Advanced Manufacturing Technology and Systems

Advanced Manufacturing Technology and Systems

Editors

Youqiang Xing
Xiuqing Hao
Duanzhi Duan

MDPI • Basel • Beijing • Wuhan • Barcelona • Belgrade • Manchester • Tokyo • Cluj • Tianjin

Editors

Youqiang Xing
School of Mechanical Engineering
Southeast University
Nanjing
China

Xiuqing Hao
College of Mechanical and Electrical Engineering
Nanjing University of Aeronautics and Astronautics
Nanjing
China

Duanzhi Duan
School of Mechanical Engineering
Xi'an Jiaotong University
Xi'an
China

Editorial Office
MDPI
St. Alban-Anlage 66
4052 Basel, Switzerland

This is a reprint of articles from the Special Issue published online in the open access journal *Micromachines* (ISSN 2072-666X) (available at: www.mdpi.com/journal/micromachines/special_issues/Advanced_Manufacturing_Technology_Systems).

For citation purposes, cite each article independently as indicated on the article page online and as indicated below:

LastName, A.A.; LastName, B.B.; LastName, C.C. Article Title. *Journal Name* **Year**, *Volume Number*, Page Range.

ISBN 978-3-0365-7791-3 (Hbk)
ISBN 978-3-0365-7790-6 (PDF)

Contents

About the Editors

Youqiang Xing

Youqiang Xing, Ph.D, is an Associate Professor at the School of Mechanical Engineering, Southeast University, China. Dr. Xing's research interests include the interrelated areas of machining processes, micro-/nano-manufacturing, surface engineering, friction and wear, etc. He has published over 80 articles and authorized 30 invention patents in China. His article citation number is more than 2000, his h-index is 26, and his h10-index is 43. He obtained the Second Prize of Natural Science of Ministry of Education, China, among others.

Xiuqing Hao

Xiuqing Hao, Ph.D, is a Professor at the College of Mechanical and Electrical Engineering, Nanjing University of Aeronautics and Astronautics, China. Dr. Hao's research interests mainly include: surface micro-texture fabrication and application, micromachining technology, surface-wettability control, etc. She has published more than 80 referred papers. She also received the Best Paper Awards at the IMCC2021 (Xi'an, China), ICFDM2020 (Jinan, China), and ICHSM2018 (Guangzhou, China) conferences. Dr. Hao has served as Principal Investigator of projects through the National Natural Science Foundation of China (NSFC), National Defense Science and Technology, Natural Science Foundation of Jiangsu Province, Aviation Science Funds, etc. In 2018, Dr. Hao was selected for the "Six Talents Peaks in Jiangsu Province". In 2019, she received the financial support of "the Natural Science Foundation of Jiangsu Province for Outstanding Young Teachers".

Duanzhi Duan

Duanzhi Duan, Ph.D, is an Associate Professor at the School of Mechanical Engineering, Xi'an Jiaotong University, China. Dr. Duan's research interests include the interrelated areas of manufacture technology of super-hard abrasive grinding wheels, precision and ultra-precision grinding technology, etc. He has published over 40 articles and authorized 20 invention patents in China. In 2018 he received support from the Postdoctoral Innovative Talent Support Program of China.

micromachines

Editorial

Editorial for the Special Issue on Advanced Manufacturing Technology and Systems

Youqiang Xing [1,*], Xiuqing Hao [2,*] and Duanzhi Duan [3,*]

[1] School of Mechanical Engineering, Southeast University, Nanjing 211189, China
[2] College of Mechanical and Electrical Engineering, Nanjing University of Aeronautics and Astronautics, Nanjing 210016, China
[3] School of Mechanical Engineering, Xi'an Jiaotong University, Xi'an 710049, China
* Correspondence: yqxing@seu.edu.cn (Y.X.); xqhao@nuaa.edu.cn (X.H.); dzduan@163.com (D.D.)

Advanced manufacturing technology and systems (AMTSs) combine the principles of mechanical engineering with design innovation to create products and processes that are better, faster and more precise [1]. The core of AMTS is the design, fabrication and application of original and effective solutions related to manufacturing machines, process integration and systems to keep up with the dynamic needs of today's ever-evolving industries [2].

Advanced manufacturing technology and systems cover a broad scope, involving manufacturing processes, machine tool design, system optimization, smart and flexible manufacturing, theoretical study and metrology [3,4]. From a fabrication point of view, AMTSs include physical, chemical, micro-/nanofabrication, machining and forming technology, additive manufacturing, non-traditional manufacturing processes, etc. From a system point of view, AMTSs include intelligent control, energy conversion for systems, optimization algorithms, smart sensors, etc. To improve the yield for mass productions, some effective testing and optimization methods are widely used.

This Special Issue comprises 19 original papers concerning recent advances in the research and development of AMTSs. Specifically, many research fields are covered as follows: machining processes (six papers), micro-/nano precision fabrication (four papers), system optimization (eight papers) and a review (one paper). These typical studies reveal the recent advances in advanced manufacturing, which are briefly summarized as follows.

Zhang et al. [5] studied the laser machining of SiC ceramics by ablation and polishing through the use of an infrared femtosecond laser, the laser ablation threshold of SiC ceramics was calculated, and the influence of pulse energy and the defocus amount of the femtosecond laser polishing of SiC ceramics were investigated. The optimal machining process parameters were obtained. Cheng et al. [6] analyzed the interfacial interaction mechanisms between a 4H-SiC wafer surface and diamond indenter during nanoscale scratching using distilled water without using an acid–base etching solution. They reported that the reaction between water and SiC on the wafer surface could be controlled. Microcracks can be avoided, and damage-free thinning of SiC wafers can be achieved by controlling the SiC–water reaction. In order to reduce the grinding force and enhance the removal rate when grinding zirconia ceramics, a nanosecond laser is used to ablate and grind the surface of a zirconia ceramic in a study conducted by Pang et al. [7]. Compared to the grinding surface without a laser-structure, a damage-free grinding surface was obtained through laser assistance. Wang et al. [8] established an ultrasonic elliptical vibration cutting (UEVC) finite element simulation model and UEVC cutting trajectory. A mechanism for the micro removal of materials in the UEVC process was obtained. Wang et al. [9] developed a water-dissolution polishing method to obtain near-damage-free KDP surfaces. They proved that the wetting characteristics of the polishing fluid should be improved during the optimization process of polishing fluid composition when using oil-based polishing

Citation: Xing, Y.; Hao, X.; Duan, D. Editorial for the Special Issue on Advanced Manufacturing Technology and Systems. *Micromachines* **2023**, *14*, 495. https://doi.org/10.3390/mi14030495

Received: 15 February 2023
Accepted: 17 February 2023
Published: 21 February 2023

fluids. Gao et al. [10] reviewed the current state of research that characterizes, estimated the effects of the process parameters on the mechanical properties, and summarized existing works on Fused Deposition Modeling (FDM). Zhang et al. [11] used atomic layer deposition (ALD) to regulate the line width of the one-dimensional grating standards with a pitch of 1000 nm, fabricated by electron beam lithography (EBL). They proved that the width of a single grating line in the standard could be regulated with great uniformity by precisely utilizing ALD. Meng et al. [12] proposed a side ohmic contact mode for the double-channel GaN/AlGaN epitaxial layer. Rectangle transmission line model (TLM) electrodes are prepared, and the specific contact resistance is tested at annealing temperatures ranging from 700 °C to 850 °C. Two sandwiched ZnO/Metal/ZnO transparent conductive thin films were deposited using magnetron sputtering technology by Lin et al. [13]. They revealed that the introduction of the Ti layer is beneficial to the overall properties of a ZnO (Ti/Cu) thin film compared to a ZnO (Cu) thin film with the same metal layer thickness, and annealing can improve the performance of the film systems. Xu et al. [14] proposed a novel chemically enhanced shear dilatancy polishing method (C-SDP). They revealed that C-SDP technology was a novel ultra-precision machining method that could achieve great surface qualities and polishing efficiency for tungsten. Du et al. [15] proposed two methods for predicting milling stability based on the composite Cotes and Simpson's 3/8 formulas. Yu et al. [16] proposed an efficient algorithm using Field Programmable Gate Array (FPGA)-accelerated You Only Look Once (YOLO) v3 based on an attention mechanism. Xie et al. [17] proposed a prognostic method based on a back-propagation neural network for corroded pipeline systems. The rationality and effectiveness of the proposed prediction models were verified. Zhou et al. [18] introduced a methodology to model, design, fabricate and optimize an air-coupled ultrasonic transducers matching system. They indicated that the self-developed air-coupled ultrasonic transducer has a 20% higher amplitude than the product available on the market. Liu et al. [19] presented an integrated piezoresistive normal force sensor. The surface micromachined normal force sensor was transferred to the readout circuit chip with a temporary stiction effect handling process. Li et al. [20] established a Multiphysics-field-coupled simulation model of electric, flow and temperature fields during the electrochemical machining (ECM) of the miniature bearing outer ring based on the gas–liquid two-phase turbulent flow model, which was important in improving the machining accuracy of the outer ring of the ECM miniature bearing. Zhao et al. [21] proposed and manufactured an all-optical fiber sensor system based on the parallel structure of gold-plated Fiber Bragg grating (FBG) and quartz FBG, which could simultaneously measure temperature and pressure. Additionally, they designed and fabricated a high-sensitivity optical fiber Mach–Zehnder refractive index sensor and an all-fiber temperature and refractive dual-parameter-sensing Michelson interferometer based on a waist-enlarged bitaper [22,23].

Author Contributions: Writing—original draft preparation, Y.X.; writing—review and editing, X.H. and D.D. All authors have read and agreed to the published version of the manuscript.

Funding: This work was supported by the National Natural Science Foundation of China (52205454, 51875285, 52005397), the Natural Science Foundation of Jiangsu Province in China (BK20190066, BK20211562), the National Postdoctoral Program for Innovative Talents (Grant No. BX20180250), the College Young Teachers Fund of the Fok Ying Tung Education Foundation (20193218210002, 171045), and the Zhishan Young Scholar Fund of Southeast University in China.

Acknowledgments: We would like to take this opportunity to thank all of the authors for submitting their papers to this Special Issue, all the reviewers for dedicating their time and helping to improve the quality of the submitted papers, and our assistant editors Ailsa Yang and Georgina Lv for their contributions to improving these submissions.

Conflicts of Interest: The author declares no conflict of interest.

References

1. Kuljanic, E. *Advanced Manufacturing Systems and Technology*; Springer: Wien, Austria, 2014.
2. Chan, F.T.S.; Chan, M.H.; Lau, H.; Ip, R.W.L. Investment Appraisal Techniques for Advanced Manufacturing Technology (AMT): ALiterature Review. *Integr. Manuf. Syst.* **2001**, *12*, 35–47. [CrossRef]
3. Qu, Y.J.; Ming, X.G.; Liu, Z.W.; Zhang, X.Y.; Hou, Z.T. Smart Manufacturing Systems: State of the Art and Future Trends. *Int. J. Adv. Manuf. Technol.* **2019**, *103*, 3751–3768. [CrossRef]
4. Wang, B.; Tao, F.; Fang, X.; Liu, C.; Liu, Y.; Freiheit, T. Smart manufacturing and Intelligent Manufacturing: A Comparative Review. *Engineering* **2021**, *7*, 738–757. [CrossRef]
5. Zhang, X.; Chen, X.; Chen, T.; Ma, G.; Zhang, W.; Huang, L. Influence of Pulse Energy and Defocus Amount on the Mechanism and Surface Characteristics of Femtosecond Laser Polishing of SiC Ceramics. *Micromachines* **2022**, *13*, 1118. [CrossRef] [PubMed]
6. Cheng, Z.; Luo, Q.; Lu, J.; Tian, Z. Understanding the Mechanisms of SiC-Water Reaction during Nanoscale Scratching without Chemical Reagents. *Micromachines* **2022**, *13*, 930. [CrossRef] [PubMed]
7. Pang, J.; Ji, X.; Niu, Y.; Chen, S. Experimental Investigation of Grinding Force and Material Removal Mechanism of Laser-Structured Zirconia Ceramics. *Micromachines* **2022**, *13*, 710. [CrossRef]
8. Wang, Z.; Pan, Y.; Zhang, Y.; Men, X.; Fu, X.; Ren, S. Study on the Material Removal Mechanism of Ultrasonic Elliptical Vibration Cutting of Medical β Titanium Alloy. *Micromachines* **2022**, *13*, 819. [CrossRef]
9. Wang, X.; Gao, H.; Deng, Q.; Wang, J.; Chen, H.; Yuan, J. Effect of Wetting Characteristics of Polishing Fluid on the Quality of Water-Dissolution Polishing of KDP Crystals. *Micromachines* **2022**, *13*, 535. [CrossRef]
10. Gao, G.; Xu, F.; Xu, J.; Tang, G.; Liu, Z. A survey of the influence of process parameters on mechanical properties of fused deposition modeling parts. *Micromachines* **2022**, *13*, 553. [CrossRef]
11. Zhang, Y.; Wang, C.; Jing, W.; Wang, S.; Zhang, Y.; Zhang, L.; Zhang, Y.; Zhu, N.; Wang, Y.; Zhao, Y.; et al. High-Precision Regulation of Nano-Grating Line width Based on ALD. *Micromachines* **2022**, *13*, 995. [CrossRef]
12. Meng, Q.; Lin, Q.; Jing, W.; Zhao, N.; Yang, P.; Lu, D. Investigation on the Effect of Annealing Temperature on the Side Ohmic Contact Characteristics for Double Channel GaN/AlGaN Epitaxial Layer. *Micromachines* **2022**, *13*, 791. [CrossRef]
13. Lin, Q.; Zhang, F.; Zhao, N.; Yang, P. Influence of annealing temperature on optical properties of sandwiched ZnO/Metal/ZnO transparent conductive thin films. *Micromachines* **2022**, *13*, 296. [CrossRef]
14. Xu, L.; Wang, L.; Chen, H.; Wang, X.; Chen, F.; Lyu, B.; Huang, W.; Zhao, W.; Yuan, J. Effects of pH Values and H_2O_2Concentrations on the Chemical Enhanced Shear Dilatancy Polishing of Tungsten. *Micromachines* **2022**, *13*, 762. [CrossRef]
15. Du, X.; Ren, P.; Zheng, J. Predicting Milling Stability Based on Composite Cotes-Based and Simpson's 3/8-Based Methods. *Micromachines* **2022**, *13*, 810. [CrossRef]
16. Yu, L.; Zhu, J.; Zhao, Q.; Wang, Z. An Efficient YOLO Algorithm with an Attention Mechanism for Vision-Based Defect Inspection Deployed on FPGA. *Micromachines* **2022**, *13*, 1058. [CrossRef]
17. Xie, M.; Li, Z.; Zhao, J.; Pei, X. A Prognostics Method Based on Back Propagation Neural Network for Corroded Pipelines. *Micromachines* **2022**, *12*, 1568. [CrossRef]
18. Zhou, J.; Bai, J.; Liu, Y. Fabrication and Modeling of Matching System for Air-Coupled Transducer. *Micromachines* **2022**, *13*, 781. [CrossRef]
19. Liu, N.; Zhong, P.; Zheng, C.; Sun, K.; Zhong, Y.; Yang, H. Integrated Piezoresistive Normal Force Sensors Fabricated Using Transfer Processes with Stiction Effect Temporary Handling. *Micromachines* **2022**, *13*, 759. [CrossRef]
20. Li, Z.; Li, W.; Cao, B. Simulation Analysis of Multi-Physical Field Coupling and Parameter Optimization of ECM Miniature Bearing Outer Ring Based on the Gas-Liquid Two-Phase Turbulent Flow Model. *Micromachines* **2022**, *13*, 902. [CrossRef]
21. Zhao, N.; Zhang, Z.; Lin, Q.; Yao, K.; Zhu, L.; Chen, Y.; Zhao, L.; Tian, B.; Yang, P.; Jiang, Z. Research on the High Temperature and High Pressure Gold-Plated Fiber Grating Dual-Parameter Sensing Measurement System. *Micromachines* **2022**, *13*, 195. [CrossRef]
22. Zhao, N.; Wang, Z.; Zhang, Z.; Lin, Q.; Yao, K.; Zhang, F.; Jiao, Y.; Zhao, L.; Tian, B.; Yamg, P.; et al. High Sensitivity Optical Fiber Mach–Zehnder Refractive Index Sensor Based on Waist-Enlarged Bitaper. *Micromachines* **2022**, *13*, 689. [CrossRef]
23. Zhao, N.; Wang, Z.; Zhang, Z.; Lin, Q.; Yao, K.; Zhu, L.; Tian, B.; Zhao, L.; Yang, P.; Jiang, Z. Simultaneous Measurement of Temperature and Refractive Index Using Michelson Interferometer Based on Waist-Enlarged Fiber Bitaper. *Micromachines* **2022**, *13*, 658. [CrossRef]

micromachines

Article

Influence of Pulse Energy and Defocus Amount on the Mechanism and Surface Characteristics of Femtosecond Laser Polishing of SiC Ceramics

Xuanhua Zhang [1,2], Xiaoxiao Chen [2,3,*], Tao Chen [1,2], Guiying Ma [2,4], Wenwu Zhang [2,3] and Lirong Huang [1,*]

1 School of Mechanical and Electrical Engineering, Jiangxi University of Science and Technology, Ganzhou 341000, China; zhangxuanhua@nimte.ac.cn (X.Z.); chentao11@nimte.ac.cn (T.C.)
2 Ningbo Institute of Materials Technology and Engineering, Chinese Academy of Sciences, Zhejiang Provincial Key Laboratory of Aero Engine Extreme Manufacturing Technology, Ningbo 315201, China; maguiying@nimte.ac.cn (G.M.); zhangwenwu@nimte.ac.cn (W.Z.)
3 University of Chinese Academy of Sciences, Beijing 100049, China
4 School of Mechanical Engineering and Mechanics, Ningbo University, Ningbo 315211, China
* Correspondence: chenxiaoxiao@nimte.ac.cn (X.C.); huanglirong@jxust.edu.cn (L.H.)

Abstract: SiC ceramics have excellent comprehensive properties and are typical hard and brittle materials that are difficult to process and are widely used in many fields. Laser polishing technology has developed into a new surface processing technology, and femtosecond laser polishing has become an important method for the precision machining of hard and brittle materials. In this paper, SiC ceramics were ablated and polished by infrared femtosecond laser, the laser ablation threshold of SiC ceramics was calculated and the influence of pulse energy and defocus amount on the surface morphology, surface roughness, polishing depth and oxidation degree of femtosecond laser polishing of SiC ceramics were investigated. The results show that when the laser repetition frequency f = 175 kHz, wavelength λ = 1064 nm and ablation time t = 9 s, the laser ablation threshold of SiC ceramics is 0.355 J/cm^2. With the increase in pulse energy, the surface roughness first decreased and then increased, and the polishing depth showed an overall upward trend. The change of defocus amount will lead to the change of the laser spot diameter. With the increase of the defocus amount, the laser spot irradiated on the workpiece surface becomes larger, and the laser energy density decreases, which results in the decrease of the laser ablation ability and polishing depth and the increase of the polished surface roughness. Periodic nano-ripple structures appeared on the laser-induced surface. Through Energy Dispersive Spectrometer (EDS) elemental analysis, it was found that there was an oxidation phenomenon in SiC ceramics polished by femtosecond laser in an air environment, and the change of pulse energy and defocus amount had insignificant effects on the degree of oxidation.

Keywords: femtosecond laser polishing; SiC ceramics; laser ablation; surface quality

Citation: Zhang, X.; Chen, X.; Chen, T.; Ma, G.; Zhang, W.; Huang, L. Influence of Pulse Energy and Defocus Amount on the Mechanism and Surface Characteristics of Femtosecond Laser Polishing of SiC Ceramics. *Micromachines* **2022**, *13*, 1118. https://doi.org/10.3390/mi13071118

Academic Editors: Youqiang Xing, Xiuqing Hao and Duanzhi Duan

Received: 10 June 2022
Accepted: 11 July 2022
Published: 15 July 2022

1. Introduction

SiC ceramics can adapt to extreme working environments due to their excellent thermal shock resistance, high strength at high temperature, low thermal expansion coefficient, excellent corrosion resistance and low density, which have huge application prospects in the fields of aerospace, space optics, semiconductors and high-temperature components [1–3]. With the gradual development of SiC ceramic–related products in the direction of high-precision and high-quality, higher requirements for the surface quality of SiC ceramics have been put forward. Due to the poor surface quality of SiC ceramics prepared by sintering technology and SiC ceramics being hard and brittle materials with high hardness and brittleness, it is very difficult to obtain high-precision SiC ceramic surfaces by traditional processing methods, which limits the application of SiC ceramics in the precision manufacturing field.

The traditional surface precision machining technologies of SiC ceramics include mechanical polishing [4], ELID grinding [5], plasma polishing [6], chemical mechanical polishing [7], magnetorheological polishing [8], etc. These processing methods have defects such as low processing quality, low processing efficiency, high cost and environmental pollution, which make it difficult to meet actual needs [9]. Laser polishing technology is non-contact polishing, which can not only effectively avoid the above defects but also has the advantages of high flexibility, easy combination with CNC technology to realize automation, a wide processing range and suitability for the surface polishing of complex parts such as planar, spherical and free-form surfaces, making it a surface precision machining method for hard and brittle materials with application prospects and development potential [10]. Considering the complex interaction mechanism between laser and materials involving complex physical and chemical processes, the laser polishing mechanism can be divided into thermal effect and photochemical decomposition (such as thermal polishing and cold polishing) [11]. Thermal polishing generally uses continuous wave lasers or medium- and long-pulse lasers. When the laser beam is radiated to the surface of material, the material absorbs laser energy, causing the local area temperature to rise rapidly. When the energy density of the laser spot reaches a certain level, the material in the irradiated area melts or evaporates, and the surface roughness is reduced by material remelting distribution or removal of the material. Domestic and foreign scientific researchers have conducted more research on laser thermal polishing, which is more suitable for polishing metal materials. Miller et al. [12] used the continuous wave laser to polish H13 die steel and found when the optimal transient combination of laser power and scanning speed was used, surface quality improvement of 83% was obtained. Ma et al. [13] used the fiber laser to polish additive-manufactured Ti-based alloy surfaces. The results showed that the surface roughness, wear resistance and microhardness after laser polishing were better than those of the original surface. Xu et al. [14] used a continuous wave laser and a nanosecond pulsed laser to polish the surface of TiAl alloy fabricated by laser deposition and studied the differences in surface morphology, microstructure, microhardness, corrosion resistance and wear resistance between the two polishing processes. Gao et al. [15] used a nanosecond laser to ablate SiC ceramics and found that when the laser energy is low, the material is removed by evaporation. When the incident laser energy is high, the material removal mechanism is liquid phase explosion, producing a splash of liquid around the ablation area. Some studies have shown that when a continuous wave laser or a medium- and long-pulse laser was used to polish hard and brittle materials, due to the large heat-affected zone, defects such as debris deposition, microcracks and oxidation were prone to occur [16]. Therefore, it is difficult for laser thermal polishing to meet the surface precision machining requirements of hard and brittle materials.

Compared with continuous wave lasers and medium- and long-pulse lasers, ultrashort-pulse lasers have the advantages of small heat-affected zone, fewer thermal defects and high machining accuracy and can more easily meet the requirements of surface polishing accuracy of SiC ceramics. Ultrashort-pulse lasers mainly include picosecond, femtosecond and attosecond lasers. Zhang et al. [17] used the overlapping parallel line scanning mode to achieve high-efficiency, large-area and high-precision polishing of alumina ceramics by picosecond laser and determined the ablation law and ablation threshold during the polishing process. Ihleman et al. [18] conducted polishing experiments on different oxide ceramics using nanosecond and femtosecond lasers. It was found that when using nanosecond pulsed laser polishing, the material removal mechanism was mainly plasma-induced ablation. When using femtosecond laser polishing, photochemical decomposition dominated. Since the pulse width of the femtosecond laser is extremely narrow, the interaction time of the femtosecond laser with the material is very short, and it hardly brings thermal effect to the surrounding materials. Therefore, femtosecond laser polishing is also called cold polishing. The action mechanism of cold polishing is that a single photon or multiple photons interact with the lattice or chemical bond of the material, and as a result, some components in the material are directly peeled off, that is, photochemically decomposed [19]. Kurita et al. [20] found that the number and size of deposited debris on the surface of SiC ceramics processed by femtosecond laser were much smaller than nanosecond laser processing. Taylor et al. [21] used femtosecond laser to polish SiC ceramics, by optimizing the polishing process parameters, and the problem of thermal oxidation on the surface of the material due to the high laser frequency was avoided. Chen et al. [22] reported a femtosecond laser polishing method for SiC ceramics and studied the influence of laser wavelength and pulse number on the surface morphology and composition formation mechanism. By fine-tuning the processing parameters, the subsurface defects were eliminated. After polishing, the subsurface structure was uniform and the friction coefficient was stable, and a high-quality polished surface was obtained. On the basis of single-laser polishing technology, researchers have carried out research on the composite polishing technology of laser and other energy fields. Wang et al. [23] used femtosecond laser-assisted chemical mechanical polishing of SiC crystals, and the corresponding composite polishing mechanism was explored. It was found that the surface quality and polishing efficiency of chemical mechanical polishing SiC crystals could be significantly improved when the laser process parameters were properly selected. Zheng et al. [24] proposed a new method of underwater femtosecond laser polishing for SiC ceramics, studied the influence of scanning trajectory and laser pulse energy on the surface morphology and polishing depth during underwater polishing and, finally, obtained a smooth polished surface.

In summary, researchers from different countries have carried out some investigations on laser polishing for die steel, alloy, SiC ceramics and other materials. At present, the related research on the ultrashort-pulse laser polishing mechanism and process for hard and brittle materials is still limited and needs to be further carried out. Femtosecond laser polishing has certain technical advantages for the surface precision machining of hard and brittle materials. However, there are many factors that affect the polished surface quality, among which the regulation of laser energy density has a significant impact on the polishing effect. This work studies the influence of laser energy density on femtosecond laser polishing of SiC ceramics under different working conditions in order to further promote the process improvement and technological progress of the laser polishing of hard and brittle materials. In this paper, the ablation and polishing experiments with SiC ceramics were carried out by infrared femtosecond laser, and the laser ablation threshold of SiC ceramics was calculated, and the influence of pulse energy and defocus amount on the surface morphology, surface roughness, polishing depth and oxidation degree of femtosecond laser polishing SiC ceramics were studied. The research results can guide the selection and optimization of process parameters for femtosecond laser polishing of SiC ceramics.

2. Materials and Methods

2.1. Original Materials

The experimental samples were SiC ceramic blocks with the size of 20 mm × 20 mm × 5 mm, and which were prepared by sintering technology. The basic performance parameters of SiC ceramics are shown in Table 1. Before the experiments, the samples were put into a beaker with pure alcohol solution, and then the ultrasonic cleaning machine was used (F-020SD, FUYANG, Shenzhen, China) to clean them for 10 min. The purpose was to remove the impurities attached to the surface of the samples and avoid the impurities affecting the experimental results. The average initial surface roughness Ra = 1.1 μm of the samples was measured by the laser scanning confocal microscope (VK-X200K, KEYENCE, Osaka, Japan) at a magnification of 1000×. Then, the initial surface morphology of the experimental samples was observed by the scanning electron microscope (QUANTA FEG 250, FEI, Hillsboro, OR, USA), and the element types and contents of the initial surface were detected by EDS (FEI, Hillsboro, OR, USA), as shown in Figure 1. There was a certain porosity on the initial surface in which the content of C and Si elements were the highest.

Table 1. The performance parameters of SiC ceramic materials.

Density	Flexural Strength	Elastic Modulus	Thermal Expansion Coefficient	Thermal Conductivity	Microhardness
3200 kg/m^3	500 MPa	420 GPa	4.2 (1 × 10^{-6}/K)	60 W/(m·K)	2500 HV

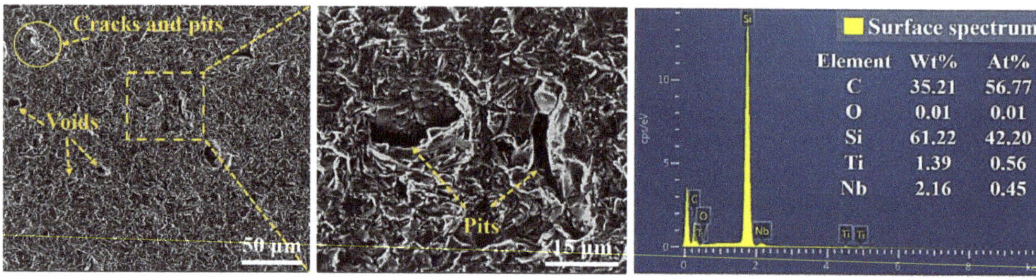

(a) (b) (c)

Figure 1. The SEM images and EDS element analysis of the initial surface. (**a**) The initial surface topography, (**b**) the partial enlarged details of the inset (**a**,**c**) the EDS elemental analysis of the inset (**b**).

2.2. Experimental Settings

In this work, experimental studies on ablation and polishing of SiC ceramics were carried out by using an infrared femtosecond laser processing system. The schematic diagrams of the infrared femtosecond laser processing system and polishing scanning trajectory are shown in Figure 2. The laser processing system is mainly composed of a femtosecond laser (HUARAY, Wuhan, China), a high-speed scanning galvanometer system, a five-axis precision motion system, an optical path system and a control system. The laser is a pulsed femtosecond laser with laser wavelength of 1064 nm, repetition frequency of 0~175 kHz, pulse width of 10~15 fs, maximum output power of 30.16 W, focusing spot diameter of 50 μm and focal length of the focusing lens of 233 mm. The cooling mode of the laser is internal circulation water cooling. In the experiments, the laser scanning confocal microscope and scanning electron microscope were used to observe the surface morphology of SiC ceramics before and after processing, the surface roughness and polishing depth were measured by VK analysis software (KEYENCE, Osaka, Japan), and elemental distribution in the machined area was detected by using EDS elemental analysis.

First, we used the laser power energy meter (OPHIR, Jerusalem, Israel) to measure the laser power, which consisted of the universal thermal power probe (50 (150) A-BB-26, OPHIR, Jerusalem, Israel) and the power meter head (VEGA, OPHIR, Jerusalem, Israel). The femtosecond laser ablation experiments were carried out on SiC ceramics under different laser powers, and the laser ablation threshold of SiC ceramics was calculated. The design of the parameters of the ablation experiments is shown in Table 2.

Table 2. The femtosecond laser ablation experimental parameters of SiC ceramics.

Number	Laser Frequency *f*/kHz	Ablation Time *t*/s	Laser Power *P*/mW
1	175	9	3920
2	175	9	4860
3	175	9	5840
4	175	9	6920
5	175	9	8060
6	175	9	9220
7	175	9	9900
8	175	9	11,180
9	175	9	11,890

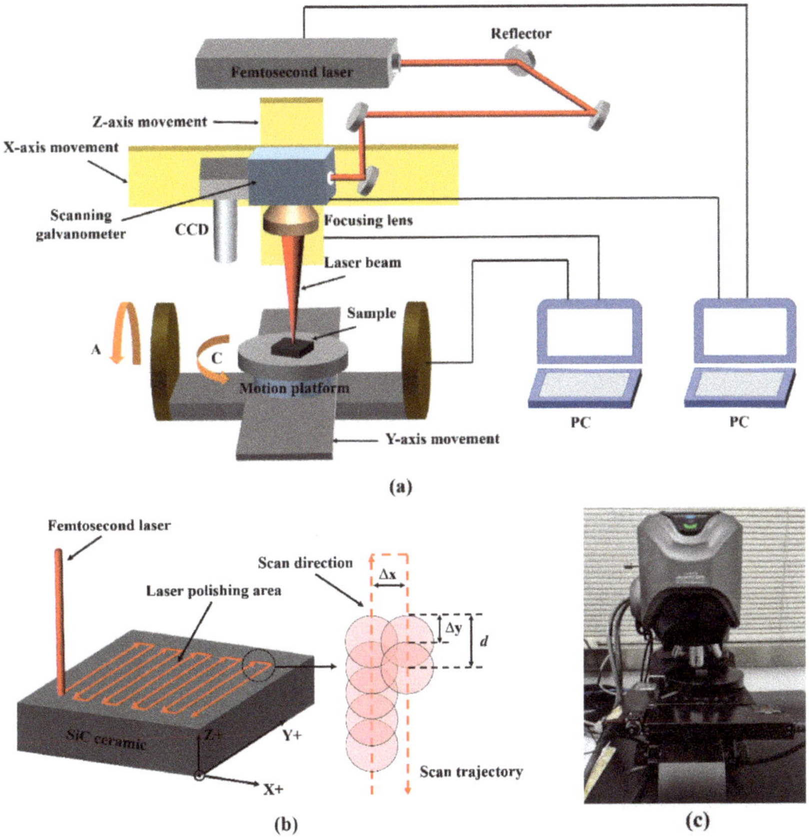

Figure 2. (**a**) The schematic diagram of infrared femtosecond laser processing system, (**b**) the schematic diagram of scanning trajectory and (**c**) the laser scanning confocal microscope.

Then, keeping other parameters unchanged, by changing the pulse energy and defocus amount, several groups of experiments were designed. The designs of polishing experimental parameters are shown in Tables 3 and 4. After the polishing experiments, the samples were ultrasonically cleaned with pure alcohol solution for 10 min, and then the polished surfaces were observed and detected with relevant testing instruments.

Table 3. Experimental parameters of femtosecond laser polishing of SiC ceramics with variable pulse energy.

Parameter	Value						
Laser frequency, f/kHz	175						
Defocus amount, h/mm	0						
Scanning spacing, Δx/μm	2.5						
Spot overlap ratio in the x-direction, ψ_x/(%)	95						
Scanning speed, v/(mm/s)	4812.5						
Spot overlap ratio in the y-direction, ψ_y/(%)	45						
Number of scanning, n	4						
Laser power, P/W	2.625	3.5	4.375	5.25	6.125	7	7.875
Pulse energy, E_P/μJ	15	20	25	30	35	40	45

Table 4. Experimental parameters of femtosecond laser polishing of SiC ceramics with variable defocus amount.

Parameter	Value					
Laser frequency, f/kHz	175					
Laser power, P/W	5.25					
Pulse energy, E_P/μJ	30					
Scanning spacing, Δx/μm	2.5					
Spot overlap ratio in the x-direction, ψ_x/(%)	95					
Scanning speed, v/(mm/s)	4812.5					
Spot overlap ratio in the y-direction, ψ_y/(%)	45					
Number of scanning, n	4					
Defocus amount, h/mm	−2	−1	0	1	2	3

The abovementioned x-direction spot overlap ratio and y-direction spot overlap ratio can be obtained by the following formulas [25]:

$$\psi_x = \frac{d - \Delta x}{d} \quad (1)$$

$$\Delta y = \frac{v}{f} \quad (2)$$

$$\psi_y = 1 - \frac{\Delta y}{d} \quad (3)$$

where d is the spot diameter (μm), Δx is the distance between adjacent scanning tracks (μm) and Δy is the distance between adjacent spots along the laser scanning direction (μm).

3. Results and Discussion

3.1. Calculation and Analysis of Femtosecond Laser Ablation Threshold of SiC Ceramics

Femtosecond laser-induced material removal is the result of the combined effect of various phenomena such as multiphoton absorption, thermal conduction, avalanche ionization, plasma expansion and liquid-phase blasting. In the case of a single pulse, the action time of the femtosecond laser is much shorter than the lattice relaxation time, and the heat energy converted by SiC ceramics after absorbing photon energy can only be conducted inside the lattice, which has almost no thermal effect on surrounding materials. In the case of multiple pulses, femtosecond laser processing also thermally affects the surrounding materials due to the thermal accumulation effect, but the thermal effect is lower than continuous wave lasers and long-pulse lasers. Figure 3 is the diagram of the action mechanism of different lasers and SiC ceramics. Observing Figure 3, it can be seen that there was recast layer on the inner wall and bottom of the ablation hole. This was because the molten material could not be discharged in time during the ablation process. After cooling and solidification, the recast layer was formed on the inner wall and bottom of the ablation hole. Related studies have shown that the recast layer formed by laser ablation of SiC ceramics is mainly composed of SiC, Si and SiO_2, and this is because SiC ceramics undergo thermal decomposition and oxidation reactions during the laser ablation process. The relevant chemical reaction equations are as follows [26,27]:

$$SiC_{(l)} = Si_{(l)} + C_{(s)} \quad (4)$$

$$Si_{(l)} + O_{2(g)} = SiO_{2(s)} \quad (5)$$

$$C_{(s)} + O_{2(g)} = CO_{2(g)} \quad (6)$$

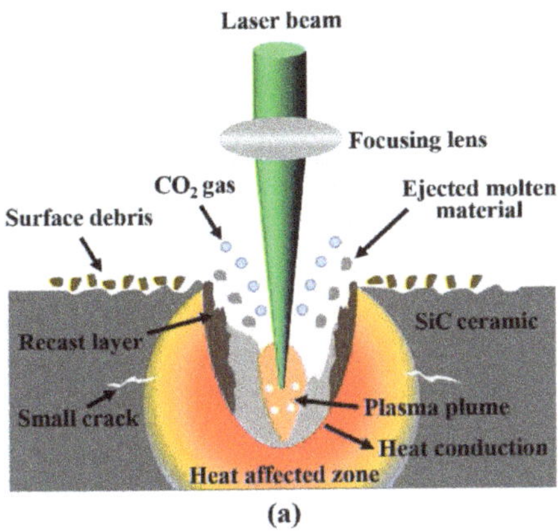

Figure 3. (**a**) The schematic diagram of nanosecond laser ablation of SiC ceramics and (**b**) the schematic diagram of femtosecond laser ablation of SiC ceramics.

The laser ablation threshold is the minimum laser energy density required to achieve material ablation removal. The calculation of the ablation threshold of SiC ceramics can guide the selection of relevant parameters in the subsequent polishing experiments, which can also avoid the phenomena of surface cracks caused by excessive laser energy or low polishing efficiency caused by too little laser energy. The equivalent diameter method was used to calculate the laser ablation threshold of SiC ceramics, which was based on the linear relationship between the square of the equivalent ablation diameter and the logarithm of the incident laser power [28,29]. Since the ablation hole was approximately circular, the width of the ablation hole was measured from the four directions of 0°, 45°, 90° and 135° along the center of the ablation hole, and then the average value of the above four width values was taken as the equivalent ablation diameter of the ablation hole.

The infrared femtosecond laser was detected by using the beam quality analyzer (SP620U, OPHIR-SPIRICON, Jerusalem, Israel), as shown in Figure 4. The femtosecond pulsed laser beam belongs to the Gaussian beams, and its laser energy density distribution obeys Gaussian distribution, which has the characteristics of high central energy density and low surrounding energy density.

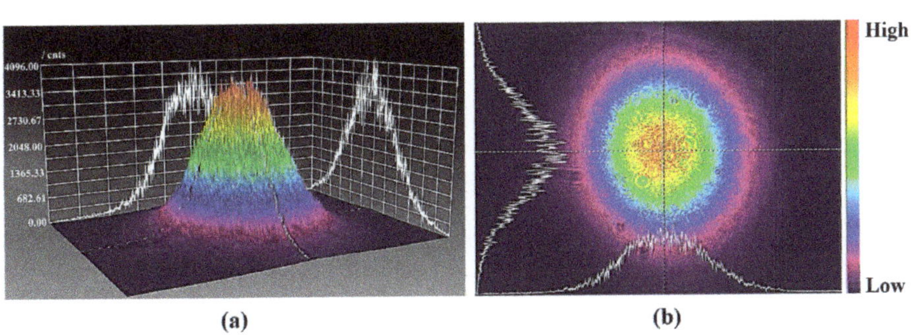

Figure 4. The schematic diagram of energy density distribution of femtosecond laser beam. (**a**) The 3D energy density distribution and (**b**) the 2D energy density distribution.

For the Gaussian beam, its energy density distribution satisfies the following relation [30]:

$$\phi(r) = \phi_0 \exp\left(\frac{-2r^2}{w_0^2}\right) \quad (7)$$

$$\phi_0 = \frac{2E_P}{\pi\, w_0^2} \tag{8}$$

where ϕ_0 represents the peak energy density of the beam center (J/cm^2), w_0 is the beam waist radius (μm), r is the distance from a point in the beam to the beam center (μm) and E_P is the laser pulse energy (μJ).

It can be seen from Equation (8) that the peak energy density of the beam center is proportional to the pulse energy. The laser pulse energy can be calculated by Equation (9):

$$E_P = \frac{P}{f} \tag{9}$$

where P is the average laser power (W), f is the laser repetition frequency (kHz). Substituting Equation (9) into Equation (8), we can obtain:

$$\phi_0 = \frac{2P}{f\,\pi\, w_0^2} \tag{10}$$

Set D as the equivalent ablation diameter of the ablation hole (μm), P_{th} as the corresponding average laser power (W) when the equivalent ablation diameter is 0 μm and ϕ_{th} as the corresponding incident laser energy density (J/cm^2) when the equivalent ablation diameter is 0 μm, which is the laser ablation threshold. According to Equations (7) and (10), we can obtain:

$$\phi_{th} = \phi_0 \exp\left(\frac{-D^2}{2w_0^2}\right) \tag{11}$$

$$D^2 = 2w_0^2\left(\ln\phi_0 - \ln\phi_{th}\right) \tag{12}$$

$$\phi_{th} = \frac{2P_{th}}{f\,\pi\, w_0^2} \tag{13}$$

Substituting Equations (10) and (13) into Equation (12), we can obtain:

$$D^2 = 2w_0^2\left(\ln P + \ln\frac{2}{\pi\, f\, w_0^2\, \phi_{th}}\right) \tag{14}$$

After the ablation experiments, the equivalent ablation diameter of the ablation hole was observed and measured by using the laser scanning confocal microscope combined with VK analysis software. The experimental data processing results are shown in Figure 5.

It can be seen from Figure 5 that the regression equation of the fitted straight line is:

$$D^2 = 1074\ln P - 6725.2 \tag{15}$$

According to the above theoretical calculation formulas, it can be obtained that when the laser repetition frequency f = 175 kHz, wavelength λ = 1064 nm and ablation time t = 9 s, the ablation threshold of the infrared femtosecond laser ablation of SiC ceramics was 0.355 J/cm^2.

3.2. The Polished Surface Quality of SiC Ceramics by Femtosecond Laser

Femtosecond laser polishing is a process in which the interaction between the laser and the material is used to remove the surface material to obtain a smoother and flatter surface. There are many factors that affect the effect of laser polishing, including the properties of the material itself, the initial surface topography, laser pulse energy, repetition frequency, scanning speed, scanning path, spot overlap rate, incident angle, defocus amount and number of scanning. In this paper, the influence of pulse energy and defocus amount on the surface characteristics of femtosecond laser polishing SiC ceramics were mainly studied. Figure 6 is the schematic diagram of the femtosecond laser polishing of SiC ceramics; due to the high energy of the femtosecond laser, the surface material is directly removed by sublimation, and the heat-affected zone is small. The laser polishing experiments were carried out in the air, and SiC ceramics chemically reacted with O_2 in the air during polishing.

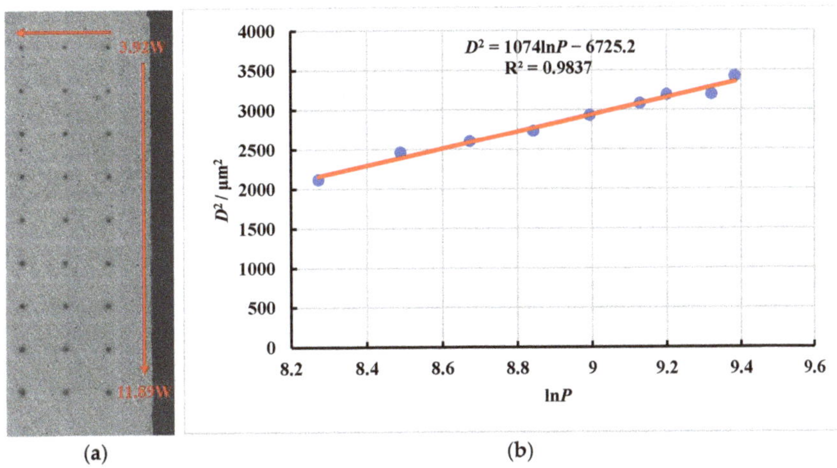

Figure 5. (**a**) The ablation hole distribution map and (**b**) the fitting line graph of the square of the equivalent ablation diameter and the logarithm of the incident laser power.

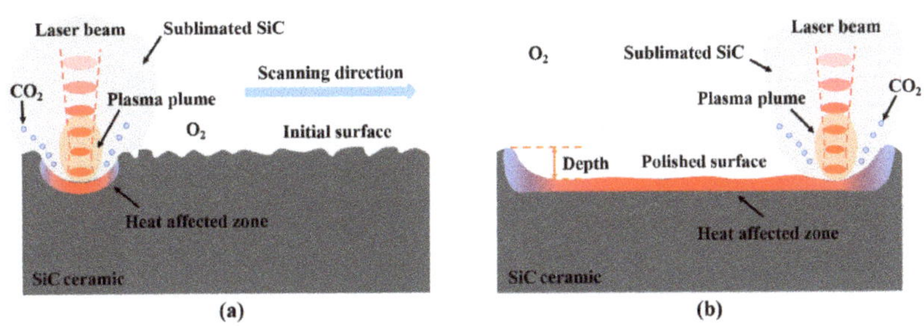

Figure 6. The schematic diagram of femtosecond laser polishing of SiC ceramics. (**a**) Before polishing and (**b**) after polishing.

3.2.1. Influence of Pulse Energy on Surface Morphology and Polishing Depth of SiC Ceramics

Controlling the repetition frequency to remain unchanged, by changing the laser power, the laser pulse energy and laser energy density were changed. According to the femtosecond laser ablation threshold of SiC ceramics calculated previously, seven different pulse energies of 15 μJ, 20 μJ, 25 μJ, 30 μJ, 35 μJ, 40 μJ and 45 μJ were selected as the experimental variable values. Femtosecond laser polishing experiments of SiC ceramics were carried out under the conditions of laser repetition frequency of 175 kHz, scanning speed of 4812.5 mm/s, scanning spacing of 2.5 μm, defocus amount of 0 mm and the number of scanning as four times. The three-dimensional morphology of the polished surface was observed by using the laser scanning confocal microscope, as shown in Figure 7. By comparing the three-dimensional morphologies of the polished surfaces under different pulse energies, the changing trend of the surface quality could be qualitatively reflected.

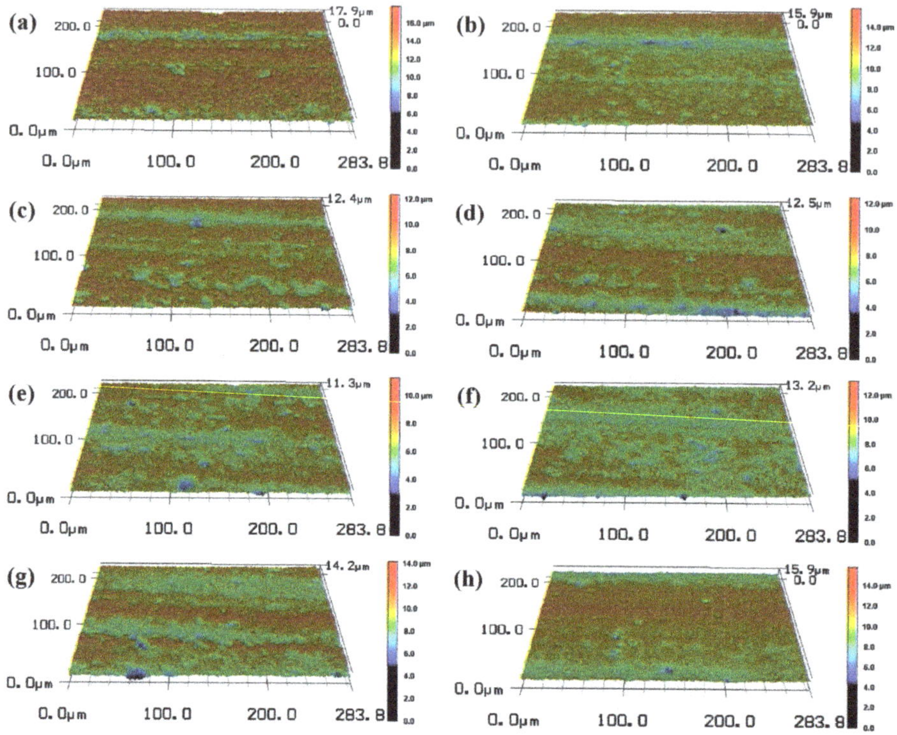

Figure 7. The three-dimensional morphologies of initial surface and the surface after polishing with different pulse energies (magnification 1000). (a) The initial surface, (b) 15 μJ, (c) 20 μJ, (d) 25 μJ, (e) 30 μJ, (f) 35 μJ, (g) 40 μJ, (h) 45 μJ.

In order to quantitatively analyze the surface characteristics, the surface roughness and polishing depth after different laser pulse energy polishing were measured by the laser scanning confocal microscope and VK analysis software, and the experimental data were fitted and analyzed, as shown in Figure 8. It can be seen from Figure 8a that the roughness of the polished surface first decreased with the increase of the laser pulse energy. When the pulse energy was 35 μJ, the lowest average surface roughness Ra = 0.664 μm was obtained, which was 39.64% lower than the initial average surface roughness Ra = 1.1 μm. With continued increase in the pulse energy, the surface roughness value showed an upward trend. The reasons for the above variation laws were that when the laser pulse energy was small, the corresponding laser energy density was low, which was not enough to completely remove the convex peaks on the surface of the material and so the surface was rough. When the pulse energy was increased to an appropriate value, the laser energy acted on the surface of the material uniformly, so a flatter polished surface can be obtained. When the pulse energy continued to increase, the laser energy was too large, the ablation was more serious, and the surface quality became poor. It can be seen from Figure 8b that the polishing depth showed an upward trend with the increase of the pulse energy, and when the pulse energy exceeded 40 μJ, the increased magnitude of polishing depth rose. This was because the greater the pulse energy, the greater the laser energy density, and the more energy absorbed by the material per unit area, the more intense the ablation between the laser and the material, and with greater amounts of material removed, the polishing depth thus gradually increased.

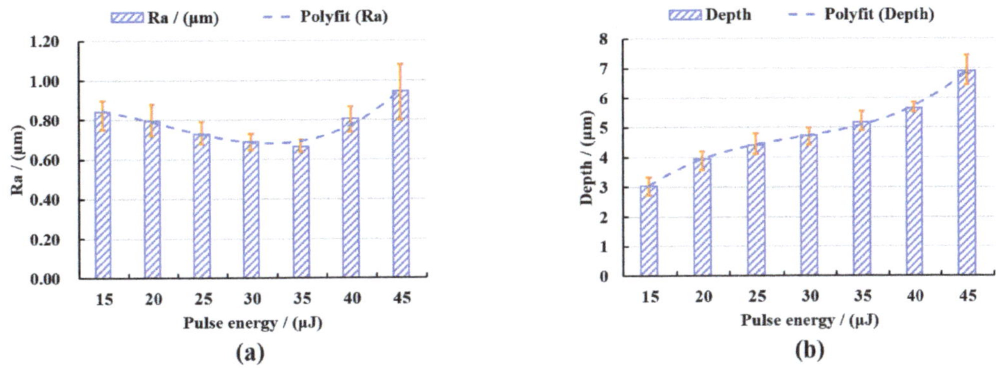

Figure 8. (**a**) The variation of surface roughness with pulse energy and (**b**) the variation of polishing depth with pulse energy.

For the purpose of further study, the influence of pulse energy on the surface morphology and oxidation degree of femtosecond laser polished SiC ceramics in air environment, scanning electron microscopy was used to observe the microscopic morphology of the polished surface, and EDS was used to detect the content of elements on the polished surface, as shown in Figures 9–11.

By observing Figures 1 and 9, it can be found that the surface quality after polishing was improved, and the initial surface wrinkle-like morphology was basically eliminated. However, at low pulse energy, defects such as initial voids, pits and cracks were still preserved, as shown in Figure 9a, and this was because the low pulse energy was not enough to completely ablate the material. With the increase of the pulse energy, the voids on the surface became smaller and less numerous. When the pulse energy was 35 μJ, the surface quality was the best, and defects such as voids and pits were basically eliminated, as shown in Figure 9c. With continued increase in the pulse energy to 45 μJ, spalling and cracks appeared on the surface, and the surface quality deteriorated. This was because the larger pulse energy improved the ablation ability of the laser, and the interaction between the laser and the material became more intense, which led to thermal stress produced on the machined surface, resulting in brittle fracture of the material, and then the phenomena of spalling and cracks appeared. At the same time, there was a small amount of microparticles on the polished surface under the action of different pulse energies, and the periodic nano-ripple structure appeared. The periodic nano-ripple structure was induced by the laser, which was related to the laser wavelength, polarization and initial surface morphology [31,32]. The EDS point scan elemental analysis of the polished surface microparticles was performed at the pulse energy of 45 μJ, as shown in Figure 10c. Compared with the initial surface element content, it can be seen that the polished surface microparticles might be splash products during the processing, and some oxidation had occurred. Observing Figures 1c and 11, it can be found that compared with the initial surface, the content of O element on the polished surface increased, indicating that there was an oxidation phenomenon in SiC ceramics polished by femtosecond laser in the air environment, and the relevant reaction equations were shown in Equations (4)–(6). Therefore, in order to eliminate the oxidation phenomenon during processing, the processing can be carried out in the environment of protective gas. It was also found that the content of each element on the polished surface under the action of different pulse energies was not much different, and the increase of pulse energy had no significant effect on the oxidation of SiC ceramics polished by femtosecond laser.

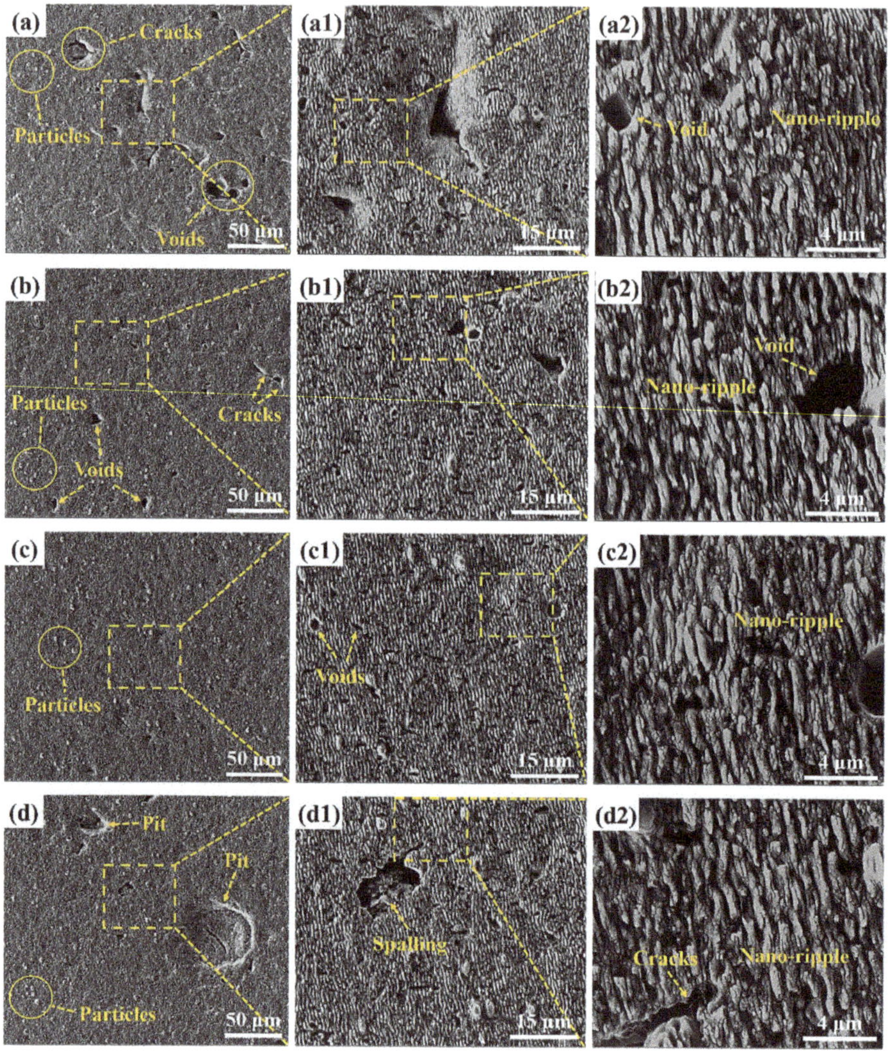

Figure 9. Polished surface morphologies and local magnifications under different pulse energies. (**a**) 20 μJ, (**b**) 25 μJ, (**c**) 35 μJ, (**d**) 45 μJ. The insets (**a1**), (**b1**), (**c1**) and (**d1**) were the partial enlarged details of the insets (**a**–**d**), respectively. The insets (**a2**), (**b2**), (**c2**) and (**d2**) were the partial enlarged details of the insets (**a1**), (**b1**), (**c1**) and (**d1**), respectively.

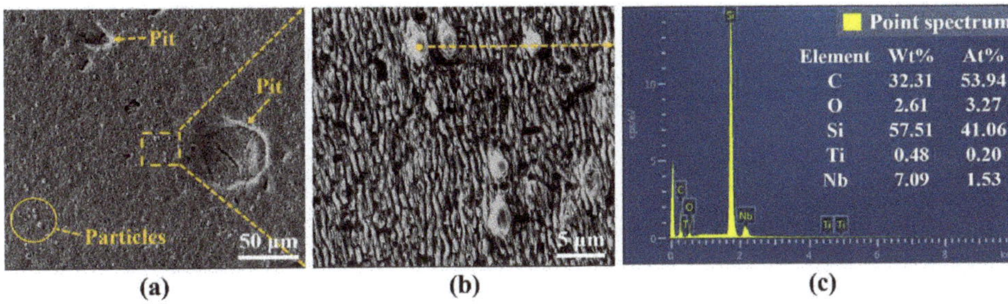

Figure 10. (**a**) The SEM image of polished surface corresponding to pulse energy of 45 μJ, (**b**) the partial enlarged details of the inset (**a**,**c**) the EDS elemental analysis of microparticles.

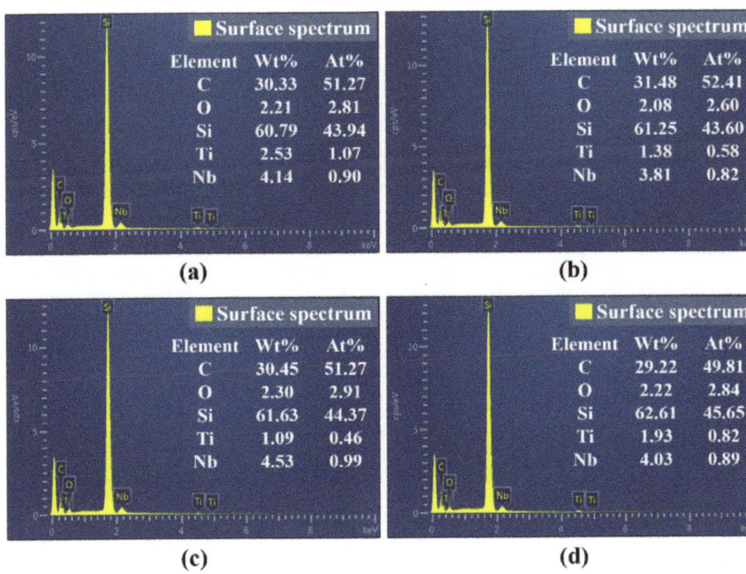

Figure 11. The elemental content of polished surfaces under different pulse energies. (**a**) 20 μJ, (**b**) 25 μJ, (**c**) 35 μJ, (**d**) 45 μJ. (Figures (**a**), (**b**), (**c**) and (**d**) in Figure 11 are the surface spectrum of figures (**a1**), (**b1**), (**c1**) and (**d1**) in Figure 9, respectively).

3.2.2. Influence of Defocus Amount on Surface Morphology and Polishing Depth of SiC Ceramics

Defocus amount refers to the distance between the focal point of the laser beam and the surface to be machined of the workpiece, also known as the focus offset distance. The change in the defocus amount directly affects the size of the laser spot, which in turn affects the laser energy density in the radiation area. According to the distance between the focus of the laser beam and the surface to be machined of the workpiece, it can be divided into three cases: positive defocus, zero defocus and negative defocus. Figure 12 showed the schematic diagram of the laser beam with positive defocus, zero defocus and negative defocus.

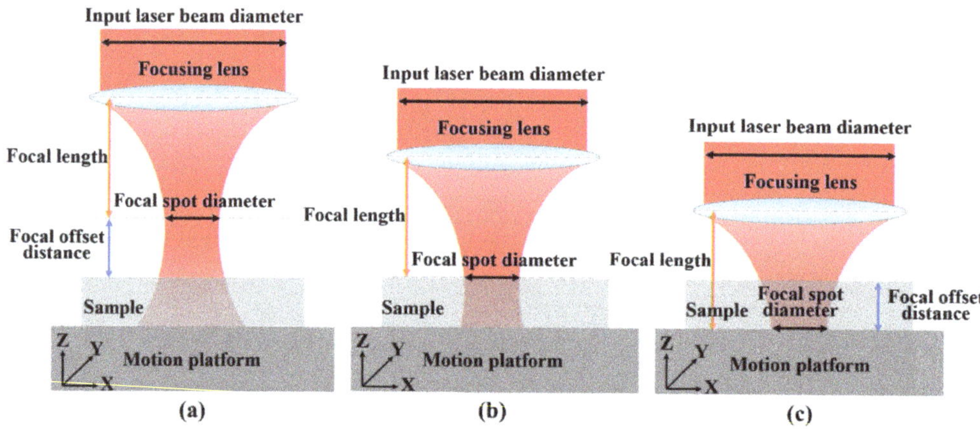

Figure 12. The schematic diagrams of three defocus situations of laser beam. (**a**) Positive defocus, (**b**) zero defocus and (**c**) negative defocus.

Six different defocus amounts of −2 mm, −1 mm, 0 mm, 1 mm, 2 mm and 3 mm were selected as the experimental variable values. Femtosecond laser polishing experiments of SiC ceramics were carried out under the conditions of laser pulse energy of 30 μJ, repetition frequency of 175 kHz, scanning speed of 4812.5 mm/s, scanning spacing of 2.5 μm and the number of scanning as four times. The three-dimensional morphology of the polished surface was observed by using the laser scanning confocal microscope, as shown in Figure 13. By comparing the three-dimensional morphologies of the polished surfaces under different defocus amounts, the changing trend of the surface quality could be qualitatively reflected.

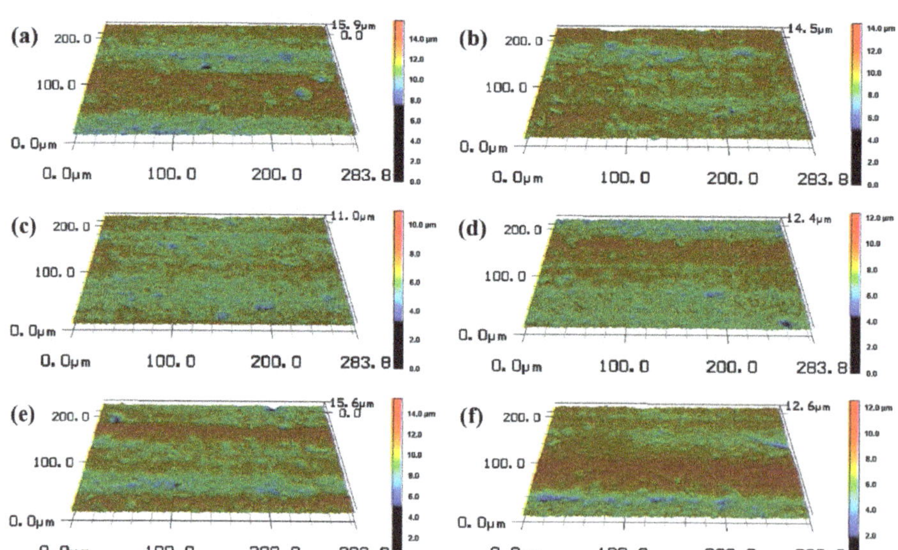

Figure 13. The three-dimensional topographies of the surface after polishing with different defocus amounts (magnification 1000). (**a**) −2 mm, (**b**) −1 mm, (**c**) 0 mm, (**d**) 1 mm, (**e**) 2 mm, (**f**) 3 mm.

In order to quantitatively analyze the surface characteristics, the surface roughness and polishing depth after different defocus amounts polishing were measured by the laser scanning confocal microscope and VK analysis software, and the experimental data were fitted and analyzed, as shown in Figure 14.

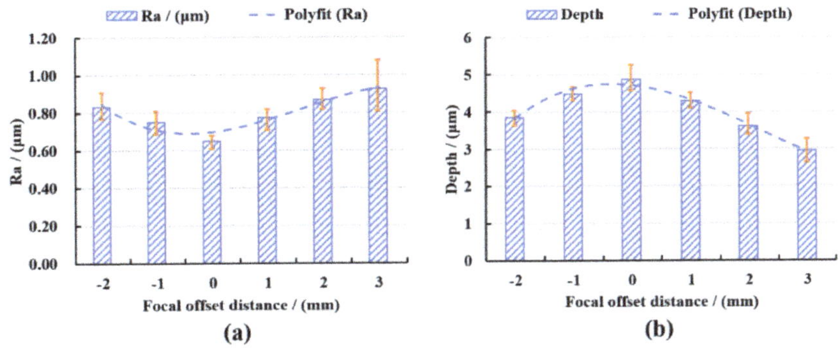

Figure 14. (**a**) The variation of surface roughness with defocus amount and (**b**) the variation of polishing depth with defocus amount.

It can be seen from Figure 14a that the polished surface roughness increased with the increase in the defocus amount whether it was positive defocus or negative defocus, and in the case of positive defocus, with the increase in the defocus amount, the increase in the surface roughness gradually slowed down. When the defocus amount was 0 mm, the lowest average surface roughness Ra = 0.652 μm was obtained, which was 40.73% lower than the initial average surface roughness Ra = 1.1 μm. The increase in the defocus amount leads to a larger laser spot irradiated on the workpiece surface, a decrease in the laser energy density and a weakening of the material removal ability. It is well-known that when other parameters remain unchanged, and the laser energy density is inversely proportional to the diameter of the laser spot, so the energy density in the laser spot irradiated on the workpiece surface gradually decreases with the increase in the defocus amount. The lower laser energy density is not enough to completely remove the convex peaks on the surface of the material, and the surface roughness is larger. It can be seen from Figure 14b that when the defocus amount was 0 mm, the maximum average polishing depth of 4.874 μm was obtained. In the positive defocus state, the polishing depth decreased with the increase in the defocus amount. This was because the larger defocus amount, the larger diameter of the laser spot, resulting in smaller laser energy density resulting in less laser energy absorbed by the material per unit area, which resulted in continuous weakening of the laser ablation processing capability, so the polishing depth gradually decreased. It can also be found in Figure 14 that when the defocus amounts were −1 mm and 1 mm, both the average roughness value of the polished surface and polishing depth were basically the same. This was because positive and negative defocus of equal magnitude had similar laser energy density and processing capacity, so the polishing results were similar for both.

In order to further study the influence of defocus amount on the surface morphology and oxidation degree of femtosecond laser polished SiC ceramics in an air environment, scanning electron microscopy was used to observe the microscopic morphology of the polished surface, and EDS was used to detect the content of elements on the polished surface, as shown in Figures 15 and 16.

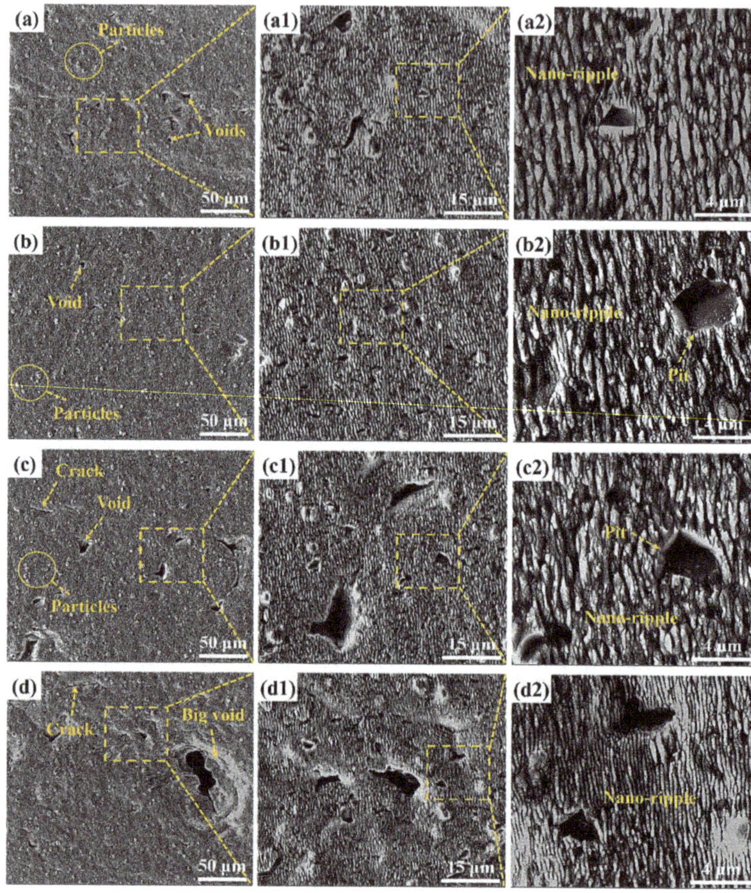

Figure 15. Polished surface morphologies and local magnifications under different defocus amounts. (**a**) −1 mm, (**b**) 0 mm, (**c**) 2 mm, (**d**) 3 mm. The insets (**a1**), (**b1**), (**c1**) and (**d1**) were the partial enlarged details of the insets (**a**), (**b**), (**c**) and (**d**), respectively. The insets (**a2**), (**b2**), (**c2**) and (**d2**) were the partial enlarged details of the insets (**a1**), (**b1**), (**c1**) and (**d1**), respectively.

It can be seen from Figure 15 that the surface defects were the least and the surface quality was the best under the zero defocus condition. Compared with the initial surface, the polished surface roughness of the abovementioned positive and negative defocus states was reduced. However, the polished surface retained the defects of the initial surface, e.g., voids, cracks and pits. With the increase of the absolute value of defocus amount, the surface defects became more obvious and the surface quality was poor. The larger the defocus amount is, the larger the spot diameter is. The smaller the laser energy density is, the more the removal of the surface material is weakened by the low energy density. At the same time, microparticles and periodic nano-ripple structures were generated on the polished surface. When the defocus amount was 3 mm, the periodic nano-ripples generated on the polished surface were significantly finer, as shown in Figure 15(d2). Observing Figures 1c and 16, it can be found that oxidation occurred during the polishing process, and the element contents of the polished surfaces with different defocus amounts were not much different. However, the content of the O element on the polished surface in the defocus states was slightly higher than that in the focus state.

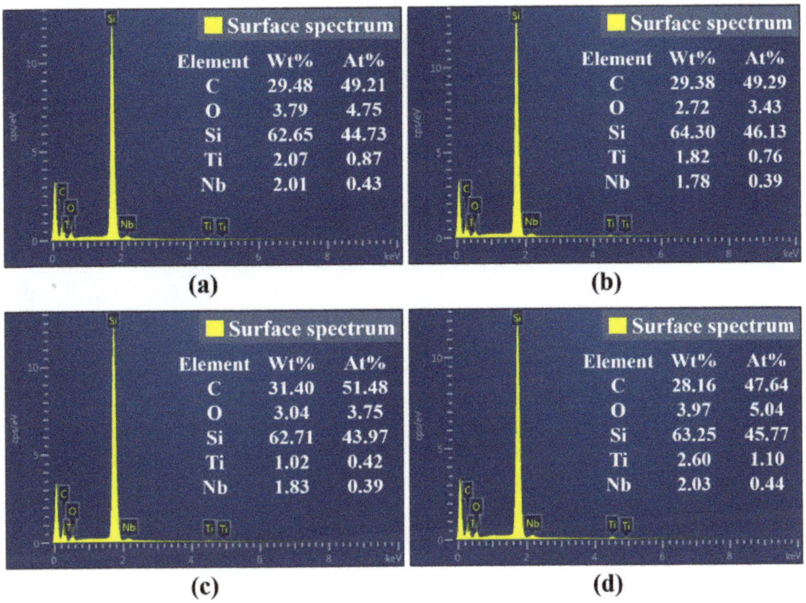

Figure 16. The elemental content of polished surfaces under different defocus amounts. (**a**) −1 mm, (**b**) 0 mm, (**c**) 2 mm, (**d**) 3 mm. (Figures (**a**), (**b**), (**c**) and (**d**) in Figure 16 are the surface spectrum of figures (**a1**), (**b1**), (**c1**) and (**d1**) in Figure 15, respectively).

4. Conclusions

In this paper, in order to further promote the process improvement and technological progress of femtosecond laser polishing of SiC ceramics, the interaction mechanism between femtosecond laser and SiC ceramics and the influence of laser energy density on femtosecond laser polishing of SiC ceramics under different working conditions were studied. The infrared femtosecond laser was used to ablate and polish SiC ceramics. Then, the equivalent diameter method was used to calculate the laser ablation threshold of SiC ceramics, and the influence of pulse energy and defocus amount on the surface morphology, surface roughness, polishing depth and oxidation degree of femtosecond laser polishing SiC ceramics were investigated. The main conclusions are as follows:

(1) Based on the linear relationship between the square of the equivalent ablation diameter and the logarithm of the incident laser power, the ablation threshold of SiC ceramics was deduced and calculated by the equivalent diameter method. It was obtained that when the laser repetition frequency f = 175 kHz, wavelength λ = 1064 nm and ablation time t = 9 s, the laser ablation threshold of SiC ceramic is 0.355 J/cm^2. The beam quality analyzer was used to verify that the laser energy density distribution of the femtosecond laser beam obeyed the Gaussian distribution, with the characteristics of high central energy density and low peripheral energy density.

(2) Compared with the initial surface, the surface quality of polished surface was improved. In the case of low pulse energy, the laser energy was not enough to completely remove the surface material, and the polished surface still retained initial surface defects such as voids, pits and cracks, so the surface roughness was high. With the increase in pulse energy, the surface roughness decreased first. When the pulse energy was 35 μJ, the polished surface quality was the best, and the lowest average surface roughness, Ra = 0.664 μm, was obtained. With continued increase in the pulse energy, spalling and cracks appeared on the polished surface, the surface quality deteriorated, and the surface roughness value showed an upward trend. The phenomena of spalling and cracks were caused by the intense ablation of the laser and the material

under the high pulse energy, and the thermal stress on the surface caused the brittle fracture of the material. The greater the pulse energy, the more energy absorbed by the material per unit area and the more the material removal, the overall increase in the trend of the polishing depth correlated with the increase in the pulse energy. There were periodic nano-ripple structures that appeared on the polished surface, and oxidation phenomenon occurred. The change of pulse energy had no significant effect on the oxidation phenomenon of the SiC ceramics polished by femtosecond laser.

(3) Under the condition of constant laser pulse energy, repetition frequency, scanning speed, scanning spacing and the number of scanning, the surface roughness increased with the increase in the absolute value of the defocus amount, and the polishing depth decreased with the increase in the absolute value of the defocus amount. With the increase in the absolute value of the defocus amount, the laser spot irradiated on the surface of the workpiece became larger, which would further reduce the laser energy density. The low laser energy density could not completely remove the surface material, the material removal amount was small, and the initial surface defects were retained, so the surface quality gradually deteriorated, and the polishing depth gradually decreased. In the case of zero defocus, the lowest average surface roughness, Ra = 0.652 μm, and the maximum average polishing depth of 4.874 μm were obtained. At the same time, the periodic nano-ripple structures were generated on the polished surface, and when the defocus amount was 3 mm, the size of the periodic nano-ripple structures generated on the polished surface was significantly smaller. A small amount of oxidation also occurred during the polishing process with variable defocus amounts, and the change of defocus amount had no significant effect on the oxidation phenomenon of the SiC ceramics polished by femtosecond laser polishing.

Author Contributions: Conceptualization, X.Z.; software, T.C.; formal analysis, X.Z., T.C. and L.H.; investigation, X.Z. and G.M.; resources, X.C. and W.Z.; data curation, X.Z., T.C. and G.M.; writing—original draft preparation, X.Z. and X.C.; writing—review and editing, X.Z. and X.C.; supervision, X.C. and L.H.; project administration, X.C. and W.Z.; funding acquisition, X.C. and W.Z. All authors have read and agreed to the published version of the manuscript.

Funding: This work was financially supported by the Natural Science Foundation of Zhejiang Province (LY20E050004), the Zhejiang Provincial Key Research and Development Program (2020C01036) and the "Science and Technology Innovation 2025" Major Project of Ningbo City (2019B10074).

Data Availability Statement: Not applicable.

Acknowledgments: The authors gratefully acknowledge Ningbo Institute of Materials Technology and Engineering, Chinese Academy of Sciences.

Conflicts of Interest: The authors declare no conflict of interest.

References

1. Willander, M.; Friesel, M.; Wahab, Q.U.; Straumal, B. Silicon carbide and diamond for high temperature device applications. *J. Mater. Sci. Mater. Electron.* **2006**, *17*, 1–25. [CrossRef]
2. Samant, A.N.; Dahotre, N.B. Laser machining of structural ceramics—A review. *J. Eur. Ceram. Soc.* **2009**, *29*, 969–993. [CrossRef]
3. Pushpakaran, B.N.; Subburaj, A.S.; Bayne, S.B.; Mookken, J. Impact of silicon carbide semiconductor technology in Photovoltaic Energy System. *Renew. Sustain. Energy Rev.* **2016**, *55*, 971–989. [CrossRef]
4. Chen, R.F.; Zuo, D.W.; Lu, W.Z.; Li, D.S.; Xu, F.; Ji, T.; Wang, M. Study on mechanical polishing for CVD diamond films of forming nucleus surface and growing surface. *Key Eng. Mater.* **2008**, *359–360*, 285–289.
5. Chen, F.; Mei, G.J.; Zhao, B.; Bie, W.B.; Li, G.X. Study on the characteristics of zirconia ceramic in three-dimensional ultrasonic vibration-assisted ELID internal grinding. *J. Mech. Sci. Technol.* **2020**, *34*, 333–344. [CrossRef]
6. Jia, G.L.; Li, B.; Zhang, J.F. Influence of SiC surface defects on materials removal in atmospheric pressure plasma polishing. *Comp. Mater. Sci.* **2018**, *146*, 26–35. [CrossRef]
7. Zhang, P.; Chen, G.M.; Ni, Z.F.; Wang, Y.G.; Teng, K.; Qian, S.H.; Bian, D.; Zhao, Y. The Effect of Cu2+ Ions and Glycine Complex on Chemical Mechanical Polishing (CMP) Performance of SiC Substrates. *Tribol. Lett.* **2021**, *69*, 94. [CrossRef]
8. Du, H.; Song, C.; Li, S.Y. Study on Surface Roughness of Modified Silicon Carbide Mirrors polished by Magnetorheological Finishing. *IOP Conf. Ser. Mater. Sci. Eng.* **2018**, *301*, 012164. [CrossRef]

9. Gu, Y.; Zhu, W.H.; Lin, J.Q.; Lu, M.M.; Kang, M.S. Subsurface Damage in Polishing Process of Silicon Carbide Ceramic. *Materials* **2018**, *11*, 506. [CrossRef]
10. Xu, G.; Dai, Y.T.; Cui, J.L.; Xiao, X.; Mei, H.P.; Li, H.L. Simulation and experiment of femtosecond laser polishing quartz material. *Integr. Ferroelectr.* **2017**, *181*, 60–69. [CrossRef]
11. Zhao, Z.Y.; Zhang, Z.J.; Yin, B.; Xu, X.M. Experimental Study on Continuous Laser Polishing Die Steel S136D. *IOP Conf. Ser. Mater. Sci. Eng.* **2020**, *758*, 012037. [CrossRef]
12. Miller, J.D.; Tutunea-Fatan, O.R.; Bordatchev, E.V. Experimental Analysis of Laser and Scanner Control Parameters During Laser Polishing of H13 Steel. *Procedia Manuf.* **2017**, *10*, 720–729. [CrossRef]
13. Ma, C.P.; Guan, Y.C.; Zhou, W. Laser polishing of additive manufactured Ti alloys. *Opt. Lasers Eng.* **2017**, *93*, 171–177. [CrossRef]
14. Xu, Z.F.; Ouyang, W.T.; Liu, Y.F.; Jiao, J.K.; Liu, Y.Z.; Zhang, W.W. Effects of laser polishing on surface morphology and mechanical properties of additive manufactured TiAl components. *J. Manuf. Process.* **2021**, *65*, 51–59. [CrossRef]
15. Gao, Y.B.; Zhou, Y.; Wu, B.X.; Tao, S.; Jacobsen, R.L.; Goodman, B. Time-Resolved Experimental Study of Silicon Carbide Ablation by Infrared Nanosecond Laser Pulses. *J. Manuf. Sci. Eng.* **2011**, *133*, 021006. [CrossRef]
16. Kim, M.; Bang, S.; Kim, D.H.; Lee, H.T.; Kim, G.H.; Ahn, S.H. Hybrid CO_2 laser-polishing process for improving material removal of silicon carbide. *Int. J. Adv. Manuf. Technol.* **2020**, *106*, 3139–3151. [CrossRef]
17. Zhang, X.M.; Ji, L.F.; Zhang, L.T.; Wang, W.H.; Yan, T.Y. Polishing of alumina ceramic to submicrometer surface roughness by picosecond laser. *Surf. Coat. Technol.* **2020**, *397*, 125962. [CrossRef]
18. Ihlemann, J.; Scholl, A.; Schmidt, H.; Wolffrottke, B. Nanosecond and Femtosecond Excimer-Laser Ablation of Oxide Ceramics. *Appl. Phys. A-Mater.* **1995**, *60*, 411–417. [CrossRef]
19. Tsai, Y.H.; Chen, B.C.; Ho, C.Y.; Chiou, Y.J.; Chen, K.H.; Chen, C.S.; Wen, M.Y. Ablation characteristics of femtosecond laser processing for nanometer-sized ceramic films. *Ceram. Int.* **2017**, *43*, S573–S577. [CrossRef]
20. Kurita, T.; Komatsuzaki, K.; Hattori, M. Advanced material processing with nano- and femto-second pulsed laser. *Int. J. Mach. Tool Manuf.* **2008**, *48*, 220–227. [CrossRef]
21. Taylor, L.L.; Qiao, J.; Qiao, J. Femtosecond laser polishing of optical materials. *Proc. SPIE* **2015**, *9633*, 96330M.
22. Chen, G.P.; Li, J.G.; Luo, H.M.; Zhou, Y.; Peng, Q.F.; Xie, X.Z.; Pan, G.S. One-step fabrication of fine surfaces via femtosecond laser on sliced SiC. *Mater. Sci. Semicond. Process.* **2021**, *132*, 105926. [CrossRef]
23. Wang, C.W.; Kurokawa, S.; Doi, T.; Yuan, J.L.; Sano, Y.; Aida, H.; Zhang, K.H.; Deng, Q.F. The Polishing Effect of SiC Substrates in Femtosecond Laser Irradiation Assisted Chemical Mechanical Polishing (CMP). *ECS J. Solid State Sci. Technol.* **2017**, *6*, P105–P112. [CrossRef]
24. Zheng, Q.Z.; Cui, J.L.; Fan, Z.J.; Mei, X.S. An experimental investigation of scan trajectory into the underwater femtosecond laser polishing SiC ceramic. *Ferroelectrics* **2020**, *563*, 77–86. [CrossRef]
25. Lin, Q.Y.; Fan, Z.J.; Wang, W.J.; Yan, Z.X.; Zheng, Q.Z.; Mei, X.S. The effect of spot overlap ratio on femtosecond laser planarization processing of SiC ceramics. *Opt. Lasers Technol.* **2020**, *129*, 106270. [CrossRef]
26. Fedorenko, L.; Medvid, A.; Yusupov, M.; Yukhimchuck, V.; Krylyuk, S.; Evtukh, A. Nanostructures on SiC surface created by laser microablation. *Appl. Surf. Sci.* **2008**, *254*, 2031–2036. [CrossRef]
27. Rudolph, P.; Brzezinka, K.W.; Wasche, R.; Kautek, W. Physical chemistry of the femtosecond and nanosecond laser-material interaction with SiC and a SiC-TiC-TiB2 composite ceramic compound. *Appl. Surf. Sci.* **2003**, *208*, 285–291. [CrossRef]
28. Kim, S.H.; Sohn, I.-B.; Jeong, S. Ablation characteristics of aluminum oxide and nitride ceramics during femtosecond laser micromachining. *Appl. Surf. Sci.* **2009**, *255*, 9717–9720. [CrossRef]
29. Mustafa, H.; Mezera, M.; Matthews, D.T.A.; Römer, G.R.B.E. Effect of surface roughness on the ultrashort pulsed laser ablation fluence threshold of zinc and steel. *Appl. Surf. Sci.* **2019**, *488*, 10–21. [CrossRef]
30. Mannion, P.T.; Magee, J.; Coyne, E.; O'Connor, G.M.; Glynn, T.J. The effect of damage accumulation behaviour on ablation thresholds and damage morphology in ultrafast laser micro-machining of common metals in air. *Appl. Surf. Sci.* **2004**, *233*, 275–287. [CrossRef]
31. Bonse, J.; Kruger, J.; Hohm, S.; Rosenfeld, A. Femtosecond laser-induced periodic surface structures. *J. Lasers Appl.* **2012**, *24*, 042006. [CrossRef]
32. Tomita, T.; Kinoshita, K.; Matsuo, S.; Hashimoto, S. Effect of surface roughening on femtosecond laser-induced ripple structures. *Appl. Phys. Lett.* **2007**, *90*, 153115. [CrossRef]

micromachines

Article

High-Precision Regulation of Nano-Grating Linewidth Based on ALD

Yaxin Zhang [1,2,†], Chenying Wang [1,3,*,†], Weixuan Jing [1,2,*], Song Wang [1,2], Yujing Zhang [1,2], Liangliang Zhang [1,2], Yijun Zhang [4], Nan Zhu [1,2], Yunxiang Wang [5], Yifan Zhao [1,4,*], Qijing Lin [1,3] and Zhuangde Jiang [1,2,3]

1 State Key Laboratory for Manufacturing Systems Engineering, International Joint Laboratory for Micro/Nano Manufacturing and Measurement Technologies, Xi'an Jiaotong University, Xi'an 710049, China; zhangyaxin@stu.xjtu.edu.cn (Y.Z.); wangsong2015@stu.xjtu.edu.cn (S.W.); zhang2020@stu.xjtu.edu.cn (Y.Z.); zhangliangliang@mail.xjtu.edu.cn (L.Z.); nan.zhu@xjtu.edu.cn (N.Z.); qjlin2015@mail.xjtu.edu.cn (Q.L.); zdjiang@xjtu.edu.cn (Z.J.)
2 School of Mechanical Engineering, Xi'an Jiaotong University, Xi'an 710049, China
3 Collaborative Innovation Center of High-End Manufacturing, Xi'an Jiaotong University, Xi'an 710049, China
4 Electronic Materials Research Laboratory, Key Laboratory of the Ministry of Education & International Center for Dielectric Research, School of Electronic Science and Engineering, Xi'an Jiaotong University, Xi'an 710049, China; zhangyj518@mail.xjtu.edu.cn
5 Suzhou Institute of Metrology, Suzhou 215128, China; wangyx@szjl.com.cn
* Correspondence: wangchenying@mail.xjtu.edu.cn (C.W.); wxjing@mail.xjtu.edu.cn (W.J.); zhaoyifan100@xjtu.edu.cn (Y.Z.)
† These authors contributed equally to this work.

Abstract: A nano-grating standard with accurate linewidth can not only calibrate the magnification of nano-measurement instruments, but can also enable comparison of linewidths. Unfortunately, it is still a challenging task to control the linewidth of nano-grating standards. Accordingly, in this paper, atomic layer deposition (ALD) was used to regulate the linewidth of the one-dimensional grating standards with a pitch of 1000 nm, fabricated by electron beam lithography (EBL). The standards were measured using an atomic force microscope (AFM) before and after ALD, and the linewidth and pitch of the grating were calculated through the gravity center method. The obtained results prove that the width of a single grating line in the standard can be regulated with great uniformity by precisely utilizing ALD. Meanwhile, the proposed method does not affect the pitch of grating, and the measurement uncertainty of standards is less than 0.16% of the pitch, thereby demonstrating a high surface quality and calibration reliability of the standards, and realizing the integration of linewidth and pitch calibration functions. Moreover, the precise and controllable fabrication method of the micro-nano periodic structure based on ALD technology has many potential applications in the fields of optoelectronic devices and biosensors.

Keywords: atomic layer deposition (ALD); linewidth regulation; micro- and nano-metrology; one-dimensional nano-grating standard

Citation: Zhang, Y.; Wang, C.; Jing, W.; Wang, S.; Zhang, Y.; Zhang, L.; Zhang, Y.; Zhu, N.; Wang, Y.; Zhao, Y.; et al. High-Precision Regulation of Nano-Grating Linewidth Based on ALD. *Micromachines* **2022**, *13*, 995. https://doi.org/10.3390/mi13070995

Academic Editors: Youqiang Xing, Xiuqing Hao and Duanzhi Duan

Received: 7 June 2022
Accepted: 23 June 2022
Published: 24 June 2022

1. Introduction

With the rapid advances in nanofabrication technology, the minimum gate length scales of transistors have been reduced to sub-10 nm [1–5], so they require precise geometric measurements, which in turn induces high demands on the accuracy of nano-measurement instruments. Therefore, it is necessary to develop nano-standards with traceability to calibrate the nano-measurement instruments to ensure the accuracy of characterization in nanofabrication, and accordingly, improve the performance of integrated circuits. In this regard, one-dimensional nano-grating standards, as one type of the important nanometric standards, are mainly used to calibrate the magnification of nano-measurement instruments. Correspondingly, a large number of research institutions and companies have developed a series of one-dimensional (1D) micro- and nano-grating standards [6–13].

Nevertheless, most nano-grating standards have a single function which provides reliable pitch calibration values. With such standards, the controllability and uniformity of the linewidth cannot be guaranteed, hence the linewidth cannot be calibrated. In addition, a grating with a constant duty cycle cannot simultaneously match the different requirements of measurement instruments with different calibration principles, for the optimal duty cycle. That is to say, the efficiency and accuracy of the calibration can be improved by making the linewidth or duty cycle of the nano-grating standard controllable.

Micro- and nano-fabrication processes such as electron beam lithography (EBL), focused ion beam (FIB) and extreme ultra-violet (EUV) are commonly used for manufacturing micro- and nano-structures. Most of the grating standards with pitch ranging between 100 and 4000 nm and fabricated using EBL have good periodicity. However, the linewidths fabricated by the EBL process have certain randomness due to the inevitable proximity effect and the instability of the current [14–16]. Therefore, despite using the same parameters, producing the same linewidth each time cannot be guaranteed. The FIB technology is a direct patterning process without a photoresist and includes several basic principles such as milling and deposition [17]. The edge and surface roughness of grating structures obtained by milling is large due to material redeposition [18]. While the structures obtained by FIB deposition have more uniform surface topography, the thickness of the deposited metal will change the designed linewidths at the same time [19]. EUV requires a mask, so the graphic size cannot be flexibly adjusted in time according to the experimental results [20]. Furthermore, none of the above processes can repatch the linewidth again after fabrication. In summary, it is not easy to precisely regulate the linewidth or duty cycle of each grating line, and likewise, it is more challenging to ensure the linewidth accuracy of nano-grating standards.

Atomic layer deposition (ALD) can precisely grow thin films of controlled thickness (from a few to tens of nanometers) on the underlying three-dimensional (3D) structures, with high accuracy, uniformity and consistency. The technique of depositing 3D conformal films on periodic structures using ALD has also been demonstrated in several papers [21–25]. To develop multifunctional grating standards with controllable linewidth and pitch, 1D nano-grating standards with a theoretical pitch of 1000 nm were fabricated in this paper using EBL, and the linewidth of shaped nano-grating was regulated by ALD. In addition, the linewidth and pitch of the 1D grating structure were measured and evaluated by an atomic force microscope (AFM) and a scanning electron microscope (SEM), which validated the feasibility and excellent performance of precise linewidth regulation via ALD, and demonstrated the high surface quality, calibration reliability, and measurement consistency of the standards.

2. Design and Fabrication

2.1. Structural Design

The 1D grating standard presented in this study is a 1.5 cm × 1.5 cm chip, whose surface structure mainly includes the calibration area and the guidance area, as shown in Figure 1a,b. The calibration area is a 1000 nm pitch 1D grating structure with an overall size of 30 μm × 60 μm (Figure 1b). Notably, the size of the 1D grating structure is very small, only 0.4% of the size of whole sample. It is difficult for the users to position the grating correctly when calibrating; therefore, a guidance area is designed at the periphery of the calibration area (Figure 1a). This area consists of a multi-level marker pattern pointing towards the center of the sample, which helps the users to identify the placement orientation of the sample and locate the calibration area rapidly, hence improving calibration efficiency greatly.

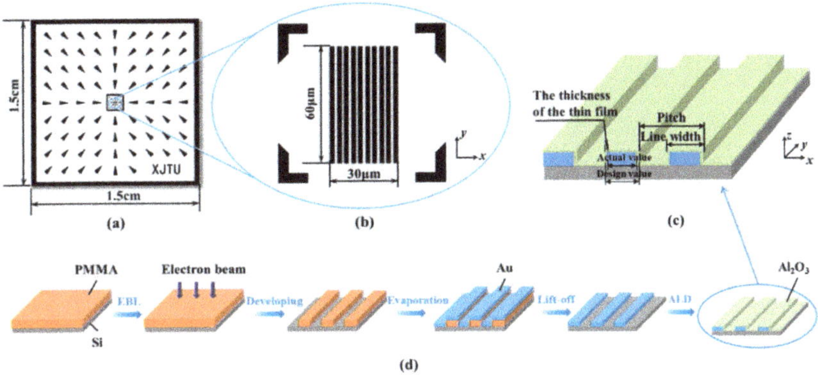

Figure 1. (**a**) Schematic of the guidance area of 1D grating standard. (**b**) Schematic of the calibration area of 1D grating standard. (**c**) Schematic of the linewidth regulation by ALD. (**d**) Schematic of the fabrication process of 1D grating standard.

2.2. Materials and Fabrication

The substrate material of the standard is Si(100) wafer, while the grating material is Cr and Au. The Au with well wear resistance, stability and conductivity can be used to calibrate the measurement instruments that require the conductivity of the material, such as SEM, and will have an excellent contrast with the substrate. The material for the linewidth regulation is an amorphous Al_2O_3 film deposited by ALD. The film growth mode corresponds to a self-limiting chemical reaction between the chemical vapor-phase precursors and the substrate surface, in the ALD process. It is worthwhile to note that the number of reacting precursors on the surface does not increase further, when the surface chemisorption reaches the saturation. As a result, ALD controls the film growth accurately by adding single atomic layers one by one until the film thickness reaches a preset value, thereby ensuring 100% uniformity and conformity of the film. Therefore, a 1D nano-grating standard with controlled linewidth can be produced by depositing an Al_2O_3 film on the surface of the grating structure with a thickness that is half the deviation of the linewidth (Figure 1c).

All the experiments were performed in a class 1000 clean room with a constant temperature of (25 ± 1) °C. The patterns of sizes 10–200 μm in the guidance area were fabricated by conventional micro-fabrication processes including ultraviolet lithography and lift-off process [26]. The specific fabrication and regulation process of the 1D grating structure is demonstrated in Figure 1d. The sample was cleaned sequentially in acetone, isopropanol (IPA) and deionized water. After that, a layer of polymethyl methacrylate (PMMA) photoresist AR-P 679 with a thickness of 100 nm was spin-coated on the substrate at 2000 rpm and baked for 2 min at 150 °C on a hot plate. Then, the one-dimensional grating structure pattern was exposed on the photoresist layer using EBL (CABL-9000C, Crestec, Hamamatsu, Japan). After the EBL process, the sample was developed in a mixture of methyl isobutyl ketone (MIBK) and IPA (1:3) for 1 min at 25 °C. Progressively, 5 nm Cr and 25 nm Au films were evaporated (TF500, Hind High Vacuum, Crawley, United Kingdom) on the sample, followed by the removal of remaining photoresist in dioxolane solution for 10 min at 25 °C. The sample was then cleaned with acetone, IPA and deionized water for 5 min each. Next, the grating structure was measured by AFM (INNOVA, Bruker, Karlsruhe, Germany) and the deviation between the designed dimension of linewidth and the actual fabricated dimension was calculated. Finally, the three-dimensional amorphous Al_2O_3 thin film was grown on the grating surface by ALD (R-200, Picosun, Masala, Finland). All the specific parameters of the ALD process have been described in our previous works [26].

To study the controllability of the modulated linewidth by ALD, three 1D grating standards with a pitch of 1000 nm, named A, B and C, were fabricated in the experiment.

By depositing Al_2O_3 films of 5, 10, and 15 nm thickness on the surfaces of samples A, B, and C, respectively, each side of the grating lines is expected to widen by 5, 10, and 15 nm, consequently increasing the width of grating lines by 10, 20, and 30 nm.

2.3. Measuremnt

The 1D nano-grating standards were measured by the AFM in the tapping mode. The scanning range was selected at the center of the grating with a size of 10 μm × 10 μm, and the number of sampling points was selected to be 256. The data measured by AFM are inevitably interspersed with some low-frequency noise signals coupled with the profile data, which can lead to bowing distortion of the measured image. Further, there exists a certain cosine error between the sample and the measurement instrument, when the sample is placed on the measurement bench. To better extract the linewidth and pitch data of the standard, linear interpolation, filtering and cosine error correction were applied to the original measurement data, to effectively reduce the tilt, bowing and other low-frequency noise, while preserving the real surface topography of the standard.

3. Results and Discussion

3.1. Results of Regulation

Firstly, in order to evaluate the linewidth uniformity of each grating line, ten grating lines in the standard were selected and their linewidths were calculated and analyzed. A length of L = 2000 nm was intercepted from each grating line, where 20 positions were chosen uniformly from top to bottom. The linewidth at each position of the grating line was the difference between the threshold line calculated by the gravity center method [27] and the two intersection points generated by left and right edges of grating line. The results of the linewidth of 10 grating lines for the three samples after ALD are shown in Figure 2a, while the position deviation curves of the linewidths are shown in Figure 2b. The inset in Figure 2a shows the AFM image of sample B after ALD. As evident from the figure, the 1D grating standard possesses well-distributed grating lines and excellent parallelism. However, there are a few particles or defects introduced by the tensile stress on Au with high ductility, during the lift-off process. It is shown in Figure 2a that the linewidths of the ten grating lines of each sample are close to each other with a slight degree of fluctuation. The deviation between the linewidths of 10 grating lines of sample A is the largest, but the maximum deviation is still only 2.1 nm (Figure 2b), which accounts for only 0.3% of the linewidth, thereby indicating the uniformity of linewidths of multiple grating lines in this structure. Hence, the user can select any grating line for the linewidth calibration, which certainly improves the repeatability of the calibration results.

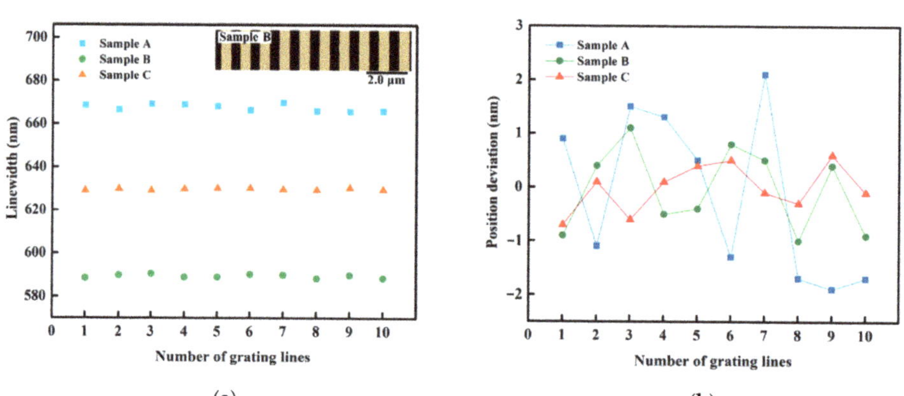

Figure 2. (**a**) Calculated linewidths of the 10 grating lines of samples A, B and C, after ALD. (**b**) The position deviation curves of the calculated linewidths.

The average value of the linewidths of 10 grating lines was taken as the linewidth calibration value of each sample. The AFM images and comparison of the linewidths of the three samples before and after ALD are shown in Figure 3. It is shown in Figure 3a,b that the height of the grating was not changed and only the grating lines were widened since ALD was growing the film simultaneously in the 3D direction. Here, the actual increase in the linewidths of three samples was 13.4, 19.6, and 29.7 nm, respectively. Evidently, the actual increase in the linewidths of samples B and C was close to the estimated value, while the actual increase in the linewidth of sample A deviated from the expected increment by 3.4 nm. This is because the linewidth uniformity of sample A is worse than the other two samples, as shown in Figure 2b. Further, the line edge roughness (LER) of the three samples was calculated according to Equation (1) [28] as LER_A = 18.9 nm, LER_B = 16.4 nm, and LER_C = 15.9 nm, respectively. Thus, the final evaluation results are likely to be disturbed by many parameters such as the selection of linewidth evaluation position, quality of grating line edge, and linewidth evaluation algorithm when measuring and calculating the linewidth of this standard.

$$\begin{cases} \overline{x} = \frac{(\sum_{i=1}^{N} x_i)}{N} \\ LER = 3\sigma = 3\sqrt{\frac{\sum_{i=1}^{N}(x_i-\overline{x})^2}{N-1}} \end{cases} \quad (1)$$

where $\overline{x}$ is the average edge of grating line, x_i is the intersection of threshold line and grating profile, N is the number of intersections, and σ is the standard deviation of line edge.

The results here reveal that the linewidth of the standard can be regulated by the ALD process, nevertheless there are some certain requirements for the standard fabrication and measurement process: (1) The regulation scheme is unidirectional since ALD can only increase the linewidth of the convex structures or decrease the linewidth of the concave structures. Thus, it is necessary to confirm the desired range of linewidths when processing the grating structure with EBL, according to the type of grating structure (convex or concave). (2) The effect of regulation is related to the linewidth uniformity and the edge straightness of grating line. The better the linewidth uniformity and edge straightness of grating line, the better the regulation performance. (3) The linewidth evaluation algorithm can be further optimized by filtering out various noises as well as the disturbances of particles at the edge of the line. The actual increase in the calculated linewidth in this case will be more reliable and can be fed back to the ALD process, to further improve the experimental parameters and form a closed-loop control.

While aiming to compare the changes in the pitch before and after the regulation, all raw data of 10 scanning lines were obtained uniformly from top to bottom along the y-direction within the scanning range of the sample, and the average pitch of each sample was calculated by the gravity center method. Meanwhile, the measurement uncertainty of each sample was evaluated according to the International Bureau of Weights and Measures (BIPM), the International Electrotechnical Commission (IEC), the International Federation of Clinical Chemistry and Laboratory Medicine (IFCC), the International Organization for Standardization (ISO), the International Union of Pure and Applied Chemistry (IUPAC), the International Union of Pure and Applied Physics (IUPAP), and the International Organization of Legal Metrology (OIML)-1993 Guide to the Expression of Uncertainty in Measurement [29], and the corresponding results are provided in Figures 3 and 4, and Table 1. The average pitch of three samples A, B and C changed by 1.8, 5.5 and 4.9 nm, respectively, before and after ALD. Theoretically, ALD should not change the pitch of the grating, and a small variation in the actual results may be caused by the fact that the quality of ALD depends on the quality of the substrate surface [30]. Hence, when there are large raised particles of several nanometers in size at the edge of grating lines, the surface of such raised particles is uniformly covered with a layer of Al_2O_3 film after ALD due to the three-dimensional conformal property of ALD. The shape of the particles is still retained, which would change the line width of the grating and affect the accuracy of the calculated data in turn. Here, the change in the pitch is small, where the maximum variation is only about 0.5% of the average pitch, which still indicates that the Al_2O_3 films deposited by ALD have a great uniformity in terms of film thickness and almost do not change the average pitch of the grating standards. As shown in Table 1, the measurement uncertainty of the standards after ALD is less than 0.16% of the average pitch, thus the calibration reliability is quite satisfactory. Certainly, the uncertainties introduced by the surface uniformity and measurement repeatability of the standards are largely minimized here compared with those before ALD, which validates that the surface quality of standard can be optimized and the measurement uncertainties can be reduced by utilizing ALD.

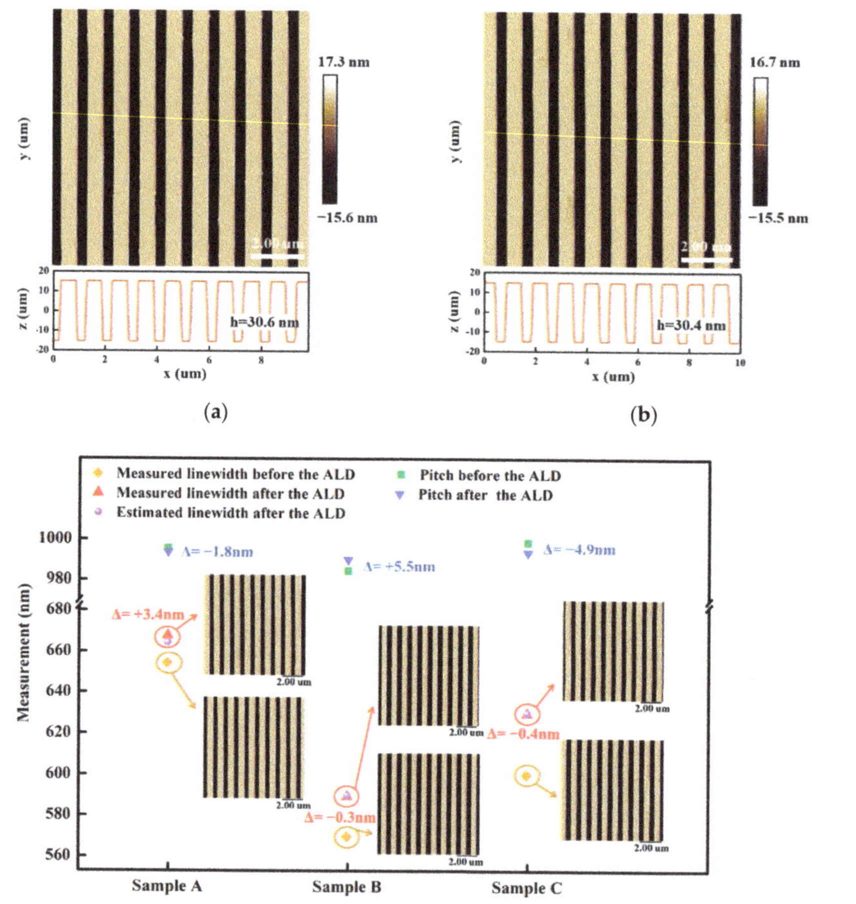

Figure 3. (**a**) The image of sample A before ALD by AFM. (**b**) The image of sample A after ALD by AFM. (**c**) Comparison of linewidths and pitches of samples A, B, and C, before and after ALD.

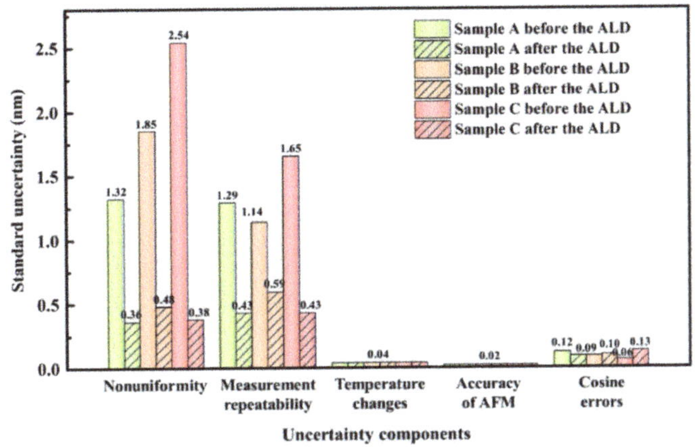

Figure 4. The standard uncertainty of major uncertainty components of samples A, B, and C, before and after ALD.

Table 1. The evaluation results of the one-dimensional grating standards.

Sample	A		B		C	
	Before ALD	After ALD	Before ALD	After ALD	Before ALD	After ALD
Average pitch (nm)	995.6	993.8	984.6	990.1	998.7	993.8
Expanded uncertainty (k [1] = 2) (nm)	3.70	1.14	4.35	1.54	6.06	1.18

[1] k is the coverage factor.

3.2. Application

To verify the calibration applicability of the sample obtained from this method in different measurement instruments, the AFM and SEM were used to measure sample B. The data obtained from the two measurement instruments were analyzed to provide a reference for the nano-geometry measurements between different instruments. The linewidth and pitch of sample B were measured by SEM system (SU8010, Hitachi, Tokyo, Japan), calculated by the gravity center method, and then evaluated for the uncertainties. The comparison of the SEM and AFM images, along with the evaluation results, is shown in Figure 5. The linewidth of the standard measured by AFM was evaluated as (589.4 $\pm$ 2.8) nm (k = 2) while the pitch was evaluated as (990.1 $\pm$ 1.5) nm (k = 2). On the other hand, the linewidth of the standard measured by SEM was evaluated to be (585.7 $\pm$ 3.1) nm (k = 2), whereas the pitch was (990.5 $\pm$ 1.8) nm (k = 2). Clearly, the pitch measured by the both instruments is very close. However, the difference in the linewidths is more obvious.

In this work, the E_n [31] was used to assess the level of agreement between the two measurements, which can be defined using Equation (2). When $|E_n| \leq 1$, the consistency of the results is good and acceptable; whereas $|E_n| > 1$ indicates a poor consistency of the results, which is unacceptable. Based on this criterion, the $|E_n|_{\text{pitch}} = 0.17$ and the $|E_n|_{\text{linewidth}} = 0.79$ were calculated for the considered sample, illustrating high agreement between the pitch and linewidth values obtained using two measurement instruments.

$$E_n = \frac{x_{AFM} - x_{SEM}}{\sqrt{U^2{}_{AFM} + U^2{}_{SEM}}} \quad (2)$$

where x_{AFM} is the value measured by AFM; x_{SEM} is the value measured by SEM; U_{AFM} is the expanded uncertainty of the result measured by AFM; U_{SEM} is the expanded uncertainty of the result measured by SEM.

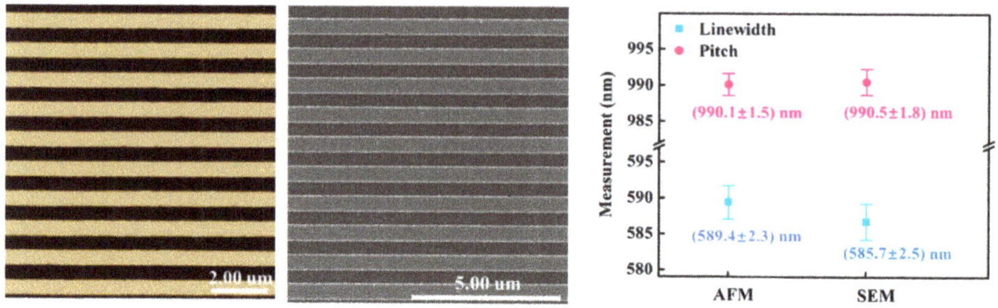

(**a**) (**b**) (**c**)

Figure 5. (**a**) The image of sample B after ALD by AFM. (**b**) The image of sample B after ALD by SEM. (**c**) Comparison of the calculation results of linewidth and pitch.

Compared with SEM, AFM can measure the 3D surface morphology of the sample more accurately, and the resolution of AFM in the horizontal and vertical directions is close to the atomic scale, hence the measurement uncertainty by AFM is lower. Meanwhile, the measurement uncertainty for SEM is higher, which is mainly introduced by the errors in the image resolution and the variation of electron beam spot diameter. However, the width of AFM probe cannot be neglected while measuring the linewidth, which can induce a spreading effect in the scanning image. As a result, the shape of AFM probe has a significant impact on the linewidth measurement, thereby resulting in a larger linewidth measured by AFM compared to SEM.

On the basis of measurement results, the sample satisfies a cross-comparison of the measurement capabilities of two measurement instruments. Excellent 3D morphology measurements were realized in the AFM, and clear edges of grating lines along with a sharp contrast with the substrate were demonstrated in the SEM. In conclusion, it can be assessed that the consistency level of results obtained by both instruments is superior in this work, based on the E_n. Accordingly, this experiment demonstrates that the samples obtained via precise linewidth regulation based on ALD can be applied to many different types of measurement instruments, where simultaneous calibration of nanoscale pitch and linewidth can be achieved, thereby enhancing calibration efficiency substantially.

4. Conclusions

In this work, we have studied the controllable regulation of the linewidth of a 1D grating standard with a pitch of 1000 nm, using ALD. The results reported herein show that the linewidth of the standard can be regulated precisely by utilizing a thin film with controllable thickness and 3D conformal structure, based on the self-limiting layer-by-layer deposition mechanism of ALD. Moreover, the better the edge straightness and linewidth uniformity of the grating line, the better the regulation performance. Evidently, the ALD process can improve the surface uniformity as well as the measurement repeatability of the standard, and restrict the measurement uncertainty of the grating standard below 0.16% of the average pitch, thus potentially guaranteeing the calibration reliability of the standard. Since the thickness of thin film grown by ALD generally does not exceed 100 nm, and the film and substrate bonding will be worse when the film is thicker. That is, the regulation value of this method for linewidth is typically less than 100 nm (film thickness is 50 nm), which requires that the deviation between the designed dimension and the actual fabricated dimension during the patterning process must not be larger than 100 nm. As a result, it is more appropriate for linewidth regulation of the grating standard with a pitch of 100 nm–10 μm.

The results acquired here from the comparisons of linewidth and pitch of the same sample by AFM and SEM are consistent; therefore, the 1D nano-grating standard with controllable pitch and linewidth can be integrated with the calibration function of grating and linewidth. It can be used not only to calibrate the magnification of the measuring instruments, but also to achieve the measurement of critical dimensions of micro- and nano-devices, thereby avoiding the repetitive errors introduced by frequent standard replacement and improving calibration efficiency significantly. Furthermore, it can also be adapted to the measurement requirements of different measurement instruments regarding the duty cycle of standard, easily and efficiently, by simply adjusting the duty cycle of the formed 1D nano-grating standard, which essentially expands the application range of the standard, and economizes the manufacturing expense.

The follow-up work will focus on further reducing the deviation between the actual and estimated regulation values of the linewidth by optimizing the parameters of the ALD process based on the molecular microscopic properties of thin film materials, and realizing precise regulation of the linewidth of 1D grating standards at the sub-nanometer scale.

Author Contributions: Conceptualization, Y.Z. (Yaxin Zhang) and C.W.; methodology, Y.Z. (Yaxin Zhang), S.W. and L.Z.; validation, Y.Z. (Yujing Zhang) and Y.W.; formal analysis, W.J. and N.Z.; investigation, Y.Z. (Yaxin Zhang); data curation, Y.Z. (Yijun Zhang); writing—original draft preparation, Y.Z. (Yaxin Zhang); writing—review and editing, C.W. and Q.L.; visualization, S.W.; supervision, W.J. and Y.Z. (Yifan Zhao); project administration, Z.J.; funding acquisition, C.W., Y.Z. (Yifan Zhao) and Z.J. All authors have read and agreed to the published version of the manuscript.

Funding: This research was funded by the National Natural Science Foundation of China (Grant Nos. 52175434, 62001366), 111 Program (No. B12016).

Conflicts of Interest: The authors declare no conflict of interest.

References

1. Xu, L.; Yang, J.; Qiu, C.; Liu, S.; Zhou, W.; Li, Q.; Shi, B.; Ma, J.; Yang, C.; Lu, J.; et al. Can Carbon Nanotube Transistors Be Scaled Down to the Sub-5 nm Gate Length? *ACS Appl. Mater. Interfaces* **2021**, *13*, 31957–31967. [CrossRef] [PubMed]
2. Loubet, N.; Hook, T.; Montanini, P.; Yeung, C.W.; Kanakasabapathy, S.; Guillom, M.; Yamashita, T.; Zhang, J.; Miao, X.; Wang, J.; et al. Stacked nanosheet gate-all-around transistor to enable scaling beyond FinFET. In Proceedings of the 2017 Symposium on VLSI Technology, Kyoto, Japan, 5–8 June 2017; IEEE: Piscataway, NJ, USA, 2017; pp. T230–T231.
3. Wu, F.; Tian, H.; Shen, Y.; Hou, Z.; Ren, J.; Gou, G.; Sun, Y.; Yang, Y.; Ren, T.L. Vertical MoS_2 transistors with sub-1-nm gate lengths. *Nature* **2022**, *603*, 259–264. [CrossRef] [PubMed]
4. Hills, G.; Lau, C.; Wright, A.; Fuller, S.; Bishop, M.D.; Srimani, T.; Kanhaiya, P.; Ho, R.; Amer, A.; Stein, Y.; et al. Modern microprocessor built from complementary carbon nanotube transistors. *Nature* **2019**, *572*, 595–602. [CrossRef] [PubMed]
5. Qiu, C.; Zhang, Z.; Xiao, M.; Yang, Y.; Zhong, D.; Peng, L.M. Scaling carbon nanotube complementary transistors to 5-nm gate lengths. *Science* **2017**, *355*, 271–276. [CrossRef]
6. Dixson, R.G.; Chernoff, D.A.; Wang, S.; Vorburger, T.V.; Tan, S.L.; Orji, N.G.; Fu, J. Multilaboratory comparison of traceable atomic force microscope measurements of a 70-nm grating pitch standard. *J. Micro-Nanolithogr. MEMS MOEMS* **2011**, *10*, 013015.
7. Vladár, A.E.; Postek, M.T. Reference Material (RM) 8820: A New Scanning Electron Microscope Scale Calibration Artifact. *Microsc. Microanal.* **2009**, *15*, 668–669. [CrossRef]
8. Raid, I.; Eifler, M.; Kusnezowa, T.; Seewig, J. Calibration of Ellipso-Height-Topometry with Nanoscale Gratings of Varying Materials. *Opt.-Int. J. Light Electron. Opt.* **2015**, *126*, 4591–4596. [CrossRef]
9. Misumi, I.; Kitta, J.I.; Fujimoto, H.; Gonda, S.; Azuma, Y.; Maeda, K.; Kurosawa, T.; Ito, Y.; Omote, K.; Nakayama, Y.; et al. 25 nm pitch comparison between a traceable x-ray diffractometer and a metrological atomic force microscope. *Meas. Sci. Technol.* **2012**, *23*, 136. [CrossRef]
10. Nakayama, Y.; Yamamoto, J.; Kawada, H. Critical dimension-scanning electron microscope magnification calibration with 25-nm pitch grating reference. *J. Micro/Nanolithogr. MEMS MOEMS* **2011**, *10*, 013021. [CrossRef]
11. Deng, X.; Dai, G.; Liu, J.; Hu, X.; Bergmann, D.; Zhao, J.; Tai, R.; Cai, X.; Li, Y.; Li, T.; et al. A new type of nanoscale reference grating manufactured by combined laser-focused atomic deposition and X-ray interference lithography and its use for calibrating a scanning electron microscope. *Ultramicroscopy* **2021**, *226*, 113293. [CrossRef]
12. Deng, X.; Liu, J.; Zhu, L.; He, P.; Cheng, X.; Li, T. Natural square ruler at nanoscale. *Appl. Phys. Express* **2018**, *11*, 075201. [CrossRef]
13. VLSI Standards Incorporated, Nanolattice Pitch Standard (NLSM). Available online: https://www.vlsistandards.com/products/dimensional/lattice_info.asp?SID=78,2022 (accessed on 27 April 2022).
14. Feng, B.; Chen, Y.; Sun, D.; Yang, Z.; Yang, B.; Li, X.; Li, T. Precision integration of grating-based polarizers onto focal plane arrays of near-infrared photovoltaic detectors for enhanced contrast polarimetric imaging. *Int. J. Extrem. Manuf.* **2021**, *3*, 035201. [CrossRef]
15. Dai, G.; Zhu, F.; Heidelmann, M.; Fritz, G.; Bayer, T.; Kalt, S.; Fluegge, J. Development and characterisation of a new linewidth reference material. *Meas. Sci. Technol.* **2015**, *26*, 115006. [CrossRef]

16. Huebner, U.; Morgenroth, W.; Boucher, R.; Meyer, M.; Mirandé, W.; Buhr, E.; Ehret, G.; Dai, G.; Dziomba, T.; Hild, R.; et al. A nanoscale linewidth/pitch standard for high-resolution optical microscopy and other microscopic techniques. *Meas. Sci. Technol.* **2007**, *18*, 422. [CrossRef]
17. Sloyan, K.; Melkonyan, H.; Apostoleris, H.; Dahlem, M.; Chiesa, M.; Al Ghaferi, A. A review of focused ion beam applications in optical fibers. *Nanotechnology* **2021**, *32*, 472004. [CrossRef] [PubMed]
18. Diddens, C.; Linz, S.J. Continuum modeling of particle redeposition during ion-beam erosion. *Eur. Phys. J. B* **2015**, *88*, 397. [CrossRef]
19. Mirmohammad, H.; Kingstedt, O.T. Theoretical considerations for transitioning the grid method technique to the microscale. *Exp. Mech.* **2021**, *61*, 753–770. [CrossRef]
20. Philipsen, V.; Luong, K.V.; Opsomer, K.; Detavernier, C.; Hendrickx, E.; Erdmann, A.; Evanschitzky, P.; Van De Kruijs, R.W.; Heidarnia-Fathabad, Z.; Scholze, F.; et al. Novel EUV mask absorber evaluation in support of next-generation EUV imaging. In *Photomask Technology 2018*; SPIE: Bellingham, WA, USA, 2018; Volume 10810, pp. 53–65.
21. Geng, G.; Zhu, W.; Pan, R.; Zhang, Z.; Gu, C.; Li, J. Precise tailoring of multiple nanostructures based on atomic layer assembly via versatile soft-templates. *Nano Today* **2021**, *38*, 101145. [CrossRef]
22. Franklin, D.; George, M.; Fraser, J.; Chanda, D. Atomic layer deposition tuning of subwavelength aluminum grating for angle-insensitive plasmonic color. *ACS Appl. Nano Mater.* **2018**, *1*, 5210–5216. [CrossRef]
23. Meng, X.; Comstock, D.J.; Fister, T.T.; Elam, J.W. Vapor-phase atomic-controllable growth of amorphous Li_2S for high-performance lithium–sulfur batteries. *ACS Nano* **2014**, *8*, 10963–10972. [CrossRef]
24. Lee, K.; Kim, H.; Kim, J.H.; Choi, D. Structural color and near-infrared tunability of ruthenium-coated anodic aluminum oxide by atomic layer deposition. *Scr. Mater.* **2020**, *187*, 125–129. [CrossRef]
25. Tamm, A.; Tarre, A.; Kozlova, J.; Rähn, M.; Jõgiaas, T.; Kahro, T.; Link, J.; Stern, R. Atomic layer deposition of superparamagnetic ruthenium-doped iron oxide thin film. *RSC Adv.* **2021**, *11*, 7521–7526. [CrossRef] [PubMed]
26. Wang, C.; Yang, S.; Jing, W.; Ren, W.; Lin, Q.; Zhang, Y.; Jiang, Z. Fabrication of nanoscale step height structure using atomic layer deposition combined with wet etching. *Chin. J. Mech. Eng.* **2016**, *29*, 91–97. [CrossRef]
27. Dai, G.; Koenders, L.; Pohlenz, F.; Dziomba, T.; Danzebrink, H.U. Accurate and traceable calibration of one-dimensional gratings. *Meas. Sci. Technol.* **2005**, *16*, 1241–1249. [CrossRef]
28. Tortonese, M.; Prochazka, J.; Konicek, P.; Schneir, J.; Smith, I.R. 100-nm-pitch standard characterization for metrology applications. In *Metrology, Inspection, and Process Control for Microlithography XVI*; SPIE: Bellingham, WA, USA, 2002; Volume 4689, pp. 558–564.
29. BIPM; EC; IFCC; ISO; IUPAC; IUPAP; OIML. *Guide to the Expression of Uncertainty in Measurement*; International Organization for Standardization: Geneva, Switzerland, 1995.
30. Wang, C.; Pu, J.; Li, L.; Jing, W.; Zhang, Y.; Zhang, Y.; Han, F.; Liu, M.; Ren, W.; Jiang, Z. Effect of the Different Substrates and the Film Thickness on the Surface Roughness of Step Structure. In Proceedings of the 2021 IEEE 16th International Conference on Nano/Micro Engineered and Molecular Systems (NEMS), Xiamen, China, 25–29 April 2021; IEEE: Piscataway, NJ, USA, 2021; pp. 47–50.
31. Misumi, I.; Dai, G.; Lu, M.; Sato, O.; Sugawara, K.; Gonda, S.; Takatsuji, T.; Danzebrink, H.U.; Koenders, L. Bilateral comparison of 25 nm pitch nanometric lateral scales for metrological scanning probe microscopes. *Meas. Sci. Technol.* **2010**, *21*, 035105. [CrossRef]

micromachines

Article

Experimental Investigation of Grinding Force and Material Removal Mechanism of Laser-Structured Zirconia Ceramics

Jingzhu Pang [1,*], Xia Ji [1,2], Yan Niu [1] and Shaojun Chen [3]

1 College of Mechanical Engineering, Donghua University, Shanghai 201620, China; jixia@dhu.edu.cn (X.J.); 2211008@mail.dhu.edu.cn (Y.N.)
2 Shanghai Collaborative Innovation Center of High Performance Fibers and Composites, Donghua University, Shanghai 201620, China
3 YoanTion Industrial Inc., Ltd., Shaoxing 312599, China; junshaochen@163.com
* Correspondence: pangjz@dhu.edu.cn

Abstract: Zirconia is a high demanded structural ceramic with desirable mechanical, thermal, and chemical properties. Poor surface integrity and limited material removal rate caused by high cutting force and wheel wear are the main problems in ceramic grinding. In order to reduce the grinding force and enhance the removal rate in grinding, zirconia ceramics are firstly ablated by laser and then be grinded. A nanosecond laser is used to ablate the surface of zirconia ceramic, the laser-ablated structures with micro pits and thermal microcracks are generated. With the input of subsequent grinding, the machinability of zirconia ceramic workpiece with laser-ablated structures changes. Grinding experiments are conducted to study the grinding force and the material remove of laser-structured zirconia ceramic. Results show that the grinding forces in tangential and normal direction are significantly reduced. Compared to the grinding surface without laser-structured, a damage-free grinding surface is obtained by laser assistance.

Keywords: laser-assisted grinding; grinding force; material removal mechanism; microcracks

Citation: Pang, J.; Ji, X.; Niu, Y.; Chen, S. Experimental Investigation of Grinding Force and Material Removal Mechanism of Laser-Structured Zirconia Ceramics. *Micromachines* **2022**, *13*, 710. https://doi.org/10.3390/mi13050710

Academic Editors: Youqiang Xing, Xiuqing Hao and Duanzhi Duan

Received: 23 March 2022
Accepted: 28 April 2022
Published: 30 April 2022

1. Introduction

Owing to the superior combination of physical and mechanical properties, advanced ceramic materials such as zirconia, silicon nitride, and alumina are widely used in precision bearing of high-grade machines, such as wind turbines, high-grade CNC machine tools, and heat settings. Laser assisted machining (LAM) mainly aims at improving the machining efficiency and cutting performance of difficult-to-machining materials, such as Al_2O_3 [1], Si_3N_4 [2] and ZrO_2 [3]. Based on the local softening mechanism, most LAM use laser to instantly heat and soften the local area of ceramic surface, and then carry out tool cutting, to obtain continuous chips and reduce the cutting force [4].

If the temperature or thermal stress of the material caused by the laser reaches the threshold, irreversible thermal damage, including cleavage, melting and vaporization will be generated [5]. The microcracks induced by thermal stress on the surface and inside will result in the reduction in material strength. Sun et al. [6] pointed out that by preheating the material before cutting, the yield strength of the material can be reduced, so that the ceramic machining mechanism can be transformed from brittle to ductile. Marinescu et al. [7] conducted laser preheating grinding on four ceramic materials, and found that laser preheating grinding can not only reduce the hardness of ceramics and improve the remove rate, but also avoid grinding cracks.

Tsai et al. [8] used laser thermal stress to realize laser assisted milling of pits on the surface of ceramic samples, pointed out that the tensile stress perpendicular to the laser scanning path is the largest, and controlled the growth of cracks. Wang et al. [9] used heat to improve the mechanical properties of quartz glass in the grinding and realized the efficient ductile grinding of quartz glass. Azarhoushang et al. [10] identified that the reduced specific grinding energy through laser structuring of workpiece is mainly due to the induced lateral cracks. Xu et al. [11] also showed that laser irradiation on zirconia ceramics prior to the grinding process led to development of thermal damages on the surface of the workpiece, which contributed to the decrease in the specific grinding energy. Ma et al. [12] built a grinding force predictive model for the LAG process that reveals the mechanism for the reduction in grinding force during LAG on zirconia ceramics. Li et al. [13] revealed the structural changes and hardness decrease enhanced the probability of plastic removal in LAG and obtained better surface integrity. Zhang et al. [3] presented a theoretical grinding force model by taking into account the three grinding stages in laser macro-micro-structured grinding (LMMSG). The model may be available to predict the grinding force of zirconia ceramics under the LMMSG condition.

Kumar et al. [14], Zhang et al. [15] proposed that laser irradiations should be carried out before processing silicon nitride ceramics to induce cracks and microcracks on the surface, reduce the strength of the material and improve the machinability. Pratap et al. [16] investigated the material removal mechanism and analyzed the damage occurrence for parallel as well as intersecting micro grinding. Rao et al. [17] revealed that adhesion and pullout of diamond grits in laser-assisted grinding were remarkable difference from those in conventional grinding due to the thermal softening of both RB-SiC specimen and bond material of the grinding wheel. Kadivar et al. [18] showed that the laser-cut depth could be predicted and choosing optimal laser parameters is possible to achieve a certain laser-cut depth. Yang et al. [19] proposed a heat flux theoretical model based on the dynamic grinding force of hard-brittle bone ductile micro-grinding to solve this bottleneck problem.

In laser-assisted grinding (LAG) process, laser thermal effect and grinding effect are successively loaded on workpiece surface, the grinding force and material removal mechanism will be different from that of grinding. However, few research has been conducted on the benefits of applying LAG on zirconia. In this paper, a LAG process is studied for applying a nanosecond laser ablation before grinding of zirconia. Zirconia workpieces will be laser-structured as the first step, then grinding experiments are carried out to study the effects of laser-generated microstructures and their corresponding parameters on grinding forces and material removal mechanism.

2. Experimental Setup and Procedures

2.1. Laser Pretreatment

The laser conditioning setup is shown in Figure 1. A nanosecond laser (poplar-355-20) is used to ablate the surface of zirconia workpiece. The maximum laser power is 20 W with a wavelength of 355 nm. Zirconia is selected as the workpiece material, of which the size is 15 mm × 15 mm × 15 mm. The main parameters of the laser are shown in Table 1. Four parameters are mainly considered in laser ablation process, such as laser frequency, laser line span, laser scan speed, and irradiation times.

The laser moves vertically and focuses the beam on the workpiece surface. The workbench is equipped with a cross slide so that the laser can follow a specified path on the workpiece surface. Direction A is the scanning direction of the laser, and direction B is the laser line span. Span S_1 and S_2 directly affect the overlap of two adjacent laser spots and have a direct effect on the laser processing efficiency and quality. The parameters of Poplar-355-20 are shown in Table 1.

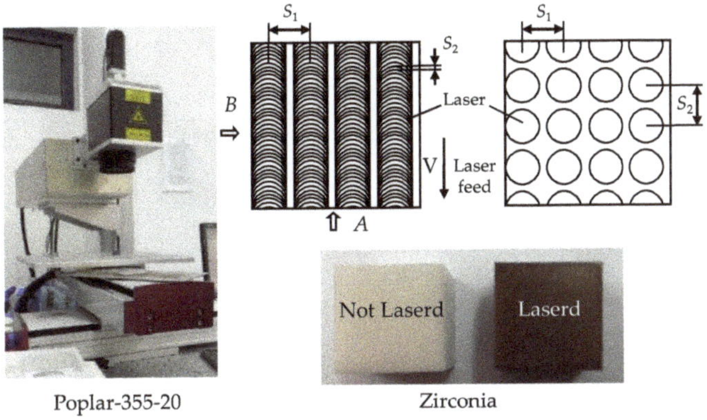

Figure 1. Experimental setup of laser pretreatment.

Table 1. Parameters of Nanosecond UV Laser system.

Characteristics	Symbol	Unit	Values
Average power	P_{ap}	[W]	0.02–20.61
Wavelength	λ	nm	355
Maximum pulse power	L_p	kW	8
Pulse energy	J_e	mJ	0.4
Pulse duration	T_D	ns	50
Pulse frequency	f	[kHz]	20~50
Scan speed	V	[mm/s]	1~5000

2.2. Grinding Force Monitoring

Grinding force experiments on the zirconia samples were carried out on a CNC machining center (ecoMill 635 V, DMG MORI, Shanghai, China) equipped with a dynamometer Kistler 9256 C, as shown in Figure 2. Normal and tangential grinding force components were measured during the grinding process by this dynamometer. The dynamometer is equipped with an amplifier Kistler 5080 A. DAQ system (LMS) was used for data acquisition (sampling frequency 6400 Hz). The diameter of electroplated diamond wheel (HS-JGPC10) is Φ10 mm, and the size of abrasive grain is about 120 μm. Grinding parameters are shown in Table 2.

Table 2. Grinding parameters on DMG 635 V.

Wheel Speed v_s [m/s]	Workpiece Speed v_w [m/s]	Cutting Depth a_p [μm]
1.57	0.0017	15

In up grinding process, the grinding wheel rotates in clockwise and the workpiece moves in directions X and Y. The workpiece moves from the left side of the grinding wheel to its right side, and the grinding force is recorded. When the material is removed by the layer thickness of 15 μm, the workpiece will return to the left, and the grinding for the second layer with the depth of 15 μm will be carried out with the same grinding parameters. In order to carry out grinding experiments with identical conditions, the grinding wheel is replaced in time to retain the same micromorphology.

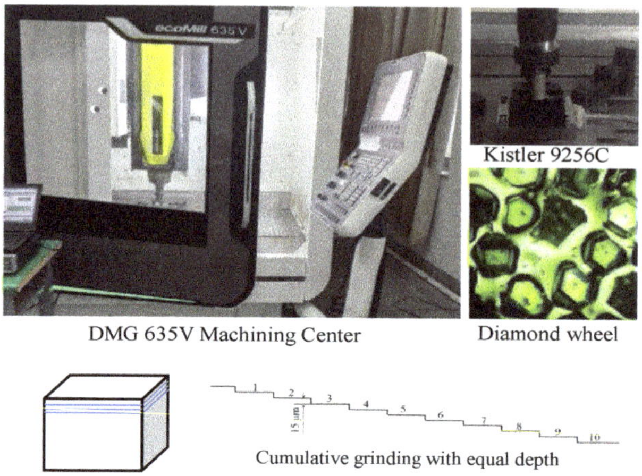

Figure 2. Experimental setup for grinding force monitoring.

2.3. High-Speed Grinding

High-speed grinding of zirconia ceramics with laser ablation was conducted on a high-speed grinding machine (MGKS1332/H, SMTW, Shanghai, China) as shown in Figure 3. A vitrified bond diamond wheel (D91 V + 2046J1SC C150E, Φ400 × 22 × 203.5) was employed. The size of the diamond grit was approximately 91 μm. The diameter of the wheel was 400 mm, and the width was 22 mm. Before up grinding, the wheel was balanced using a dynamic balancing instrument (SB-4500, SMIT, Portland, OR, USA). Zirconia workpiece is slightly inclined and installed on the fixture (the tilted angle is shown in Figure 3). Wheel speed v_s = 60 m/s, workpiece speed v_w = 0.1 m/s, and depth of cut a_p = 15 μm are used for grinding in order to obtain surfaces with different grinding removal depths under the same process conditions.

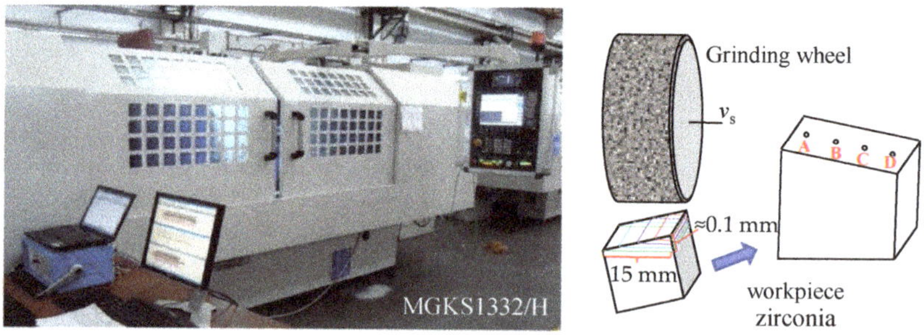

Figure 3. Experimental setup for high-speed grinding.

3. Results and Discussion

3.1. Laser Ablated Surface

The laser pretreatment parameters for zirconia ceramic workpieces are listed in Table 3. Laser spot diameter is about 0.05 mm. The laser line span S_1 is variable, which is 40 μm, 60 μm, 80 μm and 100 μm, respectively. The scan speed can be determined by the distance S_2 and the pulse frequency. If the laser pulse frequency is 49 Hz, S_2 is about 40 μm, the scan speed is about 1960 mm/min. The laser power is changed by adjusting the pulse frequency, and when the pulse frequency is 50 Hz, the laser power equals to 20 W. Three samples are processed for each laser parameters.

Table 3. Laser processing parameters.

Characteristics	Symbol	Unit	Values
Pulse frequency	f	[kHz]	20, 49
Scanning speed	V	mm/min	Determined by f and S_2
Number of scans	N	/	1, 20, 40
Laser line span	S_1	μm	40, 60, 80, 100

The surface microtopography was examined by an environment scanning electron microscope (QUANTA 250, FEI, Brno, Czech Republic). Two laser-structured workpieces with different laser parameters are illustrated in Figure 4. From Figure 4a, it is found that when the laser scanning speed is 4900 mm/min, the lasered structure of group pits are difficult to see after one scan. However, when the laser ablates 40 times at the same position of the workpiece surface, the lasered structure of group pits is clear to see, as shown in Figure 4b.

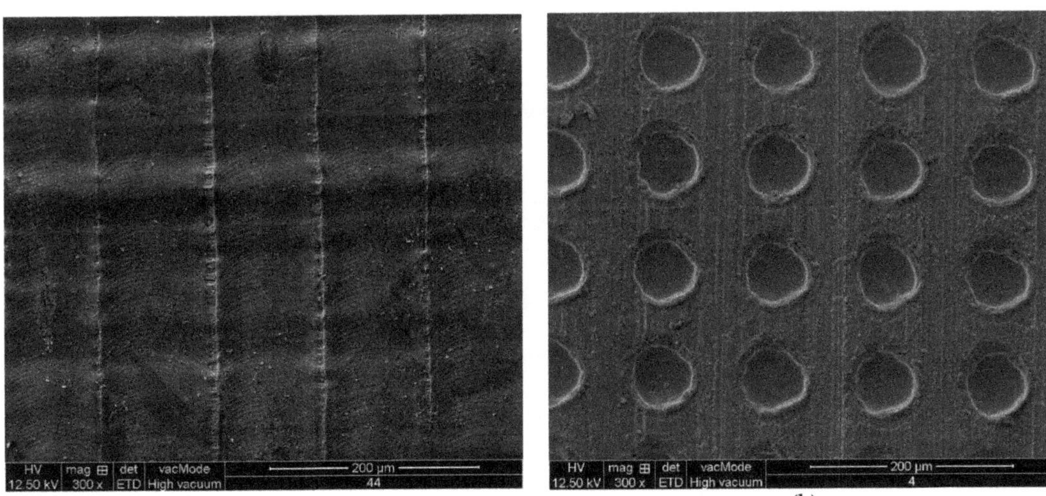

(a) (b)

Figure 4. Two laser-structured workpieces with different laser parameters (f = 49 kHz, V = 4900 mm/min, S_1 = 100 μm). (**a**) N = 1; (**b**) N = 40.

Figure 5 shows the SEM microtopography of the zirconia workpiece. From Figure 5a, we can see that the diameter of the laser spot is about 50 microns. The surface is melted by the laser, which generates heat-affected zone and thermal microcracks. The material around the pit is melted and raised with microcracks at the protrusion. As the temperature of the material increases rapidly, the material expands and compressive stress is generated. When the heat source leaves, the temperature drops quickly and the material shrinks, resulting in the tensile stress which would lead to crack initiation, material fracture and rapid reduction in stress.

From Figure 5b, it is seen that the bottom is relatively flat with lateral cracks on it. The top diameter is about 50 μm, and the depth is about 10 μm. There are no median cracks on subsurface below this pit. In Figure 5b, it is found that the propagation of intergranular microcracks leads to the microcrack network in the laser ablated area. These microcracks caused by laser is initiated by the defects on the ceramic surface and expanded by the thermal stress.

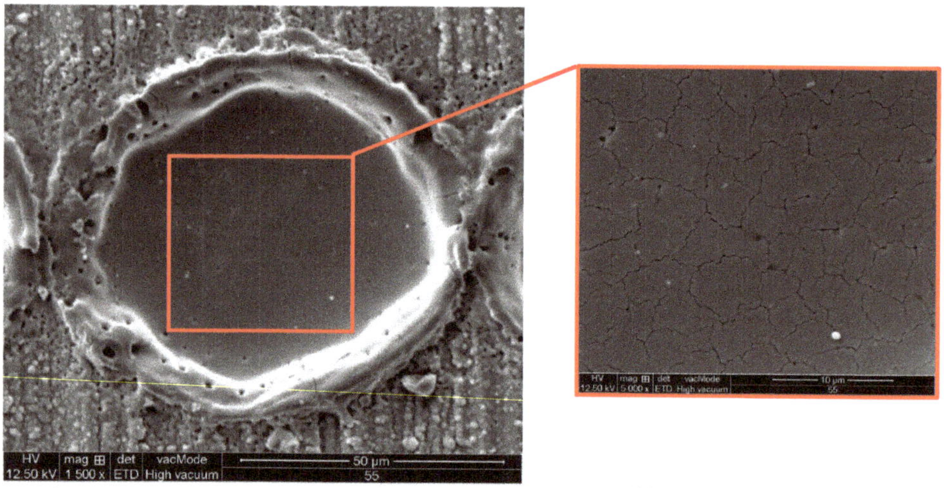

(a)

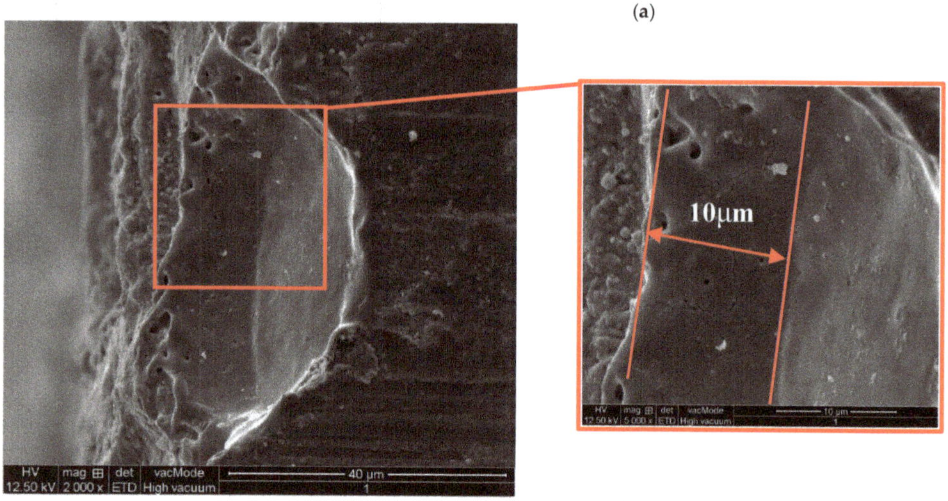

(b)

Figure 5. SEM microtopography of a laser pit on zirconia workpiece (f = 49 kHz, V = 100 mm/min, S_1 = 100 μm, N = 1). (**a**) Top View, (**b**) Section view.

As the laser moves slowly, the laser spots gradually overlapped. The total laser energy on the zirconia surface with scan speed of 100 mm/min by once (as shown in Figure 6) and with scan speed of 3920 mm/min by 40 times (as shown in Figure 5) are approximately equal, while the surface morphology is completely different. The diameter of the laminated arc formed by the laser on the workpiece surface is much larger than the spot diameter. The latter laser reprocesses some areas of the surface affected by the previous laser. The width of a single microcrack on the workpiece surface in Figure 6a,b is greater than that in Figure 5, which also shows that the depth of the microcracks will be relatively deeper. By comparing Figure 6 with Figure 5, it can be noticed that not only lateral cracks, but also median cracks are generated in slow speed scanning. In Figure 6c,d, there are median cracks, which extends to the subsurface of the material by 20~30 μm. The existence of microcracks will reduce the material strength, which is helpful for subsequent grinding.

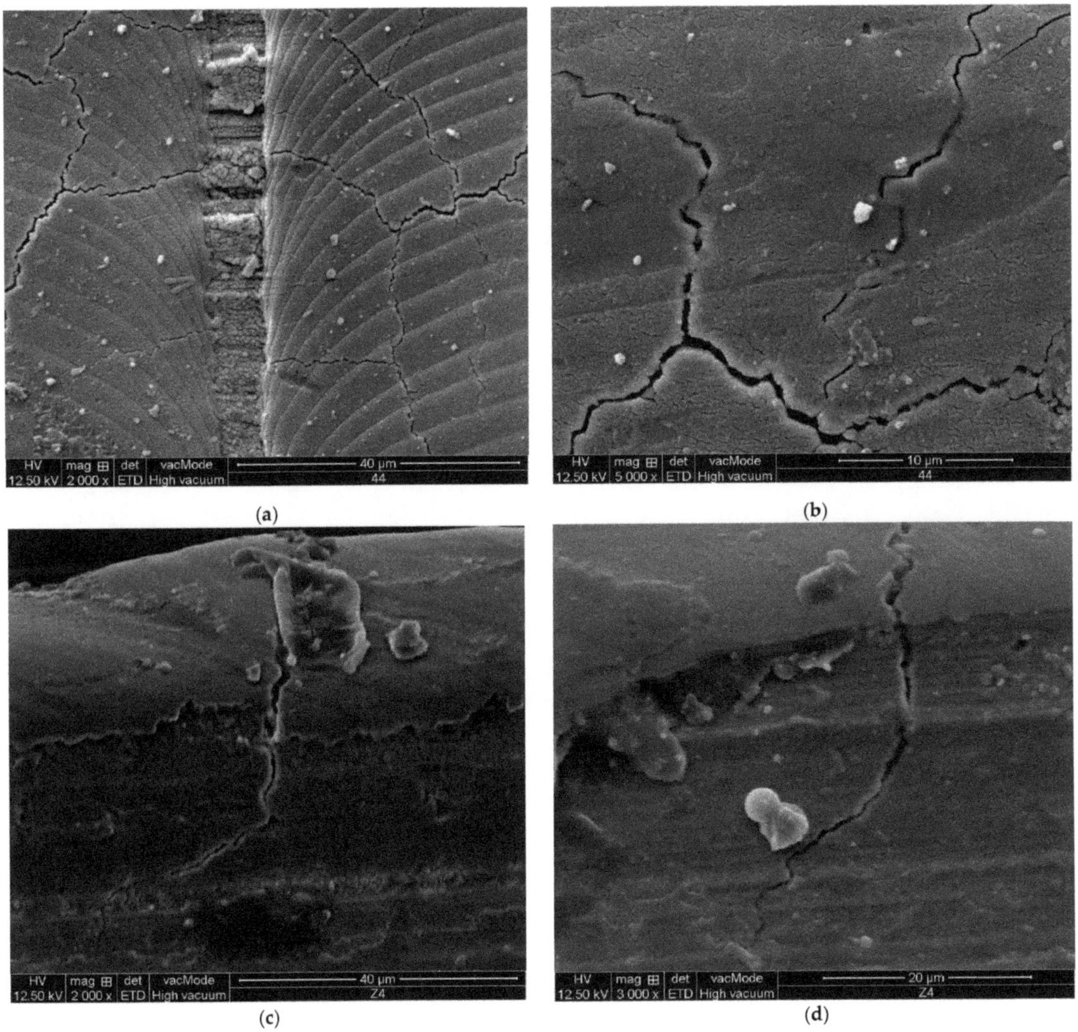

Figure 6. SEM microtopography of cracks on lasered zirconia workpiece (f = 49 kHz, V = 100 mm/min, S_1 = 100 μm, N = 1). (**a**) Lateral cracks. (**b**) Lateral cracks (Enlarged). (**c**) Median crack (Position 1). (**d**) Median crack (Position 2).

3.2. Grinding Force

Figure 7a,b show the grinding force of the workpiece without laser irradiation. The grinding wheel rotates at 3000 rpm, the workpiece approaches the wheel at a speed of 0.1 m/s, and the grinding depth is 15 μm. As the workpiece contacts with the wheel, the amplitude of the signal increases rapidly. As the workpiece leaves the grinding wheel, the amplitude of the grinding force quickly returns to zero.

The original signal is low-pass filtered, and the low-pass frequency is 10 Hz. The tangential grinding force and normal grinding force are shown in Figure 7c,d, respectively. Intercept the signals in the stable stage of grinding, which are shown in the box in Figure 7c,d. Remove the gross errors, and take their mean values as the results of grinding force. The average value of filtered normal grinding force is about 2.5 N, as the dotted line in Figure 7c,d.

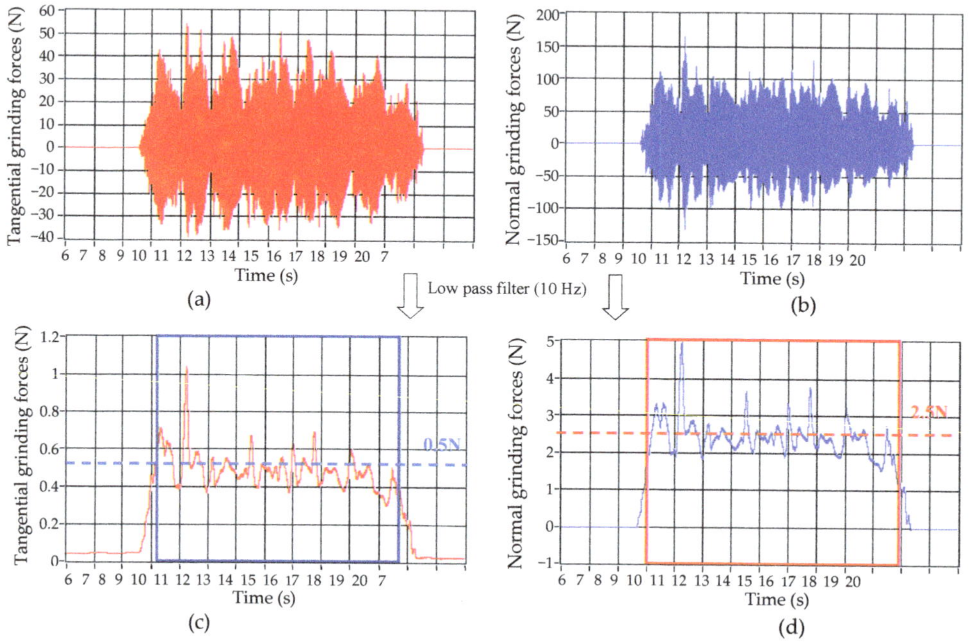

Figure 7. Grinding force (No irradiation, v_s = 3000 RPM, v_w = 0.1 m/min, a_p =15 μm). (**a**) Original signal of Tangential force. (**b**) Original signal of normal force. (**c**) Filtered signal tangential force. (**d**) Filtered signal of normal force.

The normal grinding force signals of the layer with 0–15 μm and 75–90 μm are illustrated in Figure 8. From Figure 8a,b, it is found that for the 0–15 μm layer, the filtered signals of normal grinding force fluctuate greatly, which RMS increases significantly. It is also found that there are micro pits, protrusions and microcracks on the lasered surface, and laser ablation changes the mechanical properties along the workpiece depth. The average amplitude of the filtered signal decreases significantly compared with the value in Figure 7d. From Figure 8c,d, it is found that when the laser ablation is all removed, the normal grinding force returns to about 2.5 N.

The relationship between the normal grinding force and the depth of removal is shown in Figure 9. The data in Figure 9 are the average values calculated by the grinding forces of the workpiece with the same laser parameters. It shows that the normal grinding force increases with the increasing of the grinding depths. When the distance is far away from the workpiece surface, the effect of laser irradiation is not significant. The reason for the rise of grinding force is that the unaffected ceramic materials gradually join the grinding process. The grinding resistance of materials without microcracks is greater than that of materials with laser irradiated microcracks. As the laser-affected layer generated by laser irradiation is removed completely, the grinding force stabilizes to an asymptotic value shown in Figure 7. It was noted that the maximum reduction in grinding force is more than 40%. However, the maximum falling amplitude is not at 0–15 μm layer thickness, but about at 75–90 μm layer thickness. At first, the laser ablation layer on the workpiece surface is removed. As the top layer is removed, there are the intersections of median cracks and lateral cracks on the left surface, which will lead to the decrease in its hardness. As the abrasive grains land on the cracked surface, the cracked surface is further broken and separated from the surface under the action of mechanical stress.

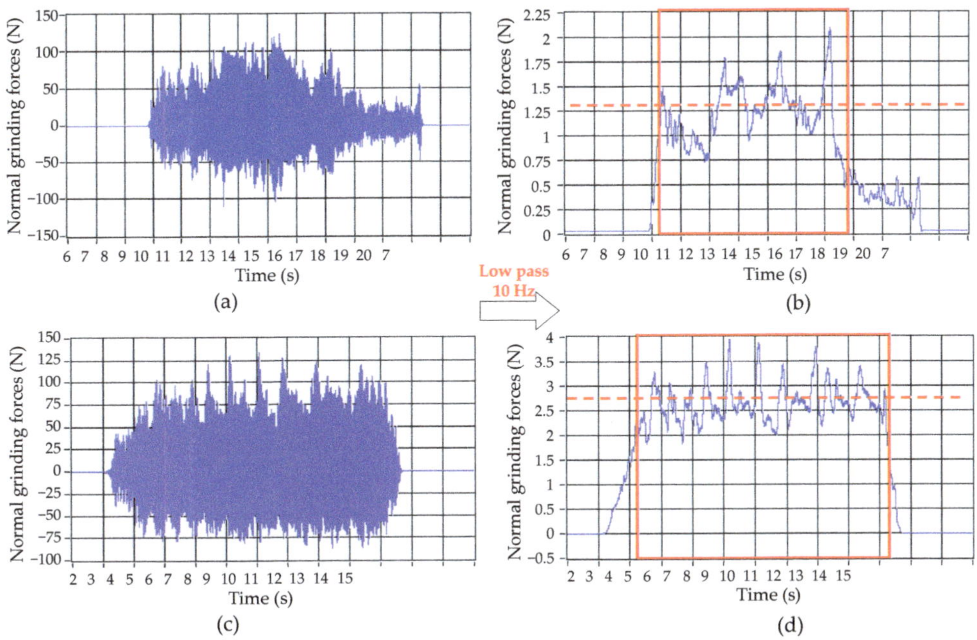

Figure 8. Normal grinding force (k = 20 kHz, V = 1600 mm/min, S_1 = 80 µm, N = 40). (**a**) Original signal (Layer: 0–15 µm). (**b**) Filtered signal (Layer: 0–15 µm). (**c**) Original signal (Layer: 75–90 µm). (**d**) Filtered signal (Layer: 75–90 µm).

From the comparison between Figure 9a,b, it is found that the overall decrease in grinding force is significant under high frequency, and its laser energy is much larger. It is also found that the smaller S_1 is, the stronger the material is affected by the laser and the deeper the ablation influence layer obtains. The trend of normal grinding force of four cases is relatively consistent under high frequency. Under low frequency, the curves of the laser line span 80 µm and 100 µm rise earlier than that of 40 µm and 60 µm, because the laser ablation influence layer of 40 µm and 60 µm is greater than that of 80 µm and 100 µm.

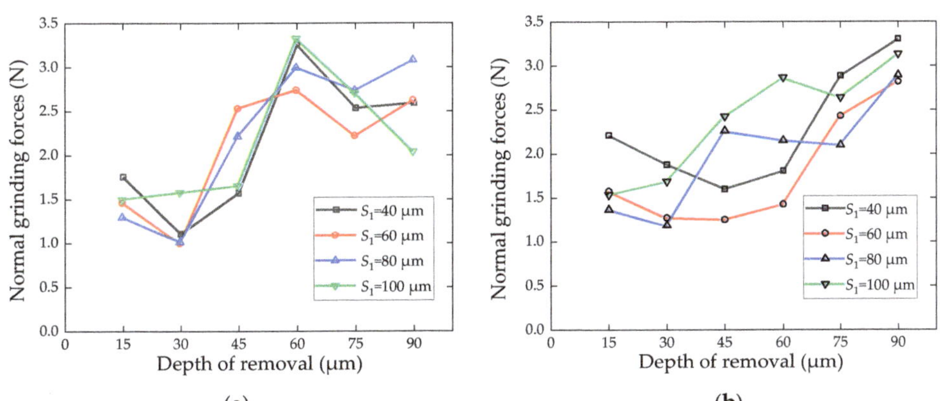

Figure 9. The relationship between the normal grinding force the total material removal thickness (N = 40). (**a**) k = 49 kHz. (**b**) k = 20 kHz.

Grinding force ratio (Ft/F_N) is an index reflecting material brittleness in traditional hard and brittle material grinding. Large brittleness leads to small force ratio. Figure 10 shows that laser ablation pretreatment can improve grinding force ratio. The surface hardness of zirconia ceramics decreases due to laser ablation, and the normal load required for abrasive grains to invade the material decreases. In the initial stage of grinding, there are many brittle fractures in material removal, which also shows that laser irradiation increases the brittleness of ceramics. The influence depth of laser ablation on the grindability of the workpiece exceeds the depth of the ablation influence layer observed in Figures 5 and 6. Compared with Figure 10a,b, it is found that high laser energy leads to high grinding force ratio during workpiece surface removal.

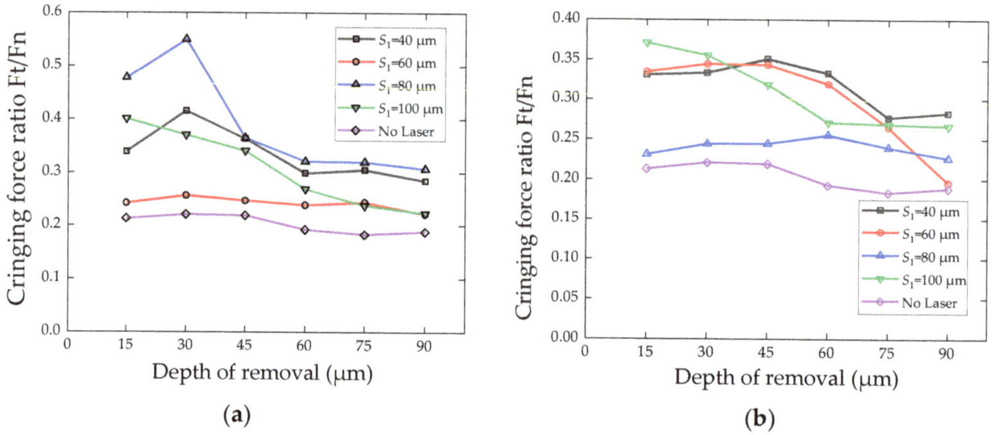

Figure 10. The relationship between the normal grinding force the total material removal thickness (N = 40). (**a**) k = 49 kHz. (**b**) k = 20 kHz.

3.3. Material Removal in Low-Speed Grinding

Figure 11 shows the SEM microtopography of the surface of a workpiece, of which the left side is the grounded area. From Figure 11a, it is found that the remelting protrusion around the original lasered pits are partially removed, and micro grinding chips are left on the surface. In grinding, when the abrasive grains ground material surface, the normal and tangential forces lead to the generation of median cracks and lateral cracks on the material surface and sub surface, respectively. From these SEM microtopography, It can be found that there are many interconnected lateral microcracks on the surface of the pits. The stress generated by the contact between abrasive grains and workpiece will lead to the generation of microcracks. However, propagation of these microcracks caused by grinding will be interrupted by the microcracks already caused by laser. The material surrounded by lateral cracks is broken into smaller fragments. As shown in Figure 11b, the remelting protrusion around the original lasered pits are all removed by grinding, and the damage of median and lateral cracks caused by the previous laser ablation does not affect the integrity of the newly created surface.

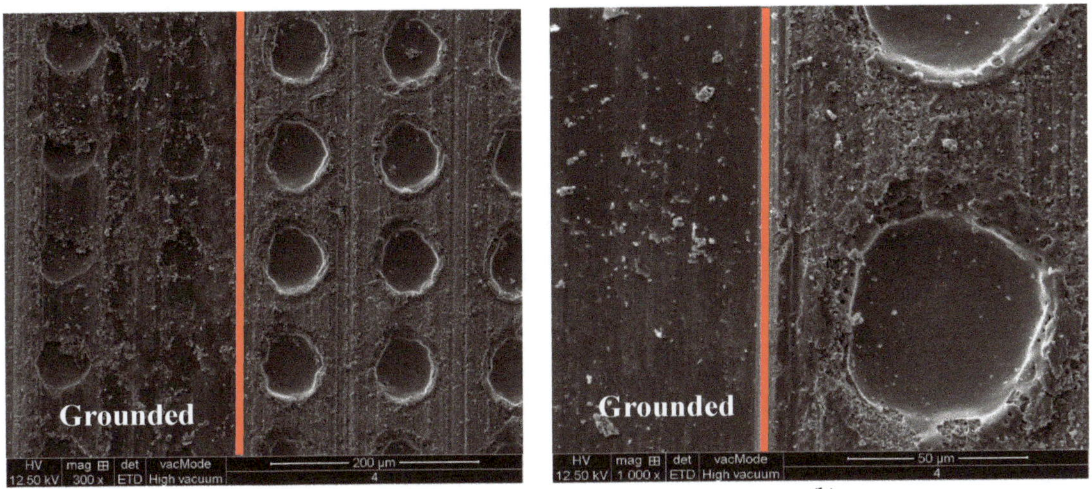

(a) (b)

Figure 11. SEM microtopography of the lasered surface which is half grounded. (**a**) Partially grounded. (**b**) All grounded.

The frequent impacts between grain and workpiece at laser pit will result in powerful mechanical shocks. As shown in Figure 12a, the size of grinding chips is mostly less than 10 μm in size, which is less than the grid size formed by the microcracks at the bottom of lasered pits or around the pits before grinding. Therefore, the new and smaller microcracks are produced by grinding mechanical stress, which improve the further crushing of the material surface. Some broken pits have not yet fallen off the surface. As the grinding process continues, these pits will be further crushed or separated from the surface, as shown in Figure 12b. With the further removal of materials, the depth of lateral cracks remaining on the newly formed grinding surface decreases.

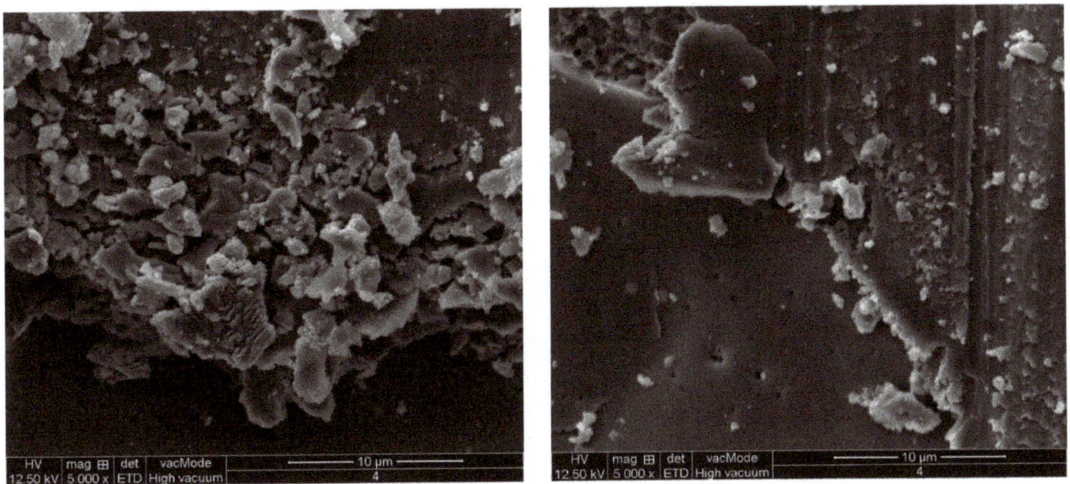

(a) (b)

Figure 12. Grinding chips and plastic scratches near lasered pits. (**a**) Chips. (**b**) Plastic scratches.

3.4. Material Removal in High-Speed Grinding

Figure 13 shows the micromorphology of the workpiece in Figure 3 which is grounded by MGKS1332/H. Figure 13a–d show the surfaces at different positions of the same workpiece after grinding, in which the material removal of surface *A* is the smallest and that of surface *D* is the largest. The layers irradiated by laser on Surface *D* are all removed. From Figure 13, it is found that there are some pits and microcracks on Surface *B* and *C*, while no microcracks are found on Surface *D*.

It is also found that there are obvious grids of laser irradiated microcracks on Surface B. In the enlarged picture of Surface *B*, material fractures and the scratches of abrasive grains can be found. Combined with the previous analysis results of grinding force, the removal of laser ablated layer should be dominated by brittle removal. As the grinding wheel speed increases, the maximum undeformed chip thickness decreases and the material is more easily removed by ploughing. The resistance of the material to crack generation and propagation increases. The speed of the grinding wheel used here is 60 m/s, and the corresponding surface characteristics show that the grinding in the laser ablated layer is in a mixed material removal mode which combined brittle material removal with ductile material removal

On Surface C, the laser-structured and influenced layer is further removed, but the grid traces left by lateral cracks still can be seen on the surface. Obvious grinding marks can be seen on Surface D, on which there is no more visible lateral cracks. There are small and shallow defects and irregularities which are formed by abrasive grains. They are typical features of ductile material removal.

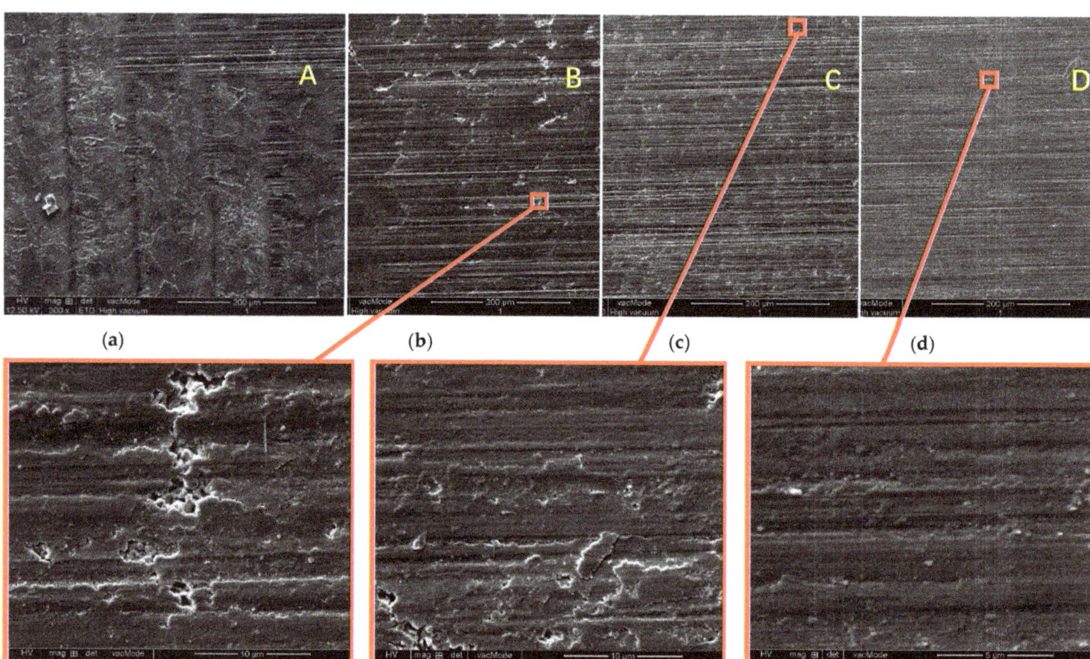

Figure 13. Micromorphology of the workpiece produced by high-speed grinding. (**a**) Surface *A*. (**b**) Surface *B*. (**c**) Surface *C*. (**d**) Surface *D*.

The above results show that the surface cracks introduced by laser ablation will improve the grindability and reduce the grinding force of zirconia ceramic, and the cracks will not expand into the matrix after the influenced layer of laser is completely removed. The surface quality produced by laser-structured grinding is consistent with that of grinding.

4. Conclusions

In this paper, different laser-structured workpieces are made to investigate the grinding force and material removal mechanism of laser-structured zirconia ceramics. The grinding force of the workpiece is experimentally studied, and the maximum reduction in grinding force is more than 40% for the laser ablated surface. The microcracks caused by the laser not only prevent the further propagation of the median cracks generated by grinding, but also improve the material removal rate. The grinding experiments show that low-speed grinding of laser ablation affected layer presents mainly brittle removal, while with the increase in grinding wheel speed, high-speed grinding presents mainly ductile removal.

Author Contributions: J.P.: methodology, formal analysis, writing—original draft preparation; X.J.: writing—review and editing, supervision, project administration, funding acquisition; Y.N.: software, validation; S.C.: Conceptualization. All authors have read and agreed to the published version of the manuscript.

Funding: This research was funded by the National Natural Science Foundation of China (Grant No. 52175384), Natural Science Foundation of Shanghai (Grant No. 21ZR1400600), and Research Project of Shanghai Collaborative Innovation Center of High Performance Fibers and Composites (Gant No. X12812101/024).

Institutional Review Board Statement: Not applicable.

Informed Consent Statement: Not applicable.

Data Availability Statement: Not applicable.

Conflicts of Interest: The authors declare no conflict of interest.

References

1. Chang, C.W.; Kuo, C.P. An investigation of laser-assisted machining of Al_2O_3 ceramics planning. *Int. J. Mach. Tools Manuf.* **2007**, *47*, 452–461. [CrossRef]
2. Lei, S.; Shin, Y.C. Experimental Investigation of Thermo-Mechanical Characteristics in Laser-Assisted Machining of silicon nitride ceramics. *J. Manuf. Sci. Eng.* **2001**, *123*, 639–646. [CrossRef]
3. Zhang, X.; Kang, Z.; Li, S.; Shi, Z.; Wen, D.; Jiang, J.; Zhang, Z. Grinding force modelling for ductile-brittle transition in laser macro-micro-structured grinding of zirconia ceramics. *Ceram. Int.* **2019**, *45*, 18487–18500. [CrossRef]
4. Dubey, A.K.; Yadava, V. Laser beam machining—A review. *Int. J. Mach. Tools Manuf.* **2008**, *48*, 609–628. [CrossRef]
5. Arami, S.; Tabatabae, M.H.; Namdar, S.F.; Chiniforush, N. Effects of Different Lasers and Particle Abrasion on Surface Characteristics of Zirconia Ceramics. *J. Dent. Tehran Univ. Med. Sci.* **2014**, *11*, 233–241.
6. Sun, S.; Brandt, M.; Dargusch, M.S. Thermally enhanced machining of hard-to-machine materials—A review. *Int. J. Mach. Tools Manuf.* **2010**, *50*, 663–680. [CrossRef]
7. Marinescu, I.D. *Handbook of Advanced Ceramics Machining*, 14th ed.; CRC Press: Boca Raton, FL, USA, 2006; pp. 293–300.
8. Tsai, C.H.; Chen, H.W. The laser shaping of ceramic by a fracture machining technique. *Int. J. Adv. Manuf. Technol.* **2004**, *23*, 342–349. [CrossRef]
9. Wang, W.; Yao, P.; Wang, J.; Huang, C.; Zhu, H.; Zou, B.; Liu, H.; Yan, J. Crack-free ductile mode grinding of fused silica under controllable dry grinding conditions. *Int. J. Mach. Tools Manuf.* **2016**, *109*, 126–136. [CrossRef]
10. Azarhoushang, B.; Soltani, B.; Daneshi, A.J.C.A. Study of the effects of laser micro structuring on grinding of silicon nitride ceramics. *CIRP Ann.* **2018**, *67*, 329–332. [CrossRef]
11. Xu, S.; Yao, Z.; Cai, H.; Wang, H. An experimental investigation of grinding force and energy in laser thermal shock-assisted grinding of zirconia ceramics. *Int. J. Adv. Manuf. Technol.* **2017**, *91*, 3299–3306. [CrossRef]
12. Ma, Z.; Wang, Q.; Chen, H.; Chen, L.; Qu, S.; Wang, Z.; Yu, T. A grinding force predictive model and experimental validation for the laser-assisted grinding (LAG) process of zirconia ceramic. *J. Mater. Process. Technol.* **2022**, *302*, 117492. [CrossRef]
13. Li, Z.; Zhang, F.; Luo, X.; Chang, W.; Fei, D. Material removal mechanism of laser-assisted grinding of RB-SiC ceramics and process optimization. *J. Eur. Ceram. Soc.* **2018**, *39*, 705–717. [CrossRef]
14. Kumar, M.; Melkote, S.; Lahoti, G. Laser-assisted microgrinding of ceramics. *CIRP Ann. Manuf. Technol.* **2011**, *60*, 367–370. [CrossRef]
15. Zhang, X.H.; Wen, D.D.; Deng, Z.H.; Li, S.; Wu, Q.P.; Jiang, J. Study on the grinding behavior of laser-structured grinding in silicon nitride ceramic. *Int. J. Adv. Manuf. Technol.* **2018**, *96*, 3081–3091. [CrossRef]
16. Pratap, A.; Patra, K.; Dyakonov, A.A.J.C.I. Experimental analysis of ductile-brittle transitions for parallel and intersecting micro-slot grinding in BK-7 glass. *Ceram. Int.* **2019**, *45*, 11013–11026. [CrossRef]

17. Rao, X.; Zhang, F.; Lu, Y.; Luo, X.; Li, C.J.C.I. Analysis of diamond wheel wear and surface integrity in laser-assisted grinding of RB-SiC ceramics. *Ceram. Intern.* **2019**, *45*, 24355–24364.
18. Kadivar, M.; Shamray, S.; Soltani, B.; Daneshi, A.; Azarhoushang, B. Laser-assisted micro-grinding of Si_3N_4. *Precis. Eng.* **2019**, *60*, 394–404. [CrossRef]
19. Yang, M.; Li, C.; Said, Z.; Zhang, Y.; Li, R.; Debnath, S.; Ali, H.; Gao, T.; Long, Y. Semiempirical heat flux model of hard-brittle bone material in ductile microgrinding. *J. Manuf. Process.* **2021**, *71*, 501–514. [CrossRef]

micromachines

Article

Study on the Material Removal Mechanism of Ultrasonic Elliptical Vibration Cutting of Medical β Titanium Alloy

Zhenda Wang [1], Yongzhi Pan [1], Yijia Zhang [1], Xiuhua Men [1], Xiuli Fu [1,*] and Shengfeng Ren [1,2]

1 School of Mechanical Engineering, University of Jinan, Jinan 250024, China; 202021100286@stu.ujn.edu.cn (Z.W.); me_panyz@ujn.edu.cn (Y.P.); 202121100307@stu.ujn.edu.cn (Y.Z.); me_menxh@ujn.edu.cn (X.M.); me_rensf@ujn.edu.cn (S.R.)
2 Linqing Institute of industry and Technology of Shandong, Liaocheng 252600, China
* Correspondence: me_fuxl@ujn.edu.cn

Abstract: For new medical β titanium implants, the surface micro texture processing technology is a difficult problem. To solve this problem, a new method of ultrasonic elliptical vibration cutting (UEVC) is adopted in this paper. The mechanism of material removal in ultrasonic elliptical vibration cutting is explored for different cutting paths. By means of simulation and experimentation, the material removal mechanism of ultrasonic elliptical vibration cutting medical β titanium alloy is revealed with respect to the aspects of cutting deformation, stress distribution, force and thermal variation, and chip formation mechanism. The results show that: (1) The cutting temperature and cutting force in the UEVC process obey the law of periodic change, and the maximum point of cutting force appears ahead of the maximum point of cutting temperature. (2) The material removal process of UEVC is a "press–shear–pull" composite cutting process. The tool squeezes the material to form the chips. Under the action of high temperature, the material is removed by adiabatic shear. (3) The difference of UEVC paths will affect the removal mode of materials and form different surface morphology. (4) For different cutting paths, compressive stress is distributed at the lowest point of the machining pit, and tensile stress is distributed at the protrusion position.

Keywords: medical β titanium alloy; ultrasonic elliptical vibration cutting; composite cutting process; material removal mechanism

Citation: Wang, Z.; Pan, Y.; Zhang, Y.; Men, X.; Fu, X.; Ren, S. Study on the Material Removal Mechanism of Ultrasonic Elliptical Vibration Cutting of Medical β Titanium Alloy. *Micromachines* **2022**, *13*, 819. https://doi.org/10.3390/mi13060819

Academic Editors: Xiuqing Hao, Duanzhi Duan and Youqiang Xing

Received: 28 April 2022
Accepted: 20 May 2022
Published: 25 May 2022

1. Introduction

The demand for medical implants such as bone implantations and bone replacements is increasing due to joint diseases and the aging population. Titanium alloy has the characteristics of being non-magnetic, corrosion resistant, and possessing high strength and high toughness. In particular, metastable β titanium alloy is a new type of medical titanium alloy. Its elastic modulus is similar to that of human bones. It can effectively avoid the problem of stress shielding. In addition, the material does not contain cytotoxic elements. Therefore, it is favored in the medical field [1,2].

To improve the biocompatibility and wear resistance of medical titanium implants, micro texture on the surface is usually processed in order to realize surface modification [3,4]. At present, laser engraving, acid etching, alkali etching, and other methods are used to process the surface micro texture of titanium implants, but it is difficult to accurately control the geometry and surface morphology of the micro texture [5–7]. Moreover, micro textured surfaces processed by laser are prone to slag, irregular shape, and complex residual stress [8–10]. Therefore, UEVC can be used to solve the problems of difficult machining and unstable surface texture processing quality in titanium implants. It also overcomes the problems of high cutting temperature, heavy tool wear, and poor machining quality in the traditional cutting process [11–14].

UEVC was first proposed by the Japanese scholars Shamoto Eiji and Moriwaki Toshimichi [15]. By applying ultrasonic excitation in two or more directions of the tool, the tool tip can cut the workpiece along an elliptical path. In this paper, ultrasonic excitation is applied in the cutting direction and cutting depth direction of the tool in order to process the pit texture on the surface of the workpiece. Lotfi et al. [16] used UEVC to texture the surface of titanium alloy. It was found that micro texture was formed on the surface of the material under the impact of certain frequencies. Friction and wear tests were carried out. The results showed that the surface micro texture could effectively improve the friction and wear properties of titanium alloy. Zhang et al. [17] processed sinusoidal, sawtooth, oblique wave and other different groove nanostructures on hardened steel by means of amplitude control in ultrasonic elliptical vibration cutting. The restrictions of vibration conditions and tool geometry on machining shape were studied, and a compensation method for amplitude control command was proposed. Yang et al. [18] prepared ordered micro/nano grating structures on the surface of aluminum alloy using the ultrasonic elliptical vibration cutting process. The influence mechanism of micro texture on surface color and optical reflectivity was studied theoretically and experimentally.

The above research verified the machinability of UEVC in the processing of material surface micro texture. However, the high-quality processing of micro texture still needs to be deeply studied with respect to its mechanism, exploring the stress state, removal mode, chip shape and so on. Ma et al. [19] studied the effect of diamond tools on the critical cutting depth of brittle materials under the condition of ultrasonic vibration by performing groove cutting tests on brittle materials. It was found that under the condition of ultrasonic vibration, diamond tools can increase the critical cutting depth for the plastic cutting of brittle materials. Liu et al. [20] conducted molecular dynamics simulation using the improved model to explore the material removal mechanism of monocrystalline silicon under EVC. The results showed that the main material removal mechanism shifts from extrusion to shear in one vibration cycle. In addition, based on stress analysis, it was found that the formation mechanism of subsurface damage in the extrusion and shear stages is different. Huang et al. [21] developed a ductile zone machining model for UEVC of brittle materials based on the plastic zone machining model with the aim of achieving the maximum cutting depth, so as to maximize the machining efficiency while ensuring the machining surface quality. Liu et al. [22] studied the effect of amplitude on machined surface integrity in high-speed ultrasonic elliptical vibration milling of titanium alloy. It was found that the surface roughness increased with increasing vibration amplitude, and the surface residual compressive stress increased with increasing vibration amplitude. Gao et al. [23] used ultrasonic elliptical vibration milling to effectively improve the quality of the machined surface. The research found that, compared with ordinary milling, high-speed ultrasonic vibration milling demonstrated a stable improvement in the tool yield and surface roughness of the machined surface. The above research on the cutting mechanism of UEVC materials mostly focused on a single material removal method. Moreover, there is still a lack of theoretical research on micro texture processing of β titanium implants, which is a difficult-to-machine material.

This paper focuses on the technical problems of β titanium implant processing. The UEVC processing method is adopted. With the help of finite element simulation, the evolution laws of chip morphology, residual stress, and maximum principal stress in the machining process under different cutting trajectories are explored. The influence mechanism of the material removal process is revealed, and a complete model of the material removal process of UEVC is established. This provides theoretical guidance for the processing of the micro texture on the surface of β titanium implants.

2. UEVC Theoretical Model

2.1. Kinematic Model

UEVC is realized by applying periodic ultrasonic excitation with the same frequency and different amplitude to the cutting direction and cutting depth direction of the tool. Finally, the machining of micro texture on the workpiece surface is realized.

As shown in Figure 1, ultrasonic excitation is applied in the X direction (cutting speed direction) and the Y direction (cutting depth direction) to establish a UEVC process under ideal conditions.

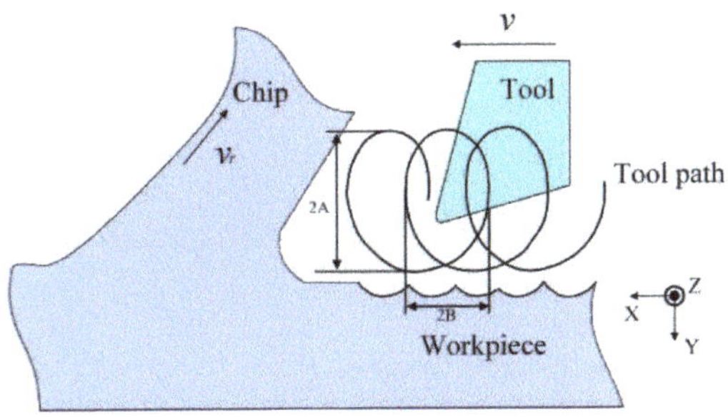

Figure 1. Schematic diagram of UEVC.

The tool tip trajectory equation is as follows:

$$x(t) = Asin(2\pi ft + \varphi) \tag{1}$$

$$y(t) = Bcos(2\pi ft) \tag{2}$$

With the cutting speed, the trajectory equation of the tool tip relative to the workpiece is as follows:

$$x(t) = Asin(2\pi ft + \varphi) + vt \tag{3}$$

$$y(t) = Bcos(2\pi ft) \tag{4}$$

where A and B are the amplitudes in X and Y directions, respectively. f is the ultrasonic vibration frequency, which is 20 kHz in this paper. φ is the phase difference of two-way sinusoidal excitation. v is the cutting speed.

The speed of the tool tip relative to the workpiece can be derived from Equations (3) and (4):

$$v_x(t) = 2\pi fAcos(2\pi ft + \varphi) + v \tag{5}$$

$$v_y(t) = -2\pi fBsin(2\pi ft) \tag{6}$$

Different cutting paths can be obtained by adjusting the parameters (A, B, f, φ and v) according to the above formula. According to Formulas (5) and (6), if $v_x(t) \geq 0$ at any time t, the tool and workpiece will not be separated, which is called non-separated ultrasonic vibration cutting. If t makes $v_x(t) < 0$, there will be a separation stage between the tool and the workpiece, which is called separated ultrasonic vibration cutting [24]. Aiming at the high-quality processing of β titanium implant surface micro texture, this paper only explores the process of separated UEVC.

2.2. Cutting Path Planning

Compared with traditional cutting, UEVC has more flexible trajectory control and more prominent advantages in the processing of material surface micro texture. Different cutting paths can be generated by adjusting the control parameters of cutting path (vibration frequency, phase difference, amplitude and cutting speed). Finally, micro textures with different shapes are processed on the material surface. The single-period trajectory can be obtained by adjusting the phase difference and amplitude using MATLAB software, as shown in Figure 2.

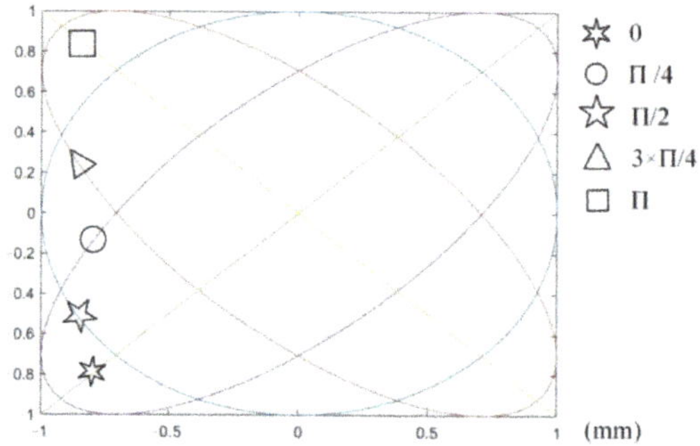

Figure 2. Single-cycle cutting trajectory.

The single-cycle trajectory parameters are adjusted in combination with the cutting speed to obtain the multi-cycle cutting trajectory under different parameters, as shown in Figure 3. In the figure, *A* and *B* in Equations (5) and (6) are taken as 0.005.

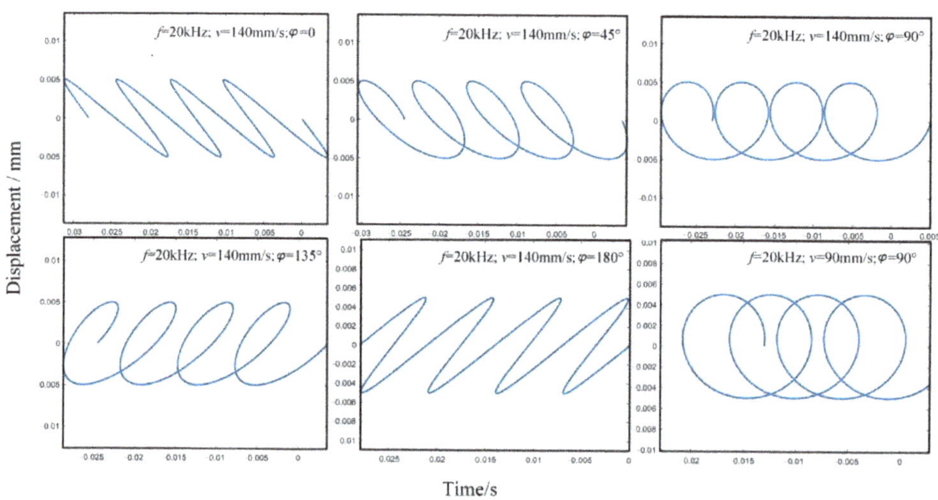

Figure 3. Variation of cutting trajectory under different parameters.

2.3. Construction of UEVC Simulation Model

Limited by experimental conditions such as ultrasonic equipment and external environment, the actual cutting trajectory obtained by adjusting parameters such as excitation voltage and frequency is often different from the theoretical trajectory. In addition, the experiment is difficult to reflect the laws of material surface stress and strain, chip formation and temperature evolution in the whole process of UEVC, so it is impossible to accurately explore the material removal mechanism in the whole process of UEVC. It is difficult to give full play to the unique advantages of UEVC. Therefore, finite element simulation is used to simulate the planned theoretical trajectory. The cutting force, cutting temperature, stress evolution and surface morphology in the cutting process are explored to reveal the UEVC material removal mechanism of medical β titanium alloy.

ABAQUS software is used to establish a 2D UEVC model, as shown in Figure 4. The amplitude and cutting depth of the UEVC process are at the micron level, and the ultrasonic frequency is set to 20 kHz. Therefore, micro-machining is used in the simulation process to more intuitively observe the material removal mode in the cutting process. The workpiece is made of metastable β titanium alloy with a size of 100 × 55 μm. The tool material is a single crystal diamond. In Figure 4, γ_0 is 5° and α is 15°. To obtain better texture morphology, and because the cutting edge of single crystal diamond tool is very sharp, the blunt circle radius of its cutting edge reaches nanometer level. Therefore, the tool model in the simulation ignores the influence of the blunt circle radius of the cutting edge. ALE (Arbitrary Lagrangian Eulerian adaptive meshing) adaptive method is used to divide the mesh to reduce the error caused by excessive distortion of the element mesh in the simulation process. The single precision offset method is used to mesh the rake face and flank of the tool. The workpiece mesh type adopts CPE4RT (four-node plane strain thermally coupled quadrilateral element) and the tool mesh type adopts CPE3T (three-node plane strain thermally coupled triangular element). The mesh size of the upper part of the workpiece is 0.0003 × 0.0003 mm. The mesh size of the lower part is 0.003 × 0.0003 mm. The lower part of the workpiece does not contact with the tool, so the lower part of the mesh adopts a large size to reduce the amount of calculation. The lower left corner of the model is taken as the reference point, and the lower half and bottom of both sides of the model are constrained in a completely fixed way.

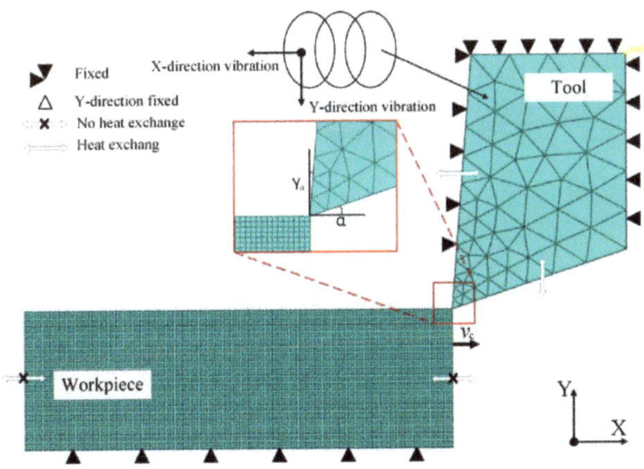

Figure 4. 2D-UEVC simulation model.

β titanium alloy is an elastic–plastic material, so the J-C (Johnson–Cook) constitutive model with simple form and convenient solution is adopted [25]. Because of the introduction of parameters such as the strain strengthening and strain rate strengthening of plastic materials, the dynamic behavior of materials in the cutting process can be accurately described. By analyzing the mesh deformation and stress distribution of materials, the material cutting process is accurately described. The mathematical expression is as follows:

$$\sigma = (A + B\varepsilon^n)(1 + C\ln\dot{\varepsilon})\left[1 - \left(\frac{T - T_r}{T_m - T_r}\right)^m\right] \tag{7}$$

Of which T, T_r and T_m are deformation temperature, room temperature and material melting point, respectively. A is the initial yield stress. B is the strain hardening modulus. n is the hardening index. C is the strain rate sensitivity coefficient. m is the thermal softening index. σ is flow stress. ε is strain. $\dot{\varepsilon}$ is the strain rate. The fracture failure criterion of β titanium alloy materials adopts J-C shear failure cumulative failure criterion.

The definition unit of the J-C fracture failure damage model is as follows:

$$d = \sum \frac{\Delta\varepsilon p}{\varepsilon f} \tag{8}$$

where d is the failure parameter, d = 0–1, initially d = 0, when d = 1, the material fails. $\triangle \varepsilon p$ is the plastic strain increment of one time step. εf is the failure strain of the current time step. Failure strain εf is as follows:

$$\varepsilon f = (d_1 + d_2 exp(d_3 \sigma^*))(1 + d_4 \ln \varepsilon^*)(1 + d_5 T^*) \tag{9}$$

where d_1, d_2, d_3, d_4 and d_5 are material parameters. σ^* is stress triaxiality. $\varepsilon^* = \varepsilon/\varepsilon_0$ is the dimensionless plastic strain rate and ε_0 is the reference plastic strain rate. T^* is dimensionless temperature.

In addition, in order to get better simulation results of chip and cutting force, the friction model in the model is set reasonably. To conform to the sliding friction form between chip and tool surface in the actual cutting process, the friction form in the simulation is defined by Coulomb model. The formula is as follows:

$$\tau_f = \mu \sigma_n \tag{10}$$

where τ_f is the friction stress. μ is Coulomb friction coefficient. σ_n is the normal compressive stress in the contact area.

Based on the established simulation model, the workpiece parameters are set as shown in Table 1. In the table, K represents the coefficient of thermal conductivity of the workpiece and C represents the specific heat of the workpiece.

Table 1. Workpiece parameter settings.

Parameter	Number
A (MPa)	1098
B (MPa)	1092
n	0.93
C	0.014
m	1.1
T_r (°C)	20
T_m (°C)	1680
d_1	−0.09
d_2	0.25
d_3	−0.5
d_4	0.014
d_5	3.87
K	7
C	546,000,000
ε_0	1
μ	0.3

The trajectory parameters are extracted in MATLAB. The extracted parameters are set as the eigenvalues of the periodic function in X and Y directions in ABAQUS, and the velocity periodic function in X and Y directions is established. The cutting path is generated in ABAQUS, as shown in Figure 5. To explore the removal mode of UEVC material under different cutting trajectories, this paper simulates and analyzes the three greatly different cutting trajectories shown in Figure 5. In the figure, A and B in Equations (5) and (6) are taken as 0.005.

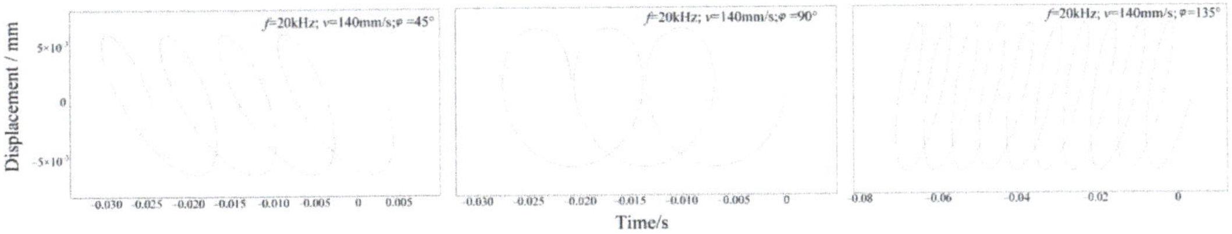

Figure 5. Simulation of cutting path.

3. UEVC Experimental Verification and Analysis

3.1. UEVC Experiment

The experimental material is a new medical titanium alloy represented by metastable β titanium alloy (Ti-25Nb-10Ta-1Zr-0.2Fe). It has the characteristics of high plasticity, high elasticity, and high strength, and has good biomedical properties [2,26]. The cutting device adopted is the Taga ultrasonic elliptical vibration cutting equipment in Japan. It was installed on the MQ-350 two-axis precision lathe for UEVC experiment. Oil cooling is used to cool and lubricate the cutting area. The workpiece is a ring sample, one end of which is fixed to the three-jaw chuck of the lathe, and its end face is processed by UEVC. The schematic diagram of cutting device and workpiece is shown in Figure 6. A single crystal diamond turning tool is selected as the UEVC tool. The rake angle of the tool is 5°, the back angle is 15°, the arc radius of the tool tip is 1 mm, and the blunt radius of the cutting edge is 0.04 μm. The experimental parameters are shown in Table 2.

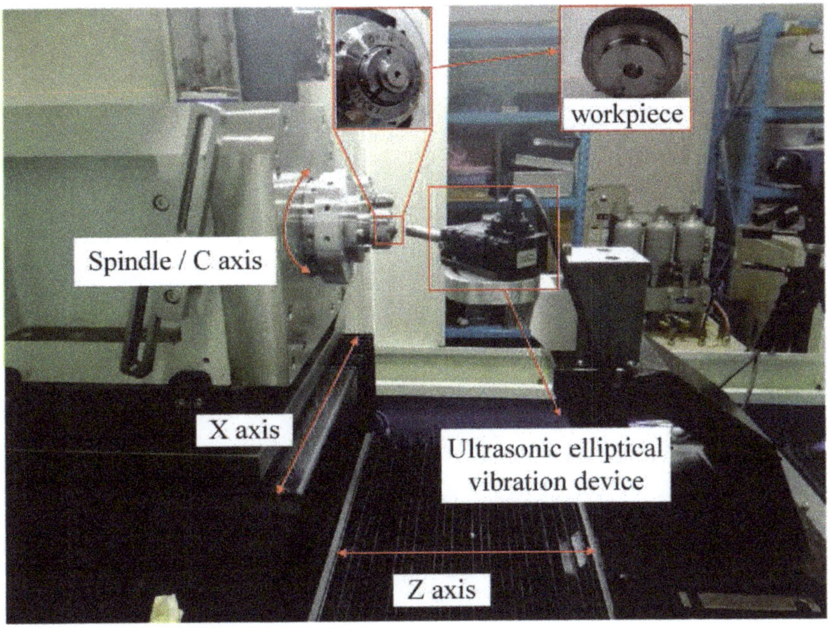

Figure 6. Schematic diagram of cutting device and workpiece.

Machining surface measurement: The surface morphology and roughness of the machined workpiece are measured by WLI-NV5000 5022S. Five areas are taken from each processed sample for inspection. The average value of five areas is taken as the surface roughness value under this machining parameter.

Table 2. Cutting parameter setting.

Group	Cutting Speed V_f (m/min)	Feed Rate F (μm/r)	Cutting Depth a_p (μm)	Ultrasonic Amplitude P-P (μm)	Ultrasonic Frequency (kHz)
1	1	20	6	4	20
2	1	25	6	4	20
3	1	20	3	4	20

3.2. Analysis of UEVC Surface Topography

Figure 7 shows a comparison of the residual height of the machined surface between UEVC experiment and simulation under the same parameters. The dotted line in the figure is the upper and lower limits of the residual height obtained by simulation, and the solid line is the surface residual height obtained by experiment. The residual height obtained by simulation can be extracted by the query function in the software, and can also be obtained by calculating the number of meshes in the figure. Because the tool model in the simulation does not consider the blunt circle radius of the cutting edge, the residual height obtained by the experiment is generally within the upper and lower limits of the residual height obtained by the simulation. Due to the errors in the experiment and the influence of machine tool vibration, the residual height obtained in the experiment fluctuates up and down in a certain range. However, the overall fluctuation trend is within the limits of the simulation results. Combined with the cutting force comparison between the experiment and simulation process under the same parameters as shown in Figure 8, it is considered that the established UEVC simulation model has high accuracy. It can be seen from Figure 8 that the error of cutting force extracted by experiment and simulation is within 20%. The cutting force obtained from simulation and experiment has periodic characteristics. In the stage of tool–workpiece separation, the cutting force is 0.

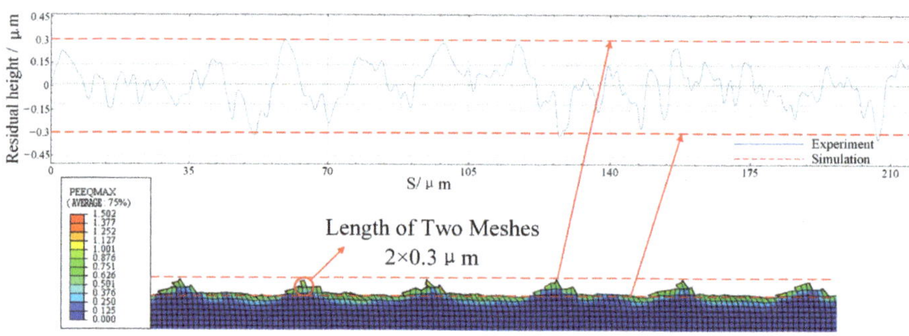

Figure 7. Comparison of machining residual height between experiment and simulation.

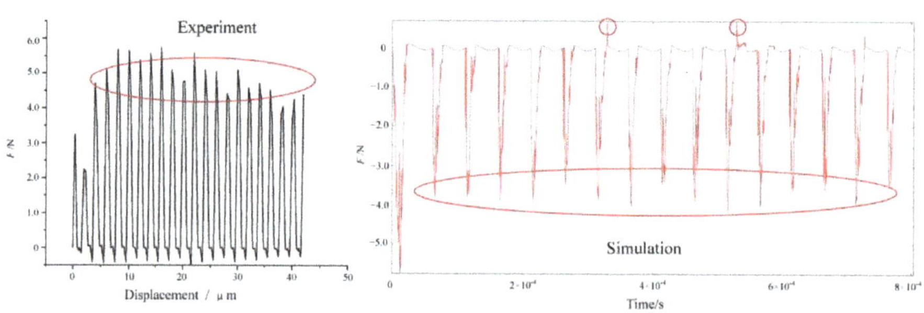

Figure 8. Comparison of cutting force between experiment and simulation.

Figure 9 shows the three-dimensional surface morphology of an area of UEVC workpiece, and the feed rate of Figure 9a is 20 μm/r, the feed rate of Figure 9b is 25 μm/r, the other parameters are consistent. It can be seen from the figure that there are evenly arranged pits on the workpiece surface processed by UEVC along the cutting direction. According to the analysis of the UEVC cutting characteristics, the pit is the result of the material removal by the periodic vibration of the tool along the cutting depth direction. By measuring the surface roughness of the workpiece, it is found that when the feed rate is 20 μm/r, the surface roughness of the material is much lower than that of the workpiece with the feed rate of 25 μm/r. The main reason for this is that when the feed rate is too large, the plowing effect of the tool on the material is more obvious. Due to the large amount of material removal and high temperature in the cutting area, the effect of UEVC on inhibiting furrow and scale thorn is gradually weakened. At the same time, with increasing cutting depth, the surface roughness also increases. Mainly due to the increase of cutting depth, the amount of material removed increases, and the contact stress between tool and workpiece increases, which makes the plastic deformation of the workpiece surface more intense.

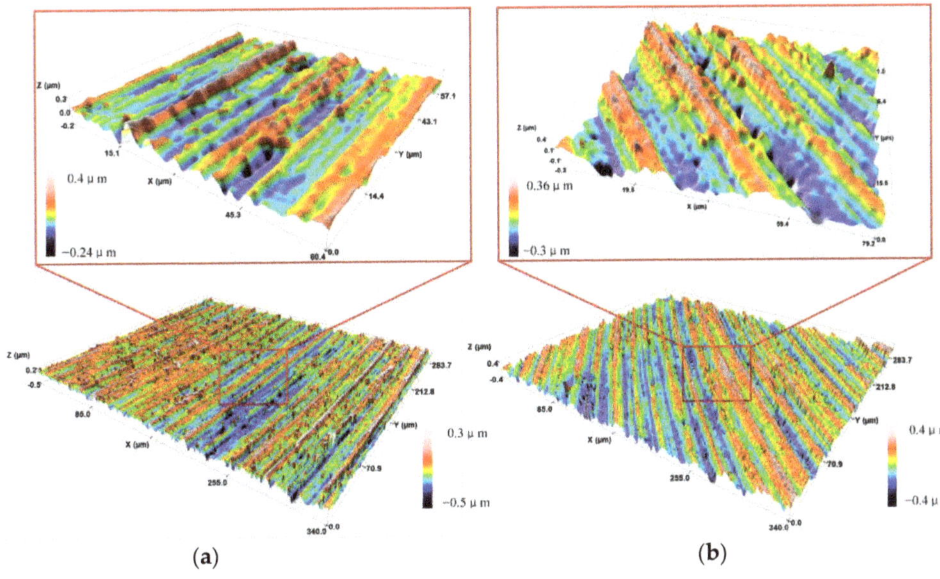

Figure 9. Three-dimensional surface morphology of UEVC workpiece. (**a**) The feed rate is 20 μm/r; (**b**) the feed rate is 25 μm/r.

4. Removal Mechanism of UEVC of Medical β Titanium Alloy

For UEVC machining of the surface micro texture of medical β titanium alloy, exploring the material removal mechanism in the cutting process can provide theoretical guidance for high-quality machining of the surface texture. Because the vibration frequency of the UEVC process reaches 20 kHz, it is difficult for human eyes to observe the material and chip formation mechanism in the cutting process. Therefore, based on the reproduction characteristics of the whole process of simulation, this paper explores the material removal mechanism in UEVC, including material cutting deformation and stress distribution, mechanical and thermal variation law, chip formation mechanism, and so on.

4.1. Analysis of UEVC Cutting Mechanism

Figure 10 shows the evolution diagram of the force–thermal evolution law of the UEVC process. The circle in Figure 10 indicates the extreme points of cutting force and cutting temperature in each cutting cycle of UEVC. It can be seen from the curve in the figure that UEVC has the characteristics of tool-workpiece cycle separation. Compared with the high-temperature process at the moment of traditional cutting, the overall temperature rise trend of the UEVC process is slower. Under the simulation parameters of v = 140 mm/s, f = 20 kHz, a_p = 0, t = 0.0008 s, the maximum temperature of the UEVC process does not exceed 120 °C. Combined with the temperature cloud diagram and the temperature change curve, it can be seen that when the tool is not separated from the workpiece, the cutting temperature at the tool tip begins to decrease. The reason for this is that at this time, the speed of the tool along the positive direction of the Y-axis gradually increases, while it decreases along the X direction, and the contact stress between the tool and the workpiece gradually decreases. At this time, the temperature is at the maximum point of the cycle, the material plasticity is strong, and the stress state between the tool and the workpiece is weakened. The resistance of the tool when cutting the workpiece decreases, the tool–workpiece interface is about to enter the separation state, and the cutting temperature begins to decrease. Based on the temperature nephogram, the maximum temperature point in the temperature curve is extracted. It can be observed that the maximum temperature point in each cutting cycle occurs at the time at which the chip is about to break or at the beginning of fracture. At this time, the contact area and contact stress between the tool and the workpiece reach the extreme point.

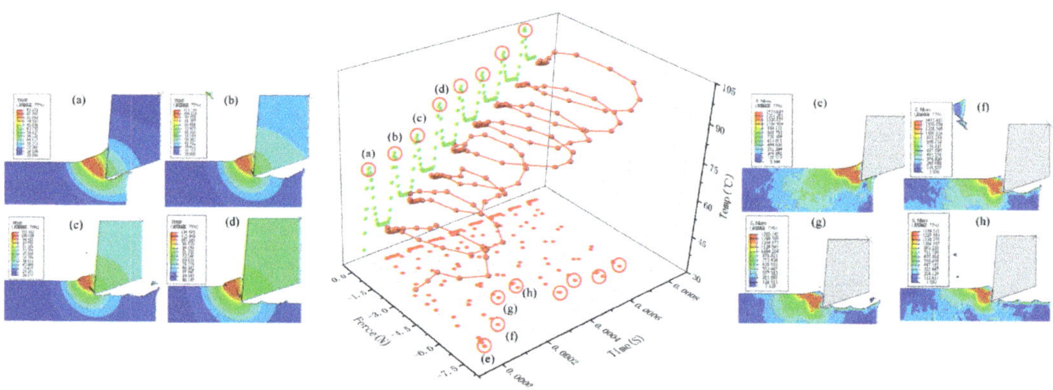

Figure 10. UEVC force–thermal evolution law (Temp-Cutting Temperature).

At the same time, it can also be found that the cutting force has a maximum point in each cutting cycle that appears before the maximum point of cutting temperature (the stress state at the maximum point of temperature is weakened). The maximum point of cutting force in each cutting cycle appears at the greatest distance between the tool tip and the workpiece along the cutting direction, that is, with the maximum value of material removal. In the first cutting cycle, the maximum value of cutting force is greater than that of the other cutting cycles. The reason for this is that the material removal is the highest in the first cutting cycle. At this time, the cutting temperature is low, and the plasticity of the material is weak, so the bonding state between grains is strong, and the cutting resistance is large [27].

4.2. Analysis of Material Removal Process

The maximum principal stress, cutting temperature and plastic deformation of the material during a cutting cycle of UEVC are extracted in the simulation, as shown in Figure 11.

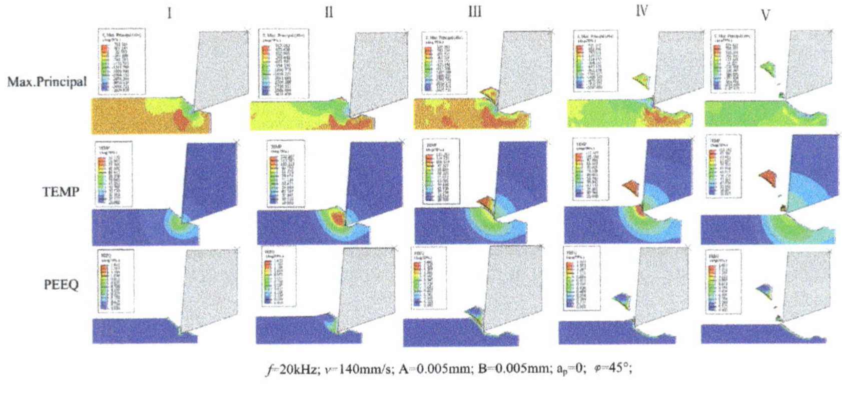

(**a**)

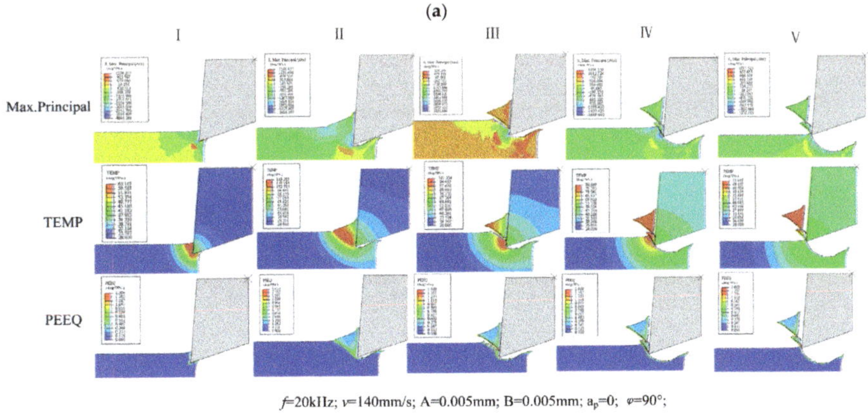

(**b**)

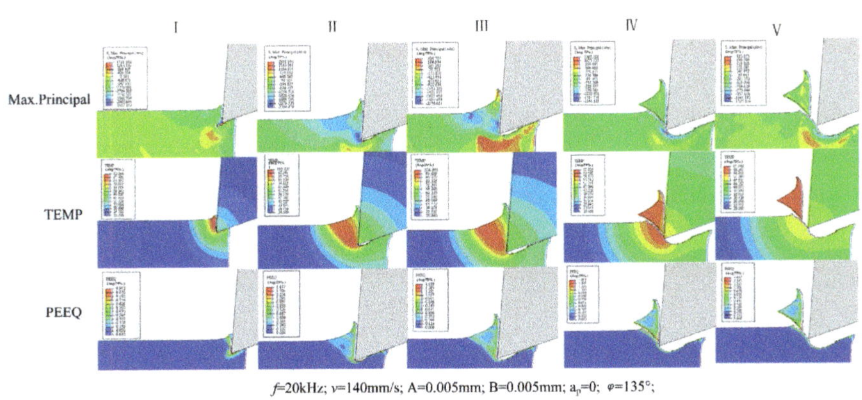

(**c**)

Figure 11. Analysis of UEVC material removal process (Max.Principal- Maximum Principal Stress, PEEQ- Plastic Strain). (**a**) Phase difference 45°; (**b**) phase difference 90°; (**c**) phase difference 135°.

It is found that under different cutting conditions and vibration parameters, the material removal process of UEVC in one cutting cycle can be divided into five stages. In the first stage, the tool just cuts into the material, at this time, the material undergoes brittle fracture, and the high temperature is mainly concentrated at the tool tip. In the second stage, the tool completely cuts into the material, and the high temperature zone is located in the material shear zone in the direction of tool cutting speed. In the third stage, the chips are about to separate, and the cutting temperature begins to drop. In the fourth stage, the chips have separated, and the tool starts to cut the material again or press the material. In the fifth stage, the tool starts to exit the cutting process and the cutting cycle ends.

Figure 11a,b show the material removal mechanism of the UEVC process when the phase difference is 45° and 90°. It can be seen that the material removal methods are basically the same under the two cutting tracks. In the first stage of the cutting cycle, the tool tip just cuts into the workpiece. At this time, the cutting temperature is low, the material plasticity is weak, and there is low plastic deformation. It can be seen from the maximum principal stress diagram that the material is brittle under the extrusion of the tool. The cutting cycle enters the second stage, and the tool completely cuts into the material. At this time, the cutting heat is mainly concentrated in the material shear zone of the tool tip along the cutting speed direction. Under the influence of high temperature, the plasticity of the material is enhanced. Under the extrusion action of the cutting tool, the material produces high plastic deformation. Under the action of high temperature and tool extrusion, chip is gradually produced on the material surface, and adiabatic shear occurs. In the third stage of the cutting cycle, the vibration of the tool along the Y direction changes from negative to positive. The tool further extrudes the gradually generated chips, and the material undergoes adiabatic shear. Then the chip breaks and the fracture surface exhibits high plastic deformation, and the cutting temperature begins to drop. Through the above three stages, it can be found that the material removal process of UEVC is a state of "press–shear–pull" composite cutting. Firstly, the tool extrudes the material to preliminarily form chips. A shear force is then applied to the formed chip. The plastic deformation at the chip fracture indicates that a certain tensile stress is generated on the surface during the fracture. Under this cutting path in Figure 11a,b, the machined surface produces certain residual protrusions under the action of the tensile stress of the chip. The tool will then perform secondary cutting on the machined surface. In the fourth stage of cutting, the tool continues to vibrate upward. The vibrating cutter carries out secondary cutting on the machined surface and produces certain chips to make the machined surface more flat. In the fifth stage of cutting, the tool starts to exit the cutting state and enter the empty cutting stage of a cutting cycle. At this stage, the tool and material are separated. At this time, the cutting area is opened, and the lubricating fluid can fully lubricate the cutting area.

Figure 11c shows the material removal mechanism of the UEVC process when the phase difference is 135°. Due to the difference of cutting path, in the first stage, the heat at the tool tip is mainly concentrated on the front face of the tool, and the heat at the back face is small. In addition, different from the UEVC process when the phase difference is 45°and 90°, there is no secondary cutting in the cutting process when the phase difference is 135°. Only one chip is produced in the whole cutting cycle, as can be seen from the analysis of the cutting trajectory in Figure 3. When the phase difference is 135°, the cutting speed of the tool along the X direction changes from negative to positive, which is earlier than the transition node in the Y direction. The tool exits the cutting state in advance. The secondary cutting of the machined surface is transformed into ironing. Different from the residual tensile stress produced by secondary cutting, there is a high residual compressive stress on the material surface. During the second stage, it can be seen that under this cutting trajectory, the compressive stress distribution in the material shear zone is more obvious. Combined with the temperature nephogram, it is found that the cutting temperature at this time is significantly lower than that at phase differences of 45° and 90°. Therefore, the plasticity of the material is weak, and the shear force required for material removal is greater.

Figure 12 shows the temperature nephogram of UEVC material removal process at different cutting speeds. It can be seen from the figure that there is no significant difference in the material removal process when the cutting speed increases from 80 mm/s to 200 mm/s. It is consistent with the removal method under the phase difference of 90° analyzed above. From the temperature nephogram, it can be seen that the cutting temperature changes significantly under different cutting speeds.

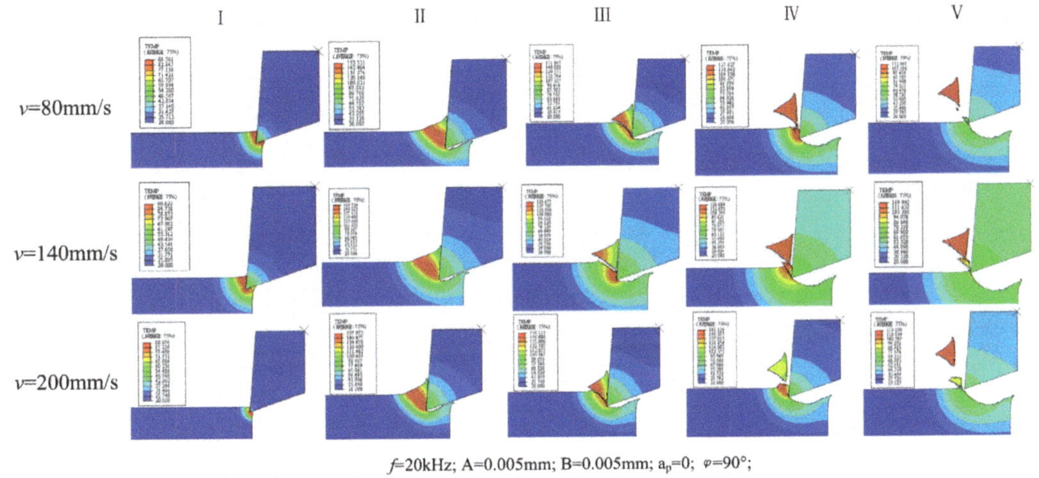

Figure 12. Material removal process at different cutting speeds.

Based on the above analysis, it can be concluded that the difference in cutting trajectory under phase difference control has a significant impact on the material removal mode. The influence of cutting speed on material removal is not obvious.

4.3. Surface Machining Quality Analysis of UEVC

The residual stress of UEVC machined surface with phase differences of 45°, 90° and 135° is extracted to obtain the evolution diagram of residual stress on the UEVC material surface, as shown in Figure 13.

Through the analysis of the machined surface topography under the two cutting trajectories in Figure 13, it can be obtained that different cutting trajectories will produce different morphologies on the material surface. For machining of different surface topographies, different cutting trajectories can be generated by adjusting the cutting parameters and ultrasonic parameters. It can be found from the curve that under the two cutting tracks, the residual stress on the machined surface is mainly compressive stress, but the tensile stress is also intermittently distributed. In addition, the tensile stress distribution range is wide when the phase difference is 135°. Compared with the surface residual stress diagram, it can be found that the tensile stress appears at the surface protrusion position, that is, the position where one effective cutting cycle (tool–workpiece contact) ends and another effective cutting cycle begins. The reason for this is that in the later stage of the effective cutting cycle, the tool begins to pull the material upward, resulting in tensile stress at the protrusion position [28]. In the next cutting cycle, the tool cuts the material using tensile stress, resulting in a small distribution range of tensile stress on the machined surface. Compared with the phase difference of 45° and 90°, the tensile stress distribution range of machined surface with 135° phase difference is wider. By analyzing the cutting path when the phase difference is 135°, it is found that the lifting stage of the tool is long.

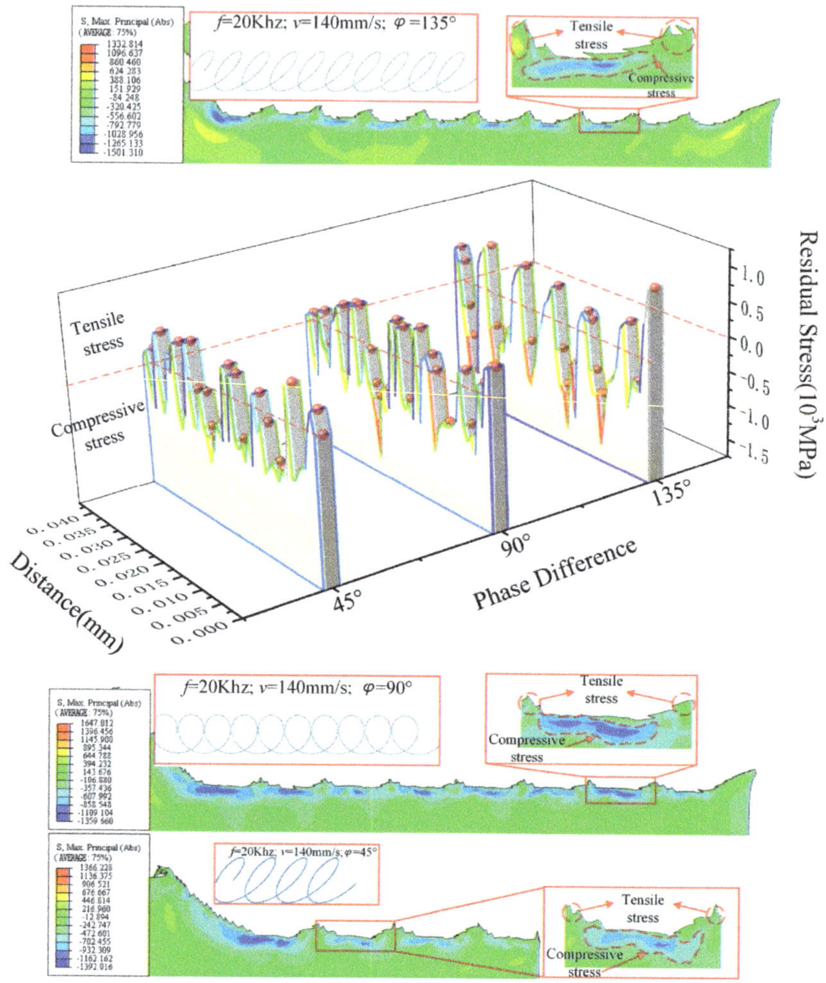

Figure 13. Evolution of surface residual stress of UEVC material.

The plastic deformation value of the UEVC machined surface is extracted to obtain the evolution diagram of plastic deformation of the UEVC material surface, as shown in Figure 14. By analyzing the plastic deformation of the machined surface, it can be concluded that the degree of plastic deformation of the machined surface with phase difference of 45° and 90° is higher than that with a phase difference of 135°. In a cutting cycle under the two cutting trajectories, the plastic deformation value of the machined surface first decreases and then increases. At the same time, it can be found that the thickness of plastic deformation layer first increases and then decreases during a cutting cycle. The main reason for this is that there is compressive stress at the lowest point of the machined surface, and the stress transmission range is wider [29]. It leads to a wider range of plastic deformation along the depth direction. The reason the plastic deformation value is opposite to the thickness of the deformation layer is the different effect of stress [30]. At the beginning of a cutting cycle, the tool cuts downward, and the material surface bears the tensile stress in the cutting direction. Therefore, high plastic deformation occurs along the cutting direction, but the effect of compressive stress along the depth direction is weak. When the tool is near the lowest point, the material surface mainly bears the compressive stress along the depth direction, which is mainly manifested in the greater depth of plastic deformation. The different performance of the two trends further shows that there is a "press–shear–pull" transformation in the removal process of UEVC material.

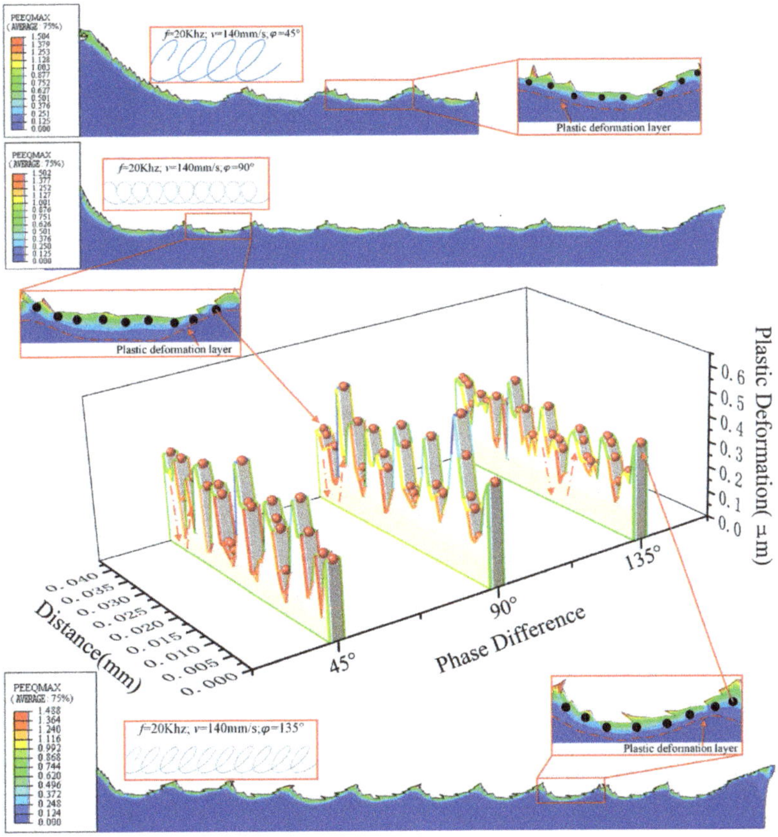

Figure 14. Evolution of surface plastic deformation of UEVC materials.

5. Conclusions

In this paper, a UEVC finite element simulation model was established, and the UEVC cutting trajectory is planned. The cutting mechanism of the UEVC process is explored through simulation analysis. Through the analysis of cutting deformation, stress distribution, force heat change law and chip formation mechanism in each stage of the UEVC process, the mechanism of micro removal of materials in the UEVC process was obtained. The results show that:

(1) The cutting temperature and cutting force in the UEVC process follow a law of periodic change. Different from traditional cutting, the cutting temperature in the UEVC process has a decreasing stage. The maximum point of cutting force in each cutting cycle is when the material removal is the greatest. The maximum point of cutting force is ahead of the maximum point of cutting temperature.

(2) The removal process of UEVC material is a "press–shear–pull" composite cutting process. When the tool cuts into the material, the material is brittle. With increasing temperature, the material undergoes plastic deformation. The tool extrudes the material to form the chips, and the material is removed by adiabatic shear under the influence of high temperature.

(3) Differences in UEVC trajectories affect the removal mode of materials and form different surface morphologies. In a cutting cycle, when the phase difference is 45° and 90°, there is secondary cutting in the cutting process. When the phase difference is 135°, the tool presses the location of chip separation following chip separation. Both methods make the machined surface more flat.

(4) For different cutting paths, compressive stress is distributed at the lowest point of the machining pit, and tensile stress is distributed at the protrusion position. The plastic deformation value of the surface layer of the machined surface first decreases and then increases, and the thickness of the surface plastic deformation layer first increases and then decreases. Compared with phase differences of 45° and 90°, the distribution range of compressive stress and plastic deformation of machined surface at 135° phase difference are smaller.

Author Contributions: Conceptualization, Z.W. and X.F.; methodology, Y.P.; software, X.M.; validation, Y.Z. and S.R.; formal analysis, Y.P.; investigation, Y.Z.; resources, X.M.; data curation, Z.W.; writing—original draft preparation, Z.W.; writing—review and editing, X.F.; visualization, X.F.; supervision, Y.P.; project administration, X.F.; funding acquisition, X.F. All authors have read and agreed to the published version of the manuscript.

Funding: This research was funded by "the General program of National Natural Science Foundation of China (5217052208), Scientific research leader studio project of Jinan (2019GXRC054), General project of Shandong Natural Science Foundation (ZR202102280460) and Intelligent Bearing Manufacturing Innovation and Entrepreneurship Community Funded project of Shandong" and "The APC was funded by 5217052208".

Conflicts of Interest: The authors declare no conflict of interest.

References

1. Zheng, X. *Research and Application of Biomedical Coating Materials*; Film and coating branch of China Silicate Society, The Chinese Ceramic Society: Jing De Zhen, Chian, 2017; Volume 1.
2. Siemers, C.; Bäker, M.; Brunke, F.; Wolter, D.; Sibum, H. Aluminum-and vanadium-free titanium alloys for application in medical engineering. In *Titanium in Medical and Dental Applications*; Woodhead Publishing: Sawston, UK, 2018; pp. 477–492.
3. Wang, P.; Yang, L.; He, F. Review on Technology of Surface Micro-nano Structure and Biological Properties on Titanium and Titanium Alloy Implant Surface. *Titan. Ind. Prog.* **2020**, *37*, 41–48.
4. Bai, Y.; Sun, X.; Wang, Y.; Zhang, X. Experimental study on the promotion of osseointegration of titanium implants with biomimetic electroactive coating. *J. Clin. Orthop. Res.* **2021**, *6*, 20–25.
5. Wang, G.; Wang, Y.; Wang, T.; Zhang, D. Review on Preparation of Micro-nano Structure on Implant Surface and Its Biocompatibility. *Surf. Technol.* **2016**, *45*, 8–18.
6. Hauschwitz, P.; Jochcová, D.; Jagdheesh, R.; Rostohar, D.; Brajer, J.; Kopeček, J.; Cimrman, M.; Smrž, M.; Mocek, T.; Lucianetti, A. Towards rapid large-scale LIPSS fabrication by 4-beam ps DLIP. *Opt. Laser Technol.* **2021**, *133*, 106532. [CrossRef]
7. Mahajan, A.; Devgan, S.; Sidhu, S.S. Surface alteration of biomedical alloys by electrical discharge treatment for enhancing the electrochemical corrosion, tribological and biological performances. *Surf. Coat. Technol.* **2021**, *405*, 126583. [CrossRef]
8. Ripoll, M.R.; Simič, R.; Brenner, J.; Podgornik, B. Friction and Lifetime of Laser Surface–Textured and MoS_2—Coated Ti6Al4V under Dry Reciprocating Sliding. *Tribol. Lett.* **2013**, *51*, 261–271. [CrossRef]
9. Kumari, R.; Scharnweber, T.; Pfleging, W.; Besser, H.; Majumdar, J.D. Laser surface textured titanium alloy (Ti–6Al–4V)—Part II—Studies on bio-compatibility. *Appl. Surf. Sci.* **2015**, *357*, 750–758. [CrossRef]
10. Wu, Z.; Xing, Y.; Huang, P.; Liu, L. Tribological properties of dimple-textured titanium alloys under dry sliding contact. *Surf. Coat. Technol.* **2017**, *309*, 21–28. [CrossRef]
11. Xie, H.; Zhang, Y.; Meng, Z. Characteristics of β titanium alloy and its application and research progress in orthopedics. *Orthop. Biomech. Mater. Clin. Study* **2013**, *10*, 29–32.
12. Jain, A.; Bajpai, V. Mechanical micro-texturing and characterization on Ti6Al4V for the improvement of surface properties. *Surf. Coat. Technol.* **2019**, *380*, 125087. [CrossRef]
13. Pratap, T.; Patra, K. Direction dependent dynamic wetting of semi-hemispherical end micro-groove textured Ti-6Al-4V surface. *Surf. Coat. Technol.* **2018**, *356*, 138–149. [CrossRef]
14. Lucia, L.; Marco, S.; Rachele, B.; Andrea, G.; Stefania, B. Anisotropy effect of additively manufactured Ti6Al4V titanium alloy on surface quality after milling. *Precis. Eng.* **2020**, *67*, 301–310.
15. Eiji, S.; Toshimichi, M. Study on Elliptical Vibration Cutting. *CIRP Ann.* **1994**, *43*, 35–38.
16. Lotfi, M.; Sajjady, S.A.; Amini, S. Wettability analysis of titanium alloy in 3D elliptical ultrasonic assisted turning. *Int. J. Lightweight Mater. Manuf.* **2019**, *2*, 235–240. [CrossRef]

17. Yang, Y.; Pan, Y.; Guo, P. Structural coloration of metallic surfaces with micro/nano-structures induced by elliptical vibration texturing. *Appl. Surf. Sci.* **2017**, *402*, 400–409. [CrossRef]
18. Liu, X.; Yu, D.; Chen, D.; Yang, S.; Wen, Y.; Xiao, Y. Self-tuned ultrasonic elliptical vibration cutting for high-efficient machining of micro-optics arrays on brittle materials. *Precis. Eng.* **2021**, *72*, 370–381. [CrossRef]
19. Ma, C.; Wang, Y. Effect of ultrasonic vibration diamond tool on critical cutting depth of brittle materials. *J. Mech. Eng.* **2005**, *6*, 198–202. [CrossRef]
20. Liu, C.; Zhang, J.; Zhang, J.; Chu, J.; Chen, X.; Xiao, J.; Xu, J. Numerical investigation on material removal mechanism in elliptical vibration cutting of single-crystal silicon. *Mater. Sci. Semicond. Processing* **2021**, *134*, 106019. [CrossRef]
21. Huang, W.; Yu, D.; Zhang, X.; Zhang, M.; Chen, D. Ductile-regime machining model for ultrasonic elliptical vibration cutting of brittle materials. *J. Manuf. Processes* **2018**, *36*, 68–76. [CrossRef]
22. Liu, J.J.; Jiang, X.G.; Gao, Z.; Zhang, M.L.; Zhang, D.Y. Investigation of the Effect of Vibration Amplitude on the Surface Integrity in High-speed Rotary Ultrasonic Elliptical Machining for Side Milling of Ti-6Al-4V. *J. Mech. Eng.* **2019**, *55*, 215–223. [CrossRef]
23. Gao, Z.; Zhang, D.; Li, Z.; Zhang, X.; Liu, J. Research on Surface Quality of Titanium Alloy Webs via High-speed Ultrasonic Elliptical Vibration Milling. *J. Mech. Eng.* **2019**, *55*, 249–256. [CrossRef]
24. Li, X.; Zhang, D. Experimental Study on the Unseparated Ultrasonic Elliptical Vibration Cutting. *J. Mech. Eng.* **2010**, *46*, 177–182. [CrossRef]
25. Ke, Z. *Application of ABAQUS Secondary Development in Constitutive Modeling of Titanium Alloy and Cutting FEA*; Tianjin University: Tianjing, China, 2016.
26. Li, J.; Tang, B. Deformation Mechanism and Microstructure Control of High Strength Metastable β Titanium Alloy. *Acta Metall. Sin.* **2021**, *57*, 1438–1454.
27. Shen, X.; Zhang, D.; Yao, C.; Tan, L. Research Progress on formation mechanism of surface integrity in titanium alloy machining. *J. Aeronaut. Mater.* **2021**, *41*, 1–16.
28. Zhu, G.; Zhang, T.; Dong, L. Summary of Research on Residual Stress of Machined Surface. *Tool Eng.* **2021**, *55*, 9–17.
29. Doan, D.Q.; Fang, T.H.; Chen, T.H. Machining mechanism and deformation behavior of high-entropy alloy under elliptical vibration cutting. *Intermetallics* **2021**, *131*, 107079. [CrossRef]
30. Tian, S.; Zhou, L.; Huang, S.; Xu, L.; Liang, S. Influence of Cutting Parameters on Titanium Alloy Finished Surface Residual Stress. *Tool Eng.* **2013**, *47*, 33–37.

micromachines

MDPI

Article

An Efficient YOLO Algorithm with an Attention Mechanism for Vision-Based Defect Inspection Deployed on FPGA

Longzhen Yu [1], Jianhua Zhu [1,*], Qian Zhao [2] and Zhixian Wang [1]

[1] College of Economics and Management, Qingdao University of Science and Technology, Qingdao 266000, China; yulongzhen@qust.edu.cn (L.Y.); wangzhixian62@126.com (Z.W.)
[2] Department of Creative Informatics, Kyushu Institute of Technology, Fukuoka 804-8550, Japan; cho@ai.kyutech.ac.jp
* Correspondence: mejhzhu@qust.edu.cn

Abstract: Industry 4.0 features intelligent manufacturing. Among them, the vision-based defect inspection algorithm is remarkable for quality control in parts manufacturing. With the help of AI and machine learning, auto-adaptive instead of manual operation is achievable in this field, and much progress has been made in recent years. In this study, considering the demand of inspection features in industrialization, we made further improvement in smart defect inspection. An efficient algorithm using Field Programmable Gate Array (FPGA)-accelerated You Only Look Once (YOLO) v3 based on an attention mechanism is proposed. First, because of the relatively fixed camera angle and defect features, an attention mechanism based on the concept of directing the focus of defect inspection is proposed. The attention mechanism consists of three improvements: (a) image preprocessing, which is to tailor images for selectively concentrating on the defect relevant things. Image preprocessing mainly includes cutting, zooming and splicing, named CZS operations. (b) Tailoring the YOLOv3 backbone network, which is to ignore invalid inspection regions in deep neural networks and optimize the network structure. (c) Data augmentation. First, two improvements can be made to efficiently reduce deep learning operations and accelerate the inspection speed, but the preprocessed images are similar and the lack of diversity will reduce network accuracy. So, (c) is added to mitigate the lack of considerable amounts of training data. Second, the algorithm is deployed on a PYNQ-Z2 FPGA board to meet the industrialization production requirements for accuracy, efficiency and extensibility. FPGA can provide a low-latency, low-cost, high-power-efficiency and flexible architecture that enables deep learning acceleration for industrial scenarios. A Xilinx Deep Neural Network Development Kit (DNNDK) converted the improved YOLOv3 to Programmable Logic (PL), which can be deployed on FPGA. The conversion process mainly consists of pruning, quantization and compilation. Experimental results showed that the algorithm had high efficiency, inspection accuracy reached 99.2%, processing speed reached 1.54 Frames per Second (FPS), and power consumption was only 10 W.

Keywords: vision; defect inspection; YOLO; FPGA; attention

Citation: Yu, L.; Zhu, J.; Zhao, Q.; Wang, Z. An Efficient YOLO Algorithm with an Attention Mechanism for Vision-Based Defect Inspection Deployed on FPGA. *Micromachines* **2022**, *13*, 1058. https://doi.org/10.3390/mi13071058

Academic Editors: Xiuqing Hao, Duanzhi Duan and Youqiang Xing

Received: 3 June 2022
Accepted: 28 June 2022
Published: 30 June 2022

1. Introduction

Manufacturing involves a large number of parts. However, installation, welding, handling and many other sectors of manufacturing inevitably cause part defects, most of which can be identified by vision. The vision-based defect inspection algorithm is crucial to ensure the quality of parts and the entire manufacturing process. Industry 4.0 features intelligent manufacturing, which means doing jobs as efficiently as possible and adapting quickly to new conditions. With the help of AI and machine learning, auto-adaptive operation is replacing manual operation in defect inspection. In particular with the emergence of cutting-edge deep learning technologies [1], the scope intelligence, accuracy, speed and efficiency of defect inspection algorithms are improved significantly [2].

Deep learning algorithms for vision-based defect inspection can be mainly divided into two types: classification-based algorithms and regression-based algorithms [3]. Algorithms based on classification are represented by Region-based Convolutional Neural Network (R-CNN) series, including R-CNN, Spatial Pyramid Pooling Networks (SPP-Net), Fast R-CNN, Region-based Fully Convolutional Networks (R-FCN), and Mask R-CNN. Based on these algorithms, Fan et al. [4], Ji et al. [5], Zhang et al. [6], Guo et al. [7], Jin et al. [8] and Cai et al. [9] have inspected surface defects of wood, gear, metal, etc. Using two-stage processing (region extraction and object classification), R-CNN algorithms generally require high computing power to achieve high accuracy but with a relatively low inspection speed. Regression-based algorithms are characterized by only one round of processing, so the speed is faster. Redmon et al. [10] proposed the well-known You Only Look Once (YOLO) algorithm, which is a representative regression-based and end-to-end model. To date, the YOLO series has evolved to include YOLOv1, YOLOv2, YOLOv3 [11], YOLOv4 [12] and YOLOv5 [13]. Furthermore, the representative regression-based algorithms also include Single Shot MultiBox Detector (SSD) [14] and CornerNet [15]. YOLOv3 is among the most widely used YOLO algorithms. Based on YOLOv3, Jing et al. [16], Li et al. [17], Huang et al. [18] and Du et al. [19] performed surface defect inspection of fabric, PCB boards, pavements, etc.

Compared with classical deep learning object detection algorithms, vision-based defect inspection can be optimized due to two characteristics: first, the recognition region on an image is predictable. As shown in Figure 1, the camera angle of the two parts is fixed. In fact, we only care about the red box region of the two photos. By identifying only this region, we can identify whether the part is defective. Other regions of the original photo can be deleted accordingly. Second, the algorithm needs to meet the deployment requirements of industrial scenarios. The indicators of efficiency must be considered, such as stability, scalability, higher speed and lower power consumption. The target system requirements for this work are as follows. Inspection accuracy should be higher than 97%, image processing speed should be higher than 1 FPS, and the power consumption of each equipment should be less than 100 W.

(a) (b)

Figure 1. Normal welding and defect welding. (**a**) Normal welding. (**b**) Defect welding.

In this work, an efficient YOLO algorithm for vision-based defect inspection with an attention mechanism deployed on FPGA is proposed. There are two main contributions.

- The camera angles of industry cameras and defect features are relatively fixed. So, an attention mechanism is proposed that is based on the concept of drawing global dependencies between the input and the output of a neural network [20], consequently directing focus on defect inspection. The improvement in attention focuses on three aspects: (1) we use image preprocessing named CZS operations for recombining the defect regions of an image and deleting useless regions. (2) We tailor the YOLOv3 backbone network, remove unnecessary recognition accuracy, and thus increase processing speed. (3) Finally, data augmentation technology is also used to further improve the accuracy of attention.
- In order to meet the requirements of industrial scenarios, the algorithm is deployed on FPGA. Deep learning generally consumes a lot of computing power, followed by high power consumption. With FPGA, we can customize hardware for accelerating large-scale computing and make the application scalable. A Xilinx PYNQ-Z2 FPGA board is used for deployment of the optimized YOLOv3. With the help of DNNDK [20], the deep learning processor unit (DPU) [21] can easily deploy the case deep learning algorithm, and involves three main steps: pruning, quantization and compilation. Case experiments showed that this results in low latency, low power consumption, extensibility, and efficiency.

2. YOLOv3 Based on an Attention Mechanism

2.1. Image CZS Preprocessing

An attention mechanism focuses on modeling the relationship between the input and the output of the algorithm, regardless of distance [22]. For defect inspection, the camera angles of industrial cameras are relatively fixed. So, we can predefine the possible defect regions and extract features only from the specific region. The self-attention mechanism calculates the sequence semantic representation by associating different positions in the sequence. We add the image preprocessing equivalent to adding a self-attention mechanism to YOLOv3 preprocessing. CZS operations are as shown in Figure 2. The blue box represents the cutting region, and the green box and the red box represent two kinds of defect markup regions. Color boxes numbered 1 to 8 in the original image on the left side correspond to the splicing regions of the image on the right side. All color boxes are the main areas of concern for defect detection. The entire process involved three steps: cut predefined regions from the original image, zoom regions to the same size and splice regions together to form a new image—named CZS operations for short.

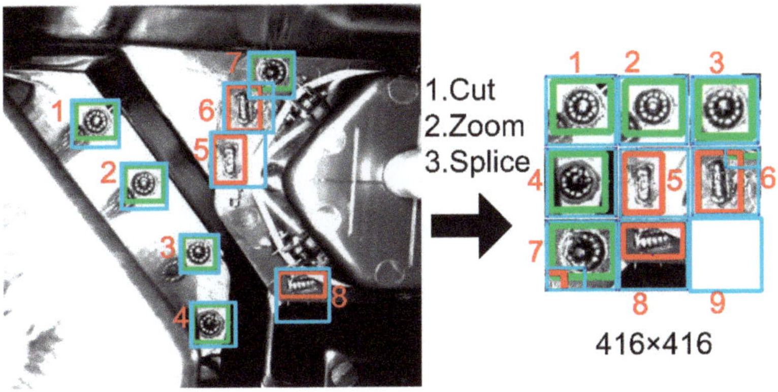

Figure 2. CZS operations.

Cutting operations take a small square box containing a defect region as a cutting region. The box is slightly larger than the smallest box containing the defect region, so as to ensure fault-tolerant positioning of the same type of images. Define the width of the original image w_s and the height h_s, respectively. The ratio of the width of the center point of the defect region to the original image is r_w, and thus the width of the center point of the defect region is $w_m = w_s \times r_w$. Similarly, define the height, x-coordinate and y-coordinate as r_h, r_x and r_y. Additionally, the height, x-coordinate and y-coordinate of the defect region's center point can be expressed as $h_m = h_s \times r_h$, $x_m = w_s \times r_x$ and $y_m = h_s \times r_y$. Define the width, height, x-coordinate and y-coordinate of the cutting box's top left corner as w_c, h_c, x_c and y_c, respectively.

$$w_c = h_c = max(w_m, h_m) \times \alpha \tag{1}$$

$$x_c = \begin{cases} x_m - \frac{w_c}{2}, x_m + \frac{w_c}{2} \leq w_s \\ w_s - w_c, x_m + \frac{w_c}{2} > w_s \end{cases} \tag{2}$$

$$y_c = \begin{cases} y_m - \frac{h_c}{2}, y_m + \frac{h_c}{2} \leq h_s \\ h_s - h_c, y_m + \frac{h_c}{2} > h_s \end{cases} \tag{3}$$

The α is the expansion coefficient, which takes a value between 1 and 2. This means that cutting box is 1- to 2-fold larger than the smallest box containing the defect region. Formulas (2) and (3) mean that when the defect region is close to the boundary of the original image, the top left corner coordinates of the cutting box should be consistent with the original image.

The zooming operation is to scale all cutting boxes on an image to the same size, so that can be fully held by a new 416 × 416 image (the standard image size processed by YOLOv3 is 416 × 416 pixel). According to YOLOv3, images will be scaled to 416 pixels in width (W) and height (H) before being processed. Suppose the number of cutting boxes on an image is N_c. So, the number of boxes that can be held in a row (N_h) or a column (N_v) of a new image is calculated as Formula (4). The target size that a cutting box should be scaled to is calculated as Formula (5). The scaling factor β is calculated as Formula (6). The *sqrt* function is used to obtain the square root of the passed argument. The *ceiling* function is used to obtain the smallest integer larger than the passed argument. The *floor* function is used to obtain the largest integer smaller than the passed argument.

$$N_h = N_v = ceiling(sqrt(N_c)) \tag{4}$$

$$w_z = h_z = floor(\frac{W}{N_h}) = floor(\frac{H}{N_v}) \tag{5}$$

$$\beta = \frac{w_z}{w_c} \tag{6}$$

The splicing operation is to combine cutting boxes from the original image into a new image after being scaled. Splicing mainly consists of two processing works; one is to map a cutting box to the splicing region; another is to map the actual defect markup region to the splicing region. For the first work, we first sort the cutting boxes from small to large according to the x_c value. If the x_c values of two cutting boxes are the same, then we sort them from small to large according to their y_c value. Suppose that a cutting box is ranked as $N^i, i = 0, \ldots, N_h$, then the width, height, x-coordinate and y-coordinate of the box's top left corner in the new image are as defined as w^i_χ, h^i_χ, x^i_χ and y^i_χ. Operator // represents the round function and % represents the remainder function.

$$n^i_h = \begin{cases} N^i//N_v + 1, N^i\%N_v \neq 0 \\ N^i//N_v, N^i\%N_v = 0 \end{cases} \tag{7}$$

$$n^i_v = \begin{cases} N^i\%N_v, N^i\%N_v \neq 0 \\ N_v, N^i\%N_v = 0 \end{cases} \tag{8}$$

$$w_{\chi}^{i} = h_{\chi}^{i} = w_{z} \tag{9}$$

$$x_{\chi}^{i} = (n_{v}^{i} - 1) \times w_{z} \tag{10}$$

$$y_{\chi}^{i} = (n_{h}^{i} - 1) \times h_{z} \tag{11}$$

For the second work, the size ratio of the width, height, x-coordinate and y-coordinate of the center point of an actual defect markup region to the new image is R_{w}^{i}, R_{h}^{i}, R_{x}^{i} and R_{y}^{i}.

$$R_{w}^{i} = \frac{w_{m} \times \beta}{W} \tag{12}$$

$$R_{h}^{i} = \frac{h_{m} \times \beta}{W} \tag{13}$$

$$R_{x}^{i} = \frac{(n_{v}^{i} - 1) \times w_{z} + (x_{m} - x_{c}) \times \beta}{W} \tag{14}$$

$$R_{y}^{i} = \frac{(n_{h}^{i} - 1) \times h_{z} + (y_{m} - y_{c}) \times \beta}{H} \tag{15}$$

In addition to the above algorithms, the new image generated may have some blank regions. We filled them with 0 or 255 values, shown as the square labelled 9 on the right-side image of Figure 2. Then, CZS preprocessing of the image is finished.

2.2. Tailoring the Backbone Network

According to the size of the inspection target, the YOLOv3 backbone network can be tailored to detect the defect regions more efficiently.

As shown in Figure 3, the backbone network of classical YOLOv3 includes 53 layers, so called Darknet-53. Among them, *Convolutional* is the convolution layer, *Residual* is the hop connection layer of residual network, *Avgpool* is pooling layer by average, and *Connected* is the full connection layer. The labels ×1, ×2, ×8, ×8 and ×4 represent repeated execution 1, 2, 8, 8 and 4 times, respectively. Note that the five repeated steps correspond to five down-sampling. Additionally, the outputs of the ×8, ×8 and ×4 down-sampling of the last three steps correspond to the classification prediction feature map (YOLO layer) at three scale resolution levels—52 × 52, 26 × 26 and 13 × 13. The final feature maps of classical YOLOv3 have three sizes, the 52 × 52 resolution has better support for detecting tiny objects, and the 13 × 13 resolution is more suitable for identifying larger objects.

Classical YOLOv3 is used for general object detection, including both large and small objects. The distance of the camera from the object of which photos are taken will also affect the size of the object to be recognized. However, the vision-based defect inspection algorithm is different from classical YOLOv3. The camera angle of industry cameras is relatively fixed, and the shape and size of the defect to be inspected are also relatively fixed. Therefore, based on the fixed shape and size of the defect, only the corresponding resolution networks need to be retained, instead of retaining all three scales (52 × 52, 26 × 26 and 13 × 13) of networks. For example, in the production site of automobile rubber and plastic parts, the visible defect commonly has a moderate size and is obviously distinguished, so the inspection network of such defects does not require a very high resolution. However, on the other hand, in the field of silicon chip solder joint quality inspection, defect inspection of welding points needs high precision. The solder joint layout on the chip is very fine and tiny, so it needs a very high-resolution network for identification. In short, the network structure can be optimized according to targeted inspection tasks. Tailoring the YOLOv3 backbone network can be based on the following formulas.

$$if(\underset{i \in N}{every}(w_{i} > \frac{W}{26} \cap h_{i} > \frac{H}{26})), tailor(yolo_{52 \times 52}) \tag{16}$$

$$if(\underset{i\in N}{every}(w_i < \frac{W}{26} \cap h_i < \frac{H}{26})), tailor(yolo_{13\times 13}) \quad (17)$$

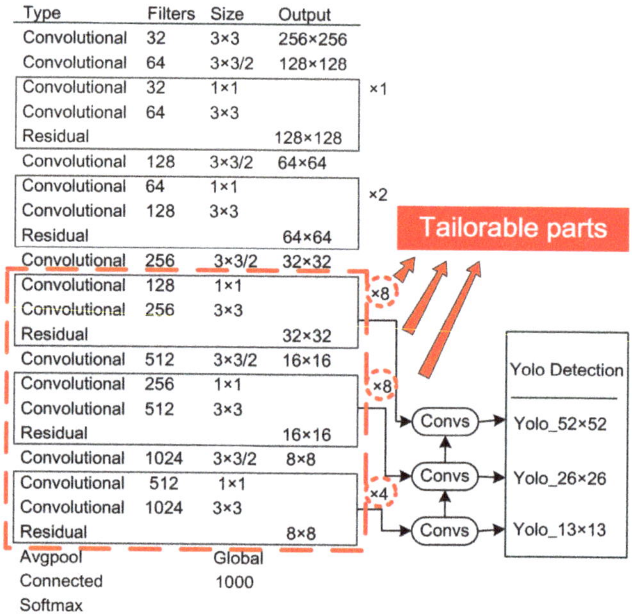

Figure 3. YOLOv3 backbone network and tailorable parts.

The *if (condition)* statement is the basic conditional control structure. This allows the *tailor* function to happen, depending on whether a given condition is true. The *tailor* function is used to delete part of the input argument network used to identify a certain scale. The $yolo_{52\times 52}$ is the smallest resolution inspection network part of YOLOv3, while the $yolo_{13\times 13}$ represents the largest resolution network part. The *every* function means every inspected target. The N represents the full amount of the inspected targets on an image. The w_i and h_i represent the width and height of a target. The W and H represent an image's width and height.

In the case study of chapter 4, the 52 × 52 tiny resolution-scale network is shrunk. That is to partially delete the third down-sampling layers of the backbone network. For classical YOLOv3, the third down-sampling layers consist of eight rounds of repetition. In our algorithm, seven rounds of repetition are tailored off, that is to delete 14 convolutional layers for all. Thus, the backbone network is condensed from 53 layers to 39 layers. Our algorithm's backbone network turns into Darknet-39.

2.3. Data Augmentation

In order to enhance attention and improve the recognition accuracy of deep learning networks, it is also necessary to implement data augmentation to expand the dataset. The whole process of data augmentation is shown in Figure 4. There are mainly two strategies.

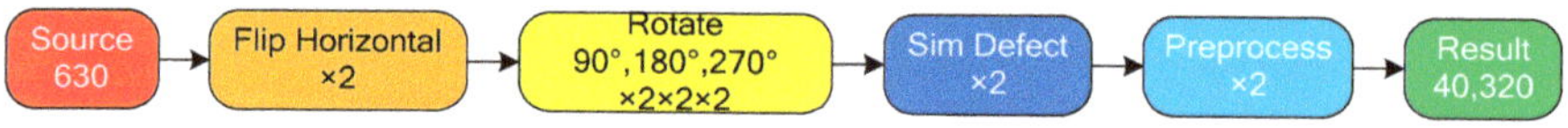

Figure 4. The process of data augmentation.

Strategy 1: It is to add random noise to the defect markup region of the original image, which changes from normal to noisy or faulty. As shown in Figure 5, the rectangular cover is used to simulate the noise or missing faults on the surface of the markup region. The position, size and color can be set randomly. From left to right, Figure 5a is normal; Figure 5b uses a rectangle to cover 1/3 region of a markup region, which is equivalent to adding some noise, so it should be ensured that the network training can recognize such markup region; Figure 5c,d completely cover one or two markup regions with rectangles to simulate the missing faults.

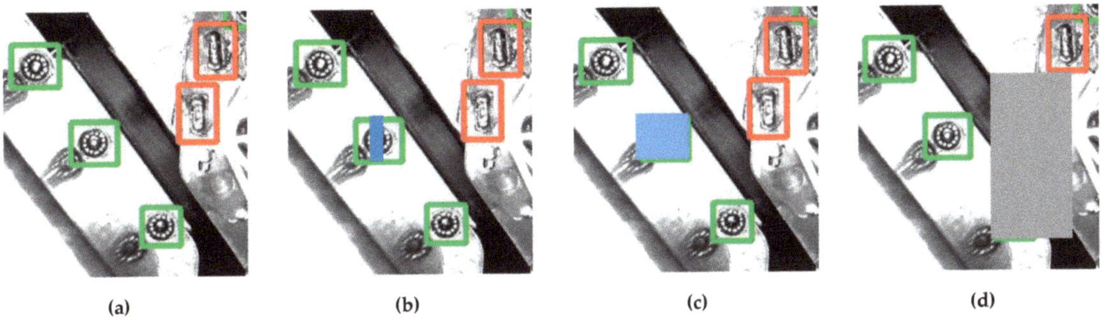

Figure 5. Simulation of noisy or missing faults. (**a**) The normal and original picture marked up with 5 inspection regions of 2 types, green box represents circular solder joint and red box represents strip solder joint. (**b**) One inspection regions was covered by a blue rectangle for 1/3. (**c**) One regions was completely covered by a blue rectangle. (**d**) Two neighbor regions was completely covered by a grey rectangle.

Strategy 2: It is to rotate the image by 90°, 180° and 270° and flip it horizontally. That is equaled to expand into 8 images by rotation and flipping, as shown in Figure 6. In the original picture, there are a total of 8 regions to be inspected, which are divided into two types, represented by red boxes and green boxes respectively.

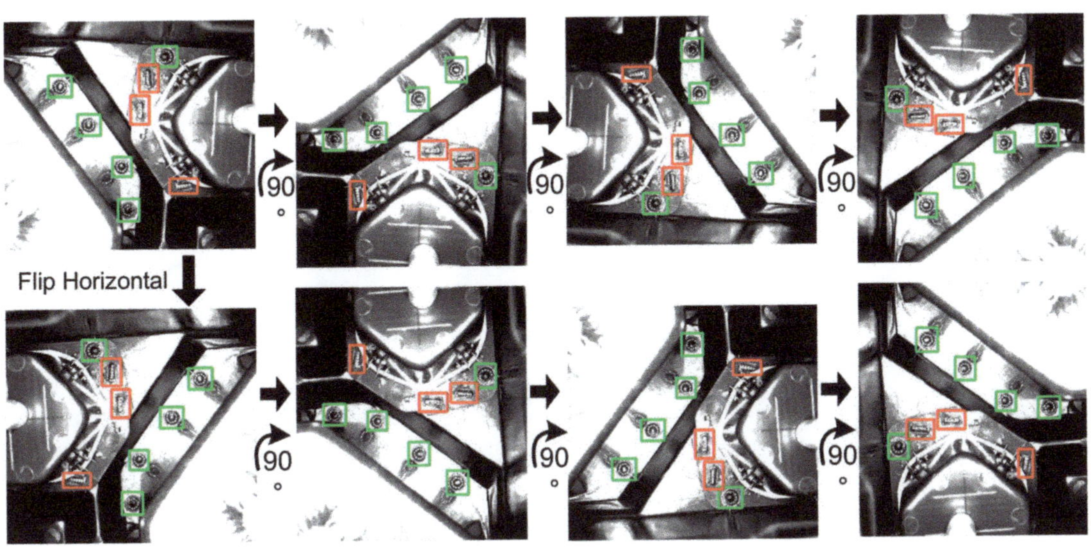

Figure 6. Rotate and flip horizontal the original image.

The use of a dataset enhancement strategy is not only to improve the quality of training, but also to effectively reduce manpower consumption. In this study, there are only 630 original photos, which can only be manually added to the defect markup region. Then, a python script can be used to automatically complete image preprocessing and dataset enhancement, so that the final dataset size for training and testing reaches 40,320 photos.

3. FPGA Deployment

3.1. Overall Framework

The vision-based defect inspection algorithm is deployed on an All Programmable System on Chip (APSoC), the Xilinx PYNQ-Z2. With its help, we can use low-power-consumption, cutting-edge, customized hardware to replace high-energy-consumption, large-footprint, non-specific-purpose and high-cost deep learning workstations. As shown in Figure 7, the whole algorithm deployed on FPGA includes two parts: image CZS preprocessing, and hardware acceleration for deep learning. Through CZS operation, only the regions to be inspected in the pictures will be retained. Refer to Figure 2 for specific explanation.

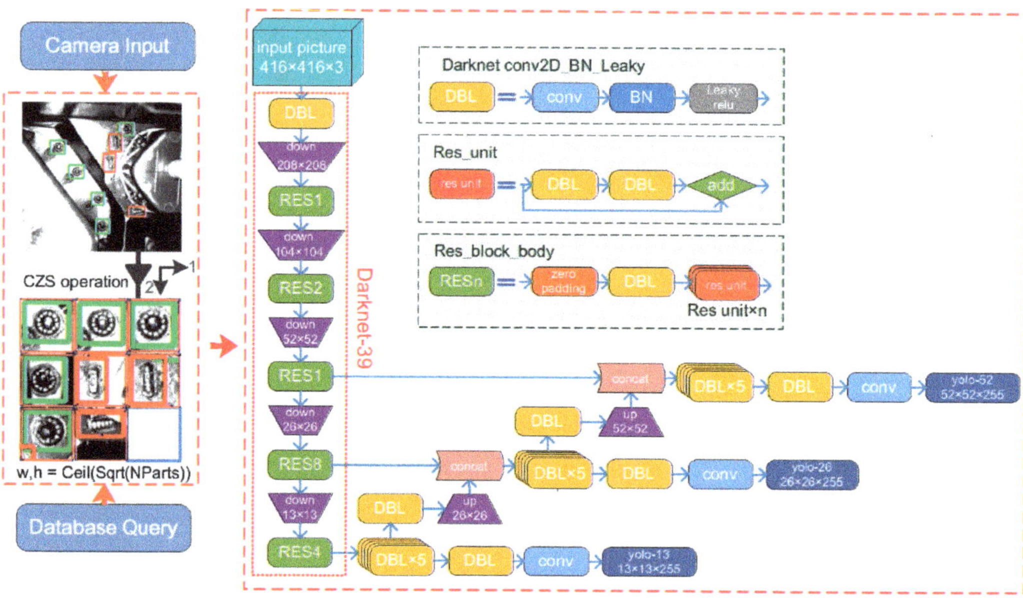

Figure 7. Overall framework of the proposed defect inspection algorithm.

At the beginning, an industry camera captures photos in FPGA. Then, programs running on the operation system of FGPA (PS, Processing System) automatically perform CZS preprocessing on those photos according to the metadata stored in the database. Preprocessed images are then transmitted to the DPU implemented by PL, which is a special customized hardware for mapping and running the darknet-39 YOLOv3 model. With it, the process of defect inspection can be accelerated and finished.

A Xilinx PYNQ-Z2 FPGA board is equipped with a ZYNQ-7020 APSoC chipset (Xilinx AMD Inc., San Jose, USA), which has both a "hard core" and a" soft core". As shown in Figure 8, the hard core and its functions are grey and green boxes, and the soft core and its functions are orange and yellow boxes. The hard core is a 650 MHz ARM Cortex-A9 dual-core processor (Arm Inc., Cambridge, UK), running an embedded Ubuntu system (Canonical Ltd., Landon, UK). This processor supports python programming for simple processing (preprocessing image, running database, etc.) and C++ programming for calling the DPU. The soft core is the PL that can be employed by the B1152 DPU architecture in accordance with the Xilinx DNNDK 3.0 (Xilinx AMD Inc., San Jose, CA, USA) framework. Deep learning algorithms can be transformed to a format that the DPU can read and execute without fully utilizing hardware resources. The efficiency of processing 416 × 416 images of YOLOv3 is approximately 3.5 FPS, and the rated power is approximately 10 W (the power information comes from the technical documents of Xilinx PYNQ-Z2), having much better power-efficient performance than a common Central Processing Unit (CPU) or a Graphics Processing Unit (GPU) chip. This can fully meet our predesigned efficiency target.

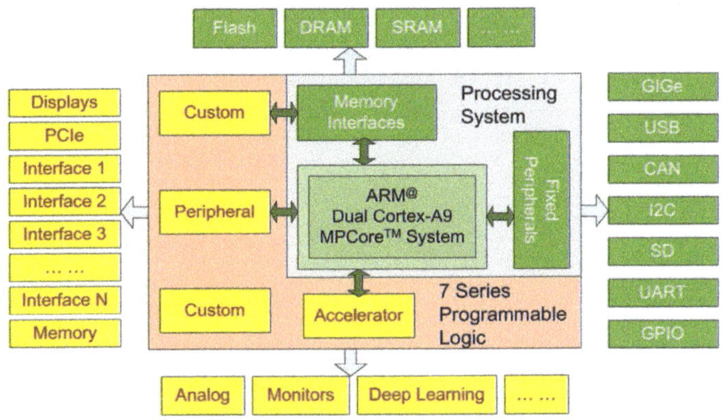

Figure 8. Xilinx PYNQ-Z2 architecture.

3.2. *Deployment*

Deployment involves two parts, host-side deployment and FPGA-side deployment. On the FPGA side, we adopt the system version of the intelligent car hydramini, the essence of which is a customized DPU platform for PYNQ-Z2. The host side is the computer side. The host side needs to install the deep learning development platform, as well as the DNNDK 3.0. DNNDK is necessary for converting a standard deep learning model into a deployable model on PYNQ-Z2. DNNDK's core functions include pruning, quantization and compiling. As shown in Figure 9, the blue flowchart represents the pruning operation, the orange flow chart represents quantization operation, and the green flowchart represents the compiler.

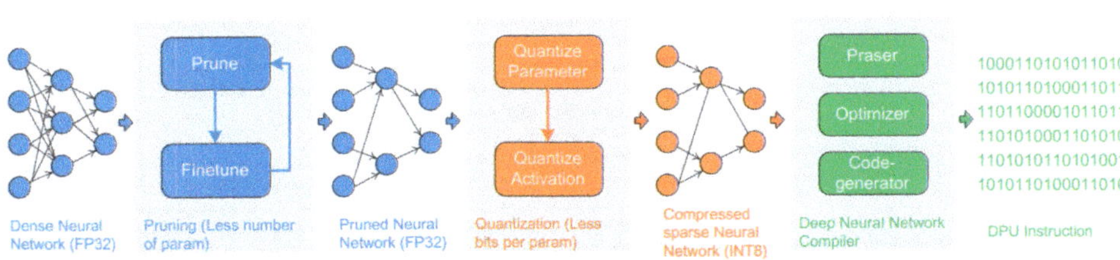

Figure 9. Pruning, quantifying, and compiling processes.

Pruning is to obtain a new condensed network from the network pretrained by PyTorch (Meta Inc., Silicon Valley, CA, USA) or Tensorflow (Google Inc., Silicon Valley, CA, USA). Pruning consists of automatically deleting some redundant branches that do not affect the network output, and replacing variable values on network nodes with constant values of the current session.

Quantization is to convert float-point values into fixed-point values. The first benefit from quantization is improving processing performance, using short bytes of data instead of long bytes. In general, 32-bit float-point values are replaced with 8-bit integers. As a result, the entire network is compressed. The second benefit lies in that the PYNQ-Z2′s Digital Signal Processing (DSP) units mainly support processing fixed-point values, the fixed-point values operations of which are specially optimized.

Compiling is the deep learning algorithm being transformed into binary instruction files that can be recognized by the DPU. The functional module consists of three parts: interpreter, optimizer and code generator. The interpreter parses the quantization model and converts it into Intermediate Representation (IR). The optimizer is responsible for optimizing IR. The code generator makes optimized IR into DPU-recognized instructions.

After the above three steps, a deep learning model file that is recognized by the PYNQ-Z2 DPU is generated. The file is deployed to PYNQ-Z2 and loaded on the core kernel process of the DPU. We can let the PYNQ-Z2 reload DPU Intellectual Property (IP) then use an executable program written by the DPU C++ library to invoke the deep learning model, including using API to read problem initialization parameters and analyze the output results. Deployment is finished. All above steps is shown in Figure 10. CZS and after-processing operations are undertaken by ARM (Arm Inc., Cambridge, UK), refer to Figure 2 for specific explanation. And defect inspection operations are undertaken by Xilinx ZYNQ-7020 (Xilinx AMD Inc., San Jose, CA, USA).

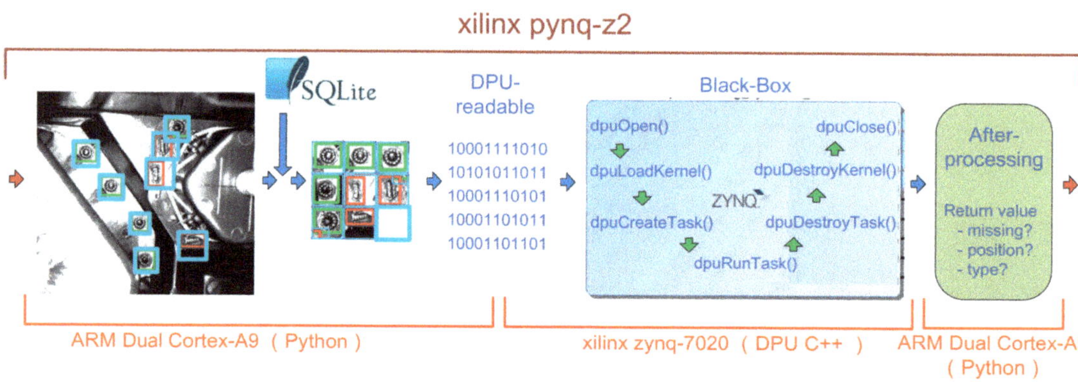

Figure 10. Process flow of PYNQ-Z2.

Access to the database and image preprocessing require a low amount of data processing resources, which are handed over to the ARM CPU for processing. In contrast, deep learning needs a lot of computing power, and FPGA is responsible for this processing work to achieve hardware acceleration. As a SoC board, PYNQ-Z2 fully realizes the seamless connection between the operating system and programmable hardware. Python and database software such as SQLite run on ARM CPU, while the DPU C++ code directly calls ZYNQ-7020 FPGA hardware resources, which optimizes the load balance of the whole process.

4. Experimental Results

4.1. Experiment Design

The case study gets experimental data from an automobile rubber and plastic parts manufacturer, which is a super class parts supplier for several well-known automobile brands in China, such as SAIC GM Wuling and FAW Volkswagen. Therefore, success in implementing the application system will have significance in the whole auto parts industry. With the help of industrial cameras with 10 fixed camera angles, we collected 630 photos compared with the original 63 sample photos. The photos have the same 1600 $\times$ 1200 pixels resolution and 8-bit depth. In addition, there are 16 types of defect markups. The ratio of normal and defective samples in the 630 pieces of original pictures is approximately 9:1. Several sample photos are shown in Figure 11, the red box on the pictures identify the area to be detected. The hardware of the host side mainly consists of a NVIDIA RTX3090 24 GB GPU (Nvidia Co., Silicon Valley, CA, USA), an AMD R9 3900X CPU (AMD Inc., Silicon Vallsey, CA, USA), 64 GB of DDR4 (Hynix Inc., Seoul, Korea) memory and a 4 TB mechanical hard disk (Seagate Technology, Scotts Valley, CA, USA). The software installed on the host side include Ubuntu, Compute Unified Device Architecture (CUDA) (Nvidia Co., Silicon Valley, CA, USA), Python (The Python Software Foundation, Wilmington, DE, USA), PyTorch, Tensorflow, Docker (Docker Inc., San Francisco, CA, USA) and DNNDK 3.0 (Xilinx AMD Inc., San Jose, CA, USA).

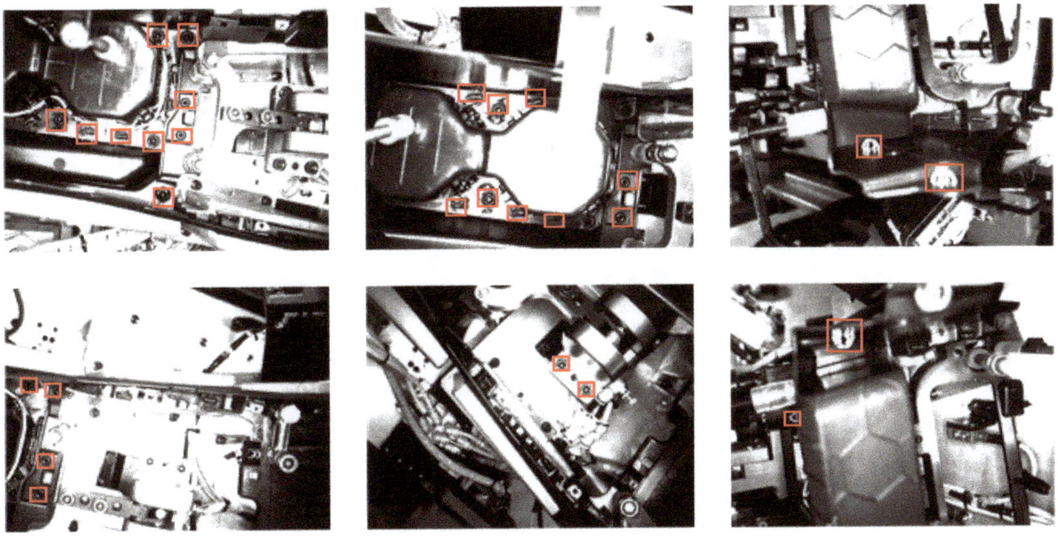

Figure 11. Several sample photos.

4.2. Host Side

On the host workstation computer side, our algorithm, the attention-based YOLOv3, is trained. The maximum number of training epochs was initially set to 300. In the training process, the training set was composed of 36139 photos after data augmentation, at a total of 11 GB. The training process took a long time, as one round of epoch took nearly 17 min. On the other hand, precision convergence was very fast. When we finished the work after 300 epochs of training, accuracy exceeded 99%.

With the pretrained model, we tested the time efficiency on the host side. The time spent is mainly divided into two parts: the loading time of the Python library is approximately 1 s, and the inspection time is approximately 0.01 s, as shown in Table 1. The loading process of the Python library is very time consuming, which can be made into a daemon, so that it is always in the loaded library state, scanning to detect changes in images, and real-time inspection.

Table 1. Comparison between our algorithm and YOLOv3.

	Training	Inspection	Accuracy
YOLOv3	28 FPS	0.014 s	95.0%
Our Algorithm	28 FPS	0.010 s	99.2%

As shown in Table 1, our algorithm's inspection time decreased by 0.004 s, and accuracy improved by 0.2%. In sum, compared with the majority of indicators, our algorithm makes an improvement to classical YOLOv3, indicating that our algorithm achieves better performance by tailoring the backbone network.

4.3. FPGA Side

The efficiency and accuracy of the algorithm on the host side fully meet the industrial requirements, so the trained neural network is moved to the FPGA side. The DNNDK environment is deployed with Docker to prune, quantify and compile the trained network; and then deploy the processed neural network files to the appropriate location on PYNQ-Z2, reload the DPU kernel and compile the executable program, which completes the migration and deployment of PYNQ-Z2. Because it is necessary to use onboard Python to call the low-performance ARM CPU for database query and image preprocessing, although the performance of programmable hardware circuit is very good, the overall timeliness of the system is lower than that of the host side. As shown in Table 2, the performance of PYNQ-Z2 will be reduced to some extent, but it is also competent for missing fault inspection. The experimental results on PYNQ-Z2 also showed satisfactory performance, similar to that of the host side. A comparison of processing times of the host workstation computer and PYNQ-Z2 is shown in Table 2.

Table 2. A comparison of processing times of workstation computer and PYNQ-Z2.

	Preprocess	**1600 × 1200**	**416 × 416**
Host	0.24 s	0.01 s	0.01 s
PYNQ-Z2	0.31 s	1.20 s	0.65 s

Although the process speed on PYNQ-Z2 is slower than that on the host side. The SoC cutting-edge equipment shows a reasonable efficiency with much lower power consumption than a workstation computer. Additionally, the comprehensive performance of our algorithm can meet the predesigned target of Section 1, as shown in Table 3.

Table 3. A comparison of processing times of different algorithms on PYNQ-Z2.

	SetImage	**RunTask**	**Deal**
No preprocess, YOLOv3	0.71 s	0.45 s	1.20 s
Preprocess, YOLOv3	0.13 s	0.45 s	0.73 s
Preprocess, our algorithm	0.13 s	0.38 s	0.65 s

4.4. Inference

In this study, the mean of average precision (mAP) and the intersection over union (IoU) were used as the main accuracy evaluation indices. In this case, the average recognition accuracy of 16 detection categories is calculated, and then the average value of these 16 average accuracies is the mAP. The closer the mAP is to 1, the better. Before training, the manually marked area to be detected is called the ground-truth bounding box. Later, the marked area detected by the model is called the predicted bounding box. The IoU is the intersection of these two regions divided by the union. The closer the IoU is to 1, the better, indicating that the model detection area is consistent with the manually marked area.

4.5. Results

This method has achieved good results on the missing faults dataset. A total of 8064 test samples are correctly classified according to type. The detection results of some examples are shown in Figures 12 and 13. This method not only has good classification accuracy, but also has good positioning, timeliness and energy consumption ratio. In the past, similar defect detection research did not achieve detection accuracy over 99%, but our algorithm not only improves accuracy, but can also be promoted at an enterprise level to complete cutting-edge detection work in a fast, cheap, stable and green way. In addition to accuracy, manufacturers are concerned with the efficiency, the stability, the scalability and the comprehensive cost performance of large-scale deployment.

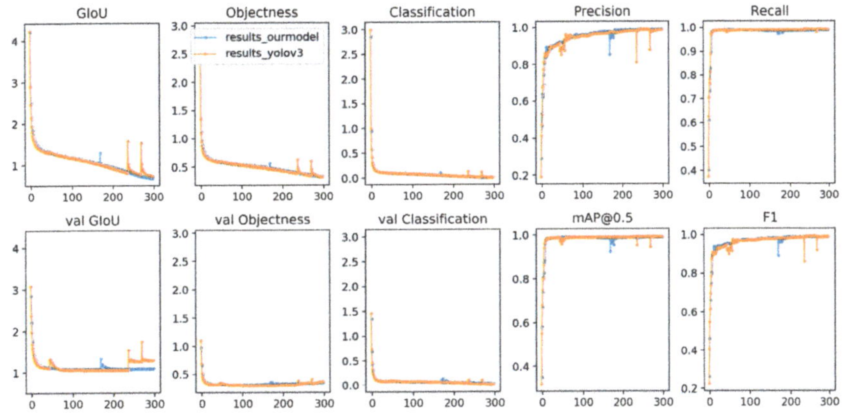

Figure 12. Comparison of our algorithm and classical YOLOv3.

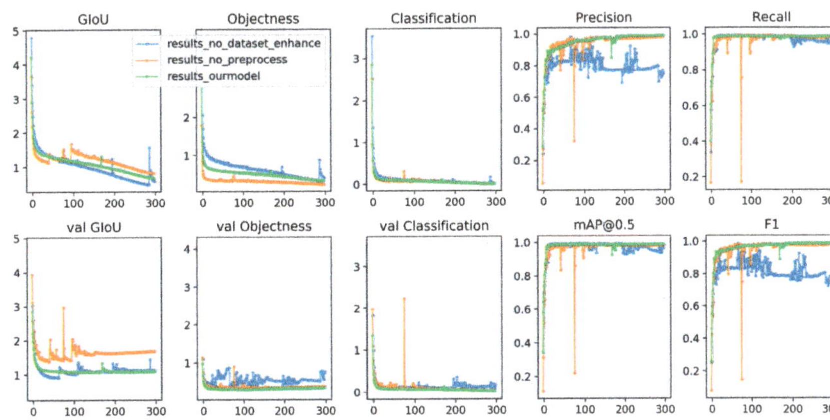

Figure 13. Effect of preprocessing and data augmentation on our algorithm.

In Figures 12 and 13, the x-coordinate axis represents training times and the y-coordinate axis represents the value of the measurement index.

The comparison results are shown in Table 4. For the first time in this study, FPGA is used to implement the deep learning missing installation detection system. The performance of the hardware scheme is moderate, and the price is moderate. Deployment is difficult, but the advantages are high stability and obvious design flexibility. A comparison of the performance of this algorithm with that of previous algorithms is shown in the table below.

Table 4. A comparison of the performance of this algorithm with that of other representative industry defect inspection studies.

	Accuracy	1/FPS	mAP_{bbox}
Nico Prappacher [23]	98%	0.153	-
Ge Liling [24]	97.2%	-	-
Ting He [25]	98.7%	0.007	-
Chunyang Xia [26]	98.4%	-	-
Junfeng Jing [16]	98%	0.046	-
Our algorithm	99.2%	0.010	0.991

The performance impact of an attention mechanism on the algorithm is shown in Table 5.

Table 5. Performance impact of an attention mechanism.

	Accuracy	FPS	mAP_{bbox}
No preprocessing	98.3%	97	0.986
No data augmentation	75.3%	97	0.948
YOLOv3	95.0%	73	0.988
Our algorithm	99.2%	97	0.991

A comparison of the main performance indices for our detection algorithm deployed on FPGA and for similar chips is shown in Table 6.

Table 6. Comparison of three representative chips.

	Disadvantage/Advantage	Power Consumption	FPS
Jetson Nano	Large heat dissipationAI performance is relatively high	10 W *	4.27
Intel NCS 2	High total cost and not SoC chipIt is easy to get started	-	1.36
PYNQ-Z2	Poor performance of its CPUHas good flexibility and scalability	10 W *	1.54

* The power consumption data are based on the official instruction document of the chip.

4.6. Further Discussion and Analysis

At the same time, some related experiments are carried out to further verify the validity of the algorithm.

Transfer learning: When training this model, weight is trained from scratch. However, some research also states that the classical YOLOv3 algorithm can be quickly developed through transfer learning. Therefore, we consider two different training schemes: (a) training the network from scratch; (b) training the network by the transfer learning method using a classical network pretrained by the Common Objects in Context (COCO) dataset. The training time of the two methods is very similar, at approximately 48 h, and the mAP of the two methods is approximately 0.991. We think that the effect of cross-dataset transfer learning is not obvious due to the large difference between the dataset of this study and the COCO dataset.

Precision analysis: Compared with the classical YOLOv3 network, our algorithm reduces 14 neural network layers. However, after 300 rounds of training, accuracy improved from 95 to 99.2%. We think that is because the optimized version has fewer network parameters and is easier to converge than the classical version.

Error analysis: Although the recognition accuracy of this method reaches 99.2%, the premise is the existing 10 types of images. When adding a new angle image, it needs automatic recognition and automatic preprocessing. Although identification did not improve, we find that accuracy cannot reach 99.2%. Therefore, it is necessary to train a new network through transfer learning.

FPGA side performance improvement: On the host side, the size of the input image has little impact on algorithm efficiency. However, on the FPGA side, there is a big difference in the recognition efficiency of different-sized images. The key of the problem is that image preprocessing is not based on a programmable hardware circuit, but is based on ARM soft core processing. The ARM CPU of PYNQ-Z2 has poor performance and a slow speed. If image preprocessing is also made into a programmable hardware circuit, FPGA hard core processing should be able to effectively improve the overall performance of the algorithm.

5. Conclusions

The traditional inspection algorithm of missing faults is limited by materials, safety, costs, etc. In this paper, a new and efficient algorithm based on an attention mechanism that can be deployed on FPGA for defects inspection is proposed. Through our algorithm, we can correctly identify the problem of missing faults in addition to achieving high precision, a fast speed and low energy consumption. The experimental results show that the accuracy of the algorithm is 99.2%, processing speed is 1.54 FPS, and energy consumption is 10 W. The algorithm can be widely deployed in the industrial field as cutting-edge equipment.

By the second quarter of 2022, there were 366, 173 and 95 articles containing YOLOv3, YOLOv4 and YOLOv5, respectively, in search titles on Web of Science. YOLOv3 is a very classical algorithm, while YOLOv4 and YOLOv5 represent an inevitable trend. In particular in the fields of robot [27] and dynamic object capture [28], YOLOv5 has made new progress, which points out the direction for future improvement in this research.

Author Contributions: Conceptualization, L.Y. and Q.Z.; methodology, L.Y., J.Z. and Q.Z.; software, L.Y. and Q.Z.; validation, J.Z.; formal analysis, L.Y. and J.Z.; investigation, L.Y.; data curation, L.Y. and Z.W.; writing—original draft preparation, L.Y.; writing—review and editing, Q.Z. and J.Z.; supervision, J.Z.; Q.Z. contributed equally to this work and should be considered co-first authors. All authors have read and agreed to the published version of the manuscript.

Funding: This research was funded by the R&D Project in Key Areas of Guangdong Province, grant number 2020B0101050001. This research was also funded by the Qingdao City Philosophy and Social Science Planning 885 Project, grant number QDSKL1801166.

Acknowledgments: The authors would like to thank the anonymous referees and journal editors for their valuable and constructive feedback.

Conflicts of Interest: The authors declare no conflict of interest.

References

1. Sengupta, S.; Basak, S.; Saikia, P.; Paul, S.; Tsalavoutis, V.; Atiah, F.; Ravi, V.; Peters, A. A review of deep learning with special emphasis on architectures, applications and recent trends. *Knowl. Based Syst.* **2019**, *194*, 105596. [CrossRef]
2. Han, J.; Zhang, D.; Cheng, G.; Liu, N.; Xu, D. Advanced deep-learning techniques for salient and category specific object detection: A survey. *IEEE Signal Processing Mag.* **2018**, *35*, 84–100. [CrossRef]
3. Liu, L.; Ouyang, W.; Wang, X.; Fieguth, P.; Chen, J.; Liu, X.; Pietikinen, M. Deep Learning for Generic Object Detection: A Survey. *Int. J. Comput. Vis.* **2020**, *128*, 261–318. [CrossRef]
4. Jianan, F.; Ying, L.; Zhongkang, H.; Qian, Z.; Luxiang, S.; Xiaolin, Z. Solid wood panel defect detection and recognition system based on faster r-cnn. *J. For. Eng.* **2019**, *4*, 112–117.
5. Weixi, J.; Meng, D.; Wei, P.; Jie, X. Research on gear appearance defect recognition based on improved faster r-cnn. *J. Syst. Simul.* **2019**, *31*, 24–31.
6. Yuyan, Z.; Yongbao, L.; Yintang, W.; Zhiwei, Z. Internal defect detection of metal three-dimensional multilayer lattice structure based on faster r-cnn. *Acta Armamentarii* **2019**, *40*, 2329–2335.
7. Wen-ming, G.; Kai, L.; Hui-fan, Q. Welding defect detection of x-ray images based on faster r-cnn model. *J. Beijing Univ. Posts Telecommun.* **2019**, *42*, 20.
8. Ying, J.; Xueying, W.; Linmao, D. Surface defect detection of inner groove in plunger brake master cylinder based on labview and mask r-cnn. *Mod. Manuf. Eng.* **2020**, *476*, 131–138.
9. Biao, C.; Kuan, S.; Jinlei, F.; Lize, Z. Research on defect detection of x-ray dr images of casting based on mask r-cnn. *Chin. J. Sci. Instrum.* **2020**, *41*, 63–71.
10. Redmon, J.; Angelova, A. Real-time grasp detection using convolutional neural networks. In Proceedings of the IEEE International Conference on Robotics & Automation, Hong Kong, China, 31 May–5 June 2014.
11. Redmon, J.; Farhadi, A. YOLOv3: An Incremental Improvement. *arXiv* **2018**, arXiv:1804.02767v1.
12. Bochkovskiy, A.; Wang, C.Y.; Liao, H.Y.M. Yolov4: Optimal speed and accuracy of object detection. *arXiv* **2020**, arXiv:2004.10934.
13. Ultralytics. ultralytics-yolov5, 1 2020. Available online: https://ultralytics.com/yolov5 (accessed on 27 June 2022).
14. Liu, W.; Anguelov, D.; Erhan, D.; Szegedy, C.; Berg, A.C. Ssd: Single shot multibox detector. In Proceedings of the European Conference on Computer Vision, Amsterdam, The Netherlands, 8–16 October 2016.
15. Law, H.; Deng, J. Cornernet: Detecting objects as paired key-points. *Int. J. Comput. Vis.* **2020**, *128*, 642–656. [CrossRef]
16. Jing, J.; Zhuo, D.; Zhang, H.; Liang, Y.; Zheng, M. Fabric defect detection using the improved YOLOv3 model. *J. Eng. Fibers Fabr.* **2020**, *15*, 155892502090826. [CrossRef]

17. Li, J.; Gu, J.; Huang, Z.; Wen, J. Application research of improved yolo v3 algorithm in pcb electronic component detection. *Appl. Sci.* **2019**, *9*, 3750. [CrossRef]
18. Huang, R.; Gu, J.; Sun, X.; Hou, Y.; Uddin, S. A rapid recognition method for electronic components based on the improved yolo-v3 network. *Electronics* **2019**, *8*, 825. [CrossRef]
19. Du, Y.; Pan, N.; Xu, Z.; Deng, F.; Kang, H. Pavement distress detection and classification based on YOLO network. *Int. J. Pavement Eng.* **2021**, *22*, 1659–1672. [CrossRef]
20. Xilinx. *Dnndk User Guide—ug1327 (v1.4)*; Xilinx: San Jose, CA, USA, 2019.
21. Xilinx. *DPU for Convolutional Neural Network v3.0—DPU ip Product Guide*; Xilinx: San Jose, CA, USA, 2019.
22. Vaswani, A.; Shazeer, N.; Parmar, N.; Uszkoreit, J.; Jones, L.; Gomez, A.N.; Kaiser, L.; Polosukhin, I. Attention is all you need. In Proceedings of the Advances in Neural Information Processing Systems 30 (NIPS 2017), Long Beach, CA, USA, 4–9 December 2017.
23. Nico, P.; Markus, B.; Gunther, B.; Frank, D.; Andreas, L. Defect Detection on Rolling Element Surface Scans Using Neural Image Segmentation. *Appl. Sci.* **2020**, *10*, 3290. [CrossRef]
24. Liling, G.; Yingjie, Z. Weld defect detection in industrial radiography based on image segmentation. *Insight* **2011**, *53*, 263–269. [CrossRef]
25. He, T.; Liu, Y.; Yu, Y.; Zhao, Q.; Hu, Z. Application of Deep Convolutional Neural Network on Feature Extraction and Detection of Wood Defects. *Measurement* **2019**, *152*, 107357. [CrossRef]
26. Xia, C.; Pan, Z.; Fei, Z.; Zhang, S.; Li, H. Vision based defects detection for Keyhole TIG welding using deep learning with visual explanation. *J. Manuf. Processes* **2020**, *56*, 845–855. [CrossRef]
27. Yang, F. A Real-Time Apple Targets Detection Method for Picking Robot Based on Improved YOLOv5. *Remote Sens.* **2021**, *13*, 1619.
28. Zhu, X.; Lyu, S.; Wang, X.; Zhao, Q. TPH-YOLOv5: Improved YOLOv5 Based on Transformer Prediction Head for Object Detection on Drone-captured Scenarios. In Proceedings of the IEEE/CVF International Conference on Computer Vision, Virtual, 11–17 October 2021.

micromachines

Article

Understanding the Mechanisms of SiC–Water Reaction during Nanoscale Scratching without Chemical Reagents

Zhihao Cheng [1], Qiufa Luo [1,*], Jing Lu [2,*] and Zige Tian [1]

[1] Institute of Manufacturing Engineering, Huaqiao University, Xiamen 361021, China; 20013080008@stu.hqu.edu.cn (Z.C.); zigetian@stu.hqu.edu.cn (Z.T.)
[2] National & Local Joint Engineering Research Center for Intelligent Manufacturing Technology of Brittle Material Products, Xiamen 361021, China
* Correspondence: qfluo2014@hqu.edu.cn (Q.L.); lujing26@hqu.edu.cn (J.L.)

Abstract: Microcracks inevitably appear on the SiC wafer surface during conventional thinning. It is generally believed that the damage-free surfaces obtained during chemical reactions are an effective means of inhibiting and eliminating microcracks. In our previous study, we found that SiC reacted with water (SiC–water reaction) to obtain a smooth surface. In this study, we analyzed the interfacial interaction mechanisms between a 4H-SiC wafer surface $(000\bar{1})$ and diamond indenter during nanoscale scratching using distilled water and without using an acid–base etching solution. To this end, experiments and ReaxFF reactive molecular dynamics simulations were performed. The results showed that amorphous SiO_2 was generated on the SiC surface under the repeated mechanical action of the diamond abrasive indenter during the nanoscale scratching process. The SiC–water reaction was mainly dependent on the load and contact state when the removal size of SiC was controlled at the nanoscale and the removal mode was controlled at the plastic stage, which was not significantly affected by temperature and speed. Therefore, the reaction between water and SiC on the wafer surface could be controlled by effectively regulating the load, speed, and contact area. Microcracks can be avoided, and damage-free thinning of SiC wafers can be achieved by controlling the SiC–water reaction on the SiC wafer surface.

Keywords: 4H-SiC wafer; amorphous silica; ReaxFF reactive molecular dynamics; plastic removal; chemical reaction

Citation: Cheng, Z.; Luo, Q.; Lu, J.; Tian, Z. Understanding the Mechanisms of SiC–Water Reaction during Nanoscale Scratching without Chemical Reagents. *Micromachines* **2022**, *13*, 930. https://doi.org/10.3390/mi13060930

Academic Editors: Youqiang Xing, Xiuqing Hao and Duanzhi Duan

Received: 24 May 2022
Accepted: 10 June 2022
Published: 11 June 2022

1. Introduction

As regards semiconductor development, silicon carbide (SiC) materials are currently known to exhibit excellent material advantages such as high forbidden bandwidth, high thermal conductivity, and high electron migration rate as compared with silicon materials [1–4]. Therefore, SiC devices are being widely used in various fields such as new energy-efficient vehicles, smart grids, and aerospace applications [5–7]. However, under high-power and extreme working conditions, the self-heating effects of SiC semiconductor materials gradually become evident and lead to poor heat dissipation and performance degradation [8]. Currently, the most effective method to solve the problem of wafer heat dissipation is the wafer back-thinning process [9,10]. Generally, SiC is considered as a typical difficult-to-machine material owing to its high mechanical hardness and chemical inertness; wafer thinning processes are constantly being improved to solve these problems [11], resulting in SiC wafers that have been thinned to less than 100 μm [12]. Therefore, ultrathin wafers with excellent performance are of remarkable significance for high-power and extreme working conditions. Generally, the wafer back-thinning process for SiC wafer involves mechanical grinding using diamond abrasives, directly resulting in wafer fragmentations and deformations owing to the formation of microcracks [13,14].

To produce thin SiC wafers and to improve their surface quality, some researchers have used fixed diamond abrasive tools for the SiC thinning process. Yam et al. [15] used a #2000 diamond wheel to grind SiC under a constant pressure, and a surface roughness Ra of 1.57 nm was achieved. Tsukimoto et al. [16] found that a 2.4 μm subsurface damage (SSD) layer existed on the SiC wafer after grinding with a #2000 diamond cup wheel, and microcracks were generated during the hard abrasive interaction and related plastic deformation and fracture. Although conventional grinding methods have achieved grinding effects, a few microcracks and SSD are still generated owing to hard abrasives with unequal protrusion heights on the surface of the tool. To improve the surface quality of SiC wafers further, Feng et al. [17] proposed a polyvinyl alcohol/phenolic resin (PVA/PF) composite sol–gel diamond wheel based on traditional grinding and obtained a high surface quality of the 4H-SiC wafer. However, the pores on the surface of the sol-gel diamond wheel were easily blocked, and the diamond wheel needed to be dressed over time. Furthermore, a few unconventional methods have been proposed for SiC thinning. Sano et al. [18] proposed a method of plasma etching with high-pressure SF6 plasma to reduce the thickness of a 2-inch wafer, and the thickness of SiC wafer could be reduced to approximately 100 μm within 20 min of plasma etching. This technology could be used as an effective method for thinning SiC wafers; however, some etch pits were not completely removed. Guan et al. [19] proposed an electric discharge grinding (EDG) method for SiC wafer thinning, which could obtain a small-sized SiC wafer with a minimum thickness of 30 μm; however, the surface roughness and subsurface quality of EDG were lower than those of the other methods. Although unconventional methods can effectively obtain ultrathin chips, the current technology is not sufficiently developed, and SSD is still a serious concern. The problem of subsurface damage introduced by SiC thinning, thus, remains unsolved, regardless of whether conventional or unconventional methods are employed. Therefore, obtaining high-quality thin wafers is difficult. Moreover, to eliminate microcracks and improve surface quality, chemical mechanical polishing (CMP), which is an effective method to obtain smooth and damage-free wafers through chemical reactions, is employed; however, the material removal rate (MRR) of SiC wafers is significantly low because of the high hardness and stable chemical properties of the SiC materials. Therefore, the production of high-performance SiC wafers is limited because CMP using acid–base chemical reagents is expensive and polluting [20–22].

In our previous studies [23–26], a green-energy and efficient semi-fixed abrasive tool for ultra-precision machining, referred to as an SG pad, was proposed. A smooth, scratch-free, and nearly damage-free SiC wafer surface (SSD of only 2 nm) could be achieved using this tool. Instead of acid, alkali, or other active substances, only distilled water was used as the coolant in the SiC processing; this effectively mitigated environmental pollution. It was found that the SiC wafer could react with deionized water and produce amorphous silicon dioxide (SiO_2) under mechanical scratching using diamond abrasives [27]. This technique could be used for the subsequent treatment of SiC thinning by eliminating microcracks and improving material removal rate via the utilization of the SiC–water reaction, which would avoid environmental pollution. Thus, SiC wafers can be obtained rapidly and efficiently by controlling the SiC–water reaction, and the conditions of the reaction should be further studied. In this regard, traditional molecular dynamics (MD) simulations based on classical force fields are used to analyze the contact behavior and material removal among atoms during nano-machining, but classical force fields cannot be used to describe reactive chemical systems [28,29]. ReaxFF reaction molecular dynamics (RMD) simulation based on the reactive force field approach can effectively simulate the formation and dissociation of chemical bonds [30–33], which has already been successfully applied to various processes based on the interaction of chemical reactions and mechanical effects, such as the interaction between water and Si [34]. Therefore, RMD simulation was used to investigate the interaction between water and SiC in this study. A method to induce a chemical reaction between SiC and water by mechanical scratching of SiC with a single diamond indenter was proposed to determine the conditions of the SiC–water reaction that can optimize the thinning process to satisfy the accuracy requirements for high-quality SiC wafers. This study aimed to regulate the scratch parameters (contact state, scratch speed, and scratch load) to control the plastic removal of 4H-SiC and realize the chemical reaction between SiC wafers and water by conducting relevant experiments and RMD simulations. The characterizations of the scratched area using transmission electron microscopy (TEM), high-resolution TEM (HRTEM), selected area electron diffraction (SAED), energy dispersive spectrometry (EDS), Raman spectroscopy, and X-ray photoelectron spectrometer (XPS) were performed after the scratching process. RMD simulations were performed to investigate the chemical reaction between the C-face $(000\bar{1})$ of 4H-SiC and water molecules under the mechanical action of a diamond indenter during nanoscale scratching.

2. Materials and Methods

2.1. Experimental Method

A friction and abrasion tester (MFT-5000, Retc, Wilmington, MA, USA) was used for scratching the SiC wafer. The size of SiC (4H-SiC-4°, $(000\bar{1})$ plane) wafers was 20 mm × 20 mm, and the original surface roughness Ra of SiC wafers was approximately 0.5 nm. The main factors of the SiC–water reaction in this experiment were determined using the control variable method, which controlled the radius of the indenter, scratching speed, and load—in that order. Diamond indenter tools were used for these experiments; these tools were made of single-crystal diamonds with radii of 0.2 mm, 0.4 mm, 0.6 mm, 1 mm, and 2 mm. The rotational speeds were set to 500 rpm, 1000 rpm, 1500 rpm, and 2000 rpm, and the eccentric radii were set to 5 mm, 6 mm, 7 mm, and 8 mm, corresponding to linear speeds of 0.262 m/s, 0.628 m/s, 1.099 m/s, and 1.625 m/s, respectively. The scratch loads were set to 0.3 N, 0.6 N, 0.9 N, 1.2 N, 1.5 N, and 1.8 N in these experiments. The processing time was set to 1 min. No chemical substance was added during the scratching process, and only distilled water was used as the coolant.

To better analyze the material removal behavior of SiC samples, the surface topographies of the scratches on SiC wafers were measured using a three-dimensional optical surface profiler (Newview 7300, ZYGO, Middlefield, CT, USA). Raman spectroscopy and XPS (K-alpha, Thermo Fisher, Waltham, MA, USA) were used to analyze the chemical composition of the scratches. TEM, HRTEM, SAED, and EDS characterizations of the wear debris extracted from the scratching-process coolant were processed using a Talos F200X G2 field-emission gun transmission electron microscope with an accelerating voltage of 200 kV.

2.2. RMD Model

The RMD simulation of scratching on the C-face of SiC with a diamond abrasive was performed to study the process of tribochemical reactions at nanoscale. The RMD simulation model of the high-speed scratching on SiC is shown in Figure 1; the workpiece size was 21.567 Å × 21.346 Å × 47.316 Å. The model was composed of a diamond abrasive tool, C-face $(000\bar{1})$ of the 4H-SiC wafer, and water molecules. The top part of the model represented a simplified diamond abrasive, and the contact interface of the diamond abrasive was divided into planar, spherical, and conical surfaces to consider the effects of different contact states. The middle part of the model represented free water molecules, and the lower part of the model represented a standard 4H-SiC lattice. The top two layers of SiC atoms used to control the loading and friction of the diamond abrasive were set as the moving layers, and the function of the bottom two layers was fixed to prevent rigid movements of the entire structure during the friction process.

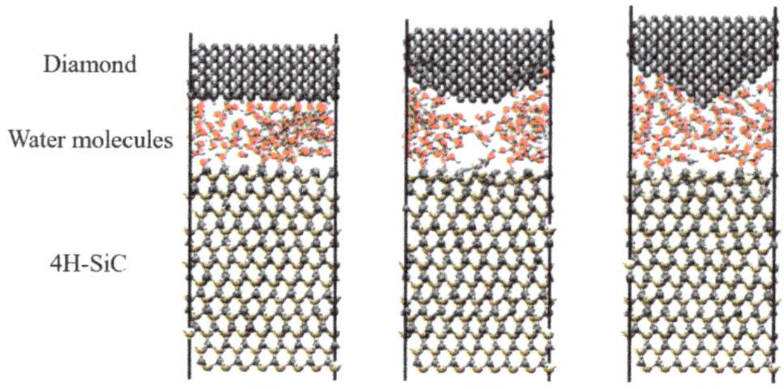

Figure 1. ReaxFF reactive molecular dynamics (RMD) simulation with (**a**) planar, (**b**) spherical, and (**c**) conical contact.

The accuracy and reliability of RMD simulations depend on the appropriate potential function selected. The ReaxFF approach was used to simulate bond breaking and bond formation during chemical reactions in the RMD simulations. Therefore, ReaxFF was used to describe the chemical reaction between SiC and H_2O molecules in the friction process. Therefore, all simulations were performed in the NVT ensemble, and the time step was set to 0.25 fs. A periodic boundary condition was applied in the x- and y-directions to reduce the influence of boundary effects, and a fixed boundary condition was applied in the z-direction. The temperature was controlled using a Nose–Hoover thermostat; the initial temperature of the simulation system was 160 °C, and the damping constant was 10 fs. The RMD simulations in this study were performed using a large-scale atomic/molecular massively parallel simulator (LAMMPS). A massively parallel simulator (MAPS) was used to construct the model, visualize the process, and analyze the results.

3. Results

3.1. Controlling the Material Removal Behaviour of SiC Wafer

Different radii were set for the indenter, scratch speeds, and loads during this experiment to reflect the mechanical action of a single diamond abrasive in SiC wafer ultra-precision machining to examine the SiC–water reaction conditions. First, the same load (0.3 N) and linear velocity (0.262 m/s) were set, and the results of the C-face $(000\bar{1})$ morphologies of 4H-SiC wafers scratched by diamond indenters having different radii are depicted in Figure 2. After scratching with a diamond indenter of 0.2 mm radius, a brittle-removal phenomenon was observed on this scratch (Figure 2a), as well as several deep pits and microcracks with micron grade depth. With increasing radius of the diamond indenter, the scratch depth decreased from micro-scale to nanoscale, and the brittle-removal mode of the SiC material gradually weakened. A scratch with no evident pits and cracks was achieved after scratching with a diamond indenter of 2 mm radius, as shown in Figure 2e, and the material removal mode of SiC changed to plastic removal.

Subsequently, a diamond indenter of 2 mm radius and a constant load of 0.3 N were controlled in this experiment, and the speeds were set sequentially to 0.262 m/s, 0.628 m/s, 1.099 m/s, and 1.625 m/s to study the effects of the scratching speed on SiC material removal. As shown in Figure 3, the depths and widths of scratches increased with an increase in speed. Furthermore, the shapes of scratches without any pits or cracks were more regular, and the extent of plastic removal of the SiC material gradually increased. Therefore, a speed of 1.625 m/s was set as a fixed parameter to deepen the extent of plastic removal of the SiC material while examining the SiC–water reaction.

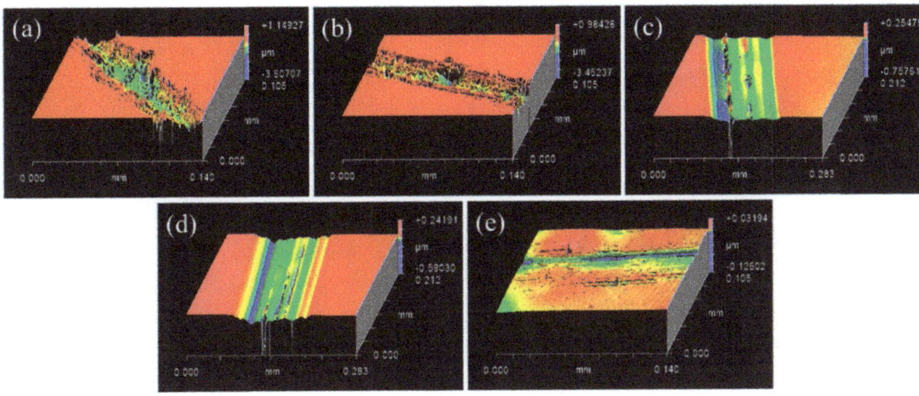

Figure 2. Surface profile of SiC scratched with 0.3 N, 0.262 m/s by different radius of indenter (**a**) 0.2 mm, (**b**) 0.4 mm, (**c**) 0.6 mm, (**d**) 1 mm, and (**e**) 2 mm.

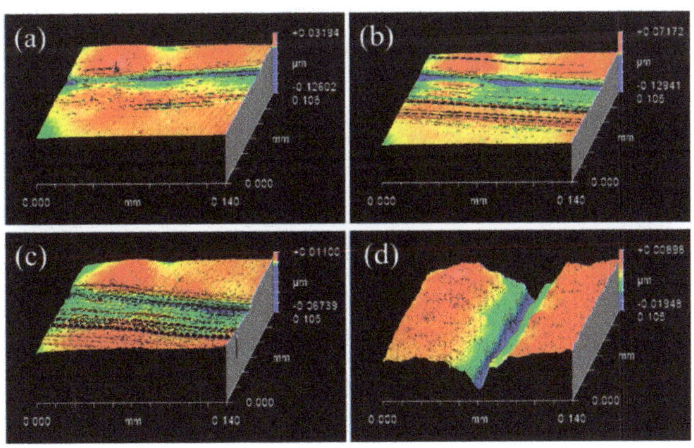

Figure 3. Surface profiles of scratched SiC after experiments with 2 mm indenter and 0.3 N at different speeds: (**a**) 0.262 m/s, (**b**) 0.628 m/s, (**c**) 1.099 m/s, and (**d**) 1.625 m/s.

Finally, a diamond indenter of 2 mm radius and a constant speed of 1.625 m/s were fixed in this test, and the loads were set to 0.3 N, 0.6 N, 0.9 N, 1.2 N, 1.5 N, and 1.8 N to study the effects of different loads on SiC material removal. The depths and widths of the scratches were increased with the increase in load shown in Figure 4a–d and when the load was between 0.3 N and 1.2 N, the depth of scratches increased from 15 nm to over 130 nm, and the extent of plastic removal of SiC material increased. However, a few pits were observed at the bottom of the scratch, which were at a depth of 400 nm, as shown in Figure 4e. The coexistence of brittle removal and plastic removal were observed when the load was 1.5 N. Nevertheless, a few deep microcracks were observed on the scratch when the load was 1.8 N, as shown in Figure 4f, when the SiC material was primarily in the brittle stage. With the increase in load, there was a large amount of fragmentation at the bottom of the scratch, gradually transitioning from the plastic stage to the brittle–plastic coexistence stage, and subsequently to the brittle stage.

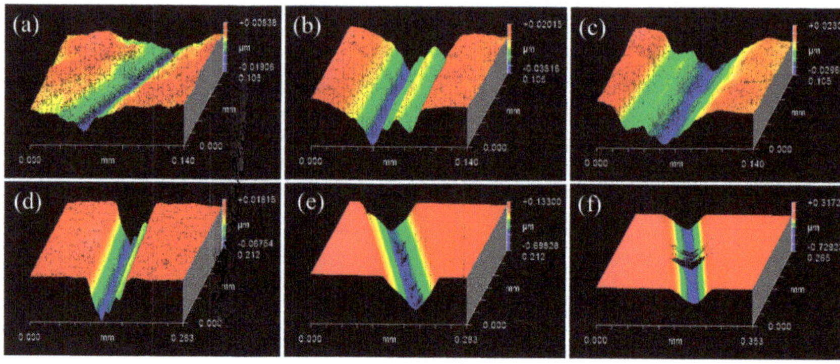

Figure 4. Scratch surface profiles of SiC scratched with a 2 mm indenter and speed of 1.625 m/s at different loads: (**a**) 0.3 N, (**b**) 0.6 N, (**c**) 0.9 N, (**d**) 1.2 N, (**e**) 1.5 N, and (**f**) 1.8 N.

3.2. Characterisations of SiC–Water Reaction on C-Face $(000\bar{1})$

As indicated by results of the afore-described experiments, when the radius of diamond indenter was 2 mm, speed was 1.625 m/s, and load was 1.2 N, a relatively uniform and regular plastic scratch without evident cracks was obtained, and wear debris was collected from the SiC wafer surface after the experiment. Therefore, the plastic scratch on the SiC surface was ultrasonically cleaned and analyzed to investigate whether the SiC–water reaction was performed under plastic conditions using Raman spectroscopy and XPS. The wear debris was examined and analyzed to further study the chemical reaction that occurred between the C-face $(000\bar{1})$ of SiC and water. The characterization results of the wear debris included TEM, HRTEM, SAED, and EDS analyses.

Figure 5 shows the Raman detection of the C-face $(000\bar{1})$ of 4H-SiC wafer under the abovementioned plastic conditions. The Raman peak of SiC was observed at 966 cm^{-1}, whereas the Raman peak of silica dioxide was at 968 cm^{-1} [35]. After fitting the original graph to split the peaks, new peaks of SiC and silica dioxide were at 963.6 cm^{-1} and 973.1 cm^{-1}, respectively, as shown in Figure 5a. This is because the oxide layer remained on the original surface of the SiC wafer after CMP; therefore, a smaller peak of silica dioxide appeared. In Figure 5b, the peaks of SiC and SiO_2 were at 970.1 cm^{-1} and 987.4 cm^{-1}, respectively. The shift of the Raman peak of SiC may be because of the influence of stress [36]. Comparison of Raman results between scratch and non-scratch regions, it could be clearly seen that the original peak was significantly broadened, and the fitted SiO_2 peak was also broadened, indicating that amorphous SiO_2 was possibly formed on the scratch [37].

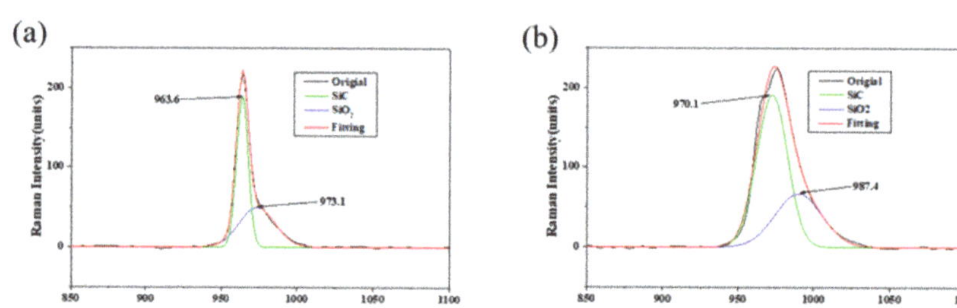

Figure 5. Detection results of debris of C-face $(000\bar{1})$ of 4H-SiC wafer after scratching under optimal conditions: (**a**) Raman detection without scratching and (**b**) Raman detection with scratching.

The composition and chemical state of the plastic scratch on the C-face $(000\bar{1})$ of 4H-SiC wafers were further characterized by XPS, as shown in Figure 6. As observed in Figure 6a, there are mainly peaks at Si 2s, Si 2p, C 1s, N 1s, O 1s, and O KLL. The N 1s peak was attributed to nitrogen doping during the preparation of 4H-SiC wafers. The fine spectra of Si 2p, C 1s, and O 1s were analyzed because the main elements on the surface of 4H-SiC wafers were Si, C, and O. There were four peaks in the C 1s spectra; the peak at 284.78 eV corresponded to SiC, whereas the other three peaks corresponded to the organic compounds adsorbed on the surface. The peak at 534.31 eV in the O 1s spectra and the peak at 103.42 eV in the Si 2p spectra corresponded to SiO_2 [38,39]. Therefore, SiO_2 existed in the scratch on the C-face $(000\bar{1})$ of 4H-SiC wafer, and the results indicated that the SiC might react with water under mechanical friction to generate SiO_2. Further analysis of the wear debris was required to verify whether the SiC–water reaction had occurred.

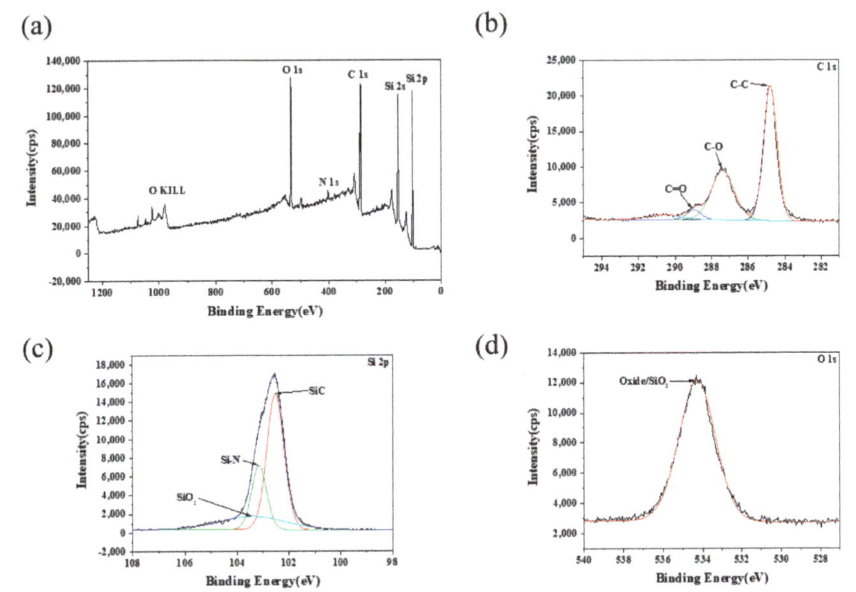

Figure 6. Detection results of debris of C-face $(000\bar{1})$ of 4H-SiC wafer 4H-SiC wafers after scratching under optimal conditions: (**a**) XPS whole spectra, (**b**) XPS fine spectra of C 1s, (**c**) XPS fine spectra of Si 2p, and (**d**) XPS fine spectra of O 1s.

TEM images, SAED patterns, HRTEM images, EDS spectra, and EDS mapping of SiC wear debris under the aforementioned conditions are shown in Figure 7. As shown in Figure 7a,b, the wear debris of the C face with a thickness of approximately 200–300 nm was spherical, and the atomic arrangement of the wear debris had no periodic variation, with no evident crystalline state. The SAED pattern without diffraction spots emerged as a halo, indicating that the wear debris was in an amorphous phase. The Mo and Cu appearing in the EDS spectra were mainly derived from the Mo mesh micro-grid supporting film, as shown in Figure 7c. The elements of the wear debris were silicon, oxygen, and carbon; however, oxygen and silicon were considerably more abundant than carbon. In addition, among the individual elements observed in the EDS mapping shown in Figure 7d, the major elements of the wear debris are silicon and oxygen. Therefore, it was confirmed that the main component of the wear debris of the C-face was amorphous SiO_2 [40,41].

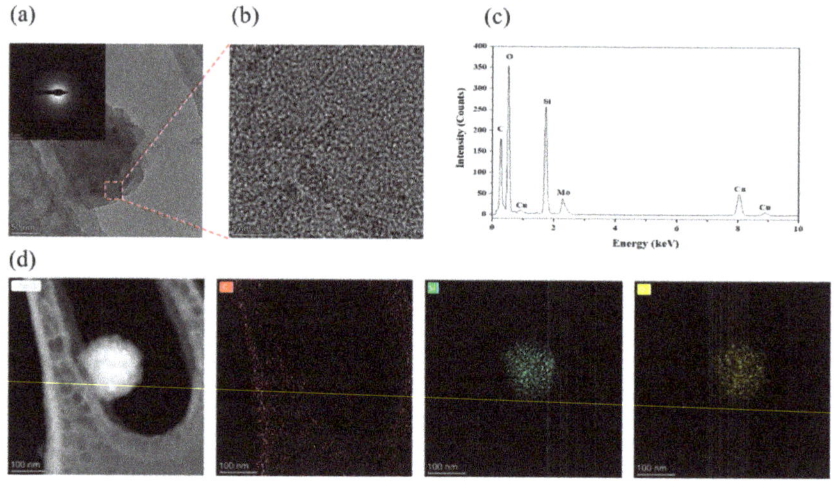

Figure 7. Detection results of debris of 4H-SiC wafers after scratching under optimal conditions: (**a**) TEM image and SAED pattern, (**b**) HRTEM image, (**c**) EDS spectra, and (**d**) EDS Mapping.

3.3. Mechanism of SiC–Water Reaction in RMD Simulation

Figure 8 shows the results of the reaction between SiC and water molecules under different contact states. The diamond indenter had a planar surface, spherical surface, and conical surface. The friction load, friction speed, and temperature were set to 60 nN, 20 m/s, and 580 °C, respectively. From different viewpoints, the results showed that the lattice structure of SiC materials was severely damaged by the conical surface, and, thus, more silicon atoms were oxidized. In this case, the layer structure of SiC materials was damaged by the spherical contact surface, which reacted with water molecules during the friction process. However, the structure of SiC materials was relatively intact under planar contact, with only the surface layer atoms forming silicon–oxygen bonds with oxygen atoms. This further indicated that the degree of chemical reaction at the interface was determined by the degree of damage to the lattice structure of SiC materials. Interfacial chemical reactions could be facilitated by contact modes that resulted in greater stress concentrations, leading to structural breakdown of SiC materials. Generally, to match the actual processing situation, the contact mode at the nanoscale was simplified to planar contact. Therefore, all simulations were performed using planar contact to investigate the effects of the remaining factors on the SiC–water reaction.

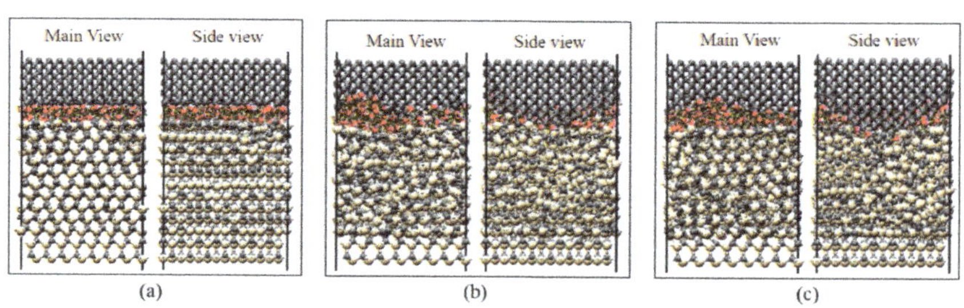

Figure 8. SiC–water reactions under different contact states when the contact interfaces were set as: (**a**) flat, (**b**) spherical, and (**c**) conical. Frictional loads, frictional velocities, and temperatures are shown as 60 nN, 20 m/s, and 580 °C.

The results of the reaction between SiC and water molecules at different friction speeds are shown in Figure 9, wherein the friction depths at a temperature of 790 °C were 0.6 nm and 0.7 nm. When the speed increased from 10 m/s to 20 m/s, the lattice structure of

SiC materials did not significantly change at the same friction depth. Conversely, at the same friction speed, when the friction depth increased from 0.6 nm to 0.7 nm, the lattice structure of SiC materials was significantly damaged, effectively promoting the SiC–water reaction. The change in friction speed had no significant effect on the chemical reaction at the interface between SiC and water molecules, as well as on the degree of damage to the lattice structure of SiC materials.

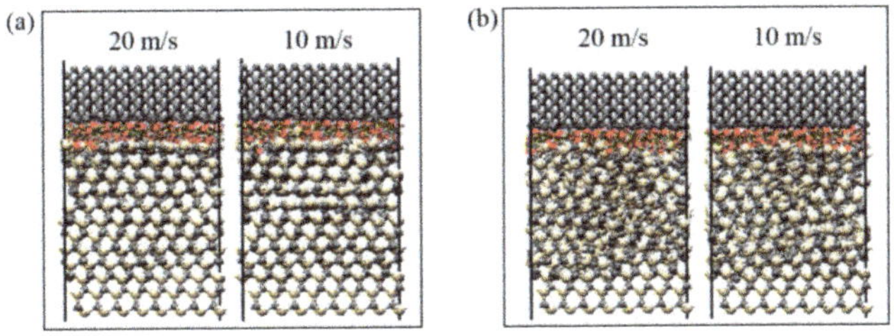

Figure 9. SiC–water reactions at different friction speeds. Friction depths and temperatures were set to: (**a**) 0.6 nm, 790 °C and (**b**) 0.7 nm, 790 °C.

The SiC–water reactions at different temperatures are shown in Figure 10, wherein the friction depths at a friction speed of 20 m/s were 0.6 nm and 0.7 nm. From the perspective of temperature, at the same friction depth, when the temperature was below 580 °C, the interfacial reaction only occurred at the contact interface between SiC materials and water molecules. At temperatures between 580 °C and 790 °C, the water molecules entered the second layer of SiC materials, starting the chemical reaction. When the temperature was further increased to 1000 °C, the lattice structure of SiC materials was significantly damaged, and the reaction occurred in the subsurface layer. As the temperature increased, the atoms at the interface moved more drastically, and the chemical reactions that occurred became more obvious.

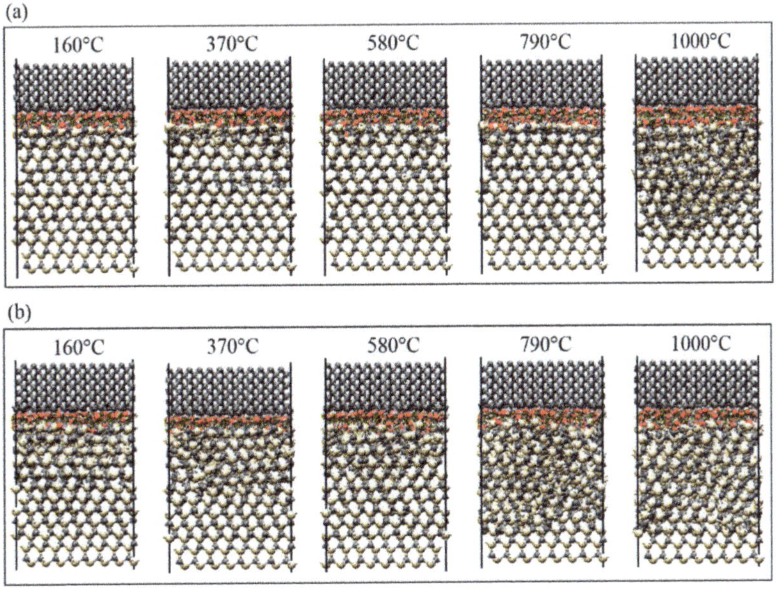

Figure 10. SiC–water reactions at different temperatures. The friction depth and friction velocity were set to: (**a**) 0.6 nm, 20 m/s and (**b**) 0.7 nm, 20 m/s.

From the point of view of friction depth (load), when the friction depth increased from 0.6 nm to 0.7 nm at the same temperature, the lattice structure of SiC materials was significantly damaged, and as the friction depth increased, the extent of SiC lattice structure damage increased and the reaction became more vigorous. Simultaneously, even at lower temperatures, as long as the friction depth was sufficiently large, the lattice structure of SiC materials would be damaged for the reaction to occur. Notably, as shown in Figure 10a, the lattice structure of SiC materials was still not significantly damaged at a temperature of 790 °C and friction depth of 0.6 nm; however, the lattice structure was already significantly disrupted at a temperature of 370 °C and friction depth of 0.7 nm, resulting in the SiC–water reaction. Therefore, the lattice structure of SiC materials was more effectively disrupted by increasing the friction depth than by increasing the temperature; thus, the reaction at the interface was significantly promoted.

4. Discussion

A regular SiC plastic scratch without evident cracks was obtained in the scratching experiments that employed a diamond indenter with a radius of 2 mm, speed of 1.625 m/s, and load of 1.2 N. In this process, only deionized water was used as the coolant without adding any acid–base etching solution. The chemical composition of the SiC plastic scratch was further analyzed using Raman spectroscopy and XPS, and the wear debris was analyzed using TEM. The chemical reaction was observed to generate amorphous silicon dioxide (the phenomenon of the chemical reaction that occurred between the diamond abrasive and SiC wafers was highly consistent with our previous work [27]). This result was highly consistent with the results of RMD simulations, which further verified that the chemical reaction at the nanoscale between SiC and water molecules occurred under mechanical action. The experimental results indicate that the SiC–water reaction occurred in the plastic removal stage of SiC wafers; thus, controlling the material removal of SiC was a prerequisite for determining the SiC–water reaction. Scratching experiments focused on the removal mode of SiC to indirectly verify the conditions of the SiC–water reaction at the macroscopic level, whereas the scratching depths at the nanoscale were controlled by precisely regulating the scratch loads, which was consistent with the scale of friction depths at the nanoscale in the RMD simulation. Therefore, both experiments and RMD simulations were combined to investigate the effects of different contact states, friction speed, friction depth (load), and different temperatures on the SiC–water reaction.

In the experiments, the brittle fracture of bulk material on the SiC surface was directly caused by the small radius indenter (a radius of 0.2 mm), indicating that the removal mode of SiC materials in this state was brittle. However, the material removal mode of SiC was plastic when SiC was scratched using a large-radius indenter (with a radius of 2 mm). In the RMD simulation, the lattice structure of SiC materials was easily damaged by the conical and spherical contacts; however, in the case of the planar contact, the lattice structure was relatively more intact, and the reaction occurred on the surface. Both the experiments and RMD simulations further illustrated that the structure of SiC was easily damaged owing to a larger stress concentration that occurred because of the difference in contact states, and the material removal mode was more likely to be brittle. Therefore, the SiC–water reaction can be effectively promoted only if the material removal form of SiC is controlled in the plastic removal phase.

As the scratching speed gradually increased (up to 1.625 m/s), the extent of plastic removal of SiC materials gradually increased during the experiment, and the plastic scratches became increasingly evident. Nevertheless, the actual experimental speed was not simulated, and the mode of material removal was less affected by high speeds in the RMD simulation. Hence, the agreement between the experimental and simulated results verified that the scratching speed had no evident effect on the material removal of SiC materials and the SiC–water reaction, whereas it was easier to observe plastic scratches with an increase in speed [42].

The load was precisely controlled within a certain range (0.3–1.5 N) in the experiment, which precisely controlled the friction depth. When the load was 1.2 N, a regular plastic scratch without any evident cracks was obtained, and the SiC–water reaction was confirmed to occur on the scratch. The SiC material changed from plastic deformation mode to brittle damage mode with an increase in load, whereas brittle cracks appeared at the bottom of the scratch when the load was greater than 1.5 N. Therefore, the brittle–plastic transformation behavior of SiC was realized in the range of 1.5–1.8 N. An increase in the friction depth (load) in the RMD simulation increased the extent of SiC lattice structure damage, and the interface reaction became significantly vigorous. Generally, the higher the temperature, the more drastic the chemical reaction is. Although it was difficult to conduct high-temperature scratching experiments owing to the limitations of the instruments and aqueous environment, RMD simulation was performed to analyze the effect of different temperatures on the SiC–water reaction. Evidently, with an increase in temperature, the damage in the lattice structure of SiC gradually increased. However, by combining the two conditions of friction depth (load) and temperature, even at a lower temperature (160 °C), the friction depth was sufficiently large; this damages the lattice structure of SiC and promotes the SiC–water reaction. This result was verified by the scratching experiment, wherein the SiC–water reaction still occurred at room temperature (20 °C) under the abovementioned conditions. Therefore, the friction depth (load) during the plastic removal stage was more important than the temperature in the mechanism of the SiC–water reaction; this indicates that the load was the most critical factor in achieving plastic removal of SiC in both experiments and RMD simulations. By effectively regulating the contact state, speed, and load, the material removal mode of SiC materials was reasonably controlled in the plastic removal phase; thus, the SiC–water reaction smoothly occurred.

The SiC-water reaction occurred under the plastic removal of SiC materials, thus controlling the reaction was primarily a matter of controlling the mode of material removal in an aqueous medium. The removal mode of SiC was achieved primarily by controlling the load and contact states, and secondarily by controlling the velocity and temperature. The lattice structure of SiC materials was directly and effectively damaged by regulating the load and contact states in an aqueous medium. Thus, the material removal form of SiC was effectively controlled in the plastic removal phase, further avoiding microcracks and brittle collapse. During the plastic phase, SiC materials on surface layer was mechanically induced to generate amorphous silicon dioxide when distilled water was used as the coolant, and a controllable SiC–water reaction was used to obtain smooth, scratch-free, and damage-free wafers during the thinning process of SiC wafers.

5. Conclusions

This study entailed experiments and RMD simulations to investigate the mechanisms of the reaction between SiC and water during nanoscale scratching in the absence of any chemical reagents. The material removal behavior of SiC at the nanoscale level was analyzed, and the mechanisms of the SiC–water reaction were summarized. The following conclusions were drawn:

(1) A regular plastic scratch with nanometer depth and almost no evident cracks was obtained using diamond abrasives to scratch 4H-SiC C face ($000\bar{1}$) without any chemical agent, and the formation of amorphous SiO_2 was clearly observed, which verified that the SiC–water reaction occurred in the SiC plastic removal stage.
(2) When the contact radius was gradually increased to 2 mm, the scratches that developed on the surface were converted from brittle to plastic removal mode, which were further concluded from the RMD simulations. The SiC crystal structure was damaged to a lesser extent owing to the larger contact radius and smaller contact stress. The plastic removal of SiC materials was not significantly affected by the increase in speed; only the removal rate of SiC materials was increased.
(3) With the increase in load, the scratch depth could be in the range of tens to hundreds of nanometers during scratching. Furthermore, the removal mode of SiC transi-

tioned from plastic to brittle and thereby achieved a controllable behavior in the plastic domain of SiC under controlled loading, which indirectly controlled the SiC–water reaction. Conversely, the controllable behavior of the SiC–water reaction at the nanoscale during RMD simulations was primarily achieved by adjusting the friction depth. Therefore, both experiments and simulations regulated the load, thereby regulating the friction depth to control the occurrence of the SiC–water reaction.

(4) The main factors controlling the SiC–water reaction were the load and contact states, and the secondary factors were the speed and temperature. The SiC–water reaction could occur smoothly at low temperatures as long as a suitable load and contact state were maintained. Therefore, microcracks could be reasonably avoided by the effective application of the SiC–water reaction in the subsequent SiC thinning process, thus improving the surface quality of SiC wafers.

Author Contributions: Experiment, data analysis, and writing, Z.C. and Q.L.; conception, material characterization and analysis, J.L. and Q.L.; software and data analysis, Z.T. All authors have read and agreed to the published version of the manuscript.

Funding: This research was funded by National Natural Science Foundation of China, grant numbers 51975222 and 52005190, and Natural Science Foundation of Fujian Province, grant numbers 2021J06027 and 2021J05060.

Conflicts of Interest: The authors declare no conflict of interest.

References

1. Yu, K.; Zhao, W.; Wu, X.; Zhuang, J.; Hu, X.; Zhang, Q.; Sun, L. In situ atomic-scale observation of monolayer graphene growth from SiC. *Nano Res.* **2018**, *11*, 2809–2820. [CrossRef]
2. Chen, L.; Wen, J.; Zhang, P.; Yu, B.; Chen, C.; Ma, T.; Lu, X.; Kim, S.H.; Qian, L. Nanomanufacturing of silicon surface with a single atomic layer precision via mechanochemical reactions. *Nat. Commun.* **2018**, *9*, 1542. [CrossRef]
3. Gao, W.; Zhang, Y.X.; Huang, P.J. Study on material removal mechanism of 6H-SiC single crystal wafer based on different nano-scratch order. *Diam. Abras. Eng.* **2022**, *41*, 92–97.
4. Deng, J.; Yan, Q.; Lu, J.; Xiong, Q.; Pan, J. Optimisation of lapping process parameters for single-crystal 4H–SiC using orthogonal experiments and grey relational analysis. *Micromachines* **2021**, *12*, 910. [CrossRef]
5. Neslen, C.L.; Mitchel, W.C.; Hengehold, R.L. Effects of process parameter variations on the removal rate in chemical mechanical polishing of 4H-SiC. *J. Electron. Mater.* **2001**, *30*, 1271–1275. [CrossRef]
6. Evans, C.J.; Paul, E.; Dornfeld, D.A.; Lucca, D.A.; Byrne, G.; Tricard, M.; Klocke, F.; Dambon, O.; Mullany, B.A. Material removal mechanisms in lapping and polishing. *CIRP Ann.- Manuf. Techn.* **2003**, *52*, 611–633. [CrossRef]
7. Endoh, N.; Akiyama, S.; Tashima, K.; Suwa, K.; Kamogawa, T.; Kohama, R.; Fukidome, H. High-quality few-layer graphene on single-crystalline SiC thin film grown on affordable wafer for device applications. *Nanomaterials* **2021**, *11*, 392. [CrossRef]
8. Royet, A.S.; Ouisse, T.; Cabon, B.; Noblanc, O.; Arnodo, C.; Brylinski, C. Self-heating effects in silicon carbide mesfets. *IEEE Trans. Electron. Devices* **2000**, *47*, 2221–2227. [CrossRef]
9. Anzalone, R.; Piluso, N.; Litrico, G.; Lorenti, S.; Arena, G.; Coffa, S.; La Via, F. Stress relaxation mechanism after thinning process on 4H-SiC substrate. *Mater. Sci. Forum* **2018**, *924*, 535–538. [CrossRef]
10. Zhao, M.; Hayakawa, S.; Nishida, Y.; Jourdain, A.; Tabuchi, T.; Leunissen, L.H. Wafer backside thinning process integrated with post-thinning clean and TSV exposure recess etch. *ECS Trans.* **2012**, *44*, 865. [CrossRef]
11. Huo, F.W.; Guo, D.M.; Kang, R.K. Nanogrinding of SiC wafers with high flatness and low subsurface damage. *Trans. Nonferrous Met. Soc. China* **2012**, *22*, 3027–3033. [CrossRef]
12. Presser, V.; Krummhauer, O.; Kailer, A.; Nickel, K.G. In situ monitoring and depth-resolved characterization of wet wear of silicon carbide. *Wear* **2011**, *271*, 2665–2672. [CrossRef]
13. Rupp, R.; Gerlach, R.; Kirchner, U.; Schlogl, A.; Kern, R. Performance of a 650 V SiC diode with reduced chip thickness. *Mater. Sci. Forum* **2012**, *717–720 Pt 2*, 921–924. [CrossRef]
14. Li, Z.; Zhang, F.; Luo, X. Subsurface damages beneath fracture pits of reaction-bonded silicon carbide after ultra-precision grinding. *Appl. Surf. Sci.* **2018**, *448*, 341–350. [CrossRef]
15. Yamaguchi, K.; Touge, M.; Nakano, T.; Watanabe, J. Study on lapping and constant-pressure grinding of single-crystal SiC. *Adv. Mater. Res.* **2009**, *76–78*, 282–287. [CrossRef]
16. Tsukimoto, S.; Ise, T.; Maruyama, G.; Hashimoto, S.; Sakurada, T.; Senzaki, J. Correlation between local strain distribution and microstructure of grinding-induced damage layers in 4H-SiC ($000\bar{1}$). *Mater. Sci. Forum* **2017**, *897*, 177–180. [CrossRef]
17. Feng, K.; Zhao, T.; Lyu, B.; Zhou, Z. Ultra-precision grinding of 4H-SiC wafer by pav/pf composite sol-gel diamond wheel. *Adv. Mech. Eng.* **2021**, *13*, 16878140211044929. [CrossRef]

18. Sano, Y.; Tajiri, K.; Inoue, Y.; Mukai, R.; Nakanishi, Y.; Matsuyama, S.; Yamauchi, K. High-speed etching of silicon carbide wafer using high-pressure sf6 plasma. *ECS J. Solid State Sci. Technol.* **2021**, *10*, 014005. [CrossRef]
19. Guan, J.; Zhao, Y. Non-contact grinding/thinning of silicon carbide wafer by pure EDM using a rotary cup wheel electrode. *Precis. Eng.* **2022**, *74*, 209–223. [CrossRef]
20. Zhou, Y.; Pan, G.; Shi, X.; Gong, H.; Luo, G.; Gu, Z. Chemical mechanical planarization (CMP) of on-axis Si-face SiC wafer using catalyst nanoparticles in slurry. *Surf. Coat. Technol.* **2014**, *251*, 48–55. [CrossRef]
21. Deng, J.Y.; Pan, J.S.; Zhang, Q.X.; Guo, X.H.; Yan, Q.S. Research progress in chemical mechanical polishing of single crystal SiC substrates. *Diam. Abras. Eng.* **2020**, *40*, 13.
22. Qi, W.; Cao, X.; Xiao, W.; Wang, Z.; Su, J. Study on the mechanism of solid-phase oxidant action in tribochemical mechanical polishing of SiC single crystal substrate. *Micromachines* **2021**, *12*, 1547. [CrossRef]
23. Lu, J.; Li, Y.; Xu, X. The effects of abrasive yielding on the polishing of SiC wafers using a semi-fixed flexible pad. *Proc. Inst. Mech. Eng. Part B J. Eng. Manuf.* **2015**, *229*, 170–177. [CrossRef]
24. Luo, Q.; Lu, J.; Xu, X. A comparative study on the material removal mechanisms of 6H-SiC polished by semi-fixed and fixed diamond abrasive tools. *Wear* **2016**, *350*, 99–106. [CrossRef]
25. Luo, Q.; Lu, J.; Xu, X.; Jiang, F. Removal mechanism of sapphire substrates (0001, $11\bar{2}0$ and $10\bar{1}0$) in mechanical planarization machining. *Ceram. Int.* **2017**, *43*, 16178–16184. [CrossRef]
26. Xu, Y.; Lu, J.; Xu, X.; Chen, C.; Lin, Y. Study on high efficient sapphire wafer processing by coupling SG-mechanical polishing and GLA-CMP. *Int. J. Mach. Tool. Manu.* **2018**, *130*, 12–19. [CrossRef]
27. Luo, Q.; Lu, J.; Tian, Z.; Jiang, F. Controllable material removal behavior of 6H-SiC wafer in nanoscale polishing. *Appl. Surf. Sci.* **2021**, *562*, 150219. [CrossRef]
28. Chrobak, D.; Tymiak, N.; Beaber, A.; Ugurlu, O.; Gerberich, W.W.; Nowak, R. Deconfinement leads to changes in the nanoscale plasticity of silicon. *Nat. Nanotechnol.* **2011**, *6*, 480–484. [CrossRef]
29. Wu, Z.; Liu, W.; Zhang, L. Revealing the deformation mechanisms of 6H-silicon carbide under nano-cutting. *Comput. Mater. Sci.* **2017**, *137*, 282–288. [CrossRef]
30. Van Duin, A.C.; Dasgupta, S.; Lorant, F.; Goddard, W.A. ReaxFF: A reactive force field for hydrocarbons. *J. Phys. Chem. A* **2001**, *105*, 9396–9409. [CrossRef]
31. Chenoweth, K.; Van Duin, A.C.; Goddard, W.A. ReaxFF reactive force field for molecular dynamics simulations of hydrocarbon oxidation. *J. Phys. Chem. A* **2008**, *112*, 1040–1053. [CrossRef]
32. Senftle, T.P.; Hong, S.; Islam, M.M.; Kylasa, S.B.; Zheng, Y.; Shin, Y.K.; Van Duin, A.C. The ReaxFF reactive force-field: Development, applications and future directions. *NPJ Comput. Mater.* **2016**, *2*, 15011. [CrossRef]
33. Wen, J.; Ma, T.; Zhang, W.; Psofogiannakis, G.; van Duin, A.C.; Chen, L.; Lu, X. Atomic insight into tribochemical wear mechanism of silicon at the Si/SiO_2 interface in aqueous environment: Molecular dynamics simulations using ReaxFF reactive force field. *Appl. Surf. Sci.* **2016**, *390*, 216–223. [CrossRef]
34. Fogarty, J.C.; Aktulga, H.M.; Grama, A.Y.; Van Duin, A.C.; Pandit, S.A. A reactive molecular dynamics simulation of the silica-water interface. *J. Chem. Phys.* **2010**, *132*, 174704. [CrossRef]
35. Roehrl, J.; Hundhausen, M.; Emtsev, K.V.; Seyller, T.; Graupner, R.; Ley, L. Raman spectra of epitaxial graphene on SiC (0001). *Appl. Phys. Lett.* **2008**, *92*, 201918. [CrossRef]
36. Ma, L.; Qiu, W.; Fan, X. Stress/strain characterization in electronic packaging by micro-Raman spectroscopy: A review. *Microelectron. Reliab.* **2021**, *118*, 114045. [CrossRef]
37. Malinovsky, V.K.; Novikov, V.N.; Surovtsev, N.V.; Shebanin, A.P. Investigation of amorphous states of SiO_2 by Raman scattering spectroscopy. *Phys. Solid State* **2000**, *42*, 65–71. [CrossRef]
38. Kusunoki, I.; Igari, Y. XPS study of a SiC film produced on Si (100) by reaction with a C_2H_2 beam. *Appl. Surf. Sci.* **1992**, *59*, 95–104. [CrossRef]
39. Ferrah, D.; Penuelas, J.; Bottela, C.; Grenet, G.; Ouerghi, A. X-ray photoelectron spectroscopy (XPS) and diffraction (XPD) study of a few layers of graphene on 6H-SiC (0001). *Surf. Sci.* **2013**, *615*, 47–56. [CrossRef]
40. Lu, J.; Wang, Y.; Luo, Q.; Xu, X. Photocatalysis assisting the mechanical polishing of a single-crystal SiC wafer utilizing an anatase TiO2-coated diamond abrasive. *Precis. Eng.* **2017**, *49*, 235–242. [CrossRef]
41. Borysiuk, J.; Bozek, R.; Grodecki, K.; Wysmołek, A.; Strupinski, W.; Stępniewski, R.; Baranowski, J.M. Transmission electron microscopy investigations of epitaxial graphene on C-terminated 4H-SiC. *J. Appl. Phys.* **2010**, *108*, 013518. [CrossRef]
42. Tian, L.; Fu, Y.; Xu, J.; Li, H.; Ding, W. The influence of speed on material removal mechanism in high speed grinding with single grit. *Int. J. Mach. Tool. Manu* **2015**, *89*, 192–201. [CrossRef]

micromachines

Article

Simulation Analysis of Multi-Physical Field Coupling and Parameter Optimization of ECM Miniature Bearing Outer Ring Based on the Gas-Liquid Two-Phase Turbulent Flow Model

Zhaolong Li [1,2,*], Wangwang Li [2] and Bingren Cao [2]

[1] Key Laboratory of Advanced Manufacturing Intelligent Technology of Ministry of Education, Harbin University of Science and Technology, Harbin 150080, China
[2] School of Mechanical and Power Engineering, Harbin University of Science and Technology, Harbin 150080, China; liwangwang19970214@163.com (W.L.); 1920110144@stu.hrbust.edu.cn (B.C.)
* Correspondence: lizhaolong@hrbust.edu.cn; Tel.: +86-0451-8639-0588

Abstract: Electrochemical machining (ECM) is an essential method for machining miniature bearing outer rings on the high-temperature-resistant nickel-based alloy GH4169. However, the influence of electrolyte temperature distribution and bubble rate distribution on electrolyte conductivity in the ECM area could not be fully considered, resulting in the simulation model not being able to accurately predict the machining accuracy of the outer ring of the miniature bearing, making it challenging to model and predict the optimal process parameters. In this paper, a multiphysics field coupled simulation model of electric, flow, and temperature fields during the ECM of the miniature bearing outer ring is established based on the gas–liquid two-phase turbulent flow model. The simulation analyzed the distribution of electrolyte temperature, bubble rate, flow rate, and current density in the machining area, and the profile change of the outer ring of the miniature bearing during the machining process. The analysis of variance and significance of machining voltage, electrolyte concentration, electrolyte inlet flow rate, and interaction on the mean error of the ECM miniature bearing outer rings was derived from the central composite design. The regression equation between the average error and the process parameters was established, and the optimal combination of process parameters for the average error was predicted, i.e., the minimum value of 0.014 mm could be achieved under the conditions of a machining voltage of 16.20 V, an electrolyte concentration of 9.29%, and an electrolyte inlet flow rate of 11.84 m/s. This is important to improve the machining accuracy of the outer ring of the ECM miniature bearing.

Keywords: ECM; gas–liquid two-phase turbulence model; miniature bearing outer ring; machining accuracy; central composite design

Citation: Li, Z.; Li, W.; Cao, B. Simulation Analysis of Multi-Physical Field Coupling and Parameter Optimization of ECM Miniature Bearing Outer Ring Based on the Gas-Liquid Two-Phase Turbulent Flow Model. *Micromachines* **2022**, *13*, 902. https://doi.org/10.3390/mi13060902

Academic Editors: Youqiang Xing, Xiuqing Hao and Duanzhi Duan

Received: 5 May 2022
Accepted: 6 June 2022
Published: 7 June 2022

1. Introduction

With the development of aerospace and automotive manufacturing and machinery production miniaturization, precision and complex internal characteristics of the small hole structure are more and more widely used, such as small holes in the reaming form of it can be applied to the outer ring of the miniature bearing [1]. GH4169 is a high-temperature-resistant nickel-based alloy material widely used to manufacture miniature bearing parts [2]. However, due to the high hardness of the alloy in the use of conventional machining methods challenging to the process, poor machining accuracy, and surface quality, the service life of miniature bearings is not long [3]. Electric discharge machining (EDM) and laser beam machining (LBM) are both thermal processes with high machining efficiency. The disadvantages are that they produce recast layers, heat-affected areas, and tensile residual stresses that also reduce the service life of micro bearings [4–7]. Electrochemical machining (ECM) is based on the principle of anodic dissolution for metal removal, independent of the hardness of the workpiece. It has become one of the main machining techniques for machining miniature bearing outer rings on the high-temperature-resistant nickel-based alloy GH4169 because of its advantage of no thermal damage [8–10]. There are still some problems to be solved. For example, the influence of electrolyte temperature distribution and bubble rate distribution on electrolyte conductivity in the ECM area cannot be fully considered, resulting in the simulation model not being able to accurately predict the machining accuracy of the outer ring of the miniature bearing. Therefore, it is difficult to model and predict the optimal process parameters.

In Deconck et al. [11–13], a simulation model for calculating the temperature distribution was developed, and it was pointed out that the electrolyte temperature distribution has an important influence on the machining accuracy. However, the model ignores the effect of electrolyte bubble rate distribution on machining accuracy and uses the laminar flow N–S model, which reduces the heat transfer effect. In Fang et al. [14], a multiphysics field coupled simulation model was proposed to predict the electrolytic machining accuracy. However, they treated the electrolyte as a single-phase flow and neglected the effect of electrolyte bubble rate distribution on machining accuracy. In Gomez-Gallegos et al. [15], a 3D coupled multiphysics field finite element model was developed to predict ECM accuracy. In Li et al. [16], a coupled model of the magnetic field, electric field, and electrolyte flow in ECM was developed to predict the accuracy of ECM. However, both of these researchers neglected the influence of the bubble rate distribution of the electrolyte in the machining area on the ECM accuracy, resulting in the simulation model's inaccurate prediction of the ECM accuracy. Kozak et al. [17] and Mayank et al. [18] pointed out that the ECM accuracy is also affected by the bubble rate distribution of the electrolyte in the machining area. Some researchers have conducted studies on the bubble rate distribution of electrolytes in the machining area. In Shimasaki et al. [19] and Zhang et al. [20], the effect of bubbles generated by the electrolyte in the machining area on the accuracy of the ECM is observed using transparent electrodes. In Chang et al. [21], a two-dimensional two-phase laminar flow quasi steady-state flow model is proposed to predict the accuracy of the ECM. In Klocke et al. [22], the flow field in the machining area is assumed to be a gas–liquid two-phase flow. The bubble rate distribution in the electrolyte is approximated using the laminar bubble flow model. The use of the laminar N–S model reduces the heat transfer effect, making the prediction accuracy of the simulation model low. In Chen et al. [23] and Zhou et al. [24], a multiphysics field coupled simulation model for the ECM of turbine blades was established based on the gas–liquid two-phase turbulent flow k-ε model, and the effects of electrolyte temperature distribution and bubble rate distribution on the accuracy of the ECM of turbine blades were fully considered. The results show that the simulation model is highly accurate in predicting the machining accuracy of turbine blades.

The complexity of the ECM makes it challenging to model and predict the optimal process parameters [25]. In addition, the random selection of process parameters or trial-and-error methods is very costly and time-consuming and does not yield the desired results. These problems can be solved by optimization techniques [26,27]. Jain et al. [28] used a genetic algorithm to optimize the ECM process parameters. Three process parameters, namely, tool cathode feed rate, electrolyte flow rate, and machining voltage were selected as input quantities and machining accuracy as output quantities, and the optimization results obtained showed significant improvement in machining accuracy. In Jegan et al. [29], the optimization of the ECM process parameters based on the particle swarm algorithm was investigated. Four process parameters, namely, machining current, machining voltage, electrolyte concentration, and tool cathode feed rate were selected as input quantities and material removal rate and surface roughness as output quantities, and the particle swarm algorithm was determined to be superior to the genetic algorithm in terms of computation time and statistical analysis. In Mehrvar et al. [30], based on the central composite design to optimize the ECM process parameters, four process parameters, namely, machining voltage, tool cathode feed rate, electrolyte flow rate, and electrolyte concentration were selected as input quantities, and the material removal rate and surface roughness were determined as output quantities. The results show that the proposed optimization method is effective and suitable.

In summary, the simulation model of ECM of the microbearing outer ring under the influence of electrolyte temperature distribution and bubble rate distribution is established based on the gas–liquid two-phase turbulent flow k-ε model to predict its machining accuracy. It is also essential to optimize the process parameters of the ECM of the miniature bearing outer ring through central composite design to improve its machining accuracy. In this paper, a multiphysics field coupled simulation model of electric, flow, and temperature fields during the ECM of the miniature bearing outer ring is established based on the gas–liquid two-phase turbulent flow model. The influence of electrolyte temperature distribution and bubble rate distribution on the accuracy of the ECM is fully considered. The process of the ECM of miniature bearing outer rings was investigated using a central composite design, and a regression equation between the mean error and the process parameters was established. The optimal combination of process parameters for the mean error was predicted.

2. Simulation Model of ECM of the Miniature Bearing Outer Ring

2.1. Geometric Model

Since the entire ECM miniature bearing outer ring is modeled as an axisymmetric figure, it is assumed that the multiphysics field coupling is the same for each cross section. Therefore, the model is simplified in two dimensions for the convenience of analysis and calculation. The geometric model of the machining area is shown in Figure 1. The part inside the red dashed box in the figure is taken for analysis, and the shape of the entire machining area can be obtained by rotating it around the tool cathode axis for one week. Boundary Γ_1 is the tool cathode without an insulating layer; boundaries Γ_2, Γ_3, Γ_4, and Γ_5 are the tool cathode with an insulating layer; boundary Γ_6 is the electrolyte inlet; boundary Γ_7 is the electrolyte outlet; boundary Γ_8 is the workpiece anode, i.e., the outer ring of the miniature bearing, and Ω is the ECM area.

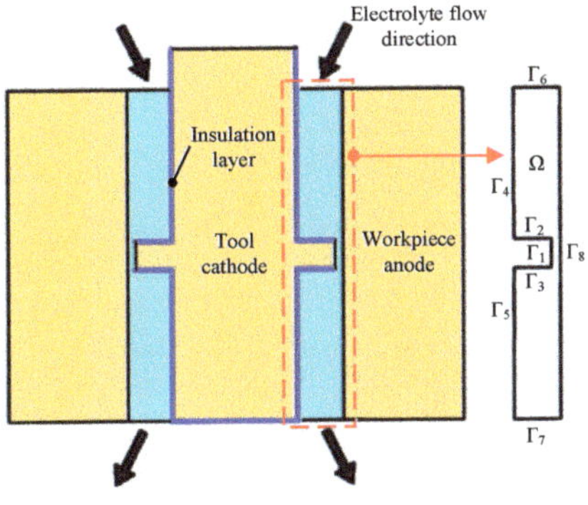

Figure 1. Geometric model of the machining area.

2.2. Mathematical Model

2.2.1. Mathematical Model of Electric Field

According to Ohm's law, the relationship between the current density in the machining area and the electric field strength and potential is:

$$E = -\nabla\varphi \tag{1}$$

$$i = \sigma E = -\sigma\nabla\varphi \tag{2}$$

where E is the electric field strength; φ is the electrolyte potential; i is the current density; σ is the electrolyte conductivity.

In the actual ECM, the speed of the ECM is usually expressed in terms of the dissolution speed in the direction normal to the metal surface of the workpiece anode:

$$v_n = \eta\omega i = -\eta\omega\sigma\nabla\varphi \tag{3}$$

where v_n is the ECM speed; η is the ECM efficiency; ω is the volumetric galvanic equivalent of the workpiece anode.

2.2.2. Mathematical Model of the Flow Field

The gas phase and solid phase products generated during the electrolysis process form a three-phase flow of gas, liquid, and solid in the machining area. In contrast, the volume ratio of the solid phase electrolysis products is minimal and has minimal effect on the electrolyte conductivity, so the flow field in the machining area can be simplified to a gas–liquid two-phase flow.

The gas–liquid two-phase flow in the ECM satisfies the conservation of mass:

$$\frac{\partial}{\partial t}(\beta_l\rho_l + \beta_g\rho_g) + \nabla\cdot(\beta_l\rho_l u_l + \beta_g\rho_g u_g) = 0 \tag{4}$$

$$\frac{\partial \beta_g\rho_g}{\partial t} + \nabla\cdot(\beta_g\rho_g u_g) = m_{lg} \tag{5}$$

$$\frac{\partial \beta_l\rho_l}{\partial t} + \nabla\cdot(\beta_l\rho_l u_l) = -m_{lg} \tag{6}$$

$$\beta_l + \beta_g = 1 \tag{7}$$

where β_g is the proportion of gas in the total volume of the two-phase flow; β_l is the proportion of liquid in the total volume of the two-phase flow; ρ_g and ρ_l are gas density and liquid density; u_g and u_l are the velocity of the gas phase and liquid phase; m_{lg} is the mass transfer rate of the liquid phase into the gas phase.

The gas–liquid two-phase flow satisfies the conservation of momentum:

$$\frac{\partial}{\partial t}(\beta_g\rho_g u_g) + \nabla\cdot(\beta_g\rho_g u_g u_g) = -\beta_g\nabla p + \nabla\cdot\tau_g + \beta_g\rho_g g + F_m \tag{8}$$

$$\frac{\partial}{\partial t}(\beta_l\rho_l u_l) + \nabla\cdot(\beta_l\rho_l u_l u_l) = -\beta_l\nabla p + \nabla\cdot\tau_l + \beta_l\rho_l g - F_m \tag{9}$$

where p is the electrolyte pressure; τ_g and τ_l are gas and liquid viscous stress tensors; F_m is the interphase force.

Assuming that hydrogen is produced only on the surface of the tool cathode in the ECM and that the pressure and temperature distributions of the gas and liquid phases are the same, Faraday's law states that:

$$m_H = k_H It = k_H iSt \tag{10}$$

where m_H is the mass of hydrogen produced; k_H is the electrochemical equivalent of the hydrogen mass; I is the current; i is the current density; S is the area of the tool cathode.

The mass transfer rate m_{lg} for the transformation of liquid phase into gas phase is the mass flux of hydrogen gas produced on the cathode per unit width of the tool, which combined with Equation (10) can be obtained after finishing:

$$Lm_{lg} = ik_H \tag{11}$$

where L is the width of the tool cathode unit.

The density of hydrogen is calculated from the ideal gas equation of state as:

$$\rho_g = \frac{p}{RT} \tag{12}$$

where R is the gas constant; T is the electrolyte temperature.

The electrolyte in the ECM region is in a turbulent state, and considering the effect of bubbles, the RANS k-ε turbulence model is used in this paper. The electrode surface near the wall is solved by the wall function as follows:

$$\frac{\partial k}{\partial t} + \nabla \cdot \left[ku - \left(\mu + \frac{\mu_T}{\sigma_k}\right)\nabla k\right] = P_k + S_k - \varepsilon \tag{13}$$

$$\frac{\partial \varepsilon}{\partial t} + \nabla \cdot \left[\varepsilon u - \left(\mu + \frac{\mu_T}{\sigma_\varepsilon}\right)\nabla \varepsilon\right] = \frac{\varepsilon}{k}(C_1 P_k + C_\varepsilon S_k - C_2 \varepsilon) \tag{14}$$

$$P_k = \frac{\mu_T}{2}\left|\nabla u + (\nabla u)^T\right|^2 \tag{15}$$

$$S_k = -\beta C_k |\nabla p|^2 \tag{16}$$

where k is the turbulent kinetic energy; ε is the turbulent dissipation rate; u is the electrolyte flow rate; μ is the electrolyte dynamic viscosity; μ_T is the turbulent viscosity coefficient; C_1, C_2, C_k, C_ε, σ_k, and σ_ε are the model constants.

2.2.3. Mathematical Model of the Temperature Field

During ECM, fluid heat transfer occurs in the machining area. The heat generated in the machining area is carried away by the thermal convection of the electrolyte in the turbulent state. According to the law of energy conservation, the expression of the convective heat transfer equation in the machining area is:

$$\rho C_p \frac{\partial T}{\partial t} + \rho C_p u \cdot \nabla T = \nabla(\lambda \nabla T) + Q \tag{17}$$

$$Q = i\nabla \varphi \tag{18}$$

where ρ is the density of the electrolyte; C_p is the specific heat capacity of the electrolyte; λ is the thermal conductivity of the electrolyte; Q is the heat generated in the ECM.

2.2.4. Multiphysics Field Coupling Model

The conductivity of the electrolyte in the actual ECM as affected by temperature and bubble rate can be expressed as:

$$\sigma = \sigma_0 (1-\beta)^m [1 + \gamma(T - T_0)] \tag{19}$$

where σ_0 is the initial conductivity of the electrolyte; β is the bubble rate; m is the bubble rate influence index; γ is the temperature correlation coefficient; T_0 is the initial temperature of the electrolyte.

Substituting Equation (19) into Equation (3), the coupling equations for the electric field, flow field, temperature field, and ECM speed are obtained:

$$v_n = -\eta\omega\sigma_0(1-\beta)^m[1 + \gamma(T - T_0)]\nabla\varphi \tag{20}$$

2.3. 2D COMSOL Multiphysics Field Coupling Simulation Model

As shown in Figure 2, the electric field, flow field, temperature field, and deformation geometry modules on the software were selected for multiphysics field coupling simulation. The coupling method is that the electric and temperature field modules were coupled through an electromagnetic heat source module, and the flow and temperature field modules were coupled through a non-isothermal flow module. The actual ECM of the outer ring of the miniature bearing can be reproduced to the maximum extent.

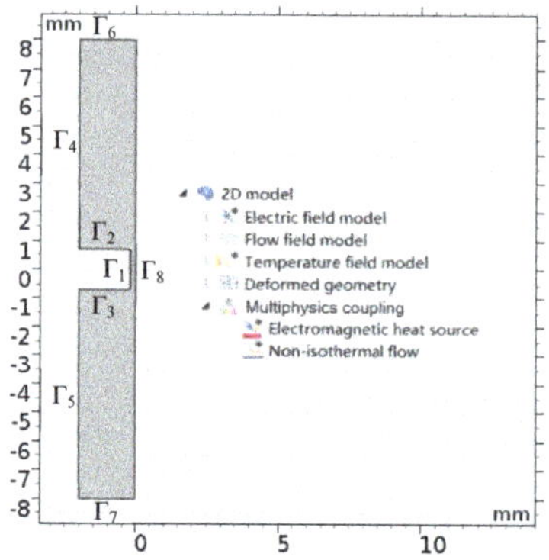

Figure 2. Simulation model and simulation module.

2.3.1. Boundary Condition Setting

The boundary conditions of the electric field module were set as follows: the boundary Γ_8 was connected to the voltage U; the boundary Γ_1 was grounded, and the boundaries Γ_2, Γ_3, Γ_4, Γ_5, Γ_6, and Γ_7 were electrically insulated. The boundary conditions of the flow field module were set as follows: boundary Γ_1 gas mass flux, electrolyte wall function; boundary Γ_6 electrolyte normal inflow velocity u_0, no gas flux; boundary Γ_7 electrolyte and gas outlet; and boundaries Γ_2, Γ_3, Γ_4, Γ_5, and Γ_8 no gas flux, electrolyte wall function. The temperature module boundary conditions were set as follows: boundaries Γ_1, Γ_2, Γ_3, Γ_4, Γ_5, and Γ_8 thermal insulation; boundary Γ_6 electrolyte initial temperature T_0; boundary Γ_7 electrolyte outflow. The boundary conditions of the deformation geometry module were set as follows: boundary Γ_8 normal mesh moving speed, i.e., electrochemical processing speed v_n.

2.3.2. Material Parameter Setting of the Simulation Model

The workpiece anode material of the simulation model was GH4169 high-temperature-resistant nickel-based alloy; the tool cathode material was titanium alloy electrode (insulating film covered with PTFE), and the electrolyte was $NaNO_3$ solution of a given concentration. The material parameters of the specific simulation model are shown in Table 1.

Table 1. Material parameter settings of the simulation model.

Simulation Parameters	Numerical Value
Specific heat capacity of electrolyte (J/kg/K)	4200
Electrolyte density (kg/m^3)	1200
Thermal conductivity of electrolyte (W/m/K)	0.64
Electrolyte power viscosity (Pa·s)	1.01×10^{-3}
The initial temperature of electrolyte (K)	293.15
Temperature correlation coefficient	0.025
Gas density (kg/m^3)	8.99×10^{-2}
Air bubble diameter (m)	1×10^{-5}
Bubble rate impact index	1.5
GH4169 volumetric electrochemical equivalent (cm^3/A/min)	0.00178

3. Simulation Analysis of ECM of the Miniature Bearing Outer Ring

The simulation conditions were as follows: machining voltage was 18 V; the electrolyte was $NaNO_3$ solution with 12% concentration, and the electrolyte inlet flow rate was 9 m/s.

3.1. Electrolyte Temperature Distribution in the Machining Area

As shown in Figure 3, the maximum electrolyte temperature in the machining area is 293.6 K, which is 0.45 K higher than the initial electrolyte temperature of 293.15 K. The maximum temperature of the electrolyte in the machining area from the initial stage of machining to the machining time of 20 s increased by 0.1 K, and the area of the thermally affected area was about 45% of the machining area. This is because more Joule heat is generated at this stage by higher current density, and the machining gap is small, resulting in poor electrolyte circulation and the accumulation of Joule heat in the direction of the electrolyte flow, causing the electrolyte temperature to increase and the area of the heat-affected area to expand. From the machining time of 20 s to the machining time of 60 s, the maximum temperature of the electrolyte in the machining area is reduced by 0.2 K, and the area of the heat-affected area is about 35% of the machining area. This is because the current density at this stage is low; the Joule heat generated is low, and the machining gap is large. The electrolyte flow is smooth, so that the electrolysis temperature is lower, and the area affected by heat is reduced.

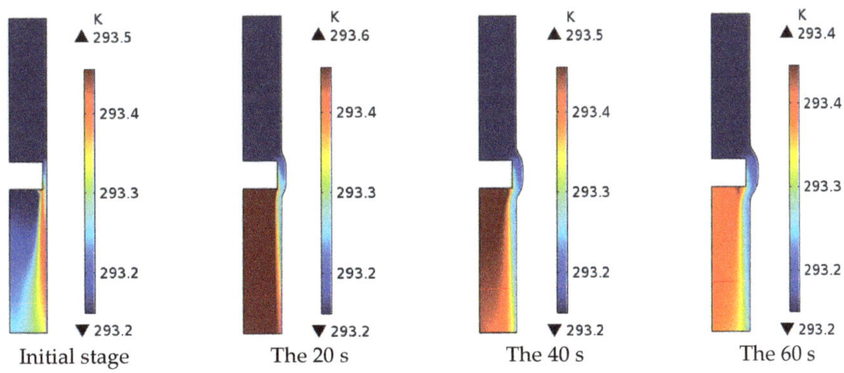

Figure 3. Electrolyte temperature distribution in the machining area.

3.2. Distribution of Electrolyte Bubble Rate in the Machining Area

As shown in Figure 4, the bubble rate of electrolytes in the machining area gradually increases as the ECM proceeds. The highest bubble rate of electrolytes in the machining area reaches 18.5% at 60 s of machining time. The highest bubble rate of electrolytes in the machining area from the initial machining stage to 20 s of machining time reached 17.3%, and the electrolyte bubbles were distributed around the boundary Γ_1 with a minimal area. This is because the high flow rate of the electrolyte at this stage can carry away the hydrogen produced at the boundary Γ_1 in time. The highest bubble rate of electrolytes in the machining area increased by 1.2% from 20 s of machining time to 60 s of machining time, and the area of electrolyte bubble distribution was about 55% of the machining area. It is because the low flow rate of electrolyte at this stage is not able to take away the hydrogen produced by the boundary Γ_1 in time, so the hydrogen accumulates along the electrolyte flow direction, causing the electrolyte bubble distribution area to expand.

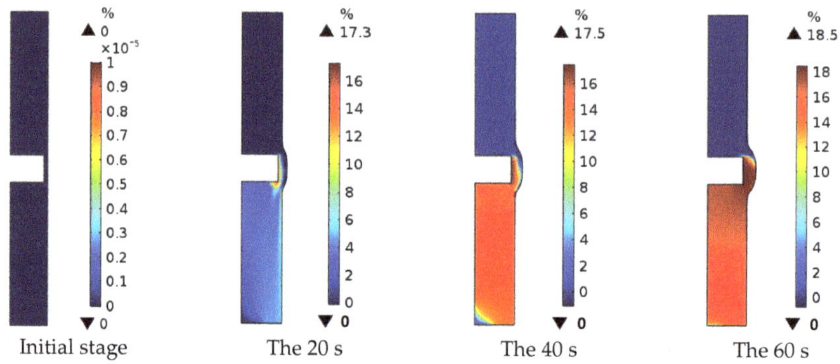

Figure 4. Distribution of electrolyte bubble rate in the machining area.

3.3. *Electrolyte Flow Rate Distribution in the Machining Area*

As shown in Figure 5, the electrolyte flow rate in the machining area at the initial stage is high, up to 106 m/s. As the ECM progresses, the electrolyte flow rate gradually decreases, and the highest electrolyte flow rate in the machining area is 55.2 m/s when the machining time is 60 s. The electrolyte flow rate decreases as the machining gap increases, while the electrolyte flow rate remains constant.

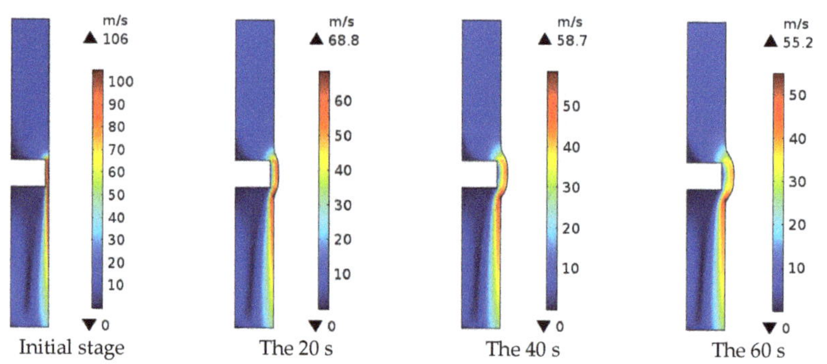

Figure 5. Distribution of electrolyte flow rate in the machining area.

3.4. *Electrolyte Current Density Distribution and Workpiece Anode Profile Changes in the Machining Area*

As shown in Figure 6a, the distance between the tool cathode and the workpiece anode gradually becomes larger. The electrolyte current density in the machining gap gradually decreases as the workpiece anode is dissolved during ECM. As shown in Figure 6b, the machining depth and height of the workpiece anode increase progressively as the ECM progresses, while the machining volume simultaneously tends to decrease. The main reason is that as the current density decreases during ECM, the speed of ECM becomes slower, and the machining volume declines simultaneously.

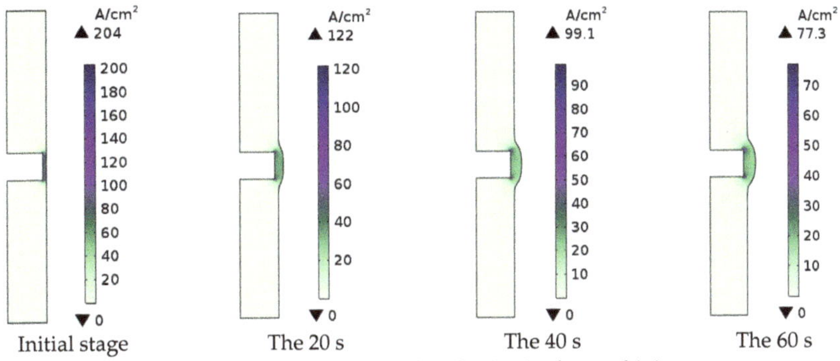

(**a**) Electrolyte current density distribution in the machining area

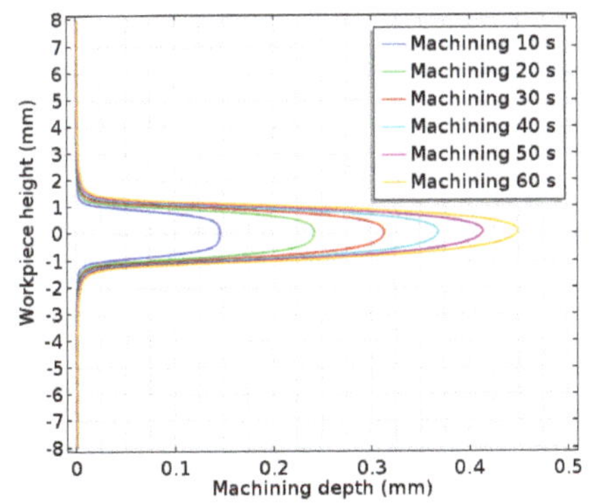

(**b**) Workpiece anode profile change curve

Figure 6. Electrolyte current density distribution in the machining area and workpiece anode profile change curve.

4. Process study of ECM of the Miniature Bearing Outer Ring

4.1. Process Evaluation Index for ECM of the Miniature Bearing Outer Ring

As shown in Figure 7, it is assumed that the equation of the standard curve of the outer ring section of the miniature bearing is: $x^2 + (y - 1.2)^2 = 4$, among them: $0 \leq x \leq 0.8$. The cross-sectional curve of the ECM miniature bearing outer ring does not precisely coincide with the standard curve, and there is a specific error, so the average error of the process evaluation index of the ECM miniature bearing outer ring is constructed:

$$\delta = \frac{1}{n}\sum_{1}^{n}|x_b - x_a| \tag{21}$$

where δ is the average error; x_a is the horizontal coordinate of the standard curve; x_b is the horizontal coordinate of the machining curve; n is the number of points taken uniformly along the vertical coordinate.

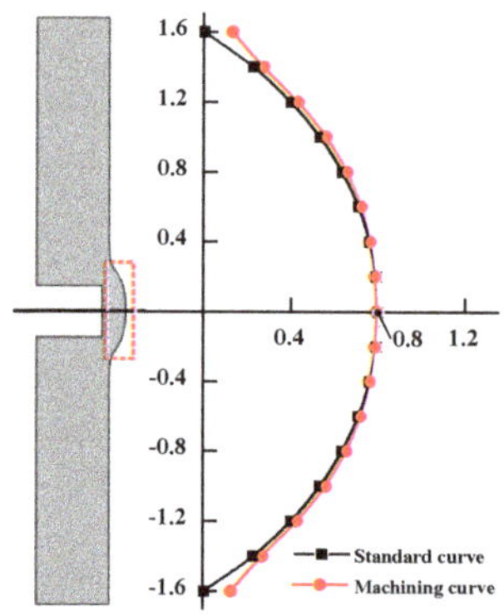

Figure 7. Standard curve and machining curve.

4.2. ECM Miniature Bearing Outer Ring Center Composite Design Solution

4.2.1. Design Solutions and Simulation Results

In this design scheme, three process parameters, namely, machining voltage, electrolyte concentration, and electrolyte inlet flow rate were selected as input quantities. One process evaluation index, namely, average error was chosen as the output quantity. The process parameters were coded with −1, 0 with 1 representing the different level values of each process parameter, where "0" represents the center point of the level value; "−1" represents the low-level value; "1" represents the high-level value. The actual and coded values of the central composite design scheme are shown in Table 2. The center composite design scheme and simulation results are shown in Table 3.

Table 2. Actual and coded values of the central composite design scheme.

Factors	Process Parameters	Codes		
		−1	0	1
A	Machining voltage U/V	12	18	24
B	Electrolyte concentration C/%	8	12	16
C	Electrolyte inlet flow rate V/m/s	6	9	12

4.2.2. Establishing the Regression Equation

According to the results given in Table 3, the regression equation between the mean error of the process evaluation index of ECM miniature bearing outer ring and the process parameters was established through multiple quadratic orthogonal regression analysis:

$$y = \alpha_0 + \sum_{i=1}^{3} \alpha_i A_i + \sum_{i=1}^{3} \alpha_{ii} A_i^2 + \sum\sum_{i<j} \alpha_{ij} A_i A_j + \gamma \quad (22)$$

where y is the output of the model; α_0 is a constant; αi is the primary term regression coefficient; α_{ii} is the quadratic term regression coefficient; α_{ij} is the interaction term regression coefficient; γ is the error estimate; A_i is the coded value of the process parameters.

4.2.3. Analysis of Variance and Significance of Each Factor

The regression analysis of the mean error of the outer ring of the ECM miniature bearing yielded the variance and significance analysis as shown in Table 4.

Table 3. Center composite design scheme and simulation results.

No.	Factor A Machining Voltage U/V	Factor B Electrolyte Concentration C/%	Factor C Electrolyte Inlet Flow Rate V/m/s	Average Error δ/mm
1	12	8	6	0.01647
2	24	8	6	0.01565
3	12	16	6	0.01636
4	24	16	6	0.03747
5	12	8	12	0.0165
6	24	8	12	0.01567
7	12	16	12	0.01635
8	24	16	12	0.03715
9	12	12	9	0.01573
10	24	12	9	0.02558
11	18	8	9	0.01602
12	18	16	9	0.02791
13	18	12	6	0.01734
14	18	12	12	0.01767
15	18	12	9	0.01756
16	18	12	9	0.01835
17	18	12	9	0.01853
18	18	12	9	0.01784
19	18	12	9	0.01712
20	18	12	9	0.01735

Table 4. Variance and significance of the mean error.

Source Items	Square and	Degree of Freedom	Average Value	F-Value	*p*-Value	Significance
Model	0.0009	9	0.0001	149.77	<0.0001	Highly significant
A	0.0003	1	0.0003	396.82	<0.0001	Highly significant
B	0.0003	1	0.0003	476.83	<0.0001	Highly significant
C	2.500×10^{-10}	1	2.500×10^{-10}	0.0004	0.9845	Not significant
AB	0.0002	1	0.0002	374.83	<0.0001	Highly significant
AC	1.280×10^{-8}	1	1.280×10^{-8}	0.0202	0.8897	Not significant
BC	1.805×10^{-8}	1	1.805×10^{-8}	0.0285	0.8692	Not significant
A^2	8.326×10^{-6}	1	8.326×10^{-6}	13.16	0.0046	Significant
B^2	0	1	0	40.43	<0.0001	Highly significant
C^2	5.467×10^{-6}	1	5.467×10^{-6}	8.64	0.0148	Significant
Residuals	6.328×10^{-6}	10	6.328×10^{-7}	——	——	——
Miss drafting	4.769×10^{-6}	5	9.537×10^{-7}	3.06	0.1226	Not significant
Pure error	1.559×10^{-6}	5	3.118×10^{-7}	——	——	——
Total	0.0009	19	——	——	——	——
$R^2 = 0.9926$		Adjusted $R^2 = 0.9860$		Predicted $R^2 = 0.9718$		Adeq Precision = 40.3255

Where at $p < 0.001$, the factor is highly significant; at $p < 0.05$, the factor is significant; at $p \geq 0.05$, the factor is not significant.

As shown in Table 4, the F-value of the model is 149.77 with a *p*-value less than 0.0001; the miss drafting F-value is 3.06, and the miss drafting *p*-value is 0.1226, indicating that the model is highly significant. The miss drafting term is not significant, showing that the model is meaningful and plausible. Simultaneously, $R^2 = 0.9926$, Adjusted $R^2 = 0.9860$, and Predicted $R^2 = 0.9718$. The three values are similar and less different from 1, indicating that the model fits relatively well throughout the regression region. Adequate precision = 40.3255 and much greater than 4 indicates the model is more realistic and reliable. The established regression model has a good response. However, factors C, AC, and BC are not significant, so by excluding these insignificant factors, the regression model can be optimized.

4.2.4. Optimization of Regression Models

As shown in Table 4, the impact of factors C, AC, and BC on the mean error was not significant, so the regression model was optimized using a stepwise elimination of insignificant factors to rerun the regression analysis. The variance and significance analysis of the mean error of the regression model after optimization are shown in Table 5.

Table 5. Variance and significance of the mean error of the optimized regression model.

Source Items	Square and	Degree of Freedom	Average Value	F-Value	p-Value	Significance
Model	0.0009	6	0.0001	290.62	<0.0001	Highly significant
A	0.0003	1	0.0003	513.35	<0.0001	Highly significant
B	0.0003	1	0.0003	616.85	<0.0001	Highly significant
AB	0.0002	1	0.0002	484.90	<0.0001	Highly significant
A^2	8.326×10^{-6}	1	8.326×10^{-6}	17.02	0.0012	Significant
B^2	0	1	0	52.30	<0.0001	Highly significant
C^2	5.467×10^{-6}	1	5.467×10^{-6}	11.18	0.0053	Significant
Residuals	6.359×10^{-6}	13	4.891×10^{-7}	——	——	——
Miss drafting	4.800×10^{-6}	8	6.000×10^{-7}	1.92	0.2441	Not significant
Pure error	1.559×10^{-6}	5	3.118×10^{-7}	——	——	——
Total	0.0009	19	——	——	——	——
R^2 = 0.9926		Adjusted R^2 = 0.9892		Predicted R^2 = 0.9813		Adeq Precision = 54.6205

Where at $p < 0.001$, the factor is highly significant; at $p < 0.05$, the factor is significant; at Podel was optimized using a stepwise.

As shown in Table 5, the F-value of the model is 290.62 with a p-value less than 0.0001; the miss drafting F-value is 1.92, and the miss drafting p-value is 0.2441, indicating that the model is highly significant. The miss drafting term is not significant, showing that the model is meaningful and plausible. Simultaneously, R^2 = 0.9926, Adjusted R^2 = 0.9892, and Predicted R^2 = 0.9813. The three values are similar and less different from 1, indicating that the model fits relatively well throughout the regression region. Adequate precision = 54.6205 and much greater than 4 indicates that the model is more realistic and reliable. The response of the established regression model is good. All the factors are significant at this time.

4.2.5. The Regression Equation of the Mean Error with the Normal Probability Distribution of the Residuals

According to the variance and significance analysis of the mean error of the optimized regression model, the calculation of each coefficient of the mean error regression equation was carried out, and the regression equation of the optimized mean error was obtained as:

$$\begin{aligned} \delta = 0.0182 + 0.0050U + 0.0055C + 0.0054U \cdot C + \\ 0.0017U^2 + 0.0030C^2 - 0.0014V^2 \end{aligned} \tag{23}$$

As shown in Figure 8, the residuals of each factor are distributed around a straight line. The residuals of each factor conform to a normal distribution, so the prediction of the mean error using Equation (23) is reliable.

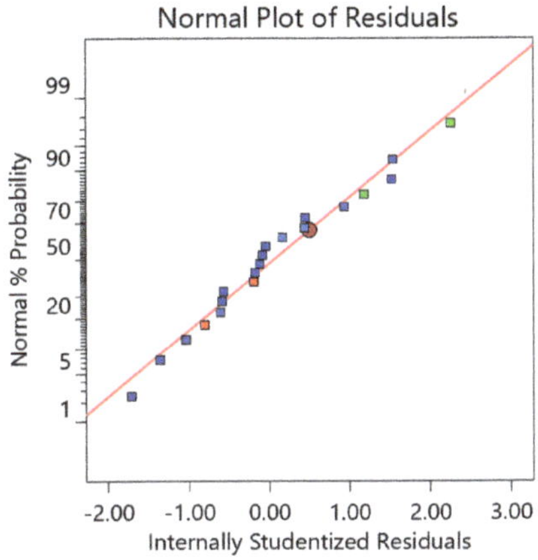

Figure 8. Normal probability distribution of residuals.

4.2.6. Response Surface Analysis

Since the effect of electrolyte inlet flow rate on the average error of the ECM of the miniature bearing outer ring is not significant, the inlet flow rate of electrolyte is taken as 9 m/s in the response surface analysis, and the effect of machining voltage and electrolyte concentration on the average error is shown in Figure 9.

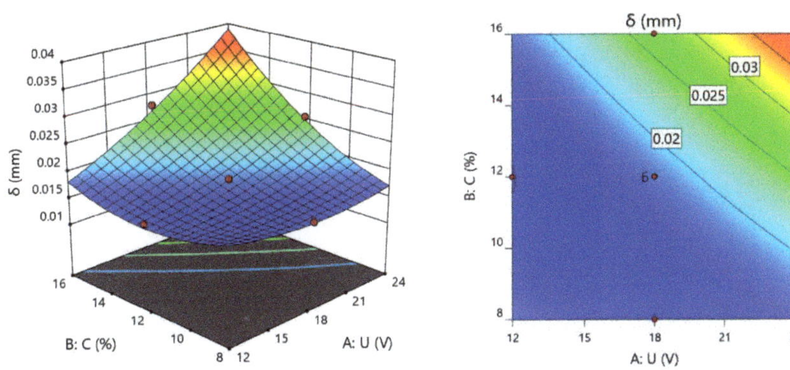

(**a**) Response surface diagram (**b**) Projected contour map

Figure 9. Effect of machining voltage and electrolyte concentration on the average error.

As shown in Figure 9, the average error increases with the rise of processing voltage and electrolyte concentration. This is because both the increase of processing voltage and electrolyte concentration will increase the dissolution rate of the metal material on the anode surface of the workpiece, and when the processing depth is 0.8 mm, the radius of curvature of the cross-sectional curve of the outer ring of the miniature bearing produced by electrolysis increases, resulting in a rise in the average error between the standard curve and the standard curve.

4.2.7. Prediction of the Best Combination of Process Parameters for the Average Error

The optimal combination of process parameters for the average error of the ECM miniature bearing outer ring was obtained using Design-Expert 12 software analysis as shown in Figure 10.

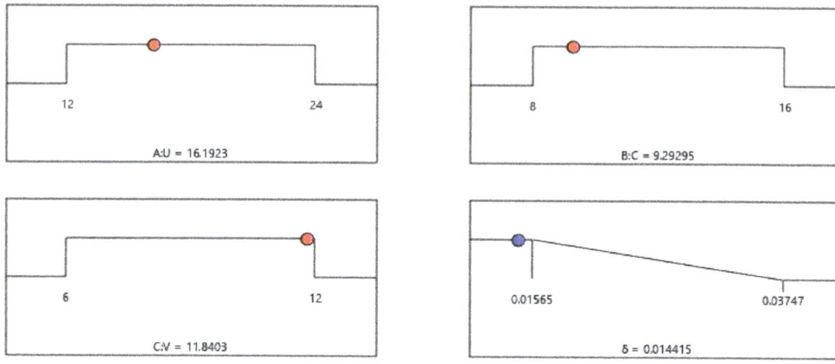

Figure 10. Optimal combination of process parameters for the average error.

As shown in Figure 10, the average error of ECM of the miniature bearing outer ring can reach the minimum value of 0.014 mm under a machining voltage of 16.20 V, an electrolyte concentration of 9.29%, and an electrolyte inlet flow rate of 11.84 m/s.

5. Conclusions

In this paper, a multiphysics field coupled simulation model of electric, flow, and temperature fields during ECM of the miniature bearing outer ring is established based on the gas–liquid two-phase turbulent flow model. The process of ECM of miniature bearing outer rings is investigated using a central composite design. Then some conclusions were drawn from this study, which are summarized as follows:

1. In the ECM progresses, the depth and height of the workpiece anode gradually increase, while the machining volume simultaneously tends to decrease. The main reason is that the current density decreases as the electrolysis process progresses, resulting in a slower machining speed and a lower machining volume in the same amount of time.
2. The impact of machining voltage and electrolyte concentration on the average error was highly significant, while the impact of the electrolyte inlet flow rate on the average error was not significant.
3. When the electrolyte concentration is 12%, the electrolyte inlet flow rate is 9 m/s, and the machining voltage is increased from 12 V to 24 V, and the average error of the ECM of the miniature bearing outer ring is increased by 62.62%. When the machining voltage is 18 V, the electrolyte inlet flow rate is 9 m/s, and the electrolyte concentration is increased from 8% to 16%, and the average error of the ECM of the miniature bearing outer ring is increased by 74.22%. When the machining voltage is 18 V, the electrolyte concentration is 12%, and the electrolyte inlet flow rate is increased from 6 m/s to 12 m/s, and the average error of ECM of the miniature bearing outer ring is increased by 1.91%.
4. The average error of the ECM of the miniature bearing outer ring can reach the minimum value of 0.014 mm under a machining voltage of 16.20 V, an electrolyte concentration of 9.29%, and an electrolyte inlet flow rate of 11.84 m/s.

Author Contributions: Conceptualization, Z.L. and W.L.; methodology, Z.L. and W.L.; validation, Z.L. and W.L.; formal analysis, Z.L. and W.L.; investigation, Z.L. and W.L.; resources, Z.L. and W.L.; data curation, Z.L.; writing—original draft preparation, Z.L.; writing—review and editing, Z.L.; visualization, Z.L.; supervision, Z.L.; project administration, Z.L. and W.L.; funding acquisition, Z.L. and W.L.; examine and proposal, B.C.; All authors have read and agreed to the published version of the manuscript.

Funding: This work was supported by the National Natural Science Foundation of China, grant number 52075134.

Data Availability Statement: Data are contained within the article.

Acknowledgments: The authors would like to thank the Key Laboratory of Advanced Manufacturing Intelligent Technology of the Ministry of Education for helpful discussions on topics related to this work. The authors are grateful to Bingren Cao for his help with the preparation of figures in this paper.

Conflicts of Interest: The authors declare no conflict of interest.

References

1. Jo, C.H.; Kim, B.H.; Chu, C.N. Micro electrochemical machining for complex internal micro features. *CIRP Ann.* **2009**, *58*, 181–184. [CrossRef]
2. Lu, X.D.; Du, J.H.; Deng, Q.; Zhuang, J.Y. Stress rupture properties of GH4169 super alloy. *J. Mater. Res. Technol.* **2014**, *3*, 107–113. [CrossRef]
3. Kong, X.; Li, B.; Jin, Z.; Geng, W. Broaching Performance of Superalloy GH4169 Based on FEM. *J. Mater. Sci. Technol.* **2011**, *27*, 1178–1184. [CrossRef]
4. Li, C.J.; Li, Y.; Tong, H.; Zhao, L.; Kong, Q.C.; Wang, Z.Q. An EDM pulse power generator and its feasible experiments for drilling film cooling holes. *Int. J. Adv. Manuf. Technol.* **2016**, *87*, 1813–1821. [CrossRef]
5. Li, C.; Xu, X.; Li, Y.; Tong, H.; Ding, S.; Kong, Q.; Zhao, L.; Ding, J. Effects of dielectric fluids on surface integrity for the recast layer in high speed EDM drilling of nickel alloy. *J. Alloys Compd.* **2019**, *783*, 95–102. [CrossRef]
6. Das, D.K.; Pollock, T.M. Femtosecond laser machining of cooling holes in thermal barrier coated CMSX4 super alloy. *J. Mater. Proc. Technol.* **2009**, *209*, 5661–5668. [CrossRef]
7. Zhang, Z.; Wang, W.; Jiang, R.; Zhang, X.; Xiong, Y.-F.; Mao, Z. Investigation on geometric precision and surface quality of microholes machined by ultrafast laser. *Opt. Laser Technol.* **2020**, *121*, 105834. [CrossRef]
8. Zhan, D.; Han, L.; Zhang, J.; He, Q.; Tian, Z.-W.; Tian, Z.-Q. Electrochemical micro/nano-machining: Principles and practices. *Chem. Soc. Rev.* **2017**, *46*, 1526–1544. [CrossRef]
9. Han, L.H.; Sartin, M.M.; Tian, Z.Q.; Zhan, D.P.; Tian, Z.W. Electrochemical nanomachining. *Curr. Opin. Electrochem.* **2020**, *22*, 80–86. [CrossRef]
10. Lohrengel, M.; Rataj, K.; Münninghoff, T. Electrochemical Machining—Mechanisms of anodic dissolution. *Electrochim. Acta* **2016**, *201*, 348–353. [CrossRef]
11. Deconinck, D.; Van Damme, S.; Deconinck, J. A temperature dependent multi-ion model for time accurate numerical simulation of the electrochemical machining process. Part I: Theoretical basis. *Electrochim. Acta* **2012**, *60*, 321–328. [CrossRef]
12. Deconinck, D.; Van Damme, S.; Deconinck, J. A temperature dependent multi-ion model for time accurate numerical simulation of the electrochemical machining process. Part II: Numerical simulation. *Electrochim. Acta* **2012**, *69*, 120–127. [CrossRef]
13. Deconinck, D.; Hoogsteen, W.; Deconinck, J. A temperature dependent multi-ion model for time accurate numerical simulation of the electrochemical machining process. Part III: Experimental validation. *Electrochim. Acta* **2013**, *103*, 161–173. [CrossRef]
14. Fang, X.; Qu, N.; Zhang, Y.; Xu, Z.; Zhu, D. Effects of pulsating electrolyte flow in electrochemical machining. *J. Mater. Process. Technol.* **2014**, *214*, 36–43. [CrossRef]
15. Gomez-Gallegos, A.; Mill, F.; Mount, A.R.; Duffield, S.; Sherlock, A. 3D multiphysics model for the simulation of electrochemical machining of stainless steel (SS316). *Int. J. Adv. Manuf. Technol.* **2018**, *95*, 2959–2972. [CrossRef]
16. Long, L.; Baoji, M.A.; Ruifeng, W.; Lingqi, D. The coupled effect of magnetic field, electric field and electrolyte motion on the material removal amount in electrochemical machining. *Int. J. Adv. Manuf. Technol.* **2017**, *91*, 2995–3006. [CrossRef]
17. Kozak, J.; Zybura-Skrabalak, M. Some Problems of Surface Roughness in Electrochemical Machining (ECM). *Procedia CIRP* **2016**, *42*, 101–106. [CrossRef]
18. Mayank, G.; Kunieda, M. Two-Phase Simulation of Electrochemical Machining. *Int. J. Electr. Mach.* **2017**, *22*, 31. [CrossRef]
19. Shimasaki, T.; Kunieda, M. Study on influences of bubbles on ECM gap phenomena using transparent electrode. *CIRP Ann.* **2016**, *65*, 225–228. [CrossRef]
20. Zhang, W.; Kitamura, T.; Kunieda, M.; Abe, K. Observation of ECM Gap Phenomena through Transparent Electrode. *Int. J. Electr. Mach.* **2014**, *19*, 40–44. [CrossRef]
21. Chang, C.; Hourng, L. Two-dimensional two-phase numerical model for tool design in electrochemical machining. *J. Appl. Electrochem.* **2001**, *31*, 145–154. [CrossRef]
22. Klocke, F.; Zeis, M.; Klink, A. Interdisciplinary modelling of the electrochemical machining process for engine blades. *CIRP Ann.* **2015**, *64*, 217–220. [CrossRef]
23. Chen, Y.; Zhou, X.; Chen, P.; Wang, Z. Electrochemical machining gap prediction with multi-physics coupling model based on two-phase turbulence flow. *Chin. J. Aeronaut.* **2020**, *33*, 1057–1063. [CrossRef]
24. Zhou, X.C.; Cao, C.Y.; Wang, H.X.; Lin, H. Study on the multi-field coupling model of electrolyte temperature distribution in electrochemical machining. *Int. J. Adv. Manuf. Technol.* **2020**, *109*, 1655–1662.
25. Hinduja, S.; Kunieda, M. Modelling of ECM and EDM processes. *CIRP Ann.* **2013**, *62*, 775–797. [CrossRef]
26. Rao, R.V.; Kalyankar, V. Optimization of modern machining processes using advanced optimization techniques: A review. *Int. J. Adv. Manuf. Technol.* **2014**, *73*, 1159–1188. [CrossRef]

27. Yusup, N.; Zain, A.M.; Hashim, S.Z.M. Evolutionary techniques in optimizing machining parameters: Review and recent applications (2007–2011). *Expert Syst. Appl.* **2012**, *39*, 9909–9927. [CrossRef]
28. Jain, N.K.; Jain, V.K. Optimization of Electro-Chemical Machining Process Parameters Using Genetic Algorithms. *Machin. Sci. Technol.* **2007**, *11*, 235–258. [CrossRef]
29. Jegan, T.M.C.; Ravindran, D. Electrochemical machining process parameter optimization using particle swarm optimization. *Comput. Intell.* **2017**, *33*, 1019–1037. [CrossRef]
30. Mehrvar, A.; Basti, A.; Jamali, A. Optimization of electrochemical machining process parameters: Combining response surface methodology and differential evolution algorithm. *Proc. Inst. Mech. Eng. Part E J. Process. Mech. Eng.* **2017**, *231*, 1114–1126. [CrossRef]

micromachines

Article

Predicting Milling Stability Based on Composite Cotes-Based and Simpson's 3/8-Based Methods

Xu Du [1], Pengfei Ren [1] and Junqiang Zheng [2,*]

1 Mechatronic Institute, Zhejiang Sci-Tech University, Hangzhou 310018, China; duxu@zstu.edu.cn (X.D.); renpengfei596@gmail.com (P.R.)
2 School of Mechanical Engineering, Hangzhou Dianzi University, Hangzhou 310018, China
* Correspondence: zhengjunqiang@hdu.edu.cn

Abstract: Avoiding chatter in milling processes is critical for obtaining machined parts with high surface quality. In this paper, we propose two methods for predicting the milling stability based on the composite Cotes and Simpson's 3/8 formulas. First, a time-delay differential equation is established, wherein the regenerative effects are considered. Subsequently, it is discretized into a series of integral equations. Based on these integral equations, a transition matrix is determined using the composite Cotes formula. Finally, the system stability is analyzed according to the Floquet theory to obtain the milling stability lobe diagrams. The simulation results demonstrate that for the single degree of freedom (single-DOF) model, the convergence speed of the composite Cotes-based method is higher than that of the semi-discrete method and the Simpson's equation method. In addition, the composite Cotes-based method demonstrates high computational efficiency. Moreover, to further improve the convergence speed, a second method based on the Simpson's 3/8 formula is proposed. The simulation results show that the Simpson's 3/8-based method has the fastest convergence speed when the radial immersion ratio is large; for the two degrees of freedom (two-DOF) model, it performs better in terms of calculation accuracy and efficiency.

Keywords: milling stability; composite Cotes-based method; Simpson's 3/8-based method; Floquet theory

Citation: Du, X.; Ren, P.; Zheng, J. Predicting Milling Stability Based on Composite Cotes-Based and Simpson's 3/8-Based Methods. *Micromachines* **2022**, *13*, 810. https://doi.org/10.3390/mi13050810

Academic Editors: Youqiang Xing, Xiuqing Hao and Duanzhi Duan

Received: 28 April 2022
Accepted: 21 May 2022
Published: 23 May 2022

1. Introduction

The three main types of vibration that occur during high-speed milling are free vibration, excited vibration, and self-excited vibration, and the regenerative chatter in self-excited vibration is the main cause of instability in the machining process [1]. Because this chatter typically leads to poor surface quality of the machined parts, aggravated tool wear, and even reduced service life of machine tools, it must be avoided to solve or overcome these problems [2]. In the analysis of chatter, dynamic milling processes that consider the delay effect are generally described using delay differential equations with respect to the time-periodic coefficients [3,4]. The chatter stability based on these differential equations is an important index for achieving high-performance machining [5].

In recent decades, experimental, analytical, and numerical methods have been studied to obtain stability lobe diagrams, which can be used to determine exact values avoiding chatter. Wu et al. [6] utilized the Lyapunov index to measure whether chatter occurred; however, they could only predict the system stability under specific parameters. Davies et al. [7] used the time-domain calculation method to predict unstable regions in the stability lobe diagrams, which was only applicable to small radial depths of cut. Altintas and Budak et al. [8] first presented a zero-order approximation method to quickly obtain the stability lobe diagrams, where the real and imaginary parts of the feature are used to determine the cutting parameters. It is worth mentioning that this method was only applicable to high radial depths of cut. Bayly et al. [9] employed the time-finite element method to calculate tool motion and utilized the weighted residual method to determine the Floquet transition matrix. Based on the delay differential equations, Insperger et al. [10] proposed the semi-discretization method (SDM) and the first-order semi-discretization method (1_{st}SDM), which are widely used as the basis for evaluating other stability analysis methods in the time domain. The SDM and 1_{st}SDM only discretized delay states and periodic coefficients. Based on direct integration, Ding et al. [11] presented a full-discretization method (FDM) to predict milling stability. Unlike the SDM and 1_{st}SDM, the FDM carried out the linear interpolation on the state terms to improve computational efficiency. Further, Insperger [12] compared the FDM with the SDM and 1_{st}SDM in terms of the convergence speed and computational efficiency. It was found that the convergence speed of the 1_{st}SDM was faster than those of the SDM and FDM, while the computational efficiency of the FDM was higher than those of the SDM and 1_{st}SDM. After the comparisons, some FDM-based methods were used to predict the milling stability. In addition to the above methods, the numerical integration methods are widely utilized to obtain the stability lobe diagrams. Ding et al. [13] approximated the integral terms with the Newton–Cotes and Gauss integral formulas to predict the milling stability. Lu et al. [14] approximated the solution of the delay differential equation with the direct integration technique. Qin et al. [15] approximated the state terms of the delay differential equation with the Chebyshev wavelets of the second kind. Moreover, Liu et al. [16] combined the numerical solution of the delay differential equation with the Simpson formula to predict milling stability (SEM), which was only suitable for large radial immersion. However, the aforementioned numerical methods cannot simultaneously guarantee fast convergence speed and high computational efficiency.

To this end, we present two numerical analysis methods, namely the composite Cotes-based method (CCM) and Simpson's 3/8-based method (S38M), in this paper to improve the convergence speed and computational efficiency at the same time. The remainder of this paper is organized as follows. Section 2 presents the mathematical model of system motion with two degrees of freedom (DOF) and the state-space. The composite Cortes-based method and its simulation and analysis for the single-DOF model are presented in Section 3. Subsequently, the Simpson's 3/8-based method and its simulation and analysis for the single-DOF and two-DOF models are presented in Section 4. Finally, the conclusions are stated in Section 5.

2. Mathematical Model

Taking down-milling as an example, the dynamic system of end-milling with two degrees of freedom is shown in Figure 1, where the workpiece is assumed to be rigid, and the milling cutter is assumed to be evenly distributed. Note that the helix angle of the milling cutter is not taken into account. Considering the regenerative effect, the mathematical model of system motion can be expressed as a second-order differential equation [17] as follows:

$$\mathbf{M}\ddot{\mathbf{q}}(t) + \mathbf{C}\dot{\mathbf{q}}(t) + \mathbf{K}\mathbf{q}(t) = -a_p\mathbf{K}_c(t)[\mathbf{q}(t) - \mathbf{q}(t-T)] \tag{1}$$

where $\mathbf{q}(t)$, $\dot{\mathbf{q}}(t)$, and $\ddot{\mathbf{q}}(t)$ represent the displacement vector, the first derivative of $\mathbf{q}(t)$ w.r.t t, and the second derivative of $\mathbf{q}(t)$ w.r.t t, respectively. $\mathbf{M}$, $\mathbf{C}$, $\mathbf{K}$, and a_p represent the mass matrix, damping matrix, stiffness matrix, and depth of cut, respectively. $T = 60/(Nv)$ represents the delay period of the delay differential equations, where N represents the cutter tooth number, and v represents the spindle speed. The radial cutting force coefficient matrix $\mathbf{K}_c(t)$ can be expressed as follows:

$$\mathbf{K}_c(t) = \begin{bmatrix} h_{xx}(t) & h_{xy}(t) \\ h_{yx}(t) & h_{yy}(t) \end{bmatrix} \tag{2}$$

where

$$h_{xx}(t) = \sum_{j=1}^{N} g(\phi_j(t))\sin(\phi_j(t))\left[K_t\cos(\phi_j(t)) + K_n\sin(\phi_j(t))\right] \tag{3}$$

$$h_{xy}(t) = \sum_{j=1}^{N} g(\phi_j(t)) \cos(\phi_j(t)) [K_t \cos(\phi_j(t)) + K_n \sin(\phi_j(t))] \tag{4}$$

$$h_{yx}(t) = \sum_{j=1}^{N} g(\phi_j(t)) \sin(\phi_j(t)) [-K_t \sin(\phi_j(t)) + K_n \cos(\phi_j(t))] \tag{5}$$

and

$$h_{yy}(t) = \sum_{j=1}^{N} g(\phi_j(t)) \cos(\phi_j(t)) [-K_t \sin(\phi_j(t)) + K_n \cos(\phi_j(t))] \tag{6}$$

where K_t and K_n represent the tangential and normal cutting force coefficients, respectively. $\phi_j(t)$ represents the angular position of tooth j, as shown in Figure 1, and is defined as follows:

$$\phi_j(t) = \frac{2\pi v}{60} t + \frac{2\pi(j-1)}{N} \tag{7}$$

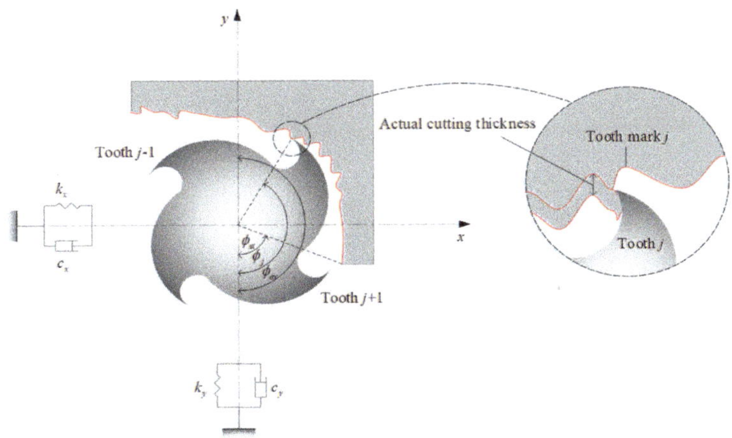

Figure 1. Dynamic system of end-milling with two degrees of freedom.

$g(\phi_j(t))$ is a piecewise function used to judge whether the tooth j is cutting or not and is defined as follows:

$$g(\phi_j(t)) = \begin{cases} 1 & \phi_{st} < \phi_j(t) < \phi_{ex} \\ 0 & \text{otherwise} \end{cases} \tag{8}$$

where ϕ_{st} and ϕ_{ex} represent the start and exit cutting angles of the tooth j, respectively. In the up-milling process, $\phi_{st} = 0$ and $\phi_{ex} = \arccos(1 - 2a/D)$; in the down-milling process, $\phi_{st} = \arccos(2a/D - 1)$ and $\phi_{ex} = \pi$, where a/D represents the radial immersion ratio.

3. Composite Cotes-Based Method

3.1. State-Space Expression

$\mathbf{x}(t)$ can be defined as $\mathbf{x}(t) = \begin{bmatrix} \mathbf{q}(t) \\ \mathbf{M}\dot{\mathbf{q}}(t) + \mathbf{C}\mathbf{q}(t)/2 \end{bmatrix}$. Then, the state expression of (1) can be expressed as follows:

$$\dot{\mathbf{x}}(t) = \mathbf{A}\mathbf{x}(t) + a_p \mathbf{B}(t)[\mathbf{x}(t) - \mathbf{x}(t-T)] \tag{9}$$

where

$$\mathbf{A} = \begin{bmatrix} -\mathbf{M}^{-1}\mathbf{C}/2 & \mathbf{M}^{-1} \\ \mathbf{C}\mathbf{M}^{-1}\mathbf{C}/4 - \mathbf{K} & -\mathbf{C}\mathbf{M}^{-1}/2 \end{bmatrix} \tag{10}$$

$$\mathbf{B}(t) = \begin{bmatrix} \mathbf{0} & \mathbf{0} \\ \mathbf{K}_c(t) & \mathbf{0} \end{bmatrix} \tag{11}$$

If $a_p\mathbf{B}(t)[\mathbf{x}(t) - \mathbf{x}(t-T)]$ is regarded as the inhomogeneous term of $\dot{\mathbf{x}}(t) = \mathbf{A}\mathbf{x}(t)$, then the following equation can be derived:

$$\mathbf{x}(t) = e^{\mathbf{A}(t-t_0)}\mathbf{x}(t_0) + a_p\int_{t_0}^{t}\left\{e^{\mathbf{A}(t-\xi)}\mathbf{B}(\xi)[\mathbf{x}(\xi) - \mathbf{x}(\xi - T)]\right\}\mathrm{d}\xi \tag{12}$$

where t_0 represents the start time. It is worth noting that $\mathbf{x}(\xi)$ is unknown such that (12) is a Volterra integral equation of the second kind.

3.2. Numerical Algorithm

To solve (9) using the composite Cotes formula, the integral equation $f(x)$ is assumed to be continuous in the interval $[c, d]$. Then, $[c, d]$ is divided into m isometric intervals; that is, the span of each interval h is equal to $(d - c)/m$. Based on the isometric nodes $u_k = c + (k - 1)/h$, where $k = 1, 2, \ldots, m + 1$, we obtain the following:

$$I_m = (d-c)\sum_{k=0}^{m} C_k^{(m)} f(u_k) \tag{13}$$

where $C_k^{(m)}$ represents the Cotes coefficients [18], expressed as follows:

$$C_k^{(m)} = \frac{(-1)^{m-k}}{nk!(n-k)}\int_0^m \prod_{j=0,j\neq k}^{m}(t-j)\mathrm{d}t, k = 0, 1, \ldots, m \tag{14}$$

It is worth noting that $C_k^{(m)}$ only depends on m. When $m = 4$, (14) can be expressed as follows:

$$C = \frac{1}{90}[7f(u_0) + 32f(u_1) + 12f(u_2) + 32f(u_3) + 7f(u_4)] \tag{15}$$

where T consists of the durations of the free and excited vibrations [15]. The durations of the free vibration and excited vibration can be defined as t_f and t_c, respectively. For the free vibration, $\mathbf{B}(\xi) = 0$, and (12) can be expressed as follows:

$$\mathbf{x}(t) = e^{\mathbf{A}(t-t_0)}\mathbf{x}(t_0) \tag{16}$$

Therefore, the state equation can be solved when $t = t_0 + t_f$, and it can be expressed as follows:

$$\mathbf{x}\left(t_0 + t_f\right) = e^{\mathbf{A}t_f}\mathbf{x}(t_0) \tag{17}$$

For the excited vibration, its time lies in $[t_0 + t_f, t_0 + T]$. The interval is divided into n isometric intervals, and the span of each interval l is equal to $(T - t_f)/n$. The discrete time can be expressed as $t_i = t_0 + t_f + (i - 1)h$, where $i = 1, 2, \ldots, n + 1$. According to the numerical integral solution of the Volterra integral equation of the second kind [19], we obtain the following:

$$\mathbf{x}(t_{i+1}) = e^{\mathbf{A}(t_{i+1}-t_i)}\mathbf{x}(t_i) + a_p\int_{t_i}^{t_{i+1}}\left\{e^{\mathbf{A}(t_{i+1}-\xi)}\mathbf{B}(\xi)[\mathbf{x}(\xi) - \mathbf{x}(\xi - t)]\right\}\mathrm{d}\xi \tag{18}$$

If $[t_i, t_{i+1}]$ is divided into four isometric intervals, then t_i, $t_{i+1/4}$, $t_{i+1/2}$, $t_{i+3/4}$, and t_{i+1} can be derived. Combining (13) and (15), the integral equation can be expressed as follows:

$$\int_{t_i}^{t_{i+1}} f(\xi)\mathrm{d}\xi = \frac{t_{i+1}-t_i}{90}[7f(t_i) + 32f(t_{i+1/4}) + 12f(t_{i+1/2}) + 32f(t_{i+3/4}) + 7f(t_{i+1})] \tag{19}$$

Upon substituting (19) into (18), we can see that $\mathbf{x}(t_{i+1/4})$, $\mathbf{x}(t_{i+1/2})$, and $\mathbf{x}(t_{i+3/4})$ cannot be solved directly. The barycentric Lagrange interpolation method [20] is introduced here to approximate $\mathbf{x}(t_{i+1/4})$, $\mathbf{x}(t_{i+1/2})$, and $\mathbf{x}(t_{i+3/4})$ and obtain the following:

$$\mathbf{x}(t_{i+1/4}) \simeq \frac{21\mathbf{x}(t_i) + 14\mathbf{x}(t_{i+1}) - 3\mathbf{x}(t_{i+2})}{32} \tag{20}$$

$$\mathbf{x}(t_{i+1/2}) \simeq \frac{3\mathbf{x}(t_i) + 6\mathbf{x}(t_{i+1}) - \mathbf{x}(t_{i+2})}{8} \tag{21}$$

$$\mathbf{x}(t_{i+3/4}) \simeq \frac{5\mathbf{x}(t_i) + 30\mathbf{x}(t_{i+1}) - 3\mathbf{x}(t_{i+2})}{32} \tag{22}$$

Substituting (19)–(22) into (18), $\mathbf{x}(t_i)$ can be expressed as follows:

$$\begin{aligned} \mathbf{x}(t_{i+1}) &= e^{\mathbf{A}h}\mathbf{x}(t_i) + \tfrac{a_p h}{6}\Big\{\tfrac{5}{2}e^{\mathbf{A}h}\mathbf{B}(t_i)(\mathbf{x}(t_i) - \mathbf{x}(t_i - T)) + 4\mathbf{B}(t_{i+1})(\mathbf{x}(t_{i+1}) - \mathbf{x}(t_{i+1} - T)) \\ &\quad -\tfrac{1}{2}e^{-\mathbf{A}h}\mathbf{B}(t_{i+2})(\mathbf{x}(t_{i+2}) - \mathbf{x}(t_{i+2} - T))\Big\} \end{aligned} \tag{23}$$

It is worth noting that $\mathbf{B}(t_{n+2})$ cannot be calculated directly using (9) when $i = n$. A Newton Cotes formula [13] is introduced here to calculate $\mathbf{x}(t_{n+1})$ and is expressed as follows:

$$\mathbf{x}(t_{n+1}) = e^{\mathbf{A}h}\mathbf{x}(t_n) + \frac{a_p h}{2}\Big[e^{\mathbf{A}h}\mathbf{B}(t_n)(\mathbf{x}(t_n) - \mathbf{x}(t_n - T)) + \mathbf{B}(t_{n+1})(\mathbf{x}(t_{n+1}) - \mathbf{x}(t_{n+1} - T))\Big] \tag{24}$$

Combining (17), (23), and (24), the transition matrix can be constructed as follows:

$$\mathbf{C}_1 = \begin{bmatrix} 0 & 0 & 0 & \cdots & 0 & 0 & 0 \\ -e^{\mathbf{A}h} & 0 & 0 & \cdots & 0 & 0 & 0 \\ 0 & -e^{\mathbf{A}h} & 0 & \cdots & 0 & 0 & 0 \\ \vdots & \vdots & \vdots & \ddots & \vdots & \vdots & \vdots \\ 0 & 0 & 0 & \cdots & 0 & 0 & 0 \\ 0 & 0 & 0 & \cdots & -e^{\mathbf{A}h} & 0 & 0 \\ 0 & 0 & 0 & \cdots & 0 & -e^{\mathbf{A}h} & 0 \end{bmatrix}_{(2n+2)\times(2n+2)} \tag{25}$$

$$\mathbf{D}_1 = \begin{bmatrix} 0 & 0 & 0 & \cdots & 0 & 0 & 0 \\ -\frac{75}{2}e^{\mathbf{A}h}\mathbf{B}_1 & -60\mathbf{B}_2 & \frac{15}{2}e^{-\mathbf{A}h}\mathbf{B}_3 & \cdots & 0 & 0 & 0 \\ 0 & -\frac{75}{2}e^{\mathbf{A}h}\mathbf{B}_2 & -60\mathbf{B}_3 & \cdots & 0 & 0 & 0 \\ \vdots & \vdots & \vdots & \ddots & \vdots & \vdots & \vdots \\ 0 & 0 & 0 & \cdots & -60\mathbf{B}_{i-1} & \frac{15}{2}e^{-\mathbf{A}h}\mathbf{B}_i & 0 \\ 0 & 0 & 0 & \cdots & -\frac{75}{2}e^{\mathbf{A}h}\mathbf{B}_{i-1} & -60\mathbf{B}_i & \frac{15}{2}e^{-\mathbf{A}h}\mathbf{B}_{i+1} \\ 0 & 0 & 0 & \cdots & 0 & 45e^{\mathbf{A}h}\mathbf{B}_i & 45\mathbf{B}_{i+1} \end{bmatrix}_{(2n+2)\times(2n+2)} \tag{26}$$

$$\mathbf{E}_1 = \begin{bmatrix} 0 & 0 & \cdots & 0 & e^{\mathbf{A}t_f} \\ 0 & 0 & \cdots & 0 & 0 \\ \vdots & \vdots & \ddots & \vdots & \vdots \\ 0 & 0 & \cdots & 0 & 0 \\ 0 & 0 & \cdots & 0 & 0 \end{bmatrix}_{(2n+2)\times(2n+2)} \tag{27}$$

where $\mathbf{B}_i$ represents $\mathbf{B}(t_i)$, $i = 1, 2, \ldots, n + 1$.

According to (25)–(27), the dynamic mapping of the discrete system can be determined as follows:

$$\left(\mathbf{I}_1 + \mathbf{C}_1 + \frac{a_p h}{90}\mathbf{D}_1\right)\begin{bmatrix} \mathbf{x}(t_1) \\ \mathbf{x}(t_2) \\ \vdots \\ \mathbf{x}(t_{n+1}) \end{bmatrix} = \left(\frac{a_p h}{90}\mathbf{D}_1 + \mathbf{E}_1\right)\begin{bmatrix} \mathbf{x}(t_1 - T) \\ \mathbf{x}(t_2 - T) \\ \vdots \\ \mathbf{x}(t_{n+1} - T) \end{bmatrix} \tag{28}$$

where $\mathbf{I}_1$ is a $(2n + 1) \times (2n + 1)$ identity matrix.

Therefore, the transition matrix of the dynamic system in a tool pass cycle is obtained as follows:

$$\mathbf{\Phi}_1 = \left(\mathbf{I}_1 + \mathbf{C}_1 + \frac{a_p h}{90}\mathbf{D}_1\right)^{-1}\left(\frac{a_p h}{90}\mathbf{D}_1 + \mathbf{E}_1\right) \tag{29}$$

It is worth noting that the chatter stability must be determined using the Floquet theory. If the modulus of any eigenvalue in $\mathbf{\Phi}_1$ exceeds 1, the system is unstable; if the moduli of all eigenvalues in $\mathbf{\Phi}_1$ are less than 1, the system is stable [21,22]. Therefore, the boundary curve between the unstable and stable regions in the lobe diagram can be used as the criterion for judging whether chatter occurs.

3.3. Simulation and Analysis

In this section, SDM [10] and SEM [16] are compared. For objective comparison, the CCM, SDM, and SEM share the same parameters and machining conditions. A benchmark example of the single-DOF milling model is utilized to validate and analyze the CCM. The single-DOF milling mathematical model is expressed as follows [23]:

$$\ddot{x}(t) + 2\zeta\omega_n\dot{x}(t) + \omega_n^2 x(t) = -\frac{a_p h(t)}{m_t}(x(t) - x(t - T)) \tag{30}$$

where ζ denotes the relative damping, ω_n denotes the angular natural frequency, and $h(t)$ denotes the specific cutting force coefficient.

The state-space for single-DOF milling mathematical model is expressed as follows:

$$\dot{\mathbf{x}}(t) = \mathbf{A}\mathbf{x}(t) + a_p\mathbf{B}(t)[\mathbf{x}(t) - \mathbf{x}(t - T)] \tag{31}$$

where

$$\mathbf{A} = \begin{bmatrix} -\zeta\omega_n & \frac{1}{m_t} \\ m_t\omega_n^2(\zeta^2 - 1) & -\zeta\omega_n \end{bmatrix} \tag{32}$$

$$\mathbf{B}(t) = \begin{bmatrix} 0 & 0 \\ -h(t) & 0 \end{bmatrix} \tag{33}$$

where $h(t)$ is equal to $h_{xx}(t)$, defined in (3).

The parameters in (32) and (33) are defined as follows: f_n = 922 Hz, ζ = 0.011, m_t = 0.3993 kg, $K_t = 6 \times 10^8$ N/m^2 and $K_t = 2 \times 10^8$ N/m^2. The adopted machining condition is down-milling. The radial immersion ratio a/D is set as 1 to avoid intermittent milling. The spindle speed is set as 5000 rpm (v = 5000 rpm). The depths of cut are set as 0.001 m, 0.0005 mm, and 0.0002 mm, respectively. All programs in this study are executed in MATLAB R2019a and run on a personal computer (AMD Ryzen 5 5600H; CPU 4.0 GHz, 16 GB). The maximum modulus of the eigenvalues of $\mathbf{\Phi}$ is labelled as $|\lambda|$, and the maximum modulus of the eigenvalues of $\mathbf{\Phi}$ with SDM is labelled as $|\lambda_0|$. In this study, $|\lambda_0|$ is treated as the exact value.

The convergence rate comparisons of the CCM, SDM, and SEM are shown in Figure 2. It is seen from the figure that when n is small, the convergence rate of the CCM is significantly higher than those of the SDM and SEM regardless of the depth of cut. As n increases, the convergence rates of the three methods approach gradually, and the convergence rates of the CCM and SEM are higher than that of SDM. The figure (a) is enlarged to observe the differences better. It is worth noting that the convergence rate of the CCM is significantly higher than those of the SDM and SEM when the discrete number changes from 75 to 100. The results demonstrate that the convergence rate of the CCM outweighs those of the SDM and SEM.

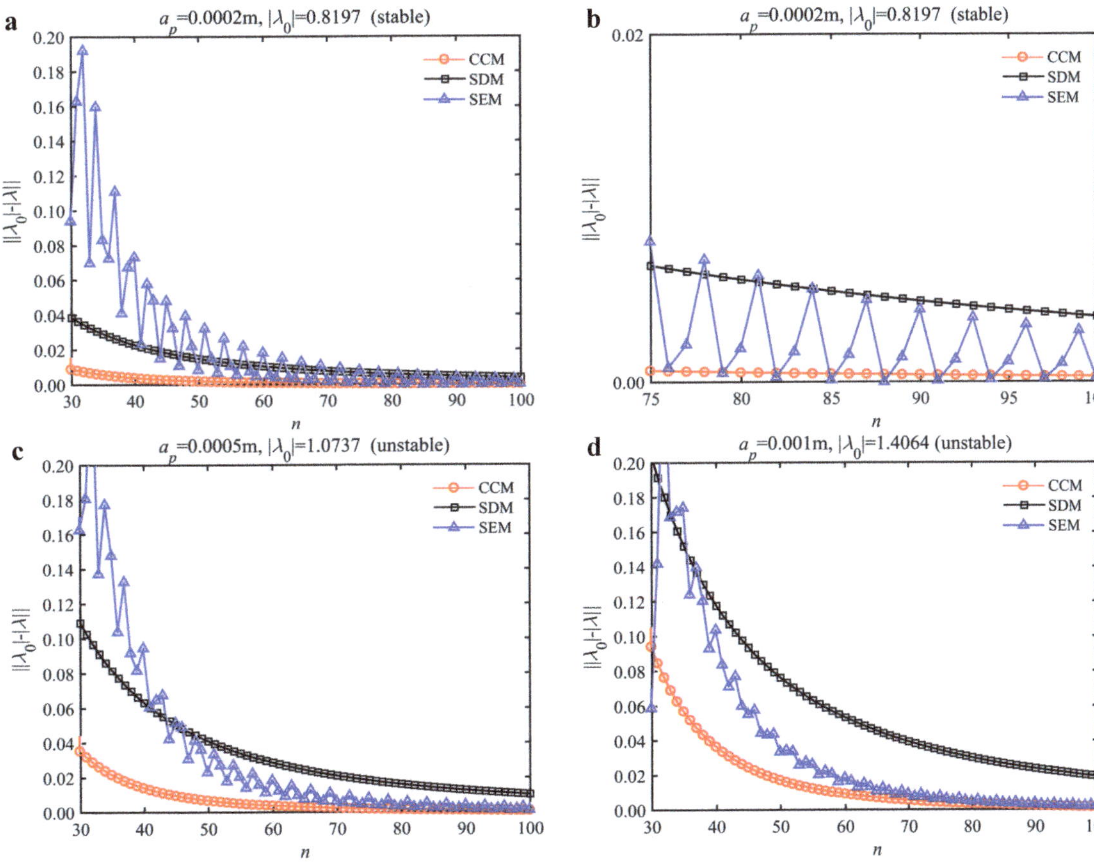

Figure 2. Convergence rate comparisons of the CCM, SDM and SEM when $a/D = 1$. (**a**) $a_p = 0.0002$ m and $n \in [30, 100]$; (**b**) $a_p = 0.0002$ m and $n \in [75, 100]$; (**c**) $a_p = 0.0005$ m and $n \in [30, 100]$; (**d**) $a_p = 0.001$ m and $n \in [30, 100]$.

The stability lobe diagrams can be used to compare the calculation accuracy among the CCM, SDM, and SEM. As a reference, the discrete number is set as 500. The other parameters are set as follows: the discrete numbers are set as 25, 40, and 55; the radial immersion ratios are set as 0.05, 0.5, and 1; the spindle speed varies from 0.5×10^4 to 2.5×10^4 rpm, and 200 equally distributed sampling points are selected within this range; the depth of cut varies from 0 to 0.01 m, and 100 equally distributed sampling points are selected within this range.

In the single-DOF system, the stability lobe diagrams obtained with the CCM, SDM, and SEM when $a/D = 1$ are given in Table 1. It can be seen that the stability lobe diagrams of the CCM are closer to the reference ones than the SDM and SEM. Note that the severe distortions appear in the stability lobe diagrams obtained with the SEM when the discrete number is 25. Overall, the CCM is more accurate than the SDM and SEM. To further compare the calculation accuracy, two indicators are introduced, i.e., the arithmetic mean of relative error (*AMRE*) and mean squared error (*MSE*) [2]. The *AMRE* and *MSE* are defined as follows [2]:

$$AMRE = \frac{1}{n}\sum_{i=1}^{n}\frac{|a_i - a_{i0}|}{a_{i0}}$$

$$MSE = \frac{1}{n}\sum_{i=1}^{n}(a_i - a_{i0})^2$$

where a_i and a_{i0} denote the predicted axial depth and reference axial depth, respectively. n denotes the discrete number. The discrete numbers are set as 25, 30, 35, 40, 45, 50, and 55. The variations of the *AMRE* and *MSE* are depicted in Figure 3. Due to the fact that the severe distortions appear in the stability lobe diagrams obtained with the SEM when the discrete number is 25, the *AMRE* and *MSE* of the SEM cannot be attained. Take n = 40 as an example. The *AMRE* obtained with the CCM, SDM, and SEM is 0.041, 0.076, and 0.154, respectively. The *MSE* obtained with the CCM, SDM, and SEM is 2.62×10^{-8}, 4.24×10^{-8}, and 1.07×10^{-7}, respectively. The results show that the CCM is closer to the reference values than the SDM and SEM.

Table 1. Stability lobe diagrams of the CCM, SDM, and SEM when a/D = 1.

n	25	40	55
CCM	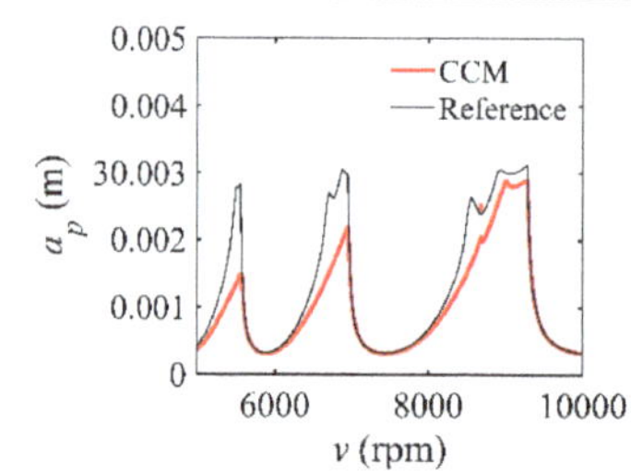	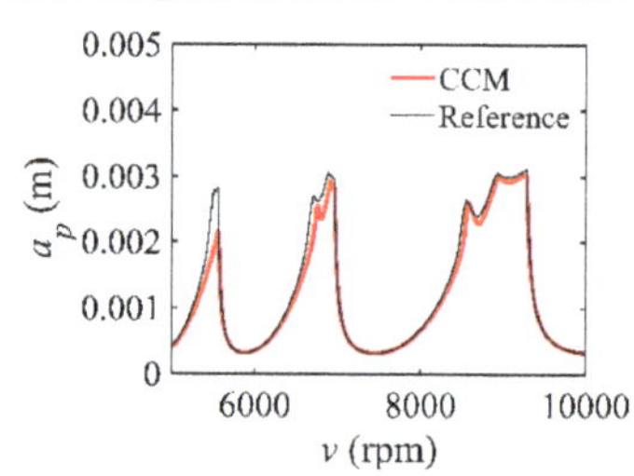	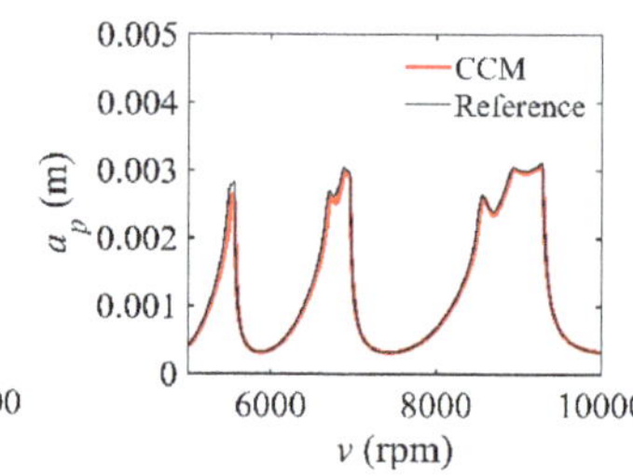
SDM	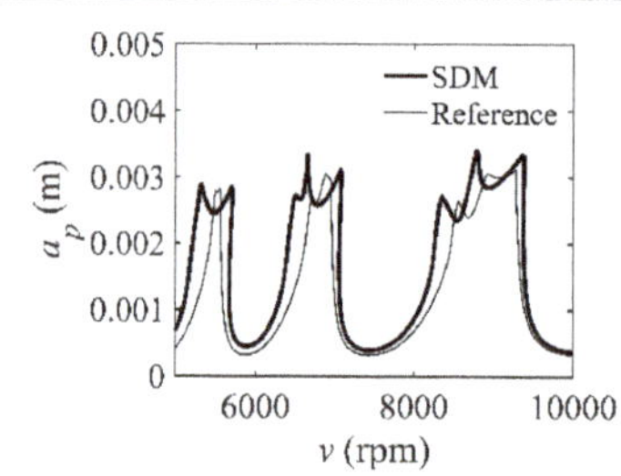	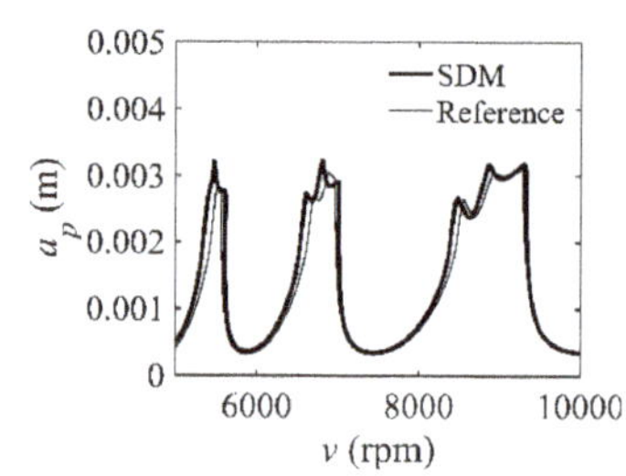	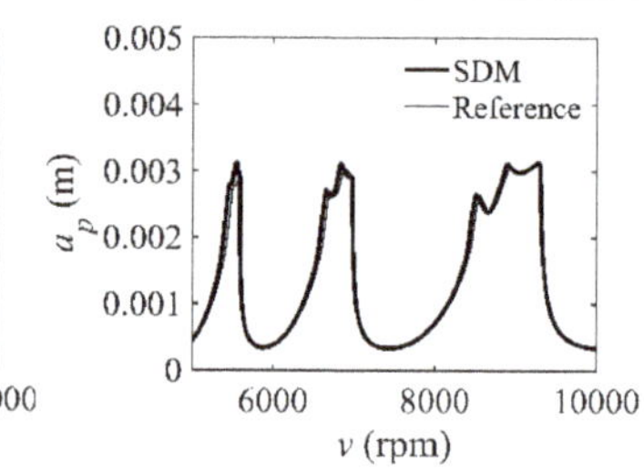
SEM	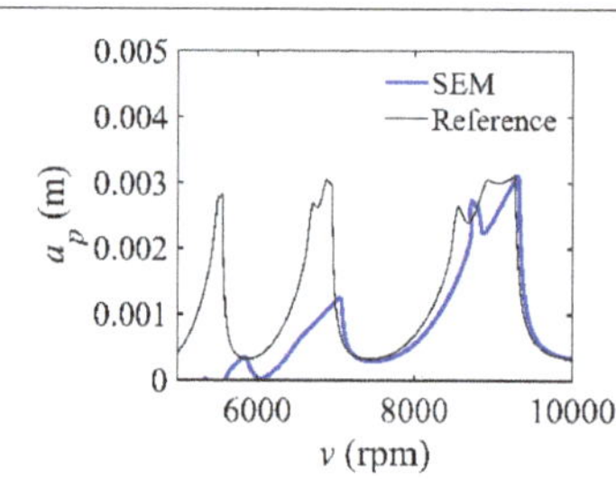	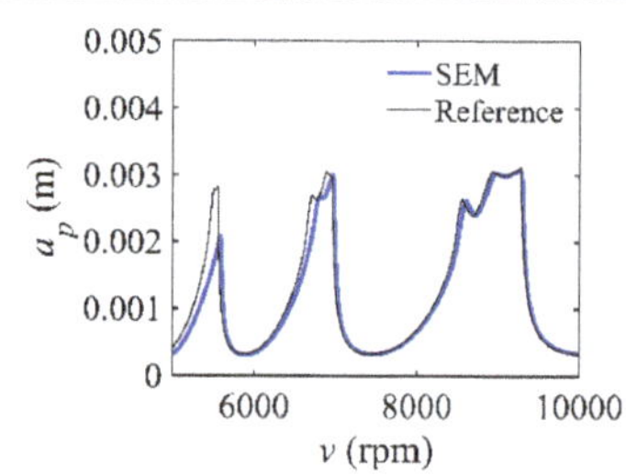	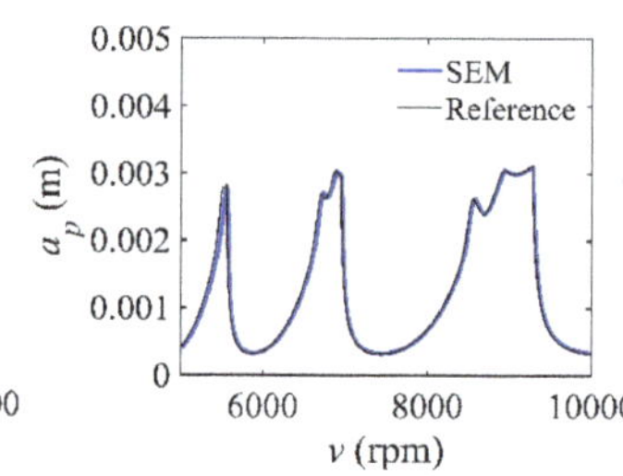

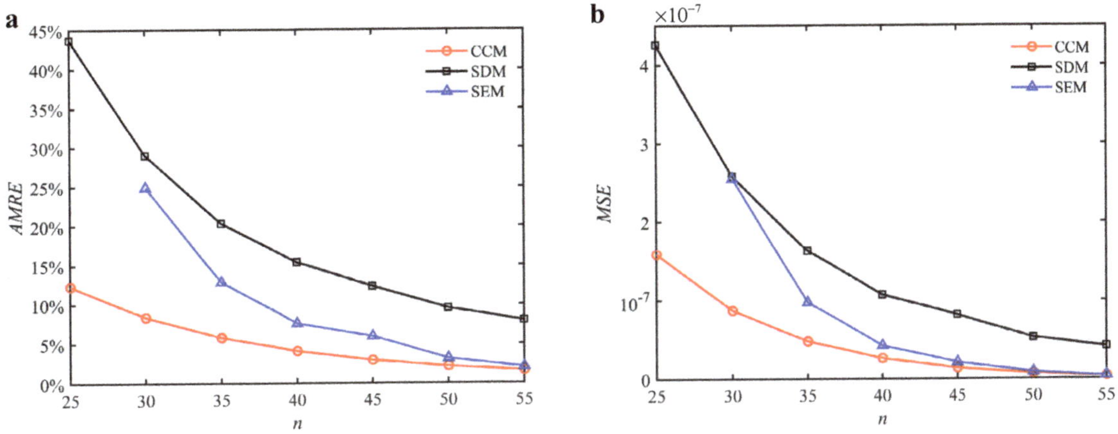

Figure 3. *AMRE* and *MSE* of the CCM, SDM, and SEM when $a/D = 1$. (**a**) *AMRE*; (**b**) *MSE*.

Furthermore, the computational efficiency is an important criterion for measuring the effectiveness of the chatter analysis methods. Figure 4 shows the computational time of the CCM, SDM, and SEM under different discrete numbers when $a/D = 1$. We observe that the computational time can be reduced from 55.63 s to 20.21 s, 62.45 s to 25.79 s, 72.06 s to 30.13 s, and 81.21 s to 35.18 s when the discrete numbers are 40, 45, 50, and 55, respectively. Furthermore, the stability lobe diagrams, *AMRE* and *MSE*, as well as the computational time when $a/D = 0.05$ and $a/D = 0.5$ are given in Tables 2 and 3 and Figures 5 and 6. It could be noted that the stability lobe diagrams obtained with the CCM are closer to the reference ones than those obtained with the SDM and SEM, and the CCM has the highest calculation accuracy. The results also demonstrate that the CCM has significant advantages in the computational efficiency.

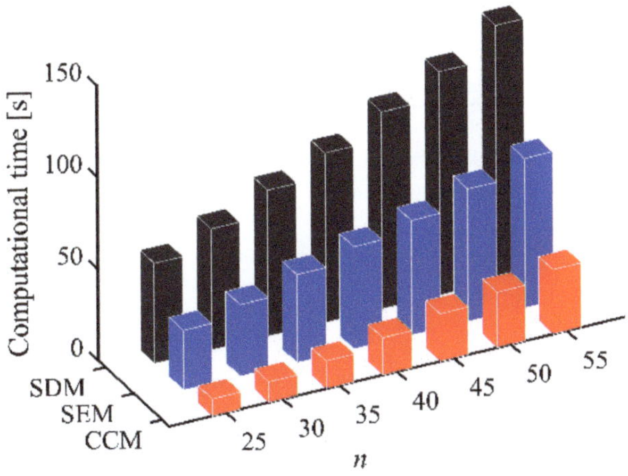

Figure 4. Computational time of the CCM, SDM, and SEM when $a/D = 1$.

Table 2. Stability lobe diagrams of the CCM, SDM, and SEM when a/D = 0.05.

n	25	40	55
CCM	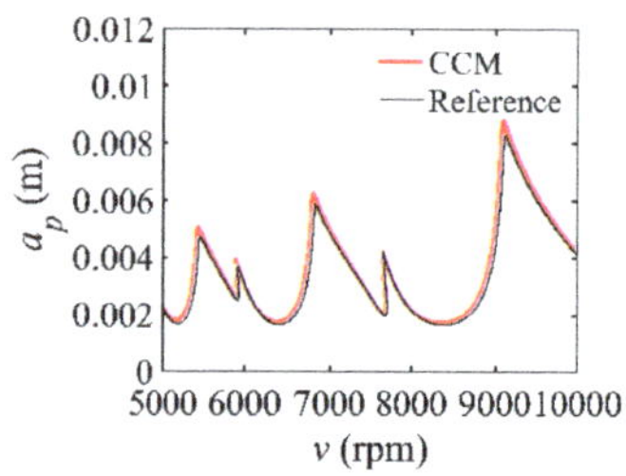	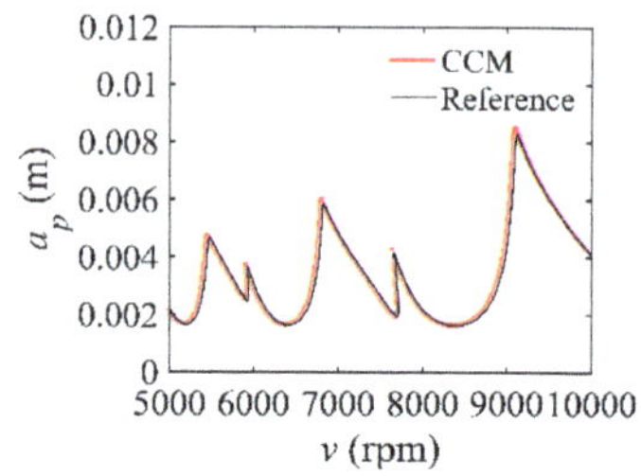	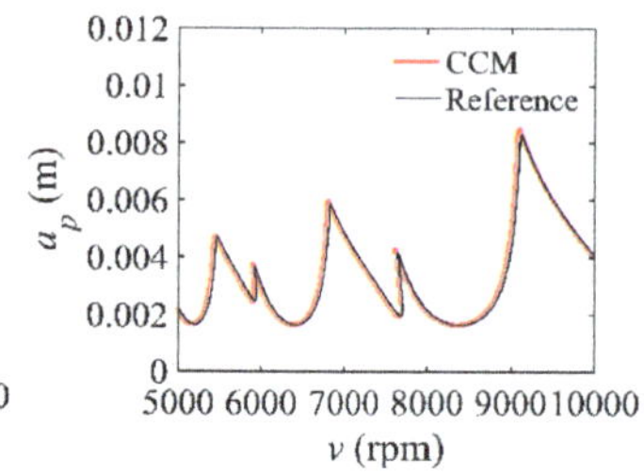
SDM	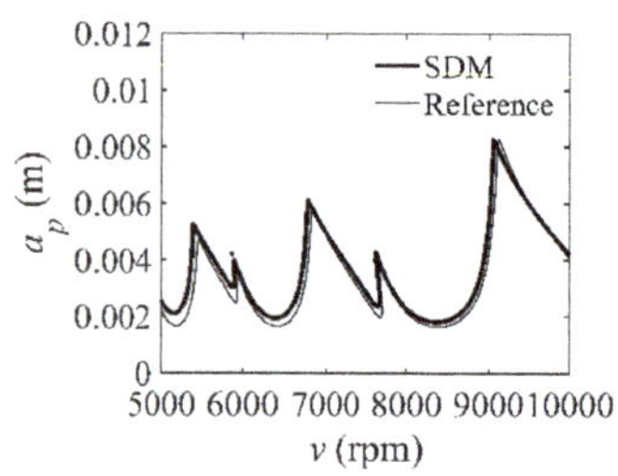	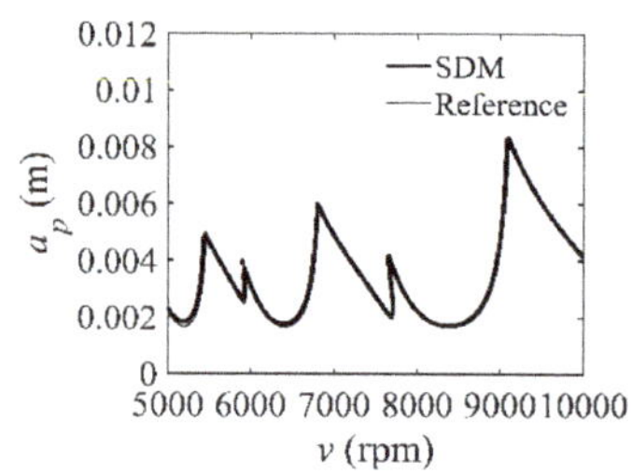	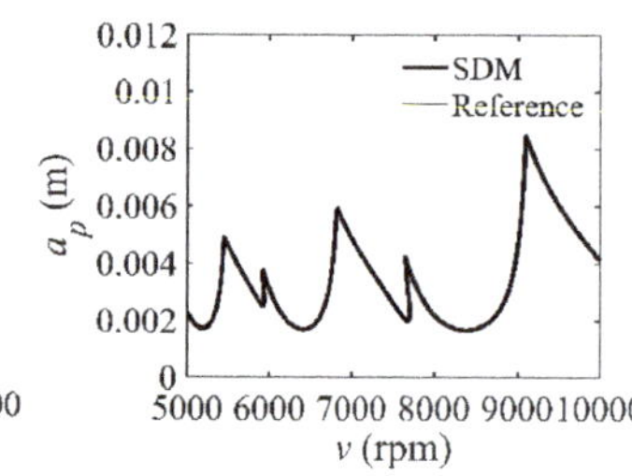
SEM	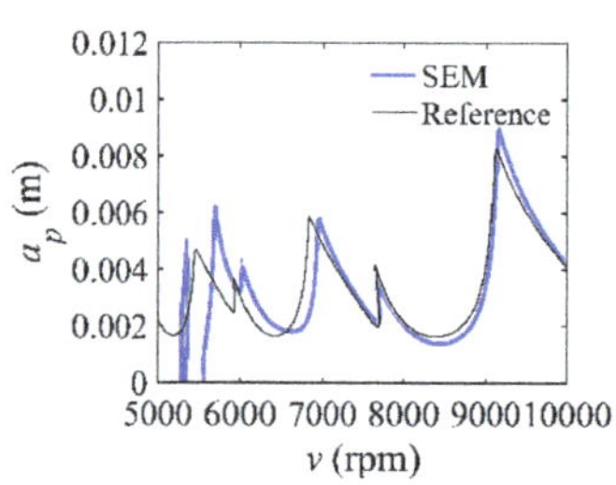	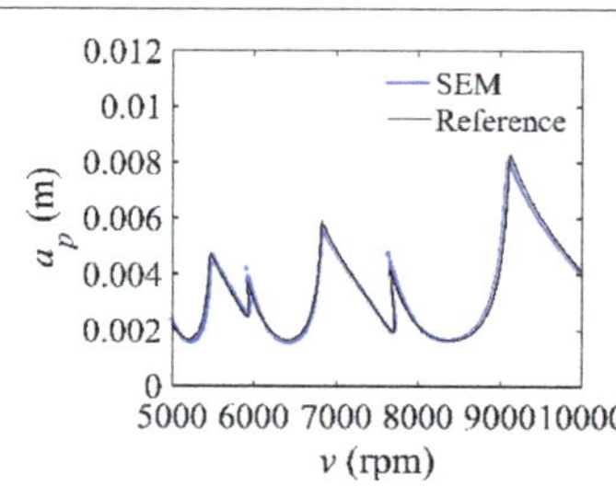	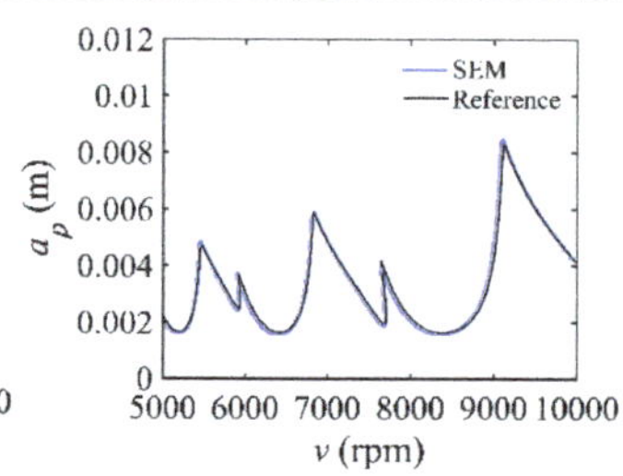

Table 3. Stability lobe diagrams of the CCM, SDM, and SEM when a/D = 0.5.

n	25	40	55
CCM	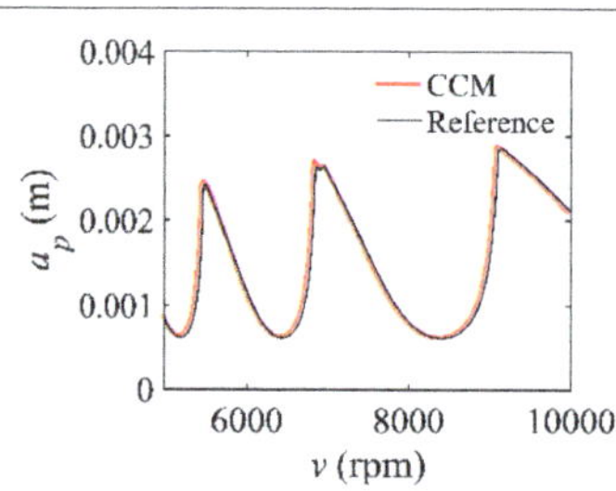	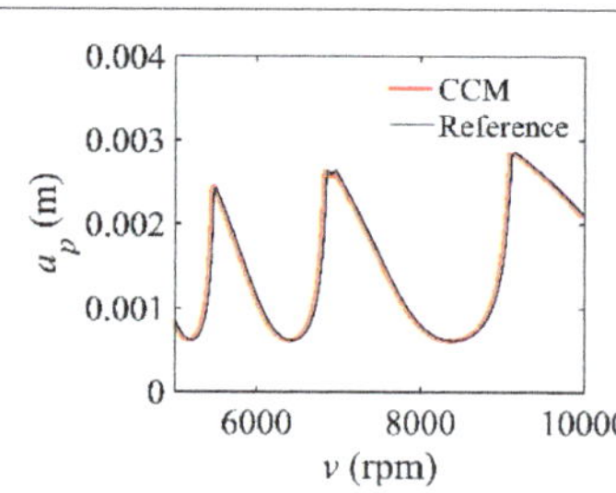	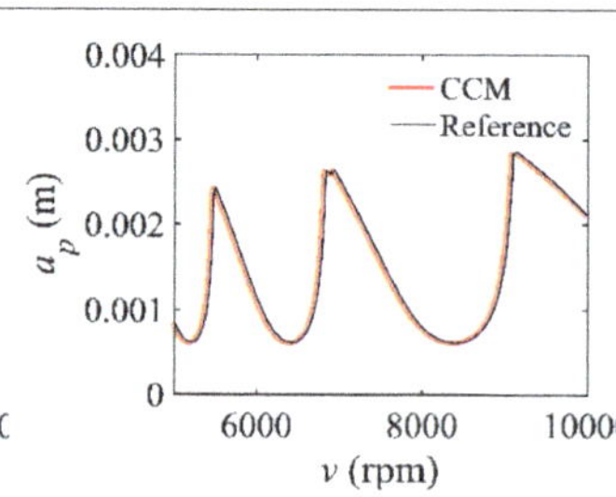

Table 3. *Cont.*

n	25	40	55
SDM	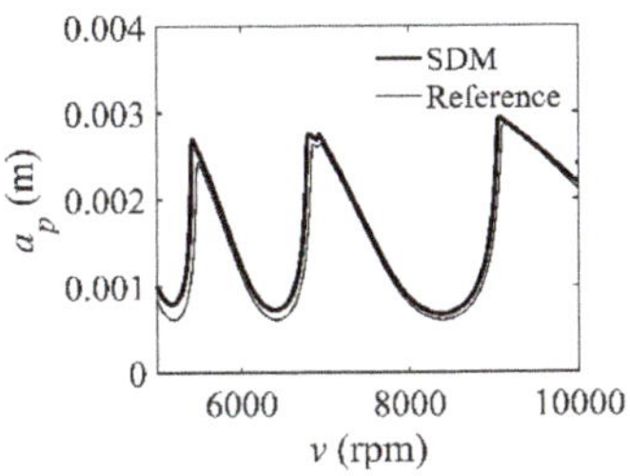	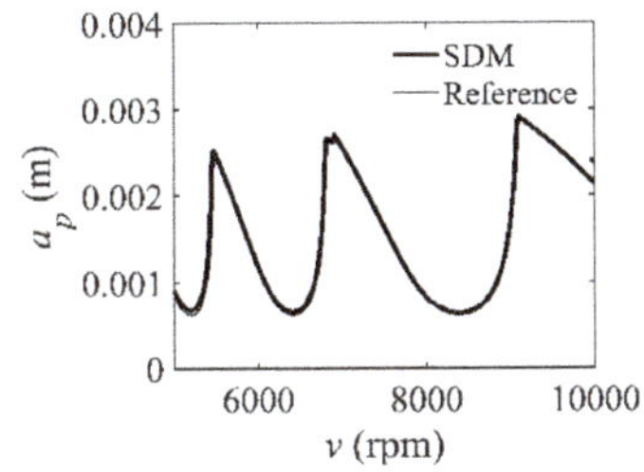	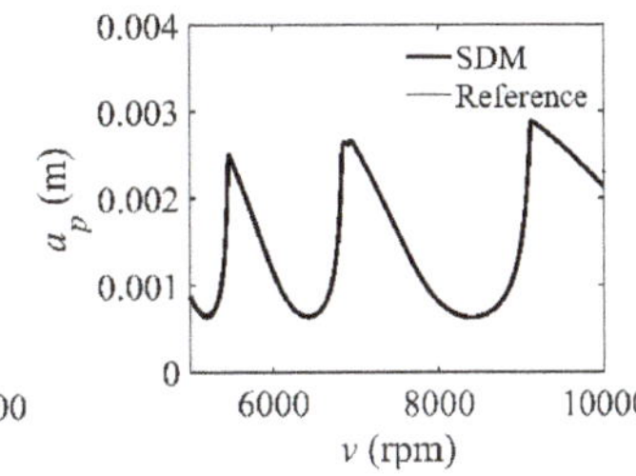
SEM	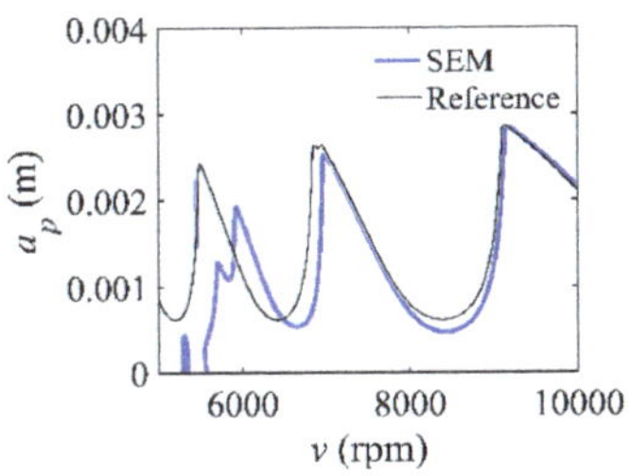	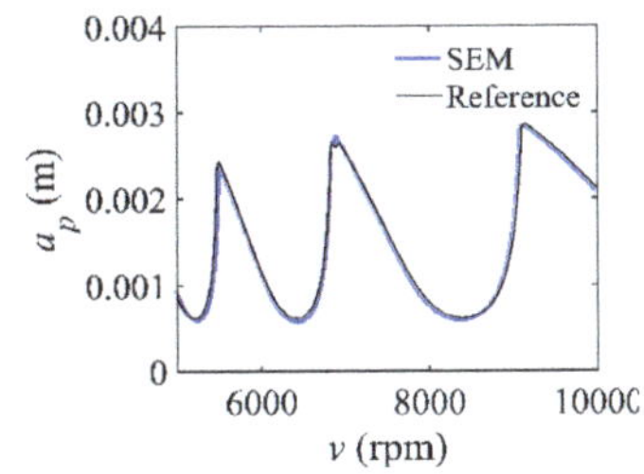	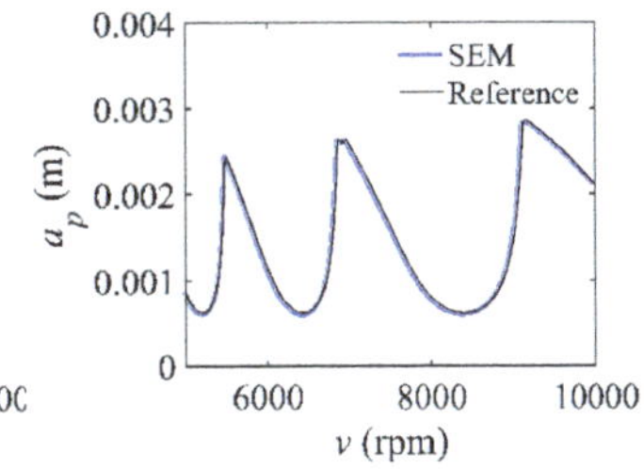

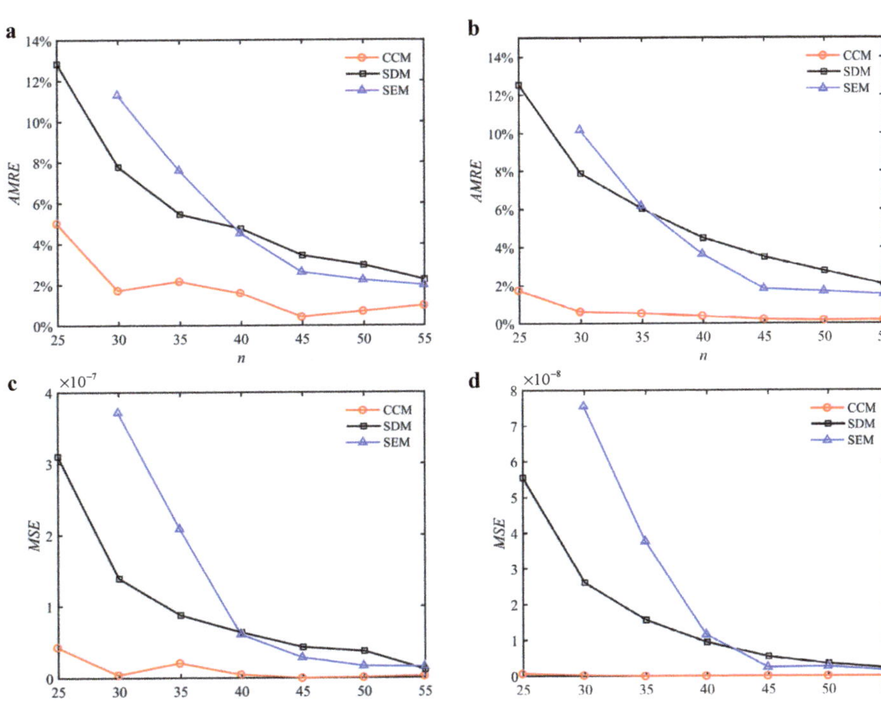

Figure 5. *AMRE* and *MSE* of the CCM, SDM, and SEM. (**a**) *AMRE* when $a/D = 0.05$; (**b**) *AMRE* when $a/D = 0.5$; (**c**) *MSE* when $a/D = 0.05$; (**d**) *MSE* when $a/D = 0.5$.

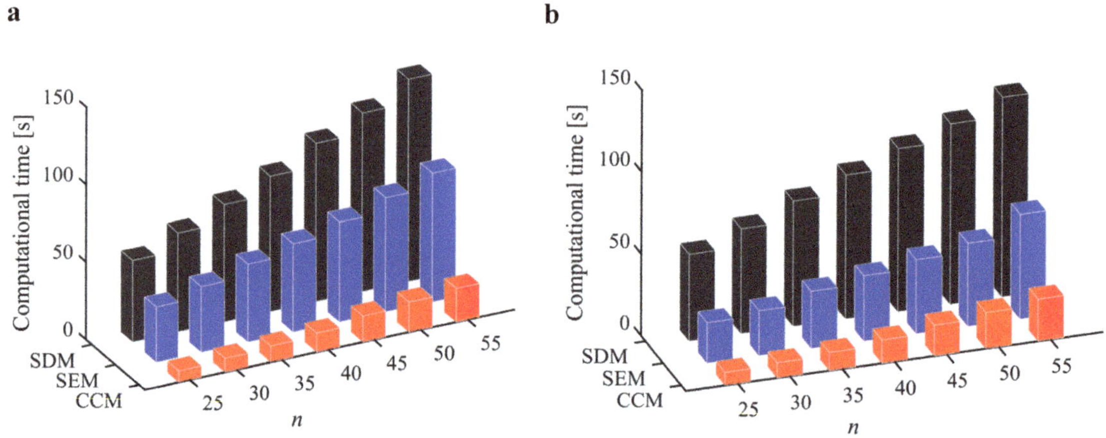

Figure 6. Computational time of the CCM, SDM, and SEM. (**a**) $a/D = 0.05$; (**b**) $a/D = 0.5$.

4. Simpson's 3/8-Based Method

4.1. State-Space Expression

To match with S38M, a new state-space expression is defined as follows [17]:

$$\dot{\mathbf{x}}(t) = \mathbf{A}\mathbf{x}(t) + a_p\mathbf{B}(t)[\mathbf{x}(t) - \mathbf{x}(t-T)] \tag{34}$$

where

$$\mathbf{A} = \begin{bmatrix} \mathbf{0} & \mathbf{I} \\ -\mathbf{M}^{-1}\mathbf{K} & -\mathbf{M}^{-1}\mathbf{C} \end{bmatrix} \tag{35}$$

and

$$\mathbf{B}(t) = \begin{bmatrix} \mathbf{0} & \mathbf{0} \\ -\mathbf{M}^{-1}\mathbf{K}_c(t) & \mathbf{0} \end{bmatrix} \tag{36}$$

Based on the numerical integration method of the Volterra integral equation of the second kind [19], the state-space equation can be deduced as follows:

$$\mathbf{x}(t) = e^{\mathbf{A}(t-t_{st})}\mathbf{x}(t_{st}) + a_p\int_{t_{st}}^{t}\left\{e^{\mathbf{A}(t-\xi)}\mathbf{B}(\xi)[\mathbf{x}(\xi) - \mathbf{x}(\xi-T)]\right\}\mathrm{d}\xi \tag{37}$$

where t_{st} represents the start time and defaults to t_1.

When $t = t_3$, we obtain the following:

$$\mathbf{x}(t_3) = e^{\mathbf{A}(t_3-t_1)}\mathbf{x}(t_1) + a_p\int_{t_{st}}^{t_1}\left\{e^{\mathbf{A}(t_1-\xi)}\mathbf{B}(\xi)[\mathbf{x}(\xi) - \mathbf{x}(\xi-T)]\right\}\mathrm{d}\xi \tag{38}$$

4.2. Numerical Algorithm

The Lagrange polynomial denoted by $P_L(x)$ [16,24] can be used to approximate $f(x)$ as follows:

$$P_L(x) = \sum_{k=0}^{L} f(x_k)\eta_{L,k}(x) \tag{39}$$

where

$$\eta_{L,k}(x) = \prod_{\substack{0 \le m \le k \\ m \ne L}} \frac{x - x_m}{x_L - x_m} \tag{40}$$

It is worth noting that x_i, x_{i+1}, and x_{i+2} only exist when $L = 2$, and then, the Simpson's 3/8 formula (41) can be derived using $P_2(x)$.

$$\int_{x_i}^{x_{i+2}} f(x)\mathrm{d}x = \frac{h}{3}(f(x_i) + 4f(x_{i+1}) + f(x_{i+2})) \tag{41}$$

Based on (41), (38) can be rewritten as follows:

$$\begin{aligned} \mathbf{x}(t_3) = \; & e^{2\mathbf{A}h}\mathbf{x}(t_1) + \tfrac{a_p h}{3}\Big\{e^{2\mathbf{A}h}\mathbf{B}(t_1)[\mathbf{x}(t_1) - \mathbf{x}(t_1 - T)] + 4e^{\mathbf{A}h}\mathbf{B}(t_2)[\mathbf{x}(t_2) - \mathbf{x}(t_2 - T)] \\ & + \mathbf{B}(t_3)[\mathbf{x}(t_3) - \mathbf{x}(t_3 - T)]\Big\} \end{aligned} \tag{42}$$

which can be expressed as follows:

$$\begin{aligned} \mathbf{x}(t_{i+2}) & = e^{2\mathbf{A}h}\mathbf{x}(t_i) + \tfrac{a_p h}{3}\Big\{e^{2\mathbf{A}h}\mathbf{B}(t_i)[\mathbf{x}(t_i) - \mathbf{x}(t_i - T)] + 4e^{\mathbf{A}h}\mathbf{B}(t_{i+1})[\mathbf{x}(t_{i+1}) - \mathbf{x}(t_{i+1} - T)] \\ & + \mathbf{B}(t_{i+2})[\mathbf{x}(t_{i+2}) - \mathbf{x}(t_{i+2} - T)]\Big\} \end{aligned} \tag{43}$$

As $\mathbf{B}(t_{i+2})$ cannot be solved directly when $i = n$, the classical numerical integration method is introduced to determine $\mathbf{x}(t_{n+1})$ as follows:

$$\mathbf{x}(t_{n+1}) = e^{\mathbf{A}h}\mathbf{x}(t_n) + \frac{a_p h}{2}\Big\{e^{\mathbf{A}h}\mathbf{B}(t_n)[\mathbf{x}(t_n) - \mathbf{x}(t_n - T)] + \mathbf{B}(t_{n+1})[\mathbf{x}(t_{n+1}) - \mathbf{x}(t_{n+1} - T)]\Big\} \tag{44}$$

Combining (17), (43), and (44), the transition matrix can be constructed as follows:

$$\mathbf{C}_2 = \begin{bmatrix} 0 & 0 & 0 & \cdots & 0 & 0 & 0 \\ -e^{2\mathbf{A}h} & 0 & 0 & \cdots & 0 & 0 & 0 \\ 0 & -e^{2\mathbf{A}h} & 0 & \cdots & 0 & 0 & 0 \\ \vdots & \vdots & \vdots & \ddots & \vdots & \vdots & \vdots \\ 0 & 0 & 0 & \cdots & 0 & 0 & 0 \\ 0 & 0 & 0 & \cdots & -e^{2\mathbf{A}h} & 0 & 0 \\ 0 & 0 & 0 & \cdots & 0 & -e^{\mathbf{A}h} & 0 \end{bmatrix}_{(2n+2)\times(2n+2)} \tag{45}$$

$$\mathbf{D}_2 = \begin{bmatrix} 0 & 0 & 0 & \cdots & 0 & 0 & 0 \\ -e^{2\mathbf{A}h}\mathbf{B}_1 & -4e^{\mathbf{A}h}\mathbf{B}_2 & -\mathbf{B}_3 & \cdots & 0 & 0 & 0 \\ 0 & -e^{2\mathbf{A}h}\mathbf{B}_2 & -4e^{\mathbf{A}h}\mathbf{B}_3 & \cdots & 0 & 0 & 0 \\ \vdots & \vdots & \vdots & \ddots & \vdots & \vdots & \vdots \\ 0 & 0 & 0 & \cdots & -4e^{\mathbf{A}h}\mathbf{B}_{i-1} & -\mathbf{B}_i & 0 \\ 0 & 0 & 0 & \cdots & -e^{2\mathbf{A}h}\mathbf{B}_{i-1} & -4e^{\mathbf{A}h}\mathbf{B}_i & -\mathbf{B}_{i+1} \\ 0 & 0 & 0 & \cdots & 0 & \frac{3}{2}e^{\mathbf{A}h}\mathbf{B}_i & \frac{3}{2}\mathbf{B}_{i+1} \end{bmatrix}_{(2n+2)\times(2n+2)} \tag{46}$$

$$\mathbf{E}_2 = \begin{bmatrix} 0 & 0 & \cdots & 0 & e^{\mathbf{A}t_f} \\ 0 & 0 & \cdots & 0 & 0 \\ \vdots & \vdots & \ddots & \vdots & \vdots \\ 0 & 0 & \cdots & 0 & 0 \\ 0 & 0 & \cdots & 0 & 0 \end{bmatrix}_{(2n+2)\times(2n+2)} \tag{47}$$

$$\mathbf{I}_2 = \begin{bmatrix} 1 & 0 & 0 & 0 & \cdots & 0 & 0 & 0 & 0 \\ 0 & 1 & 0 & 0 & \cdots & 0 & 0 & 0 & 0 \\ 0 & 0 & 0 & 0 & 1 & 0 & 0 & 0 & 0 \\ 0 & 0 & 0 & 0 & \cdots & 1 & 0 & 0 & 0 \\ \vdots & \vdots & \vdots & \vdots & \ddots & \vdots & \vdots & \vdots & \vdots \\ 0 & 0 & 0 & 0 & \cdots & 0 & 0 & 1 & 0 \\ 0 & 0 & 0 & 0 & \cdots & 0 & 0 & 0 & 1 \\ 0 & 0 & 0 & 0 & \cdots & 0 & 0 & 1 & 0 \\ 0 & 0 & 0 & 0 & \cdots & 0 & 0 & 0 & 1 \end{bmatrix}_{(2n+2)\times(2n+2)} \tag{48}$$

According to (45)–(48), the dynamic mapping of the discrete system can be determined as follows:

$$\left(\mathbf{I}_2 + \mathbf{C}_2 + \frac{a_p h}{3}\mathbf{D}_2\right)\begin{bmatrix} \mathbf{x}(t_1) \\ \mathbf{x}(t_2) \\ \vdots \\ \mathbf{x}(t_{n+1}) \end{bmatrix} = \left(\frac{a_p h}{3}\mathbf{D}_2 + \mathbf{E}_2\right)\begin{bmatrix} \mathbf{x}(t_1 - T) \\ \mathbf{x}(t_2 - T) \\ \vdots \\ \mathbf{x}(t_{n+1} - T) \end{bmatrix} \tag{49}$$

Therefore, the transition matrix of the dynamic system in a tool pass cycle is obtained as follows:

$$\mathbf{\Phi}_2 = \left(\mathbf{I}_2 + \mathbf{C}_2 + \frac{a_p h}{3}\mathbf{D}_2\right)^{-1}\left(\frac{a_p h}{3}\mathbf{D}_2 + \mathbf{E}_2\right) \tag{50}$$

4.3. Simulation and Analysis

4.3.1. Single-DOF Milling Model

A benchmark example of the single-DOF milling model is utilized to validate and analyze S38M. Its state-space expression is introduced as (31), where

$$\mathbf{A} = \begin{bmatrix} 0 & 1 \\ -\omega_n^2 & -2\zeta\omega_n \end{bmatrix} \tag{51}$$

$$\mathbf{B}(t) = \begin{bmatrix} 0 & 0 \\ -\frac{h(t)}{m_t} & 0 \end{bmatrix} \tag{52}$$

To analyze the convergence rate of S38M, the radial immersion ratio a/D is set as 1 to avoid intermittent milling. The depths of cut are set as 0.0003 m, 0.0008 m, and 0.0015 m. Figure 7 shows the convergence rate comparisons of the S38M, CCM, SDM, and SEM when the spindle speed is 8000 rpm. It is worth noting that the S38M exhibits the highest convergence rate compared with other chatter analysis methods. According to the enlarged figures on the right, the S38M has a good convergence rate even when the discrete number is large. Additionally, the results show that the convergence rate of the SEM fluctuates evidently, and the SEM is not as stable as the other methods. To further analyze the convergence rate of the S38M, the spindle speed is raised to 10,000 rpm, and the other parameters remain unchanged. Figure 8 shows the resultant convergence rates. It can be seen that these results are consistent with the previous results.

The low-immersion condition can be utilized to verify the stability of the convergence rate [25]. The radial immersion ratio is set as 0.05; that is, a/D = 0.05. The depths of cut are set as 0.0001 m, 0.0002 m, and 0.0003 m. The spindle speed was set to 5000 rpm. Figure 9 shows the convergence rate comparisons of S38M, CCM, SDM, and SEM under low immersion. The convergence rates arranged from high to low are in the following order: S38M, CCM, SDM, and SEM. Moreover, it is seen that the convergence rate of SEM fluctuated remarkably.

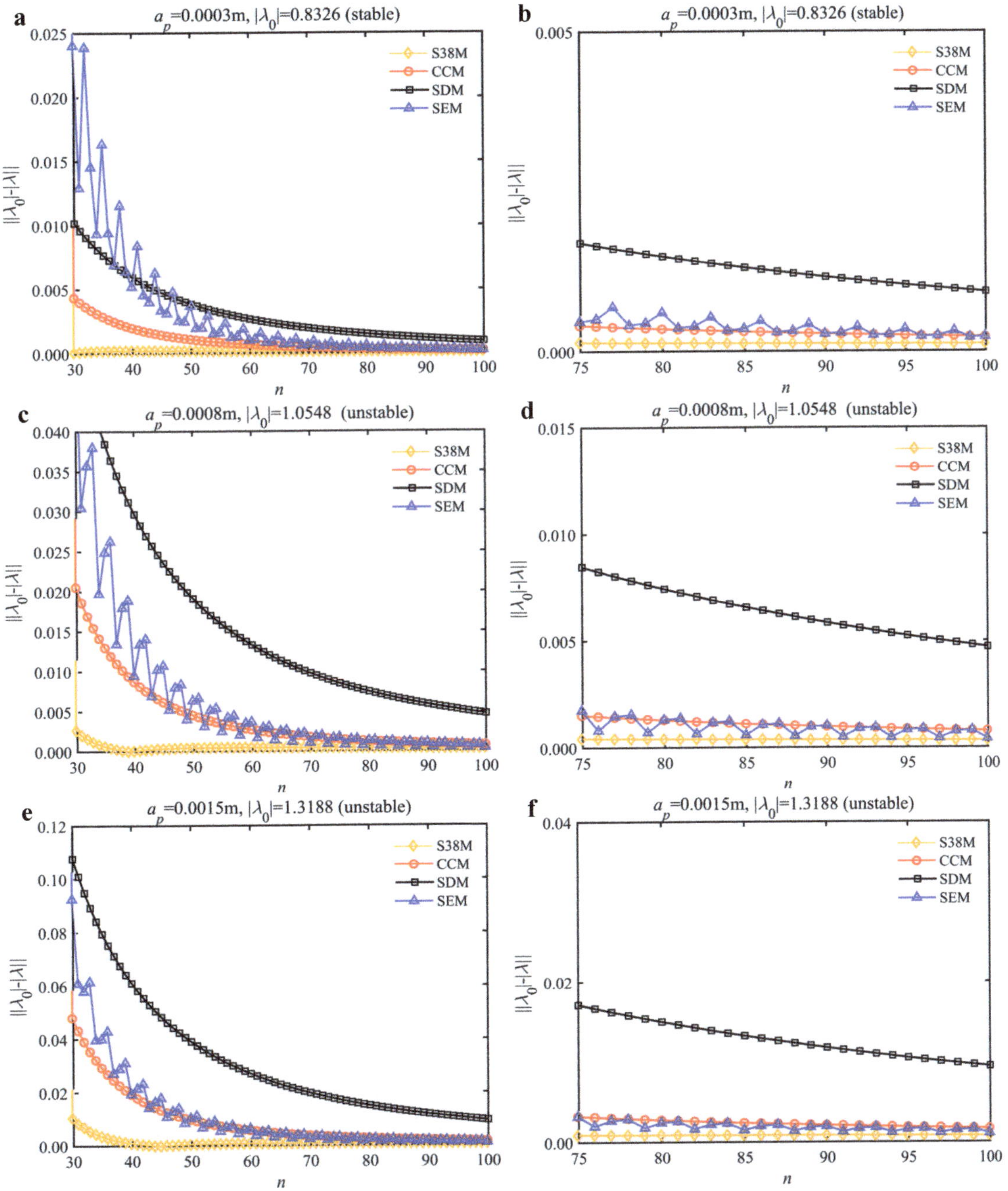

Figure 7. Convergence rate comparisons of the S38M, CCM, SDM, and SEM when v = 8000 rpm and $a/D = 1$. (**a**) a_p = 0.0003 m and $n \in$ [30, 100]; (**b**) a_p = 0.0003 m and $n \in$ [75, 100]; (**c**) a_p = 0.0008 m and $n \in$ [30, 100]; (**d**) a_p = 0.0008 m and $n \in$ [75, 100]; (**e**) a_p = 0.0015 m and $n \in$ [30, 100]; (**f**) a_p = 0.0015 m and $n \in$ [75, 100].

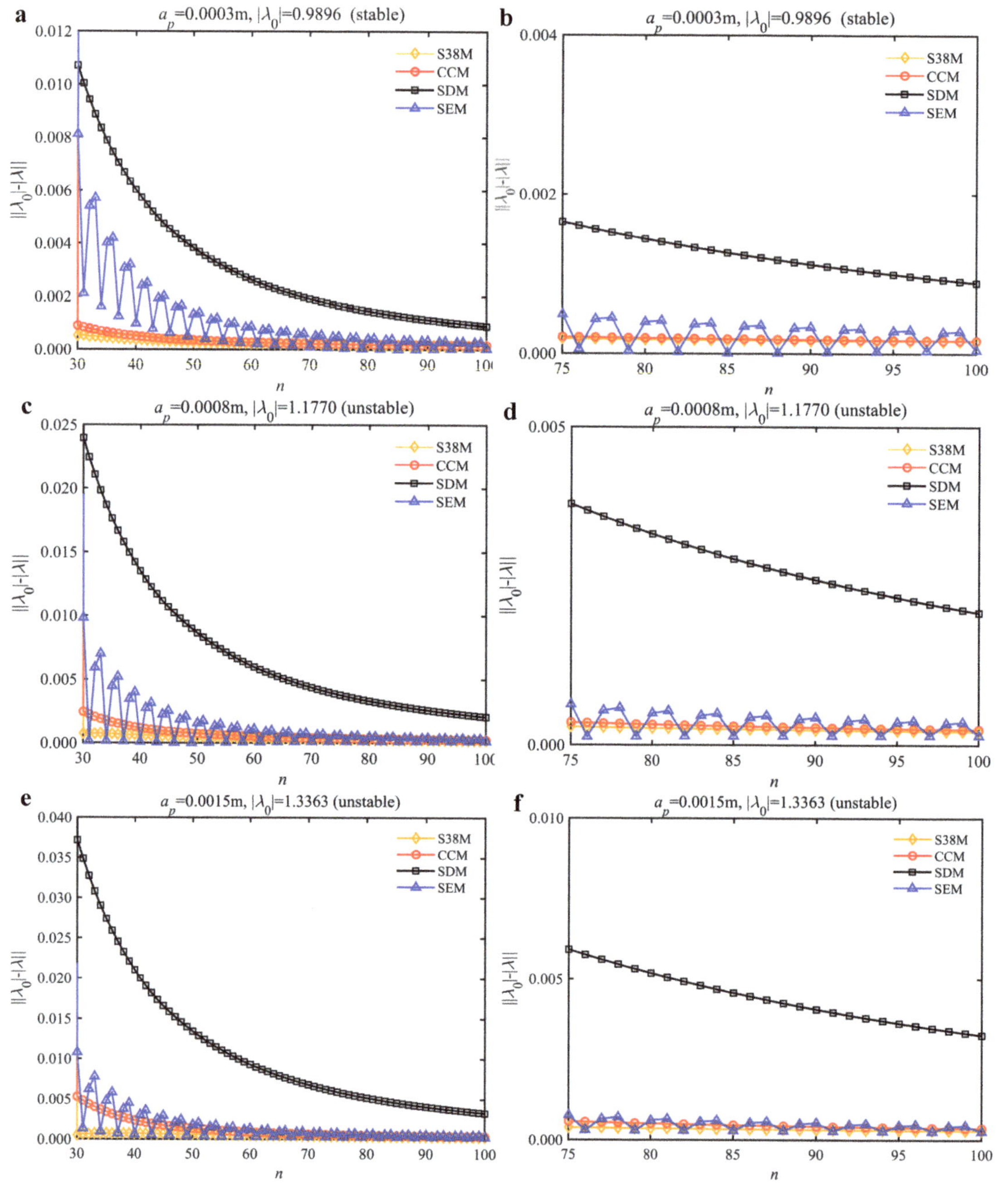

Figure 8. Convergence rate comparisons of the S38M, CCM, SDM, and SEM when v = 10,000 rpm and a/D = 1. (**a**) a_p = 0.0003 m and $n \in [30, 100]$; (**b**) a_p = 0.0003 m and $n \in [75, 100]$; (**c**) a_p = 0.0008 m and $n \in [30, 100]$; (**d**) a_p = 0.0008 m and $n \in [75, 100]$; (**e**) a_p = 0.0015 m and $n \in [30, 100]$; (**f**) a_p = 0.0015 m and $n \in [75, 100]$.

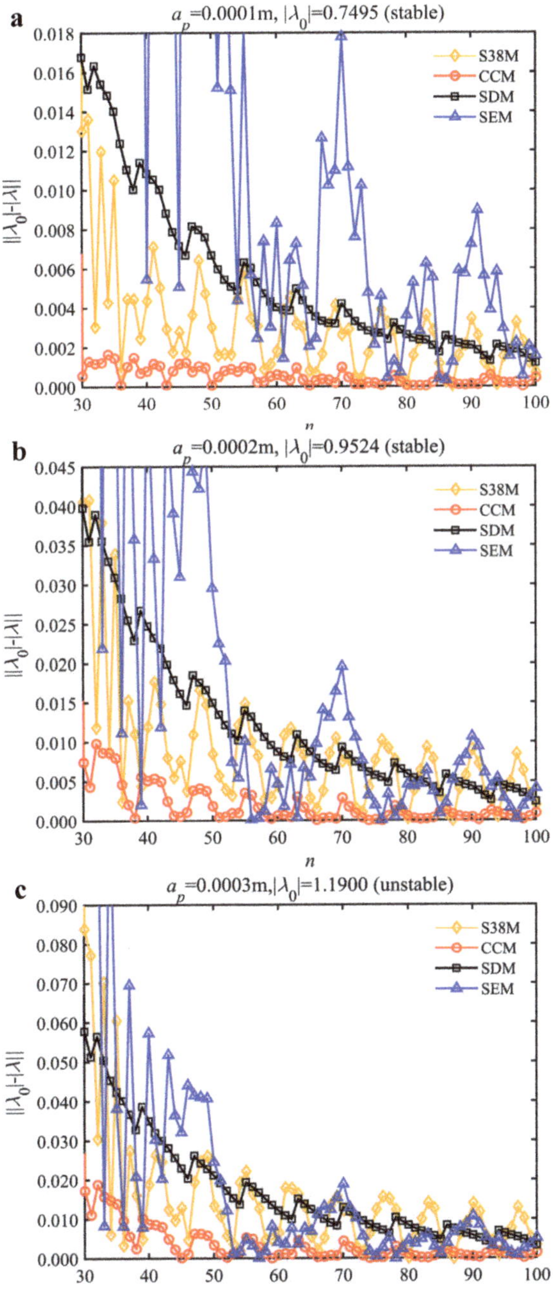

Figure 9. Convergence rate comparisons of the S38M, CCM, SDM, and SEM when and a/D = 0.05. (**a**) a_p = 0.0001 m; (**b**) a_p = 0.0002 m; (**c**) a_p = 0.0003 m.

The parameters in Section 3.2 are adopted for the S38M. Table 4 lists the computation time of the S38M. It is seen from the table that the S38M has a high calculation accuracy when a/D = 1. However, as a/D decreases, the calculation accuracy of the S38M is not as accurate as that of the CCM. The reason lies in that the interpolation of the S38M, which only uses three points, cannot reach a high calculation accuracy, while the interpolation of the CCM uses five points to interpolate, which made it calculate the smaller interval. The *AMRE* and *MSE* are shown in Figure 10. Notice that the calculation accuracy of the S38M and CCM is higher than those of the SDM and SEM when a/D = 0.05, and the calculation accuracy of the S38M is higher than those of the CCM, SDM, and SEM when a/D = 1. The reason for the former lies in that some large fluctuations exist in the S38M. The computational time of the above methods is compared in Figure 11. It is observed from the figure that compared with the SDM and SEM, the computational time can be significantly decreased with the S38M and CCM. Although the computational time of the S38M is almost equal to that of the CCM, the S38M has a higher convergence rate when the radial immersion ratio was large.

As described above, the fluctuations appear in Figure 9. It is found that for the S38M, the convergence rate is often a local maximum when the discrete number is an odd one; the convergence rate is often a local minimum when the discrete number is an even one. To further study this problem, the discrete numbers are set as 30, 40, 50, 60, 70, and 80. The corresponding *AMRE* and *MSE* are plotted in Figure 12. The results show that the S38M attains a better computational accuracy when the discrete number is an even one.

Table 4. Stability lobe diagrams of the S38M.

n	25	40	55
a/D = 0.05	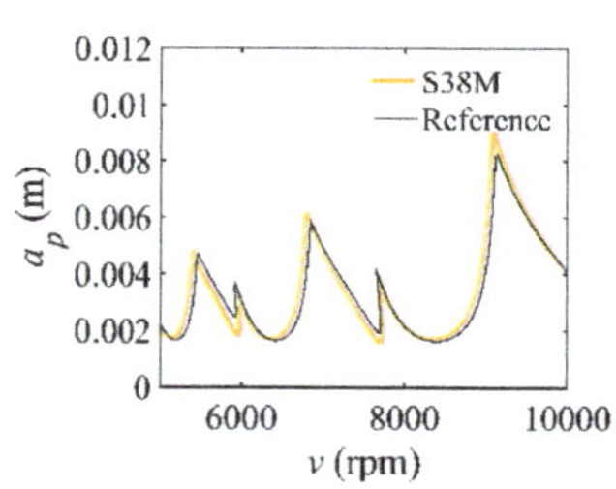	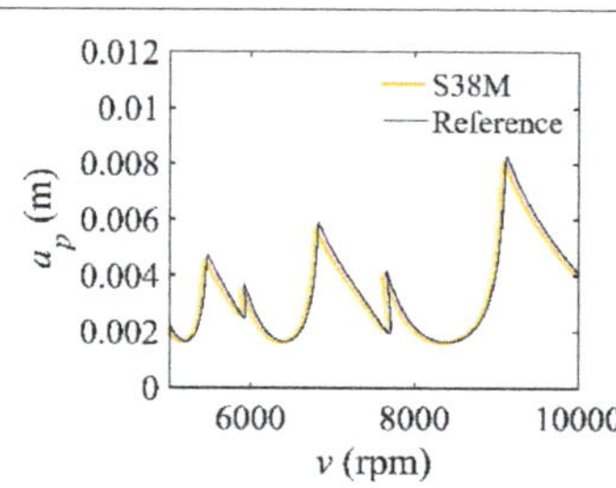	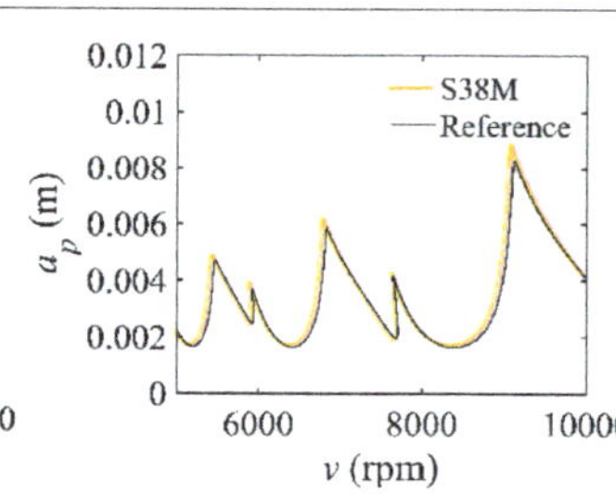
a/D = 0.5	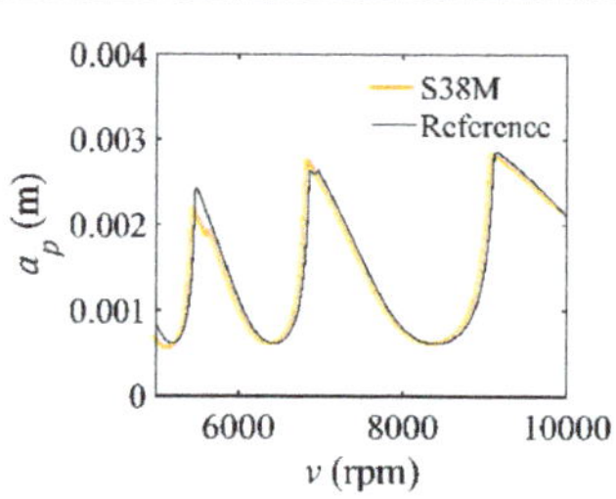	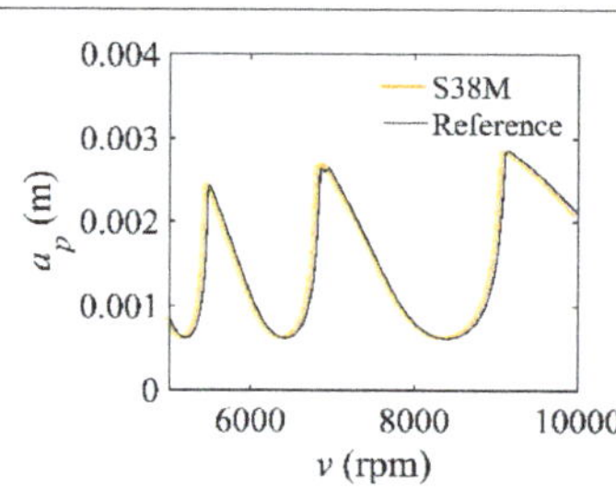	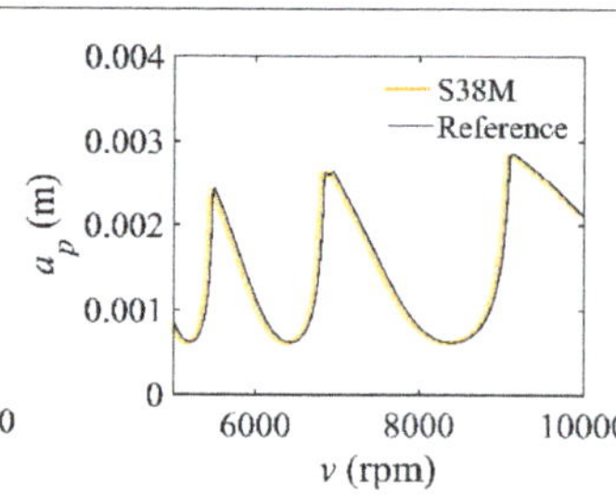
a/D = 1	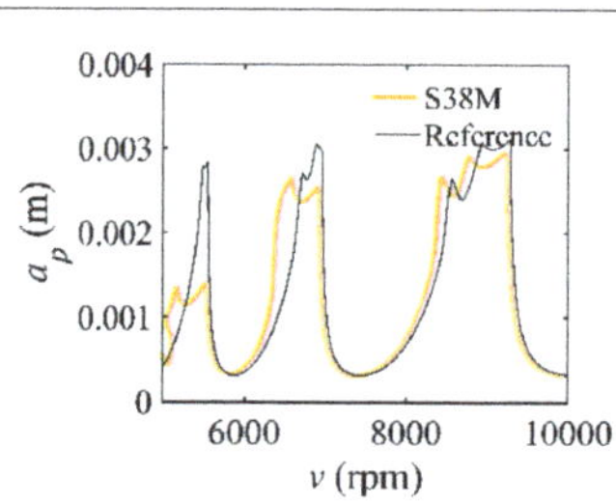	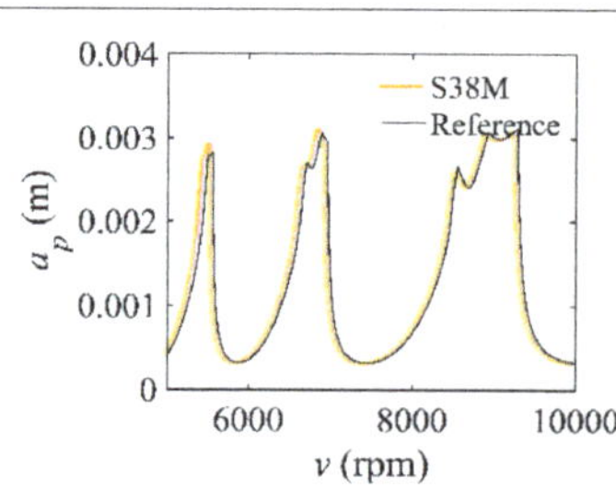	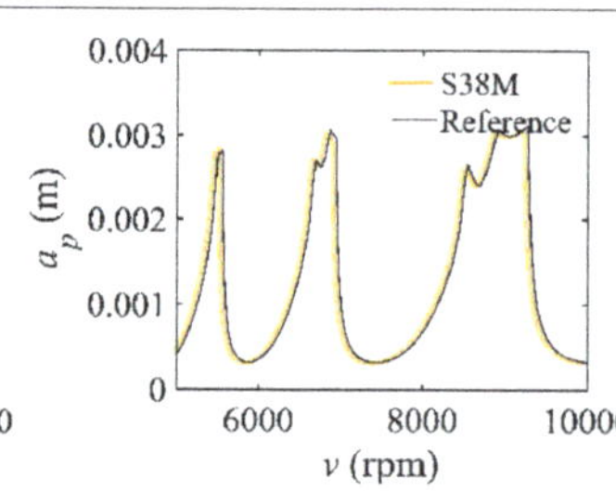

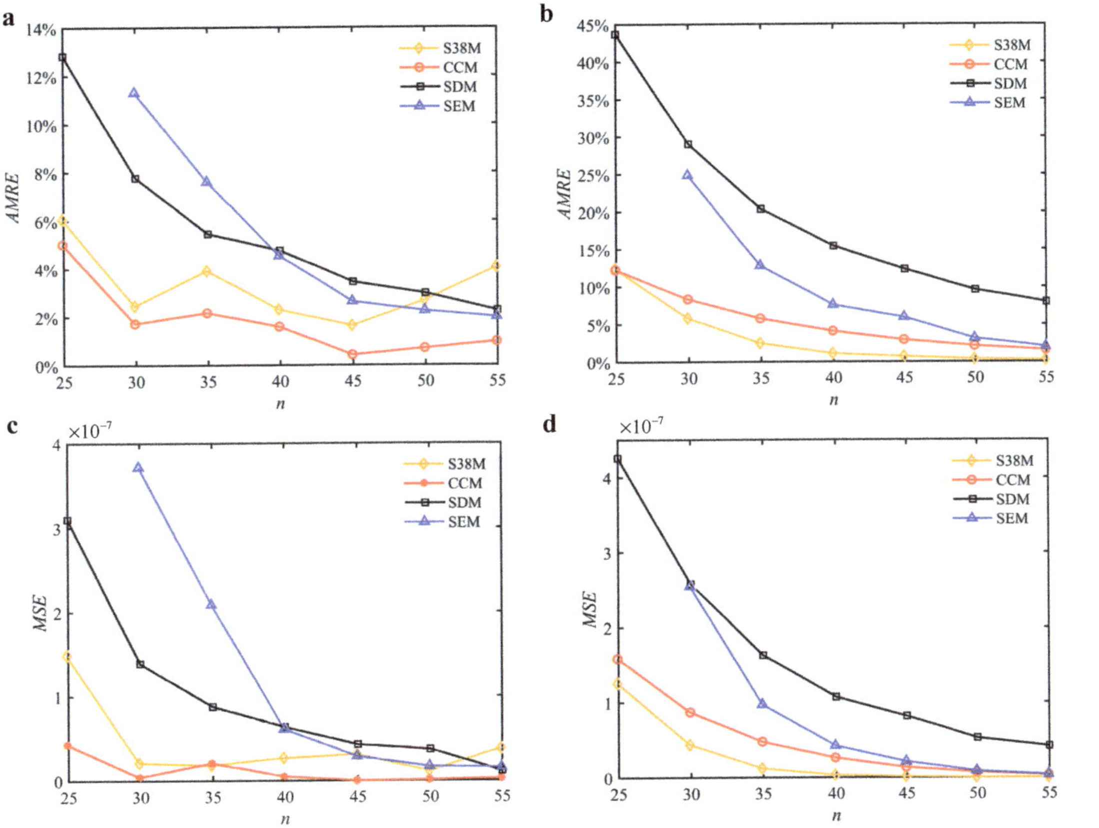

Figure 10. *AMRE* and *MSE* of the S38M, CCM, SDM, and SEM. (**a**) *AMRE* when $a/D = 0.05$; (**b**) *AMRE* when $a/D = 1$; (**c**) *MSE* when $a/D = 0.05$; (**d**) *MSE* when $a/D = 1$.

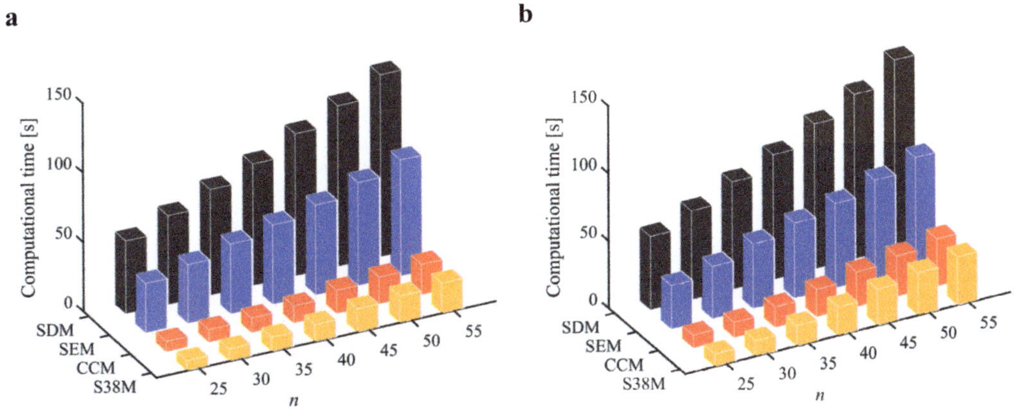

Figure 11. Computational time of the S38M, CCM, SDM, and SEM. (**a**) $a/D = 0.05$; (**b**) $a/D = 1$.

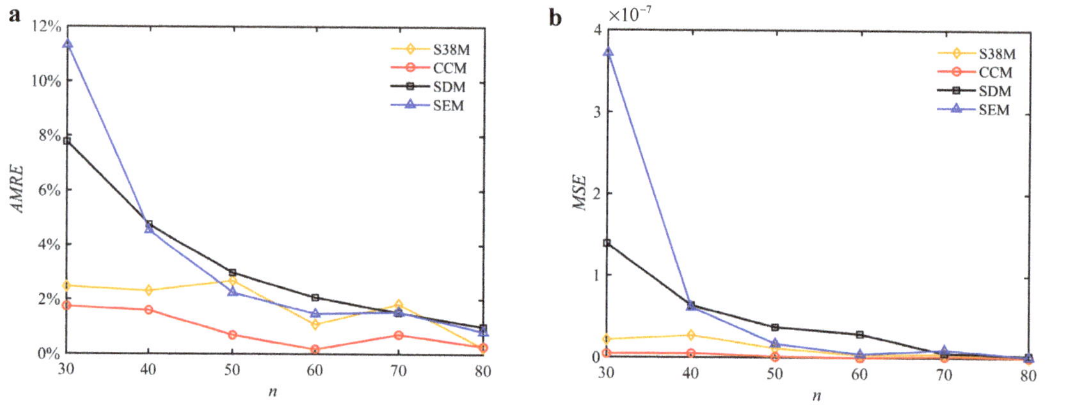

Figure 12. *AMRE* and *MSE* of the S38M for the fluctuation analysis when $a/D = 0.05$. (**a**) *AMRE*; (**b**) *MSE*.

4.3.2. Two-DOF Milling Model

The two-DOF milling mathematical model is expressed as follows [17]:

$$\begin{aligned}&\begin{bmatrix} m_t & 0 \\ 0 & m_t \end{bmatrix}\begin{bmatrix} \ddot{x}(t) \\ \ddot{y}(t) \end{bmatrix} + \begin{bmatrix} 2m_t\zeta\omega_n & 0 \\ 0 & 2m_t\zeta\omega_n \end{bmatrix}\begin{bmatrix} \dot{x}(t) \\ \dot{y}(t) \end{bmatrix} + \begin{bmatrix} m_t\omega_n^2 & 0 \\ 0 & m_t\omega_n^2 \end{bmatrix}\begin{bmatrix} x(t) \\ y(t) \end{bmatrix} \\ &= -a_p\begin{bmatrix} h_{xx}(t) & h_{xy}(t) \\ h_{yx}(t) & h_{yy}(t) \end{bmatrix} \times \left\{\begin{bmatrix} x(t) \\ y(t) \end{bmatrix} - \begin{bmatrix} x(t-T) \\ y(t-T) \end{bmatrix}\right\}\end{aligned} \tag{53}$$

where the tool is symmetrical by default. m_t, ζ, and ω_n in the x-direction are identical to those in the y-direction. $h_{xx}(t)$, $h_{xy}(t)$, $h_{yx}(t)$, and $h_{yy}(t)$ are the same as those defined in (3)–(6).

A new state vector is defined as $\dot{\mathbf{r}}(t) = [x(t), y(t), \dot{x}(t), \dot{y}(t)]^T$. Then, (53) can be expressed as follows:

$$\dot{\mathbf{r}}(t) = \mathbf{A}\mathbf{r}(t) + a_p\mathbf{B}(t)[\mathbf{r}(t) - \mathbf{r}(t-T)] \tag{54}$$

where

$$\mathbf{A} = \begin{bmatrix} 0 & 0 & 1 & 0 \\ 0 & 0 & 0 & 1 \\ -\omega_n^2 & 0 & -2\zeta\omega_n & 0 \\ 0 & -\omega_n^2 & 0 & -2\zeta\omega_n \end{bmatrix} \tag{55}$$

and

$$\mathbf{B} = \begin{bmatrix} 0 & 0 & 0 & 0 \\ 0 & 0 & 0 & 0 \\ -h_{xx}(t)/m_t & -h_{xy}(t)/m_t & 0 & 0 \\ -h_{yx}(t)/m_t & -h_{yy}(t)/m_t & 0 & 0 \end{bmatrix} \tag{56}$$

The following deduction is the same as that in Section 4.1.

The system parameters of the two-DOF milling model are identical to those of the single-DOF model. The radial immersion ratios are set as 0.05, 0.5, and 1. The figures shown in Table 5 are obtained over a 100×50-sized grid of the system parameters, that is, the spindle speed and the depth of cut. The thin, black, solid line denotes the reference stability lobe diagrams obtained using the SDM when the discrete number is 200 [26]. We observe that the accuracy of the stability lobe diagrams of the S38M is better than that of SDM. Additionally, the computation time is remarkably saved using the S38M.

Table 5. Comparisons of S38M and SDM for the two-DOF milling model.

	S38M	SDM
$a/D = 0.05$	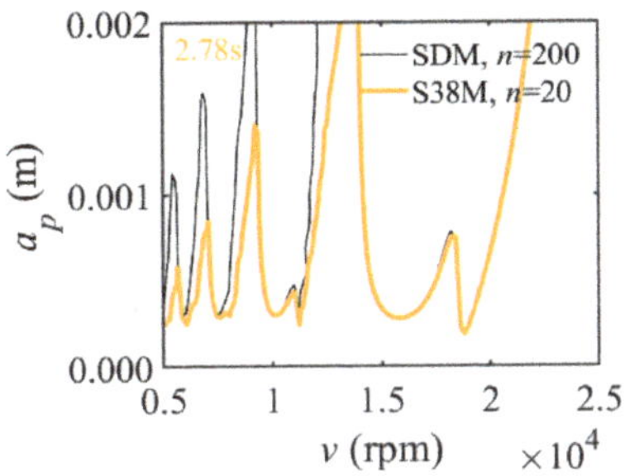	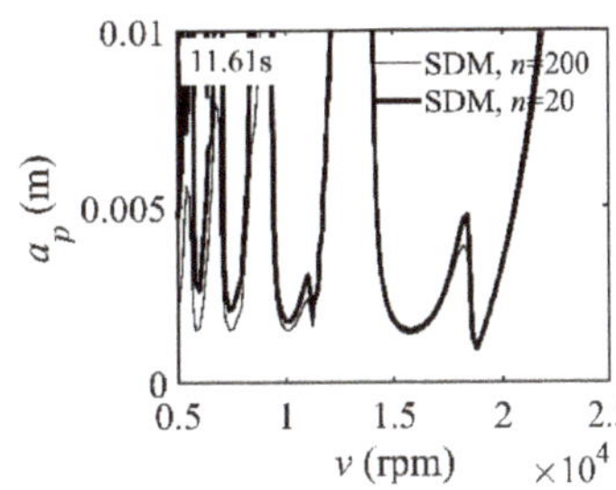
$a/D = 0.5$	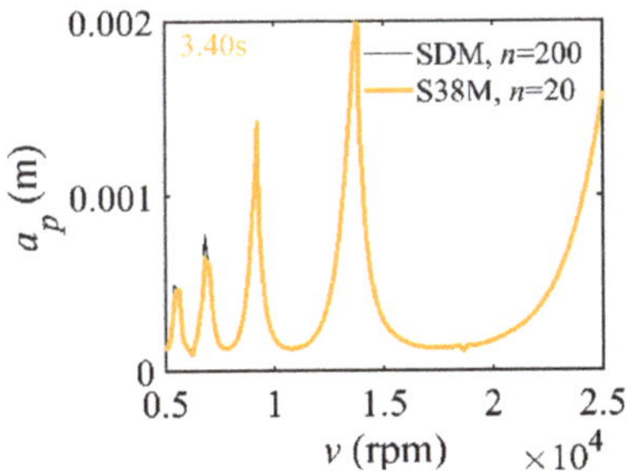	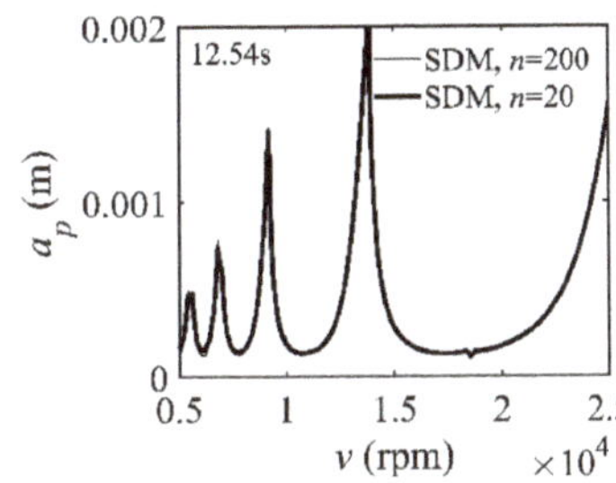
$a/D = 1$	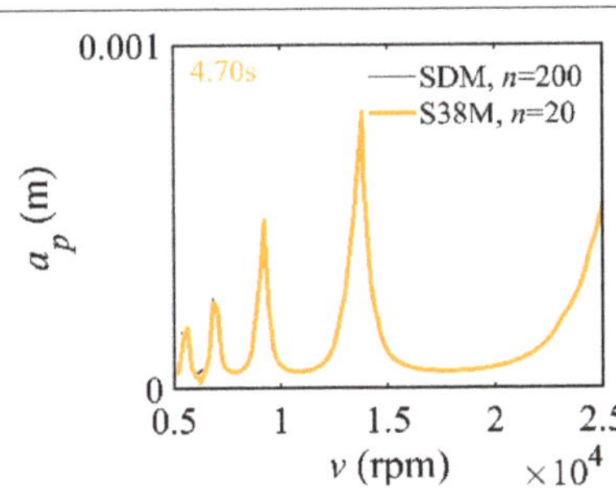	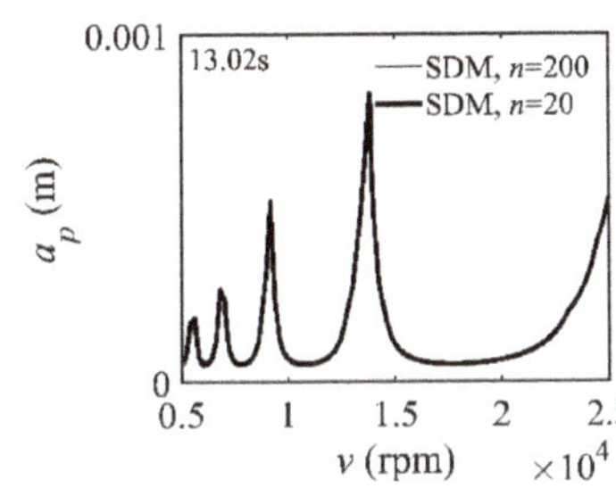

5. Conclusions

In this study, we focused on the prediction of milling stability using composite Cotes-based and Simpson's 3/8-based methods. First, the composite Cotes-based method is proposed for preventing chatter in milling processes. The three steps for obtaining the milling stability lobe diagrams are as follows: (1) establish the time-delay differential equation while considering the regenerative effects; (2) determine the transition matrix using the integral equations; (3) analyze the system stability according to the Floquet theory. To measure the calculation accuracy, the *AMRE* and *MSE* are adopted in our work. The results demonstrate that for the single-DOF model, when the discrete number is 40, the *AMRE* and *MSE* can be respectively reduced from 0.076 to 0.041 and from 4.24×10^{-8} to 2.62×10^{-8}, and the calculation time can be reduced from 55.63s to 20.21s by using the proposed composite Cotes-based method. In addition, the Simpson's 3/8-based method is proposed to further improve the calculation efficiency. The results demonstrate that for the single-DOF model, the proposed Simpson's 3/8-based method significantly improves the convergence speed while sharing the same computation time with the composite Cotes-based method when the radial immersion ratio is large; for the two-DOF model, the accuracy of the stability lobe diagrams of the Simpson's 3/8-based method are better than those of the SDM, and the computation time is remarkably saved using the Simpson's 3/8-based method.

Author Contributions: Conceptualization, X.D. and J.Z.; methodology, X.D.; software, P.R.; validation, X.D., P.R. and J.Z.; formal analysis, P.R.; investigation, X.D.; resources, X.D.; data curation, P.R.; writing—original draft preparation, X.D.; writing—review and editing, J.Z.; visualization, X.D.; supervision, J.Z.; project administration, J.Z.; funding acquisition, J.Z. All authors have read and agreed to the published version of the manuscript.

Funding: This research was funded by [National Natural Science Foundation of China] grant number [52005142] and [Open Fund of State Key Laboratory of Robotics and System] grant number [SKLRS-2021-KF-08]. The authors would like to thank the Mechatronic Institute, Zhejiang Sci-Tech University.

Institutional Review Board Statement: Not applicable.

Informed Consent Statement: Not applicable.

Data Availability Statement: The data presented in this study are available on request from the corresponding author.

Conflicts of Interest: The authors declare no conflict of interest.

References

1. Ma, J.; Li, Y.; Zhang, D.; Zhao, B.; Wang, G.; Pang, X. A Novel Updated Full-Discretization Method for Prediction of Milling Stability. *Micromachines* **2022**, *13*, 160. [CrossRef] [PubMed]
2. Yan, Z.; Zhang, C.; Jia, J.; Ma, B.; Jiang, X.; Wang, D.; Zhu, T. High-order semi-discretization methods for stability analysis in milling based on precise integration. *Precis. Eng.* **2021**, *73*, 71–92. [CrossRef]
3. Catania, G.; Mancinelli, N. Theoretical–experimental modeling of milling machines for the prediction of chatter vibration. *Int. J. Mach. Tools Manuf.* **2011**, *51*, 339–348. [CrossRef]
4. Fu, Z.; Zhang, X.; Wang, X.; Yang, W. Analytical modeling of chatter vibration in orthogonal cutting using a predictive force model. *Int. J. Mech. Sci.* **2014**, *88*, 145–153. [CrossRef]
5. Gradišek, J.; Kalveram, M.; Insperger, T.; Weinert, K.; Stepan, G.; Govekar, E.; Grabec, I. On stability prediction for milling. *Int. J. Mach. Tools Manuf.* **2005**, *45*, 769–781. [CrossRef]
6. Wu, S.; Li, R.; Liu, X.; Yang, L.; Zhu, M. Experimental Study of Thin Wall Milling Chatter Stability Nonlinear Criterion. *Procedia CIRP* **2016**, *56*, 422–427. [CrossRef]
7. Davies, M.; Pratt, J.; Dutterer, B.; Burns, T. The Stability of Low Radial Immersion Milling. *CIRP Ann.* **2000**, *49*, 37–40. [CrossRef]
8. Altintaş, Y.; Budak, E. Analytical Prediction of Stability Lobes in Milling. *CIRP Ann.* **1995**, *44*, 357–362. [CrossRef]
9. Bayly, P.V.; Mann, B.P.; Schmitz, T.L.; Peters, D.A.; Stepan, G.; Insperger, T. Effects of Radial Immersion and Cutting Direction on Chatter Instability in End-Milling. *Manufacturing* **2002**, *3641*, 351–363. [CrossRef]
10. Insperger, T.; Stépán, G. Updated semi-discretization method for periodic delay-differential equations with discrete delay. *Int. J. Numer. Methods Eng.* **2010**, *61*, 117–141. [CrossRef]
11. Ding, Y.; Zhu, L.; Zhang, X.; Ding, H. A full-discretization method for prediction of milling stability. *Int. J. Mach. Tools Manuf.* **2010**, *50*, 502–509. [CrossRef]
12. Insperger, T. Full-discretization and semi-discretization for milling stability prediction: Some comments. *Int. J. Mach. Tools Manuf.* **2010**, *50*, 658–662. [CrossRef]
13. Ding, Y.; Zhu, L.; Zhang, X.; Ding, H. Numerical Integration Method for Prediction of Milling Stability. *J. Manuf. Sci. Eng.* **2011**, *133*, 031005. [CrossRef]
14. Lu, Y.; Ding, Y.; Peng, Z.; Chen, Z.C.; Zhu, L. A spline-based method for stability analysis of milling processes. *Int. J. Adv. Manuf. Technol.* **2017**, *89*, 2571–2586. [CrossRef]
15. Qin, C.; Tao, J.; Shi, H.; Xiao, D.; Li, B.; Liu, C. A novel Chebyshev-wavelet-based approach for accurate and fast prediction of milling stability. *Precis. Eng.* **2020**, *62*, 244–255. [CrossRef]
16. Liu, C.; Tang, D.; Li, S.; Ding, G. Simpson's 3/8–based method stability analysis for milling processes. *Int. J. Adv. Manuf. Technol.* **2021**, *114*, 671–682. [CrossRef]
17. Niu, J.; Ding, Y.; Zhu, L.; Ding, H. Runge–Kutta methods for a semi-analytical prediction of milling stability. *Nonlinear Dyn.* **2013**, *76*, 289–304. [CrossRef]
18. Considine, G.D.; Kulik, P.H. (Eds.) Newton-Cotes formula. In *Van Nostrand's Scientific Encyclopedia*; John Wiley & Sons: Chichester, UK, 2005. [CrossRef]
19. Odibat, Z.M. Differential transform method for solving Volterra integral equation with separable kernels. *Math. Comput. Model.* **2008**, *48*, 1144–1149. [CrossRef]
20. Berrut, J.P.; Trefethen, L.N. Barycentric lagrange Interpolation. *Siam Rev.* **2004**, *46*, 501. [CrossRef]

21. Hajdu, D.; Borgioli, F.; Michiels, W.; Insperger, T.; Stepan, G. Robust stability of milling operations based on pseudospectral approach. *Int. J. Mach. Tools Manuf.* **2019**, *149*, 103516. [CrossRef]
22. Qin, C.; Tao, J.; Liu, C. A predictor-corrector-based holistic-discretization method for accurate and efficient milling stability analysis. *Int. J. Adv. Manuf. Technol.* **2018**, *96*, 2043–2054. [CrossRef]
23. Niu, J.; Ding, Y.; Zhu, L.; Ding, H. Stability Analysis of Milling Processes With Periodic Spindle Speed Variation Via the Variable-Step Numerical Integration Method. *J. Manuf. Sci. Eng.* **2016**, *138*, 114501. [CrossRef]
24. Zhang, Z.; Li, H.; Meng, G.; Liu, C. A novel approach for the prediction of the milling stability based on the Simpson method. *Int. J. Mach. Tools Manuf.* **2015**, *99*, 43–47. [CrossRef]
25. Merdol, S.D.; Altintas, Y. Multi Frequency Solution of Chatter Stability for Low Immersion Milling. *J. Manuf. Sci. Eng.* **2004**, *126*, 459–466. [CrossRef]
26. Tang, X.; Peng, F.; Yan, R.; Gong, Y.; Li, Y.; Jiang, L. Accurate and efficient prediction of milling stability with updated full-discretization method. *Int. J. Adv. Manuf. Technol.* **2016**, *88*, 2357–2368. [CrossRef]

micromachines

MDPI

Article

Investigation on the Effect of Annealing Temperature on the Side Ohmic Contact Characteristics for Double Channel GaN/AlGaN Epitaxial Layer

Qingzhi Meng [1,†], Qijing Lin [1,2,3,4,*,†], Weixuan Jing [1], Na Zhao [1], Ping Yang [1] and Dejiang Lu [1]

1 State Key Laboratory for Manufacturing Systems Engineering, Xi'an Jiaotong University, Xi'an 710049, China; qzmeng2016@foxmail.com (Q.M.); wxjing@mail.xjtu.edu.cn (W.J.); zn2015@stu.xjtu.edu.cn (N.Z.); ipe@xjtu.edu.cn (P.Y.); djlu@xjtu.edu.cn (D.L.)

2 Collaborative Innovation Center of High-End Manufacturing Equipment, Xi'an Jiaotong University, Xi'an 710054, China

3 School of Mechanical and Manufacturing Engineering, Xiamen Institute of Technology, Xiamen 361021, China

4 Chongqing Key Laboratory of Micro-Nano Systems and Intelligent Sensing, Chongqing Academician Workstation, Chongqing 2011 Collaborative Innovation Center of Micro/Nano Sensing and Intelligent Ecological Internet of Things, Chongqing Technology and Business University, Chongqing 400067, China

* Correspondence: qjlin2015@xjtu.edu.cn

† These authors contributed equally to this work.

Abstract: A side ohmic contact mode for the double channel GaN/AlGaN epitaxial layer is proposed in this paper. Rectangle transmission line model (TLM) electrodes are prepared, and the specific contact resistance is tested at the annealing temperatures from 700 °C to 850 °C. The results show that the minimum specific contact resistance is $2.58 \times 10^{-7}\ \Omega\cdot cm^2$ at the annealing temperature of 750 °C, which is three to four times lower than the surface contact mode. Scanning electron microscope (SEM), energy dispersive spectrometer (EDS), and atomic force microscope (AFM) were carried out for the analysis of the morphology, element composition, and the height fluctuation at the contact edge. With the increase in the annealing temperature, the specific contact resistance decreases due to the alloying of electrodes and the raised number of N vacancies. However, when the annealing temperature exceeds 800 °C, the state of the stress in the electrode films transforms from compressive stress to tensile stress. Besides, the volume expansion of metal electrode film and the increase in the roughness at the contact edge leads to the degradation of the side ohmic contact characteristics.

Keywords: side ohmic contact; transmission line model; specific contact resistance; annealing temperatures

Citation: Meng, Q.; Lin, Q.; Jing, W.; Zhao, N.; Yang, P.; Lu, D. Investigation on the Effect of Annealing Temperature on the Side Ohmic Contact Characteristics for Double Channel GaN/AlGaN Epitaxial Layer. *Micromachines* **2022**, *13*, 791. https://doi.org/10.3390/mi13050791

Academic Editors: Youqiang Xing, Xiuqing Hao and Duanzhi Duan

Received: 3 April 2022
Accepted: 16 May 2022
Published: 19 May 2022

1. Introduction

With the development of microwave communication technology, higher requirements are put forward for the frequency and power characteristics of microelectronic devices. GaN-based devices, such as GaN/AlGaN high electron mobility transistors (HEMT) and GaN diodes, have been widely used as power devices because of their high breakdown voltage [1–3] and operation frequency [4–6]. Although GaN devices have many advantages, they are still suffering from some problems such as short channel effect [7–9] and current collapse effect [10–12]. To solve these problems, some researchers proposed double channel GaN HEMTs [13–18], which contain two GaN/AlGaN heterojunction conductive channels. For devices with such structures, the current collapse effect is significantly weakened, as the lower barrier layer is far from the surface of the epitaxial layer. Besides, since double channel GaN HEMTs have a stronger confinement ability of carriers, they have higher breakdown voltage, and the carrier transport capacity is almost twice that of single channel devices [16,17].

However, the introduction of an additional channel will simultaneously involve some defects. For GaN HEMTs, the ohmic contact between a source/drain electrode and epitaxial layer is very important, as it will greatly affect the output characteristics of the device. When a positive bias is applied between source and drain, electrons firstly tunneling from the source to the conductive channel are then driven to the drain terminal. In this process, the tunneling probability, depending on the height of the barrier, plays an important role. Especially for double channel GaN HEMTs, the lower channel and source electrode are separated by two AlGaN barrier layers and a GaN channel layer, which increases the barrier height and decreases the tunneling probability of electrons. To overcome this problem, researchers have put forward the side ohmic contact mode. Metal electrodes are deposited on the side of the epitaxial layer to put electrodes directly in contact with the 2DEG conductive channel, which greatly reduces the contact resistance and improves the tunneling probability of electrons. This contact mode has already been used in single channel GaN HEMTs [19–21]. Luan et al. investigated the polarization Coulomb field (PCF) scattering in AlGaN/AlN/GaN HEMT [19] and InGaN/AlN/GaN HEMT [20] with side ohmic contact and found that the PCF scattering was greatly weakened in side ohmic contact mode compared with normal surface contact mode, and the two-dimensional electron gas (2DEG) density was also enhanced. Wang et al. [21] studied the effect of the etching process on the side ohmic contact performance and revealed that the carrier transport is more efficient in side ohmic contact than the surface contact mode. It is believed this contact mode could also be utilized in double channel GaN HEMTs to improve the tunneling probability, especially in the lower channel, to decrease the specific contact resistivity. Furthermore, if the side ohmic contact mode can improve the output characteristics of GaN HEMTs in the RF circuit, the power amplification performance would be promoted [22]. Some photodetectors [23–25] or sensors [26–28] based on GaN HEMTs also have high requirements for their output characteristics. For instance, the background noise voltage of GaN HEMT terahertz detector is related to the resistance of the source to drain conductive channel, and the outstanding ohmic contact characteristics can obtain smaller noise equivalent power [23]. For a GaN HEMT hydrogen sensor, the variation of drain current I_{ds} depends on the concentration of hydrogen [28], so the better the ohmic contact of drain and source electrodes is, the higher the sensitivity of the sensor will be. Therefore, it is necessary to study the side ohmic contact characteristics between the double channel GaN/AlGaN epitaxial layer and metal electrodes to improve the output performance of double channel GaN HEMTs. For the conventional surface contact mode, the annealing condition for the ohmic contact between a metal electrode and GaN substrate is, basically, explicitly to rapidly anneal for 30 s at the temperature of 850 °C. However, the mechanism of the side contact mode is different from the surface contact mode, the annealing temperature for the formation of ohmic contact is different. Besides, as the depth of the groove structure of side ohmic contact mode for the double channel GaN/AlGaN epitaxial layer is deeper than single channel, it may cause the lift-off of the electrode from the side of the epitaxial layer and the degradation of the ohmic contact, which is a challenge for the preparation of the side ohmic contact of the double channel GaN/AlGaN epitaxial layer and/or double channel GaN HEMT. The motivation of this paper is to explore the optimal annealing temperature of side ohmic contact mode for the double channel GaN epitaxial layer to obtain the lowest specific contact resistance between a metal electrode and the epitaxial layer.

In this paper, side ohmic contact electrodes were prepared on the double channel GaN/AlGaN epitaxial layer by magnetron sputtering technology, and they were annealed at temperatures from 700 °C to 850 °C. The specific contact resistance was tested by the transmission line model (TLM). The analysis of metal electrode films was carried out by a scanning electron microscope (SEM), an energy dispersive spectrometer (EDS), and an atomic force microscope (AFM), and the side ohmic contact performance at different annealing temperature was investigated.

2. The Measurement Method of the Specific Contact Resistance

TLM is the most commonly used method to measure the specific contact resistance. The TLM method can be divided into circular electrode TLM and rectangular electrode TLM method. For circular electrode TLM method, the metal electrode can be directly deposited onto the bulk material without the formation of the mesa, but the calculation of the specific contact resistance is complicated, while, for rectangular electrode TLM method, it is a simple and accurate way to calculate the specific contact resistance, even though an isolated mesa needs to be prepared. The schematic diagram of the rectangular electrode TLM structure is shown in Figure 1. To make electrons transfer along the horizontal direction, a rectangular isolation mesa is firstly prepared on the bulk semiconductor, and then, several rectangular metal electrodes with arithmetic sequence distance are deposited on the isolation mesa.

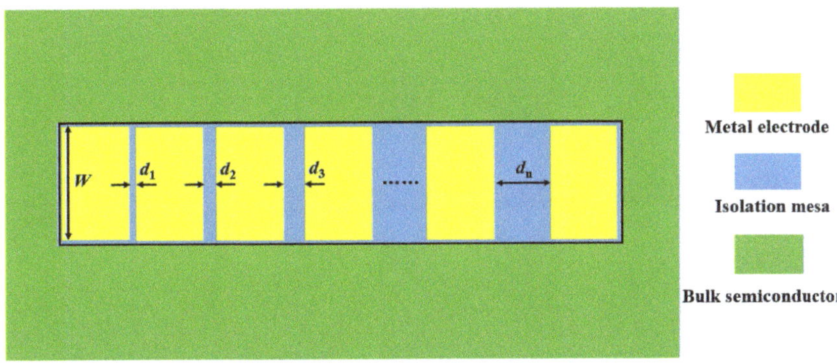

Figure 1. The schematic diagram of the rectangular electrode TLM structure.

According to the TLM theory, the total resistance between each adjacent electrode in Figure 1 is given as:

$$R_{tot} = 2R_c + R_{SH}\frac{d}{W} \quad (1)$$

where R_{tot} is the total resistance, R_{SH} is the sheet resistance of the semiconductor, $L_T = \sqrt{\rho_c/R_{SH}}$ is the transmission length, W is the width of electrodes, and d is the distance between adjacent electrodes. By measuring the I-V characteristics of electrodes with different distances, the curve of R_{tot} versus d can be plotted, as with Figure 2, where the slope of the curve is R_{SH}/W, and the intercept is $2R_{SH}L_T/W$. The R_{SH} and the L_T can be obtained, and the specific contact resistance ρ_c is calculated as $\rho_c = L_T{}^2R_{SH}$.

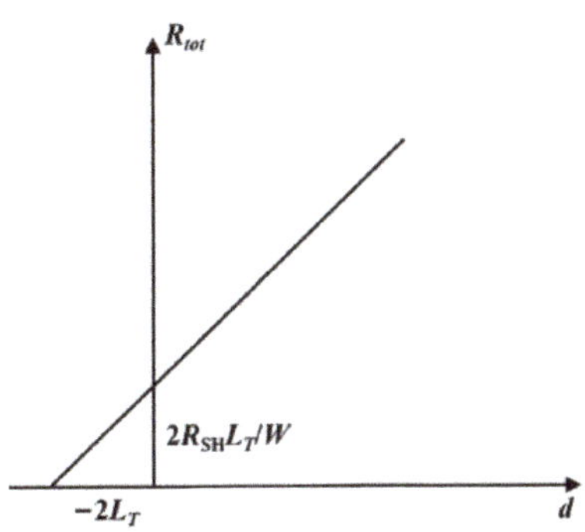

Figure 2. The curve of R_{tot} versus d. The slope of the curve is R_{SH}/W, and the intercept is $2R_{SH}L_T/W$.

3. Experimental Procedure

3.1. The Growth of Double Channel GaN/AlGaN Epitaxial Layer

Figure 3 shows the diagram of double channel GaN/AlGaN structures. GaN/AlGaN epitaxial layers were grown by metal organic chemical vapor deposition (MOCVD) technology. Triethylgallium (TEGa), trimethyl aluminium (TMAl), and ammonia (NH_3) were used as the sources of Ga, Al, and N, respectively. High-purity hydrogen (H_2) was used as the carrier gas. A 50 nm AlN nucleation layer was firstly deposited on c-axis plane (0001) sapphire substrates, and then, a 1.5µm GaN channel layer was grown, followed by a 1 nm AlN spacer and 23 nm $Al_{0.25}GaN$ barrier layer to form the lower channel. Afterwards, the upper channel was formed by the growth of 40 nm GaN channel layer, 1 nm AlN spacer, and 23 nm $Al_{0.25}GaN$ barrier layer. Finally, a 2 nm GaN cap layer is grown on the top of the epitaxial layer surface. Detailed results of the material parameters and electrical parameters for the epitaxial layer are in our previous work [29]. The average electron mobility of the epitaxial layers is 1815 $cm^2/V{\cdot}s^{-1}$, and the sheet density of carriers is $8.247 \times 10^{12}/cm^2$.

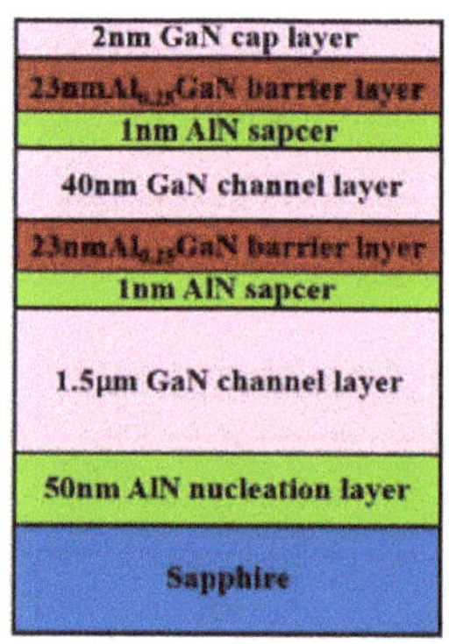

Figure 3. Diagram of the double channel epitaxial layer.

3.2. The Prepartion of TLM Electrodes

The cross section of the side ohmic contact mode is shown in Figure 4. Compared with the conventional surface contact mode, electrodes are deposited on the side of the conductive channel, instead of the surface of the epitaxial layer, to reduce the contact barrier between the electrode and the channel. For the I-V characteristics measurement of TLM electrodes, the distance between each electrode is in an arithmetic sequence.

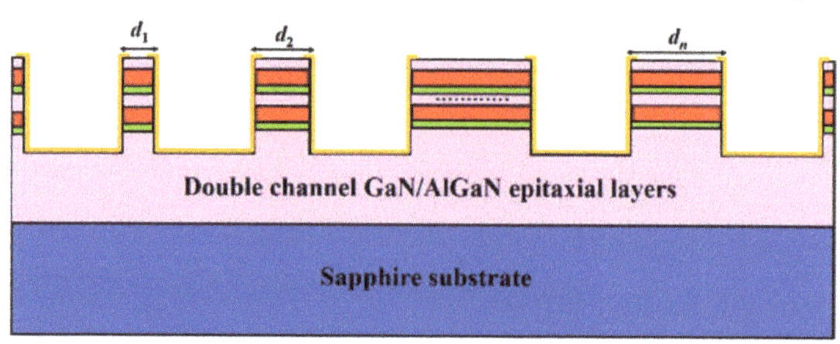

Figure 4. The cross section of the side ohmic contact mode. The pink region refers to the GaN layer, the red region refers to the AlGaN layer, and the green region refers to the AlN layer.

The isolation mesa and the groove structures were etched by inductive coupled plasma (ICP) (Corial, 210 L, Bernin, France) technology. BCl_3, Cl_2, and Ar were used as the etching gases. Main parameters used in the etching process are shown in Table 1. The Ti/Al/Ni/Au (20 nm/120 nm/40 nm/50 nm) metal film electrodes were deposited on the side of the channel using magnetron sputtering (Denton, Dsicovery635, Beijing, China) technology. Then, the samples were annealed in the rapid annealing furnace (UNTIMP, RTP-100, Pfaffenhofen, Bavaria, Germany) with N_2 atmosphere for 30 s at the temperature of 700 °C, 750 °C, 800 °C, and 850 °C, respectively. Finally, the I-V characteristics were tested by a semiconductor device analyzer (KEYSIGHT, B1500A, Santa Rosa, CA, USA). The prepared TLM electrode samples are shown in Figure 5.

Table 1. Main parameters of the ICP etching process.

Cl_2 Flow (sccm)	BCl_3 Flow (sccm)	Ar Flow (sccm)	RF Power (W)	ICP Power (W)	Etching Time (min)	Etching Depth (nm)
5	10	50	50	1500	5	120

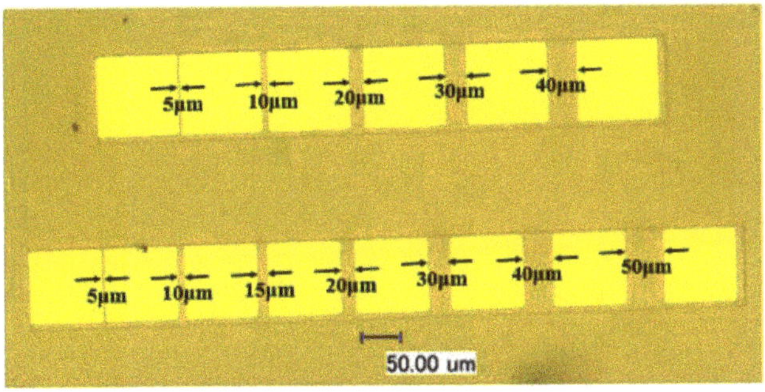

Figure 5. The top view of the prepared TLM electrodes samples.

4. Results and Discussion

Firstly, the I-V characteristics of the unannealed samples were tested. Figure 6 shows the I-V characteristic curves of the surface ohmic contact mode and side ohmic contact mode. It can be seen that, for the surface contact mode, the contact mode between the metal electrodes and epitaxial layer is Schottky contact before annealing, and the ohmic contact has not been formed yet, while for the side contact mode, the I-V curves have already shown the features of ohmic contact. This is because, in the surface contact mode, there is no alloying between the electrodes and the surface of the epitaxial layer when not annealed, and the barrier between electrodes and the channel is too high for electrons to jump over. However, metal depositing on the side of the epitaxial layer can make the electrode directly contact the conductive channel, which significantly reduces the contact barrier. After annealing at a suitable temperature, the contact resistance between the electrode and the side of the epitaxial layer will be further reduced, leading to the formation of a good ohmic contact.

The I-V characteristics curves of the side ohmic contact samples at the annealing temperature, from 700 °C to 850 °C, are shown in Figure 7a. The current rises linearly with the increase in voltage, which proves that the side contact mode can form the ohmic contact between metal electrodes and the conductive channel after rapid annealing. Figure 7b shows the total resistance R_{tot} with the change of temperature from 700 °C to 850 °C. The R_{tot} is minimum at the annealing temperature of 750 °C, after which the R_{tot} begins to increase with the annealing temperature. From these results, it can be preliminarily determined that 750 °C is the optimal annealing temperature. After measuring the I-V characteristics of side ohmic contact electrodes with different spacing, the R_{tot}-d curves were plotted in Figure 8a, from which the specific contact resistance ρ_c can be extracted by Equation (1). Figure 8b shows the calculated ρ_c at a different annealing temperature, where the minimum ρ_c is as low as 2.58×10^{-7} $\Omega{\cdot}cm^2$ at the annealing temperature of 750 °C. This value is three to four times lower than the specific contact resistance of 1×10^{-6} $\Omega{\cdot}cm^2$ in our previous prepared GaN HEMT device [23], as well as most of the reported surface contact mode [30–33]. The results show that, for a double channel GaN/AlGaN epitaxial layer, the side ohmic contact mode can obtain smaller specific contact resistance than surface contact mode (e.g., 3.29×10^{6} $\Omega \cdot cm^2$ in [30], 2.54×10^{-6} $\Omega \cdot cm^2$ in [31], 1.35×10^{-6} $\Omega \cdot cm^2$ in [32], and 3.7×10^{-6} $\Omega \cdot cm^2$ in [33]).

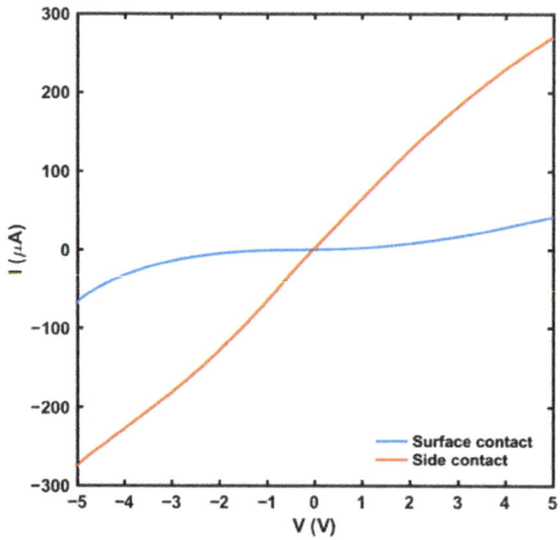

Figure 6. The I-V characteristics of the surface contact and side contact mode with the electrode distance of 40 μm when not annealed.

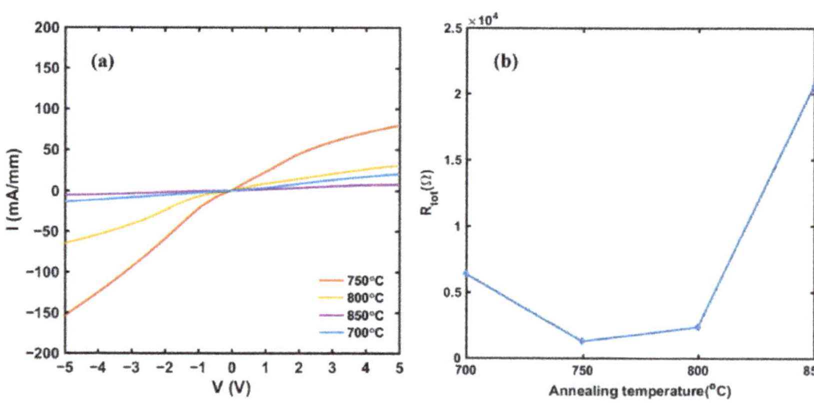

Figure 7. (**a**) I-V characteristics of side ohmic contact TLM electrodes, with a distance of 40 μm at different annealing temperatures. (**b**) The total resistance of side ohmic contact TLM electrodes with the change of annealing temperature.

In order to further investigate the ohmic contact effect of TLM electrodes at different annealing temperatures, the morphology of TLM electrodes was observed. The ohmic contact is formed on the side of the epitaxial layer, so we mainly focused on the boundary of electrode and mesa, as shown in the marked region of Figure 9. Figure 10 shows the planar SEM images of the electrode edge (marked region in Figure 9). When the annealing temperature is 700 °C, the metal electrode well contacts the side on the channel, and the edge morphology is flat and smooth. When the annealing temperature is 750 °C, some small grains are formed on the contact edge. These grains are the products of the reaction of metal layers at high temperatures, which indicates the formation of the alloying. When the annealing temperature is 800 °C, the number of grains becomes more. When the annealing temperature reaches 850 °C, there cracks appear at the contact edge and even cause the lift-off of the electrode. From Figure 10a–d, it can be noted that the stress in the Ti/Al/Ni/Au films changes with the increasing of the annealing temperature, which leads to the variation in the ohmic contact characteristics. Films initially sputtered at room temperature show a state of compressive stress due to the shot peening action of the bombarding ions and neutrals [34]. With the increase of the annealing temperature, Ti reacts with GaN/AlGaN and generates TiN to create N vacancies and form the ohmic contact. However, the Al also reacts with Ni and produces $NiAl_X$, which makes the stress move into the tensile state and weaken the film adhesion. This may be the reason why the ohmic contact characteristics first improve and then degenerate with the increase in annealing temperature. If the annealing temperature continues to rise, microcracks, together with the volume expansion of the interface, will make the electrode fall off the side of the epitaxial layer (Figure 10d).

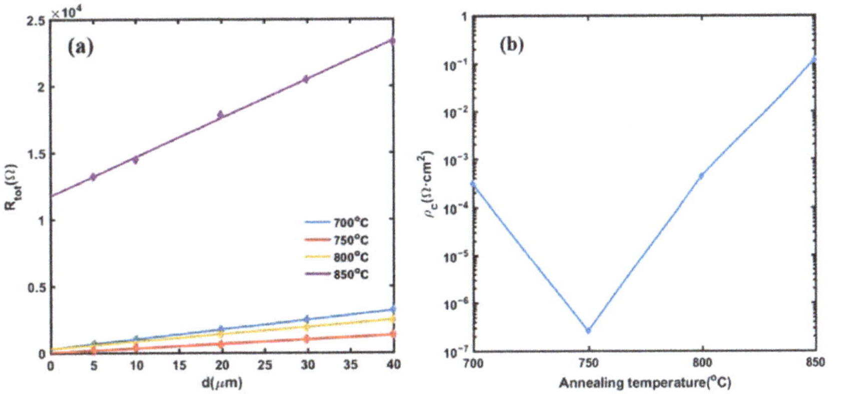

Figure 8. (**a**) The total resistance with the change of the distance between side ohmic contact TLM electrodes. (**b**) The specific contact resistance with the change of annealing temperature.

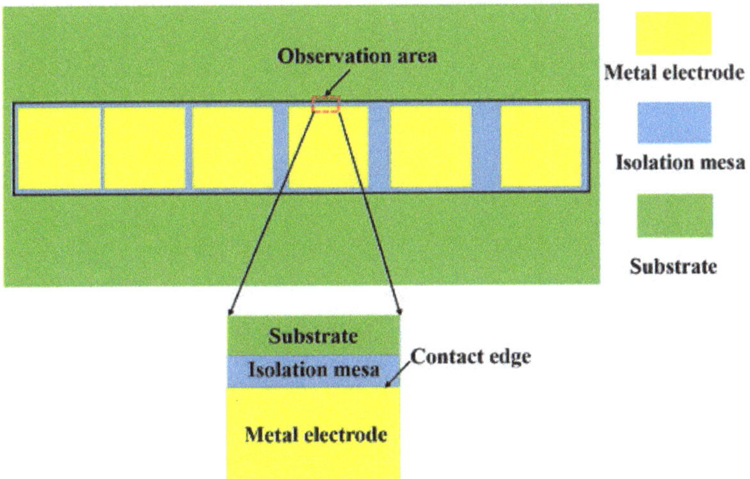

Figure 9. Schematic diagram of electrode edge area scanned by SEM.

To evaluate the composition of the annealed metal films, the EDS spectra was carried out simultaneously with the observation of the SEM. The side contact region that we mainly focused on was taken for the EDS spectra scanning (in the red circle area), and the results were shown in Figure 11. At the annealing temperature of 700 °C (Figure 11a), the Au component is dominant, as the Au layer is on the surface of metal layers, and no cracks or diffusion of metal layers appear. The Ga and N element is small but distinguishable because the AlGaN barrier is beyond the metal layers, and the annealing temperature is not high enough to make N element diffuse into the metal layer and generate N vacancies, so the specific contact resistance is large. When the annealing temperature reaches 750 °C (Figure 11b), the content of N is 20.76%, which is much higher than that of the 700 °C annealing temperature. The N element includes the compounds in the GaN/AlGaN epitaxial layer and TiN formed by high temperature annealing. The increase in N content indicates that N element diffuses into the upper metal layer and creates TiN, and the ohmic contacts are well formatted. When the annealing temperature reaches 800 °C (Figure 11c), the content of N increases to 30.74%, but the content of Au decreases sharply from 21.49% to 8.1%. This indicates the decrease in adhesion and the expansion of volume causes the lift-off of the Au film and the degradation of ohmic contact characteristics. When the annealing temperature is 850 °C (Figure 11d), the content of Au decreases to 2.82%, and the specific contact resistance increases significantly.

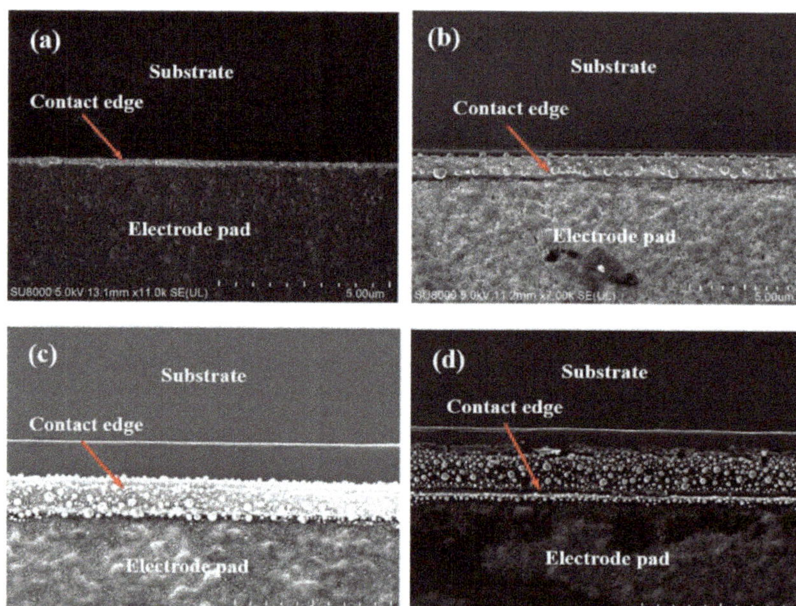

Figure 10. The SEM images of TLM electrode morphology at the annealing temperature of (**a**) 700 °C, (**b**) 750 °C, (**c**) 800 °C, and (**d**) 850 °C.

To further study the morphology variation of the contact between the metal electrode and the side of the epitaxial layer after high temperature annealing, the AFM was carried out for the quantitative analysis of the height fluctuation at the contact edge. From the three-dimensional image of the contact edge morphology in Figure 12a,b, it can basically be seen that, with the increasing of the annealing temperature, the morphology of the contact edge turns to be more and more rough, and cracks even appear at the annealing temperature of 850 °C. Figure 13 plots the variation of the step height (Figure 13a) and the root mean square roughness (RMS) (Figure 13b) of the contact edge with annealing temperature. At the annealing temperature of 700 °C, the step between the electrode and the mesa is clearly distinguishable, and the sectional measurement results show that the height of the stage is 187 nm, which is a little lower than the total thickness of the deposited metal layers. At the annealing temperature range of 700–800 °C, the step height decreases with the annealing temperature, while the RMS increases with the annealing temperature. This is possibly due to the change of the stress state and the volume expansion of metal films at high temperatures and leads to the reduction in the step height. On the other hand, with the increasing of the annealing temperature, the alloying of metals becomes more and more obvious and creates grains on the side, which makes the roughness increase. When the annealing temperature reaches 850 °C, metal layers fall off from the side of the epitaxial and form cracks at the edge of the electrode, so the step becomes more obvious. The above results reveal that, for the double channel GaN/AlGaN epitaxial layer, the metal electrode of Ti/Al/Ni/Au system can obtain the optimal side ohmic contact characteristics at the annealing temperature of 750 °C. This kind of side ohmic contact mode is, hopefully, further used in the preparation of source and drain electrodes of double channel GaN HEMTs to improve their output characteristic.

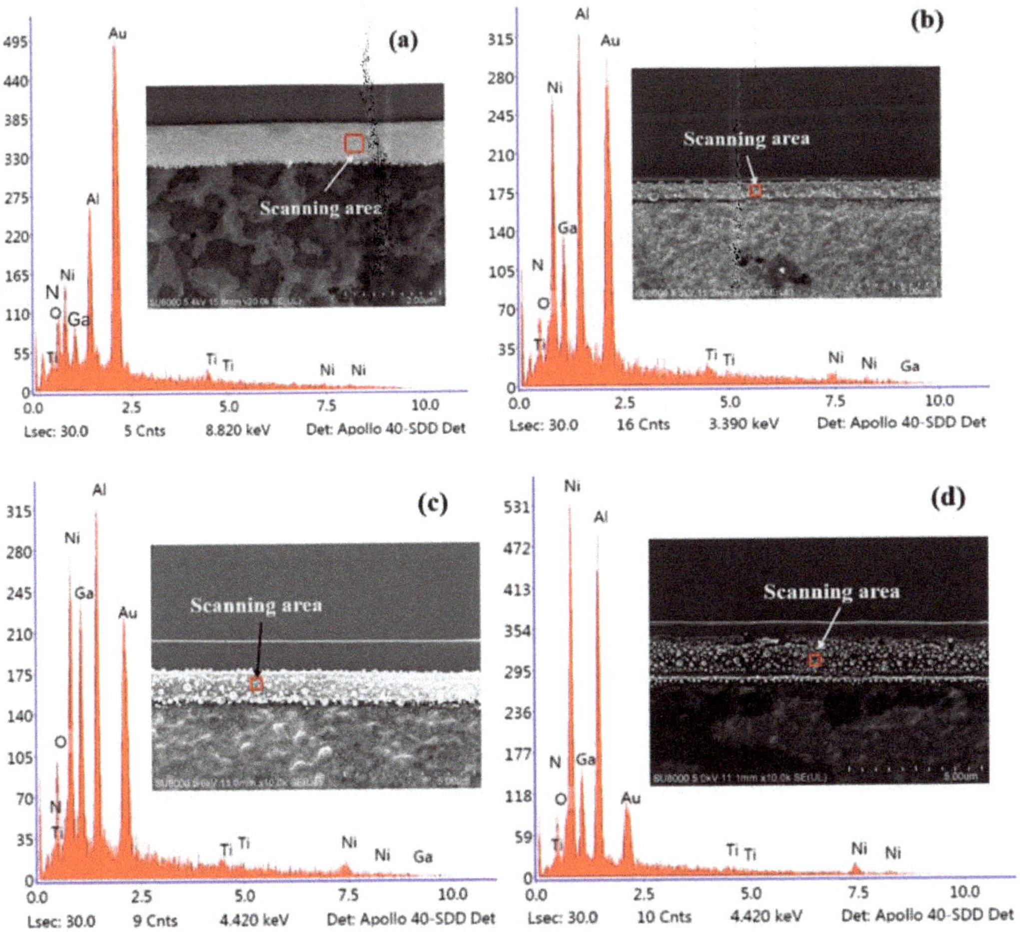

Figure 11. The EDS spectra of TLM electrodes at the annealing temperature of (**a**) 700 °C, (**b**) 750 °C, (**c**) 800 °C, and (**d**) 850 °C.

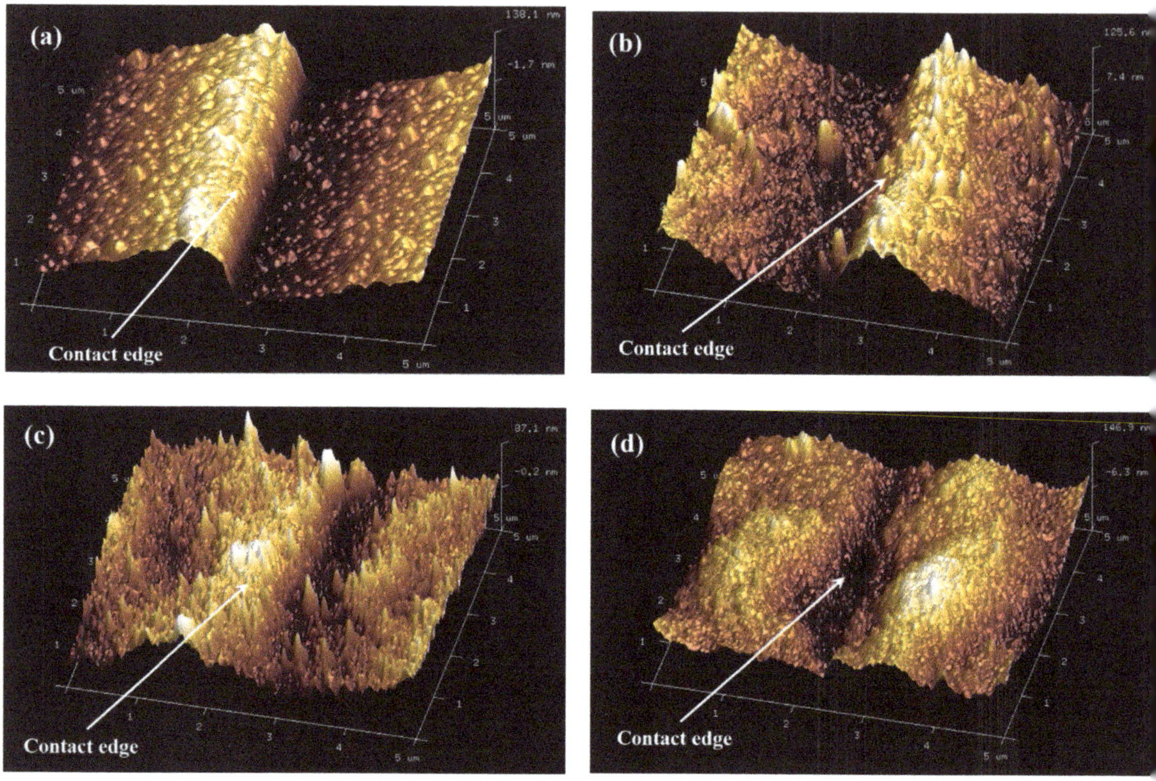

Figure 12. The morphology of the contact edge between the metal electrode and the side of the epitaxial layer, measured by AFM with the scanning area of 5 × 5 μm, after annealing at the temperature of (**a**) 700 °C, (**b**) 750 °C, (**c**) 800 °C, and (**d**) 850 °C.

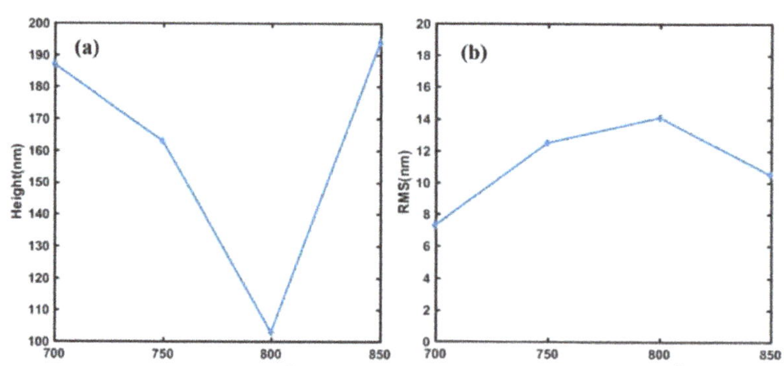

Figure 13. The step height (**a**) and the RMS (**b**) at the contact edge of the electrode and the epitaxial layer at the annealing temperature from 700 °C to 850 °C.

5. Conclusions

A side ohmic contact mode for a double channel GaN/AlGaN epitaxial layer is proposed in this paper. The TLM electrodes with a side ohmic contact mode were prepared and annealed at temperatures from 700 °C to 850 °C. The results of specific contact resistance, alone with the analysis of SEM, EDS, and AMF, show that side ohmic contact mode for the double channel GaN/AlGaN epitaxial layer can obtain smaller specific contact resistance than the conventional surface contact mode at a lower annealing temperature. The calculation results of the specific contact resistance show that the minimum specific contact resistance of 2.58×10^{-7} $\Omega \cdot cm^2$ is obtained at the annealing temperature of 750 °C. The observation of SEM, together with the EDS spectra, show that with the increase in annealing temperature, the state of the stress in the electrode films changes from compressive stress to tensile stress, leading to the volume expansion in the electrode films, and the composition of N will increase because of the creation of TiN and the N vacancies. When the annealing temperature is above 800 °C, cracks and lift-off of electrodes appear at the contact edge, which leads to a degradation of the ohmic contact performance. The AFM results reveal that, as the high temperature annealing causes the volume expansion of the metal layers, the step height decreases with the raise of the annealing temperature, while the RMS at the contact edge raises with the increase in annealing temperature, as the alloying of metal layers and the formation of grains at the contact edge begin. The experimental results prove that the proposed side ohmic mode has potential for application in double channel GaN/AlGaN HEMTs.

Author Contributions: Conceptualization, Q.M. and Q.L.; methodology, Q.M. and D.L.; validation, Q.M., Q.L., W.J., N.Z. and D.L.; data curation, N.Z.; writing—original draft preparation, Q.M.; writing—review and editing, Q.M., Q.L. and P.Y. All authors have read and agreed to the published version of the manuscript.

Funding: This research was funded by the National Key Research and Development Program (Grant No. 2021YFB3201800), the National Natural Science Foundation of China (Nos. 51720105016, 51890884), the Natural Science Foundation of Chongqing City (no. cstc2021jcyjmsxmX0223).

Data Availability Statement: Not applicable.

Acknowledgments: The authors the support from the International Joint Laboratory for Micro/Nano manufacturing and measurement technologies.

Conflicts of Interest: The authors declare no conflict of interest.

References

1. Kozak, J.P.; Zhang, R.; Song, Q.; Liu, J.; Zhang, Y. True Breakdown Voltage and Overvoltage Margin of GaN Power HEMTs in Hard Switching. *IEEE Electron Device Lett.* **2021**, *44*, 505–508. [CrossRef]
2. Kim, J.G.; Cho, C.; Kim, E.; Hwang, J.S.; Lee, J.H. High Breakdown Voltage and Low-Current Dispersion in AlGaN/GaN HEMTs with High-Quality AlN Buffer Layer. *IEEE Trans. Electron Devices* **2021**, *68*, 1513–1517. [CrossRef]
3. Liu, Y.; Yu, Q.; Du, J. Simulation design of a high-breakdown-voltage p-GaN-gate GaN HEMT with a hybrid AlGaN buffer layer for power electronics applications. *J. Comput. Electron.* **2020**, *19*, 1527–1537. [CrossRef]
4. Palacios, P.; Wei, M.; Zweipfennig, T.; Hamed, A.; Negra, R. AlGaN/GaN high electron mobility transistor oscillator for high temperature and high frequency. *Electron. Lett.* **2021**, *57*, 148–150. [CrossRef]
5. Sowmya, K.; Balamurugan, N.B.; Parvathy, V. A 2-D Modeling of Fe Doped Dual Material Gate AlGaN/AlN/GaN High Electron Mobility Transistors for High Frequency Applications. *AEU—Int. J. Electron. Commun.* **2019**, *103*, 46–56. [CrossRef]
6. Reza, S.; Klein, B.A.; Baca, A.G.; Armstrong, A.M.; Allerman, A.A.; Douglas, E.A.; Kaplar, R.J. High-frequency, high-power performance of AlGaN-channel high-electron-mobility transistors: An RF simulation study. *Jpn. J. Appl. Phys.* **2019**, *58*, SCCD04. [CrossRef]
7. Shi, Y.; Chen, W.; Fu, Z.; Chen, S.; Zhang, B. Influence of the Acceptor-Type Trap on the Threshold Voltage of the Short-Channel GaN MOS-HEMT. *IEEE J. Electron Devices Soc.* **2021**, *9*, 606–611. [CrossRef]
8. Pinchbeck, J.; Lee, K.B.; Jiang, S.; Houston, P. Dual metal gate AlGaN/GaN high electron mobility transistors with improved transconductance and reduced short channel effects. *J. Phys. D Appl. Phys.* **2021**, *54*, 105104. [CrossRef]
9. Kato, D.; Kajiwara, Y.; Mukai, A.; Ono, H.; Shindome, A.; Tajima, J.; Hikosaka, T.; Kuraguchi, M.; Nunoue, S. Suppression of short-channel effects in normally-off GaN MOSFETs with deep recessed-gate structures. *Jpn. J. Appl. Phys.* **2020**, *59*, SGGD13. [CrossRef]
10. Sharma, N.; Periasamy, C.; Chaturvedi, N.; Chaturvedi, N. Trapping Effects on Leakage and Current Collapse in AlGaN/GaN HEMTs. *J. Electron. Mater.* **2020**, *49*, 5687–5697. [CrossRef]
11. Lee, J.H.; Kim, J.G.; Kang, H.S.; Lee, J.H. Suppression of current dispersion in AlGaN/GaN MISHFETs with in-situ AlN passivation layer. *Solid State Electron.* **2021**, *178*, 107984. [CrossRef]

12. Tang, X.; Zhang, Z.; Wei, J.; Li, B.; Wang, J.; Chen, K.J. Photon emission and current-collapse suppression of AlGaN/GaN field-effect transistors with photonic–ohmic drain at high temperatures. *Appl. Phys. Express* **2018**, *11*, 071003. [CrossRef]
13. Gaska, R.; Shur, M.S.; Fjeldly, T.A.; Bykhovski, A.D. Two-channel AlGaN/GaN heterostructure field effect transistor for high power applications. *J. Appl. Phys.* **1999**, *85*, 3009–3011. [CrossRef]
14. Chu, R.; Zhou, Y.; Liu, J.; Wang, D.; Chen, K.; Lau, K. AlGaN-GaN double-channel HEMTs. *IEEE Trans. Electron Devices* **2005**, *52*, 438–446. [CrossRef]
15. Wang, X.; Hu, W.; Chen, X.; Lu, W. The Study of Self-Heating and Hot-Electron Effects for AlGaN/GaN Double-Channel HEMTs. *IEEE Trans. Electron Devices* **2012**, *59*, 1393–1401. [CrossRef]
16. Chang, Y.; Wang, J.J.; Shen, G.L. Improving Off-State Breakdown Voltage of a Double-Channel AlGaN/GaN HEMT with Air-Bridge Field Plate and Slant Field Plate. *Solid State Electron. Lett.* **2020**, *2*, 92–97. [CrossRef]
17. Palacios, T.; Chini, A.; Buttari, D.; Heikman, S.; Chakraborty, A.; Keller, S.; DenBaars, S.; Mishra, U. Use of double-channel heterostructures to improve the access resistance and linearity in GaN-based HEMTs. *IEEE Trans. Electron Devices* **2006**, *53*, 562–565. [CrossRef]
18. Chugh, N.; Kumar, M.; Haldar, S.; Bhattacharya, M.; Gupta, R.S. Applicability of Field Plate in Double Channel GaN HEMT for Radio-Frequency and Power-Electronic Applications. *Silicon* **2021**, *14*, 1029–1038. [CrossRef]
19. Luan, C.; Lin, Z.; Lv, Y.; Meng, L.; Yu, Y.; Cao, Z.; Chen, H.; Wang, Z. Influence of the side-Ohmic contact processing on the polarization Coulomb field scattering in AlGaN/AlN/GaN heterostructure field-effect transistors. *Appl. Phys. Lett.* **2012**, *101*, 113501. [CrossRef]
20. Luan, C.; Lin, Z.; Lv, Y.; Feng, Z.; Zhao, J.; Yang, Q.; Yang, M. Enhanced effect of side-Ohmic contact processing on the 2DEG electron density and electron mobility of $In_{0.17}$ $Al_{0.83}$ N/AlN/GaN heterostructure field-effect transistors. *Appl. Phys. A* **2014**, *116*, 2065–2075. [CrossRef]
21. Liang, W.; Kim, D.H.; Adesida, L. Direct contact mechanism of Ohmic metallization to AlGaN/GaN heterostructures via Ohmic area recess etching. *Appl. Phys. Lett.* **2009**, *95*, 172107.
22. Zhang, X.; Tan, R.B.; Sun, J.; Li, X.; Zhou, Y.; Lu, L.; Qin, H. Investigation of high sensitivity radio-frequency readout circuit based on AlGaN/GaN high electron mobility transistor. *Chin. Phys. B* **2015**, *24*, 105201. [CrossRef]
23. Meng, Q.; Lin, Q.; Han, F.; Jing, W.; Wang, Y.; Jiang, Z. A Terahertz Detector Based on Double-Channel GaN/AlGaN High Electronic Mobility Transistor. *Materials* **2021**, *14*, 6193. [CrossRef] [PubMed]
24. Lee, C.; Chiu, Y. Photoelectrochemical passivated ZnO-based nanorod structured glucose biosensors using gate-recessed AlGaN/GaN ion-sensitive field-effect-transistors. *Sensor. Actuat. B Chem.* **2015**, *210*, 756–761. [CrossRef]
25. Dogar, S.; Khan, W.; Kim, S.D. Ultraviolet photoresponse of ZnO nanostructured AlGaN/GaN HEMTs. *Mat. Sci. Semicon. Proc.* **2016**, *44*, 71–77. [CrossRef]
26. Thapa, R.; Alur, S.; Kim, K.; Tong, F.; Sharma, Y.; Kim, M.; Ahyi, C.; Dai, J.; Hong, J.W.; Bozack, M.; et al. Biofunctionalized AlGaN/GaN high electron mobility transistor for DNA hybridization detection. *Appl. Phys. Lett.* **2012**, *100*, 232109. [CrossRef]
27. Kang, B.; Wang, H.; Lele, T.P.; Tseng, Y.; Ren, F.; Pearton, S.J.; Johnson, J.W.; Rajagopal, P.; Roberts, J.C.; Piner, E.L. Prostate specific antigen detection using AlGaN/GaN high electron mobility transistors. *Appl. Phys. Lett.* **2007**, *91*, 112106. [CrossRef]
28. Guo, Z.; Lai, W.; Hao, Z.; Luo, Y. Modeling and experimental study on sensing response of an AlGaN/GaN HEMT-based hydrogen sensor. *Sens. Actuators B Chem.* **2013**, *176*, 241–247. [CrossRef]
29. Meng, Q.; Lin, Q.; Jing, W.; Mao, Q.; Zhao, L.; Fang, X.; Dong, T.; Jiang, Z. Characterization of the Electrical Properties of a Double Heterostructure GaN/AlGaN Epitaxial Layer with an AlGaN Interlayer. *J. Electron. Mater.* **2021**, *50*, 2521–2529. [CrossRef]
30. Li, P.; Wei, S.; Kang, X.; Zheng, Y.; Zhang, J.; Wu, H.; Wei, K.; Yan, J.; Liu, X. Optimization of Oxygen Plasma Treatment on Ohmic Contact for AlGaN/GaN HEMTs on High-Resistivity Si Substrate. *Electronics* **2021**, *10*, 855. [CrossRef]
31. Yoshida, T.; Egawa, T. Improvement of Au-Free, Ti/Al/W Ohmic Contact on AlGaN/GaN Heterostructure Featuring a Thin-Ti Layer and Low Temperature Annealing. *Phys. Status Solidi A* **2018**, *215*, 1700825. [CrossRef]
32. Wong, Y.Y.; Chen, Y.K.; Maa, J.S.; Yu, H.W.; Tu, Y.Y.; Dee, C.F.; Yap, C.C.; Chang, E.Y. Low resistance copper-based ohmic contact for AlGaN/GaN high electron mobility transistors. *Appl. Phys. Lett.* **2013**, *103*, 152104. [CrossRef]
33. Wang, L.; Mohammed, F.M.; Ofuonye, B.; Adesida, I. Ohmic contacts to n~+-GaN capped AlGaN/AlN/GaN high electron mobility transistors. *Appl. Phys. Lett.* **2007**, *91*, 012113. [CrossRef]
34. Lin, Q.; Yang, S.; Jing, W.; Li, C.; Wang, C.; Jiang, Z. Buckling and Delamination of Ti/Cu/Si Thin Film During Annealing. *J. Electron. Mater.* **2014**, *43*, 3351–3356. [CrossRef]

micromachines

MDPI

Article

Fabrication and Modeling of Matching System for Air-Coupled Transducer

Jinjie Zhou, Jiaqi Bai and Yao Liu *

School of Mechanical Engineering, North University of China, Taiyuan 030051, China; zhoujinjiechina@126.com (J.Z.); baihuahua6@outlook.com (J.B.)
* Correspondence: liuyao@nuc.edu.cn; Tel.: +86-19834530812

Abstract: The tremendous acoustic impedance difference between the piezoelectric composite and air prevents the ultrasonic transition, resulting in low amplitude for the received signal for the composite defect detection using an air-coupled transducer. The matching system, which includes the matching layers and bonding layers attached to the piezoelectric composite, can reduce the acoustic impedance difference and benefit the acoustic transition. In this paper, the fabrication method and modeling for the matching layers are proposed to optimize the transducer performance. The effects of bonding layer material on the transducer performance are also discussed. Experiments were conducted for modeling validation. The proposed model can predict the matching layer acoustic properties with an error of less than 11%. The bonding layer using the same material as the first matching layer can help to increase the sensitivity by about 33% compared to the traditional epoxy bonding. The optimized air-coupled ultrasonic transducer, based on the results of this study, has a 1283 mV amplitude in the air, which is 56% higher than commercially available transducers, and can identify the defects in two typical non-metallic composite materials easily.

Keywords: air-coupled acoustic transducer; matching layer; acoustic impedance; bonding layer

Citation: Zhou, J.; Bai, J.; Liu, Y. Fabrication and Modeling of Matching System for Air-Coupled Transducer. *Micromachines* **2022**, *13*, 781. https://doi.org/10.3390/mi13050781

Academic Editors: Xiuqing Hao, Duanzhi Duan and Youqiang Xing

Received: 26 April 2022
Accepted: 14 May 2022
Published: 17 May 2022

1. Introduction

The air-coupled acoustic transducer uses air as the medium to detect the defect in aerospace composites, foods, drugs, etc. [1,2], in which the coupling agent is prohibited or caution used. An air-coupled acoustic transducer commonly uses the piezoelectric composite as the acoustic wave source. When the acoustic wave propagates from the piezoelectric composite to the air directly, a nearly total reflection occurs on the interface due to the tremendous acoustic impedance discrepancy between the piezoelectric composite and air. The high reflection ratio of the acoustic wave limits the energy into the air and tested material, resulting in a low signal amplitude [3,4]. In order to solve this problem, the transition/matching layers are often attached to the piezoelectric composite for the acoustic transmission [5]. The properties of the matching layer, including the geometric and acoustic parameters, ultimately determine the acoustic transmission process [6].

The study of matching layers for the air-coupled transducer has been reported. Toda et al. [7] conducted impedance matching by adjusting the air space and reflectivity by insertion a reflective layer between the transducer and the propagation medium. Kelly et al. [8] added a porous material with extremely low impedance and the low-density rubber material as the matching layer, which induced the amplitude of the received signal to increase by 30 dB compared with the unmatched one. Tomas et al. [9,10] proposed a better matching configuration by studying the acoustic properties of polyethersulfone and other materials to obtain better sensitivity and bandwidth transducer. Saito et al. [11] optimized the acoustic impedance of the matching layer by using the transmission line model, which was proved by experiments to increase the sensitivity by 20 dB by using silicon rubber and thermoplastic hollow microspheres mixture as a matching layer. Botun et al. [12] developed an air-coupled ultrasonic transducer with a simple structure and high sensitivity by using honeycomb polypropylene iron electret film as a matching layer. Kazys et al. [13,14] used low impedance polystyrene foam to improve the efficiency, bandwidth, and radiation pulse waveform of the PMN-32PT crystal transducer. Guo et al. [15] analyzed the effect of the matching layer material on the vibration mode shape of the transducer and found the resin epoxy can improve the transmission ratio. Wang et al. [16] used the superposition of two low impedance matching layers to improve the sensitivity of the designed transducer, which can detect microcracks. Wu et al. [17] developed a single matched layer air-coupled ultrasonic transducer using a hollow polymer microspheres/epoxy resin system, increasing the transducer sensitivity by 20.9 dB. Song [18] introduced the periodic subwavelength apertures that employ coupled resonances to enhance the efficiency and bandwidth of the non-contact ultrasonic transducers. Based on the literature review above, the matching layer properties, including the material and acoustic properties, have a significant influence on the sensitivity of the air-coupled transducer. However, the modeling of the matching layer properties and the bonding material effects on the matching layers is currently missing.

In this study, a double-layer matching system on the air-coupled transducer was investigated. The matching theory of the acoustic impedance is firstly introduced. Then modeling to predict the density, acoustic transmission speed, and the acoustic impedance of the matching layer is proposed based on the raw material. Experiments to validate the modeling results and reveal bonding material effects on the transducer sensitivity are introduced. Lastly, the optimized transducer was applied to detect the defects in two typical non-metallic materials to prove the feasibility.

2. Matching Theory and Modeling of Acoustic Impedance

2.1. Matching Theory of Acoustic Impedance

A material acoustic impedance Z can be calculated by the density ρ and acoustic propagation speed c_L in the material, as shown in Equation (1) [19].

$$Z = \rho c_L \tag{1}$$

Assuming the piezoelectric composite, medium, and tested sample are half infinite. The acoustic impedance of piezoelectric composite, the Nth matching layer, and medium are represented by Z_0, Z_n, and Z_L, respectively. The theoretical acoustic impedance for single matching layer Z_1 can be calculated by Equation (2) [19]:

$$Z_1 = \sqrt{Z_0 Z_L} \tag{2}$$

The theoretical acoustic impedances for double-layer matching can be calculated as Equations (3) and (4) [19] for the first and second matching layers.

$$Z_1 = \sqrt[4]{Z_0^3 Z_L} \tag{3}$$

$$Z_2 = \sqrt[4]{Z_0 Z_L^3} \tag{4}$$

where Z_1 and Z_2 are the acoustic impedances for the first and second matching layers.

The ultrasonic wave generated by the piezoelectric composite propagates as a simple harmonic wave to the matching layer. In order to ensure a continuous ultrasonic oscillation, the matching layer thickness should be a quarter of the wavelength, which can also reduce the ultrasonic attenuation. Based on the theory above, the matching layer thickness should be:

$$b = \frac{c}{4f} \tag{5}$$

where c is the acoustic propagation speed, f is the frequency of the acoustic, and b is the thickness of the matching layer.

2.2. Modeling of Matching Layer Acoustic Impedance

Assuming the matching layer consists of the N components, with the weight of m_1, m_2, m_3, … , and m_N. The corresponding densities for the N components are ρ_1, ρ_2, ρ_3, … , and ρ_N, respectively. Generally, the matching layer uses the curing process to form the solid shape from liquid, which may cause the volume change. The volume variation for the N components, described by the shrinkage or expansion ratio, are k_1, k_2, k_3, … , and k_N, where the negative value means the expansion and the positive value means shrinkage. The volume of the matching layer after solidification is

$$V = \sum_{i=1}^{N} (1 - k_i) \frac{m_i}{\rho_i} \tag{6}$$

The density of the matching layer is

$$\rho = \frac{\sum_{i=1}^{N} m_i}{V} = \frac{\sum_{i=1}^{N} m_i}{\sum_{i=1}^{N} (1 - k_i) \frac{m_i}{\rho_i}} \tag{7}$$

In order to model the acoustic velocity of the matching layer, the acoustic velocity in each pure component in solid mode should be measured in advance. Assuming the acoustic velocity in N components are v_1, v_2, v_3, … , and v_N. Based on the probability statistics theory, when the matching layer thickness is b, the average prorogation distance, in statistics, of acoustic in the ith component is

$$b_i = b \frac{(1 - k_i) m_i / \rho_i}{V} \tag{8}$$

The propagation time in the ith component is $t_i = b_i / v_i$, and the total time used to cross the matching layer is

$$t = \sum_{i=1}^{N} \frac{b_i}{v_i} \tag{9}$$

So the average acoustic velocity in the matching layer is

$$v = \frac{b}{t} = \frac{\sum_{i=1}^{N} (1 - k_i) \frac{m_i}{\rho_i}}{\sum_{i=1}^{N} (1 - k_i) \frac{m_i}{\rho_i v_i}} \tag{10}$$

Based on the Equation (1), the acoustic impedance of the matching layer is

$$Z = \rho v = \frac{\sum_{i=1}^{N} m_i}{\sum_{i=1}^{N} (1 - k_i) \frac{m_i}{\rho_i}} \frac{\sum_{i=1}^{N} (1 - k_i) \frac{m_i}{\rho_i}}{\sum_{i=1}^{N} (1 - k_i) \frac{m_i}{\rho_i v_i}} = \frac{\sum_{i=1}^{N} m_i}{\sum_{i=1}^{N} (1 - k_i) \frac{m_i}{\rho_i v_i}} \tag{11}$$

Based on Equation (11), with the mass, density, shrinkage/expansion ratio, and acoustic velocity of each component, the acoustic impedance of the matching layer can be calculated.

3. Fabrication of Transducer

From the literature [8,11,13,17], the first matching layer is commonly fabricated by the mixture of hollow glass microspheres with epoxy, and the second matching layer is often made of microcellular foam polypropylene. Their acoustic impedances can be adjusted by changing the ratio between hollow glass microspheres and epoxy and the foaming rate, respectively. In this study, the matching layer used the same raw materials.

3.1. 1-3 Piezoelectric Composite

The piezoelectric composite used in this study is the 1-3 type, which consists of one dimension piezoelectric ceramic column and a 3D polymer structure. The embed polymer helps to reduce the density and acoustic impedances of the piezoelectric composite due to the low density and acoustic impedances. The reduced acoustic impedance of the composite makes it an ideal material for an air-coupled piezoelectric transducer [9,20] to ensure more wave energy can propagate into the air. The 1-3 type piezoelectric ceramic/polymer has the following advantages for the air-coupled transducer:

(1) Low acoustic impedance, which is between the pure piezoelectric ceramic and polymer, and easier to achieve the impedance matching.

(2) High electromechanical coupling factor, which is almost the same as the longitudinal electromechanical coupling coefficient of piezoelectric ceramic when the composite is working under the thickness mode.

(3) The mechanical quality factor is low, which is suitable for the broadband transducer.

(4) Weak lateral coupling effects due to the separation of the polymer, which is suitable for the longitudinal transducer.

For the machining of piezoelectric composite, the solid piezoelectric ceramic need to dice or wire saw some kerfs on the surface to separate the ceramic block into the small individual columns. Then the polymer is immersed and cured on the kerf to connect the individual columns to a whole part. After the curing, the top and bottom of the composite surface are ground to remove the uncut ceramic layer and the excess polymer layer. Then continuing grinding is conducted to obtain the specified thickness of the composite.

3.2. Fabrication of the First Matching Layer

The first matching layer consists of hollow glass microspheres, epoxy, curing agent, and diluent. The density of the hollow glass microsphere changes with the diameter; the smaller the diameter of the hollow glass microsphere, the larger density is. The density and acoustic velocity of the first matching layer can be adjusted by changing the hollow glass microsphere's diameters and weight ratios. The fabrication method is mold casting in a vacuum chamber, which is shown in Figure 1:

Step 1. The epoxy and curing agents are weighted in 6:1, suggested by the vendor (Shanghai Aotun Chemical Technology Co., LTD, Shanghai, China), by analytical balance and poured into a glass beaker for a mixture. Then the hollow glass microspheres are weighted based on the designed weight ratio and added to the beaker. If needed, the diluent weighted in the ratio of the epoxy and curing agent is added to assist the air bubble release.

Step 2. After all raw materials are put in the beaker, a stirring of about 5 min is taken for the mixture. The vacuum-pumping process under the pressure of −0.1 Mpa is applied for 5 min to help the evacuation of air bubbles.

Step 3. The mixture is then poured into the rectangular mold, followed by another 5 min vacuum to remove the bubbles and moisture.

Step 4. Curing on a thermostat at 60 °C temperature for 12 h.

Step 5. After curing, the composite is taken from the mold and cut into the same size as the 1-3 piezoelectric composite.

Step 6. The density and acoustic velocity of the matching layer are determined by using drainage and pulse insertion

Step 7. Cut the matching layer thickness to a quarter of the acoustic wavelength from the calculation.

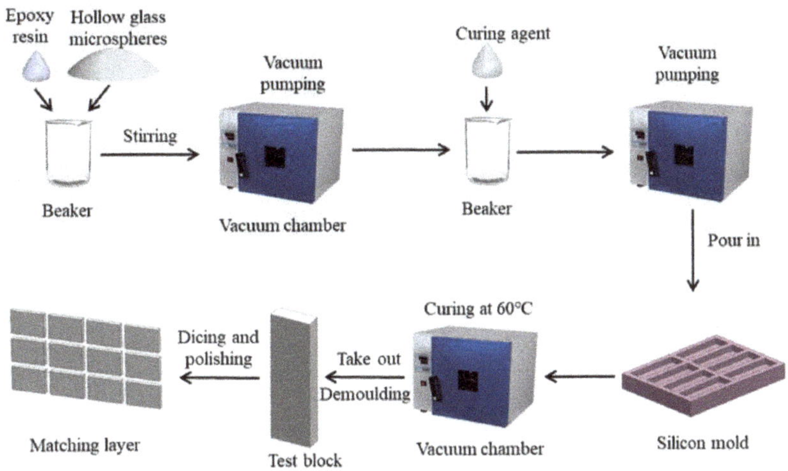

Figure 1. The steps to fabricate the matching layer.

3.3. *Fabrication of Second Matching Layer*

The second matching layer needs a low theoretical acoustic impedance and is cost-effectively machined to the designed thickness. Aerogel has low acoustic impedance, but its machinability and bonding properties are poor [21]. In this study, the second matching layer is made from polypropylene foam. It has a low density and a high bubble ratio leading to a low acoustic impedance, and can be considered as a composite of the polypropylene and air microsphere. The foaming rate controls the matching layer acoustic properties.

3.4. *Transducer Assembly*

Figure 2 shows the transducer design in this study. The front and rear surfaces of 1-3 piezoelectric composite are coated with metal films as the electrodes. On the front surface of 1-3 piezoelectric composite, the first and second matching layers are bonded to form the core of the transducer. A standard BNC connector is screwed to the housing through the front cover. BNC connector links the wire of the 1-3 piezoelectric composite to receive and send the electrical signal from the outside device.

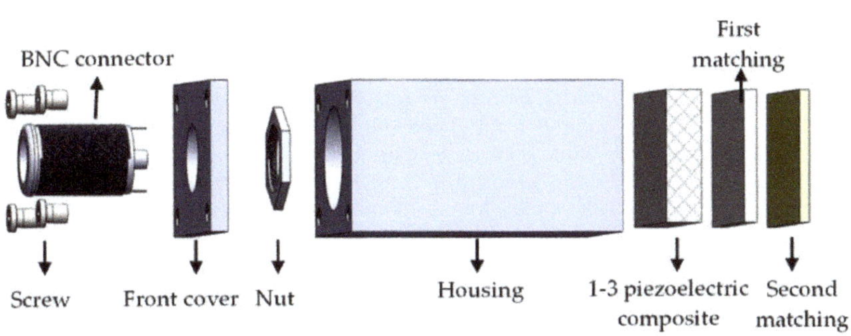

Figure 2. Structure of air-coupled transducer.

4. Experiment Setup

The PZT-4 ceramics (Shandong Weifang Jude Electronics Co. Ltd., Shanghai, China) was used as the piezoelectric composite due to the high electromechanical coupling factor, which can generate more ultrasonic waves with a provided energy. A 20 × 14 × 8 mm cuboid PZT-4 was cut by the dicing machine (SYJ-400, Shenyang Kejing, Shenyang, China) to create the 0.4 mm width and 7.8 mm depth kerf grid on the surface. The parallel distance of kerfs was 2.4 mm which can create 2 mm × 2 mm PZT-4 columns on the surface after dicing. The epoxy (E51, Shanghai Aotun Chemical Technology Co., LTD, Shanghai, China) was poured on the PZT-4 to fill the kerfs and connect the columns. The composite was put in a vacuum chamber to remove air from the composite. After 24 h curing, the composite was ground by a surface grinder (M7230H, Hangzhou grinding machine Co. Ltd., Hangzhou, China) to remove the uncut layer on the bottom and epoxy on the front surface. The ultimate thickness of the composite controlled by grinding was 7.6 mm to match the 200 kHz resonance frequency. The density and acoustic propagation velocity of the 1-3 type piezoelectric composite were measured by the drainage and pulse insertion method, which were 5501.81 kg/m^3 and 3521 m/s, respectively. Based on Equation (1), the 1-3 piezoelectric composite acoustic impedance was 19.37 MRayl. Table 1 summarizes the 1-3 piezoelectric composite properties. The front and bottom surfaces 1-3 piezoelectric composite were coated with 50 nm silver via chemical vapor deposition to generate the positive and negative electrodes. The electrode surfaces were welded to two copper wires, as shown in Figure 3a.

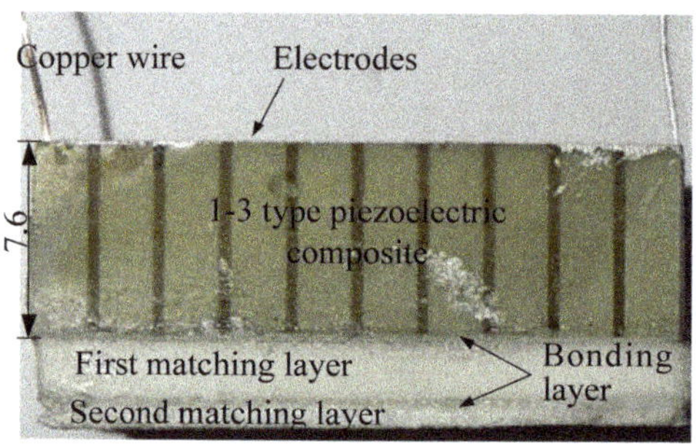

(**a**)

(**b**)

Figure 3. Air-coupled ultrasonic transducer. (**a**) 1-3 piezoelectric composite with double matching layer and (**b**) self-developed transducer.

From the matching theory mentioned in Section 2, for the double-layer matching system, the acoustic impedance of the first and second matching layers should be 1.32 and 0.0062 MRayl, respectively. In order to achieve the acoustic impedance needed for the first layer, a series of matching layers were fabricated. The first layer consisted of the hollow glass microspheres and E51 epoxy resin. In order to identify the effects of fabrication parameters on the matching layer's acoustic properties, especially in acoustic impedances, the matching layers with different glass microsphere sizes and weight ratios were made. Table 2 shows the parameters of the hollow glass microspheres used in this study, which have an average diameter of 100, 85, and 70 μm. The glass microsphere weight ratios of 10%, 15%, 20%, and 30% were experimental studied to help the modeling process. In order to facilitate the mixture and curing process, the diluent (butyl glycidyl ether) was added, if necessary, when the bubble could not be expelled completely. After the fabrication process, the matching layer's density and acoustic velocity were measured to calculate the acoustic impedances. The experiment results were used to validate the proposed matching layer modeling. The exact weight ratios of the first and second matching layers based on the calculated theoretical acoustic impedances were provided. After the determination of the recipe of the matching layer, the first and second matching layers were fabricated, and the acoustic parameters were tested.

Table 1. List of experiment parameters.

Material	Property	Value
1-3 type piezoelectric composite	Density, ρ_o (kg/m^3)	5501.81
	Acoustic velocity, c_L (m/s)	3521
	Acoustic impedance, Z_o (MRayl)	19.37
First matching layer	Hollow glass microspheres	BR20, BR40, and BR60
	Ratio of glass microspheres in weight, k	10%, 15%, 20%, and 30%
	Density of E51 resin, (kg/m^3)	1168.33
	Density of curing agent, (kg/m^3)	1036
	Density of diluent, (kg/m^3)	907
	Acoustic velocity of pure E51 + curing agent, (m/s)	3124
	Acoustic velocity of diluent in liquid, (m/s)	1040
	Shrinkage ratio of pure E51 + curing agent	2%
Second matching layer	Density, ρ_2 (kg/m^3)	67.13
	Acoustic velocity, c_{L2} (m/s)	738
	Acoustic impedance, Z_2 (MRayl)	0.049
Bonding layer	Material	Non, epoxy, epoxy/glass microsphere composite

Table 2. Hollow glass bead parameter index.

Hollow Glass Microsphere	Average Diameter, d (μm)	Real Density, ρ_R (kg/m^3)	Bulk Density, ρ_B (kg/cm^3)	Wall Thickness (μm)	Acoustic Velocity (m/s)
BR20	100	200	120	0.5–1	
BR40	85	400	240	1–2	2280
BR60	70	600	390	1.5–3.5	

The influence of the bonding layer on acoustic propagation and transducer performance is also discussed in this study. In order to connect the 1-3 piezoelectric composite, first matching layer, and second matching layer, the bonding layer should be added to the contact surfaces. Table 1 also lists three types of the bonding layer to be used in the experiment to identify the potential effect.

The first matching layer, bonding layer, and second matching layer, after determination, were attached to the surface of 1-3 piezoelectric composite to form the core of the transducer, as seen in Figure 3a. An aluminum housing enclosed the core, and a BNC connector was on the front surface of the transducer to link the 1-3 piezoelectric composite and outside device. Figure 3b. shows the transducer by using the 1-3 piezoelectric composite, first matching layer, bonding layer, and second matching layer fabrication in this study.

In order to validate the proposed matching layer modeling and study the effect of bonding layer properties on the transducer performance, a testing platform was built, as presented in Figure 4. The self-developed 200 kHz transducers and a commercially available one (0.2 K 14 × 20 N-TX, Japan Probe Co. Ltd., Japan) were put on the two ends of a linear stage to emit and receive the acoustic signal. Air and testing plate were in the middle of the transducers to act as the medium and inspected material. Both transducers are connected with ultrasonic instruments and computers to control the testing process. The exciting signals output from the self-developed ultrasonic instruments were 200 kHz Hanning windowed three-cycle sine bursts with V_{pp} of 138 V. The received signals were filtered and enveloped to obtain the amplitude. The comparisons of amplitude were conducted to evaluate the performance of the transducer. In this study, the plate was not used first to validate the influence of different matching layers and bonding layers on the transducer sensitive for the modeling and theory validation. Two types of plates (CFRP and PVC foam) with and without artificial defects were put in the middle of the transducers to emphasize the practical application of the transducer. The sizes for the CFRP and PVC foam were 540 mm × 500 mm × 5 mm and 200 mm × 200 mm × 30 mm, respectively.

The artificial defects were created by milling at the bottom of the tested materials with the diameter and depth of Φ 5 × 3 and Φ 10 × 8 mm for the CFRP and PVC foam, respectively.

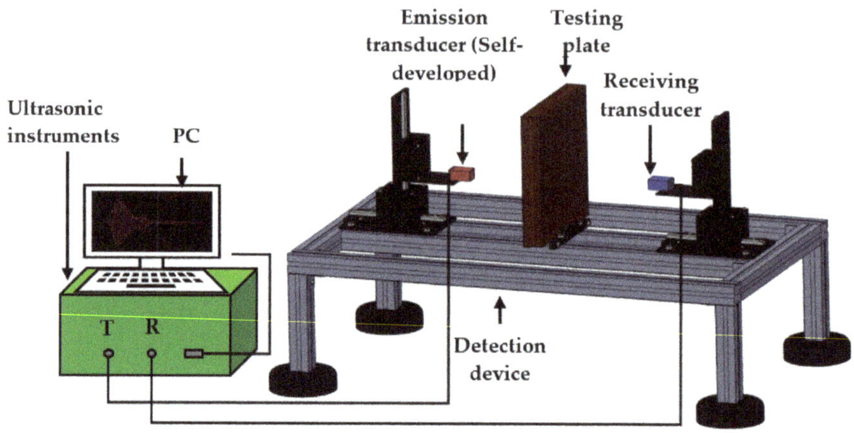

Figure 4. Testing platform.

5. Results and Discussions

5.1. Effect of Material Properties on Acoustic Impedance

Table 3 lists the measured acoustic properties of the first matching layers with different glass microsphere diameters and weight ratios. The data of the BR20 glass microsphere in the 30% weight ratio are not reported due to the high viscosity, which prevented the air bubble extraction even with the diluent and failed the homogeneous acoustic properties of the matching layer. Based on the results in Table 3, the higher the hollow glass microsphere weight ratio is, the lower the density and longitudinal wave velocity are. From the microstructure of the first matching layer, the higher hollow glass microsphere weight ratio means that more space is filled by the air inside of the microsphere and longer time the acoustic wave takes to propagate in the glass microsphere, which decreases the overall density and acoustic velocity. With the decrease in the hollow glass microsphere diameter under the same weight ratio, both the density and velocity increase due to the increase in the glass weight percentage.

Table 3. First matching layer acoustic properties. (Exp-experiment, Mod-modeling).

Model-*R*c(%)-*R*d(%)	Density, ρ_1 (kg/m^3)			Longitudinal Wave Velocity, c_{L1} (m/s)			Acoustic Impedance, Z_1 (MRayl)		
	Exp	Mod	Error	Exp	Mod	Error	Exp	Mod	Error
BR20-10-0	813	788	3.0%	2770	2726	1.6%	2.25	2.15	4.5%
BR20-15-15	695	695	0.0%	2367	2357	0.4%	1.64	1.64	0.1%
BR20-20-15	609	614	0.8%	2114	2344	10.9% *	1.29	1.44	11.6% *
BR40-10-0	954	982	2.9%	2890	2864	0.9%	2.76	2.81	1.9%
BR40-15-0	929	908	2.2%	2841	2774	2.4%	2.64	2.52	4.6%
BR40-20-0	840	845	0.6%	2820	2701	4.2%	2.37	2.28	3.7%
BR40-30-0	742	742	0.1%	2770	2590	6.5%	2.05	1.92	6.3%
BR60-10-0	1059	1069	0.9%	2962	2931	1.1%	3.14	3.13	0.2%
BR60-15-0	1015	1025	1.0%	2857	2853	0.1%	2.9	2.92	0.8%
BR60-20-0	977	984	0.7%	2807	2786	0.8%	2.74	2.74	0.0%
BR60-30-0	896	911	1.6%	2686	2673	0.5%	2.41	2.43	1.0%

* The high error is due to the bubble and inhomogeneous microstructure in the matching layer.

The first matching layer was cut into the design shape to be pasted on the 1-3 piezoelectric composite. The surface after cutting was observed by the microscopy to check the microstructure. The white microsphere and epoxy matrix can be identified in all figures.

Some white dots, marked by the arrows, are air bubbles. With the increase in the microsphere weight ratio (R_c), the number of bubbles increased, based on the observation in Figure 5a. Figure 5b,c show a similar trend. The higher R_c increases the viscosity of the mixture, which makes the bubbles hard to escape. The material flowing in the high viscosity environment is also constrained, which makes the matching layer easy to inhomogeneous. The inhomogeneous microstructure of the matching layer will affect the longitudinal wave velocity, which can be another possible reason for the experiments BR20-20-15. The bubble and inhomogeneous material also enlarge the error between the theoretical model and experiments results, indicated in Table 3, which makes the acoustic properties of the matching layer difficult to predict. The diluent can improve the bubble release, as seen in the 15% and 20% R_c in Figure 5a and Table 3.

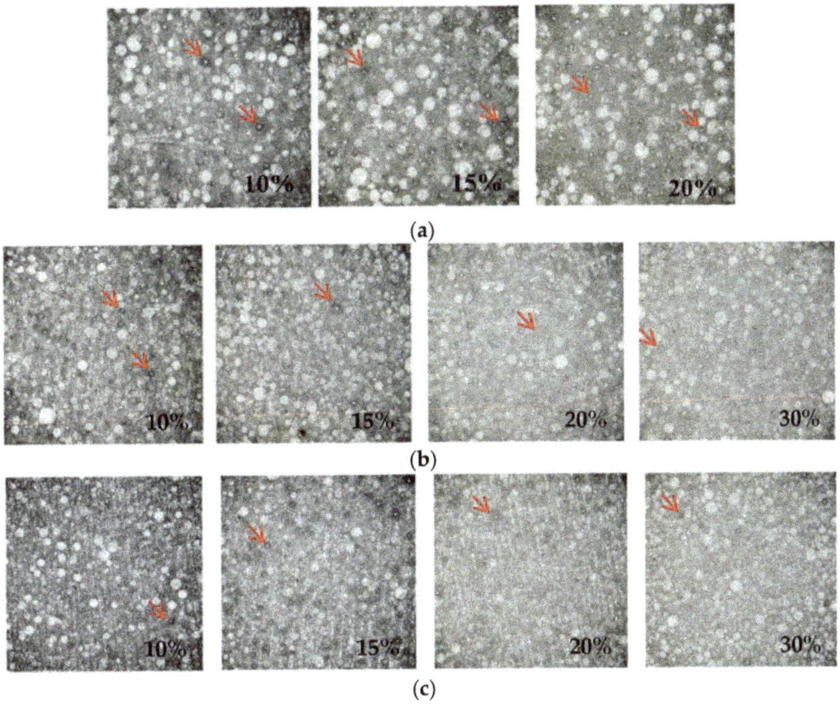

Figure 5. Microstructure of the epoxy/hollow glass microsphere composites. (**a**) BR20, (**b**) BR40, and (**c**) BR60 microsphere. (The value on the figure is the weight ratio R_c, the red arrows are the air bubble).

5.2. Validation of the Matching Layer Modeling

Table 3 also provides the modeling results of density, longitudinal wave velocity, and acoustic impedance. The proposed density model can predict all experiment density results with an error of less than 3%, which proves the correctness and accuracy of the modeling. For the longitudinal wave velocity, all experiments are modeled in the error of less than 6.5%, except BR20-20-15, which has an error of up to 10.9%. The acoustic impedance prediction results share a similar trend with the longitudinal wave velocity, which has high accuracy with an error of less than 6.3% if the BR20-20-15 is not considered. As mentioned above, in BR20, the viscosity of the composite increases tremendously when the weight ratio is larger than 15%, and the diluent has to be used. When 15% diluent is added in BR-20-15-15, the matching layer can be formed. However, when the weight ratio of hollow glass microspheres increases to 30%, the microsphere cannot be stirred to mix with the resin even with 15% diluent. More diluent, up to 50%, had been tried, which still can obtain many visible bubbles when the composite is cured. The reason for the relatively poor accuracy in BR20-20-15 maybe result from the inhomogeneous of the material and microbubble inside, which has been confirmed by Figure 5.

After verification of the matching layer modeling in this study, the best material components needed to form the theoretical acoustic impedance are predicted. Based on the proposed modeling, the theoretical first matching layer can be made by BR20 with a 19% weight ratio and 15% diluent. The fabricated sample from the modeling has a density of 625.35 kg/m^3 and 2186 m/s acoustic velocity. The calculated acoustic impedance is 1.36 MRayl, which is a 1.4% error from the theoretical value. The second matching layer, which can be considered as the composition of polypropylene and air bubble, needs an expansion rate of 80 based on the modeling in this study. The density, acoustic velocity, and acoustic impedance of the fabricated second layer are 67.13 kg/m^3, 738 m/s, and 0.049 MRayl, respectively. The larger acoustic impedance error results from the inaccuracy of the expansion rate, which is hard to control.

To further verify the modeling results on the sensitivity, several matching layers were used to make the transducer to test the signal amplitude in the air by using the testing platform in Figure 4. Due to the small acoustic impedance in the second matching layer, the accurate controlling of acoustic impedance is difficult. Thus, only the first matching layer acoustic impedance changes in the experiments. The second matching layer used the 0.049 MRayl samples. All matching materials were sliced into 20 mm $\times$ 14 mm blocks with a quarter wavelength of the acoustic in matching material thickness and pasted on the 1-3 composite with the E51 epoxy to fabricate the transducer. From the modeling result, the first layer in BR20-19-15 has the acoustic impedance closest to the theoretical value. The other two matching layers with BR20-15-15 and BR20-20-15, with higher and lower acoustic impedance, respectively, were also tested. Moreover, another purchased transducer from Japan Probe was also used to compare the performance of the self-developed with the existing industry transducer. The received signal results are shown in Figure 6. The BR20-19-15 (seen in Figure 6b) has the highest amplitude, up to 961.4 mV, which is about 10% higher than BR 20-15-15 (Figure 6a) and BR 20-20-15 (Figure 6c), which are 848.2 and 886.4 mV, respectively. This phenomenon confirms that the higher or lower acoustic impedance compared to the theoretical acoustic impedance will result in a weaker amplitude and reduce the performance of the air-coupled transducer. Compared to the Japan Probe transducer with the same 200 kHz resonant frequency, the self-developed transducer has a 20% higher amplitude, which further proves the significance of this study.

5.3. Effect of Bonding Layer on Transducer Performance

The acoustic impedances of each layer with different bonding systems are shown in Table 4, and the amplitude signals are provided in Figure 7. Without the bonding layer, the 1-3 piezoelectric composite, first matching layer, and second matching layer are connected by the air, and the transducer has an amplitude of 111.2 mV, as demonstrated in Figure 7a. This amplitude is only 11.5% of the transducer bonded by the epoxy, as provided in Figure 6b. After changing the bonding material to the hollow glass microsphere/epoxy composite, the amplitude increased to 1283 mV, 33% higher than that of the epoxy bonding. This result can be explained by the acoustic impedance distribution in Table 4.

When the acoustic wave propagates from one homogenous material to another, both the reflection and refraction happen in the interface. The reflected acoustic wave returned to the first material, and the refracted one prorogates into the second material. The larger discrepancy between the two materials, the smaller percentage of the acoustic can propagate into the second material. For the transducer with two matching layers, there are five interfaces to be penetrated. When connected by air, most acoustic waves from the 1-3 piezoelectric reflect the composite, and only a few propagate into the air. When the acoustic wave in the air propagates to the first matching layer, most of the energy reflects into the air again, which happens several times in the air bonded transducer. Thus, little energy can pass through the transducer and be received by the receiving transducer. For the transducer bonded by the epoxy, the discrepancy is reduced, especially between the 1-3 composite and the first matching layer. This reduced acoustic impedance difference allows more energy to pass through the transducer. For the transducer bonded by the hollow glass microsphere and epoxy composite, even if the difference between the 1-3 piezoelectric composite and the first matching layer increased, the interface layer decreased to three to reduce the energy dissipation.

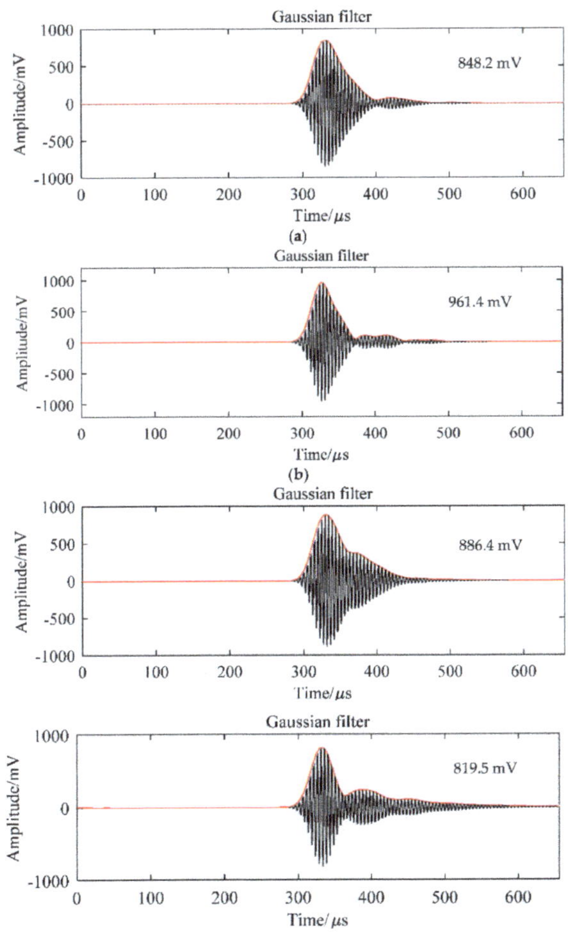

Figure 6. Time–domain filtered and enveloped signal of (**a**) BR20-15-15, (**b**) BR20-19-15, (**c**) BR20-15-15, and (**d**) transducer from Japan Probe Co., LTD.

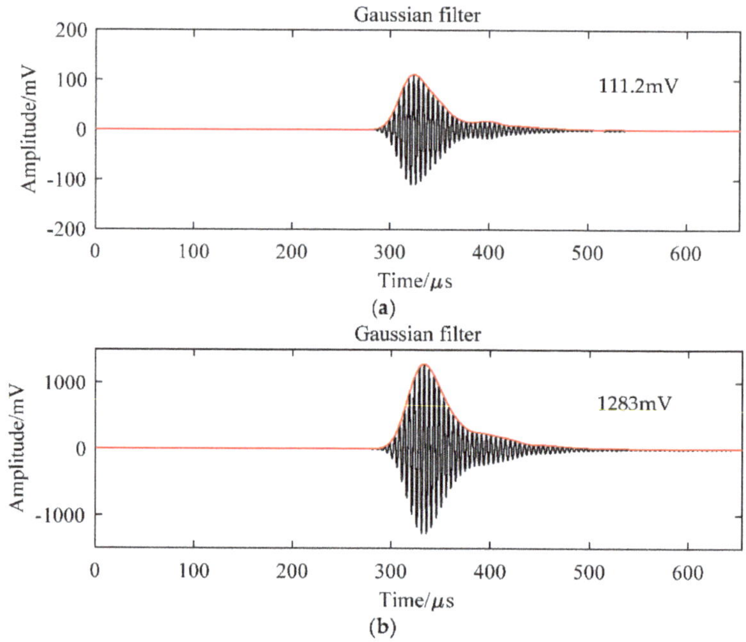

Figure 7. Time–domain filtered and enveloped signal. (**a**) Without bonding layer, (**b**) with the hollow glass beads/epoxy resin system.

Table 4. Acoustic impedance distribution of the transducer with three different bonding materials.

Structure Layer	Acoustic Impedance (MRayl)		
	Air	**Epoxy**	**Hollow Glass Microsphere and Epoxy Composite**
1-3 piezoelectric composite	19.37	19.37	19.37
First bonding layer	0.000425	3	1.36
First matching layer	1.36	1.36	1.36
Second bonding layer	0.000425	3	1.36
Second matching layer	0.049	0.049	0.049
Air	0.000425	0.000425	0.000425

5.4. Practical Application Test of Self-Developed Air-Coupled Ultrasonic Transducer

The received signals from the air-coupled ultrasonic transducer after penetrating the testing plate are shown in Figure 8. When an ultrasonic wave propagates in PVC foam and CFRP plate without defects, the peak-to-peak amplitudes of received signals are respectively 363.2 and 83.6 mV, as seen in Figure 8a,b. When the defects in PVC foam and CFRP plates locate in the ultrasonic propagation path, the peak-to-peak amplitude of received signals were 155.8 and 54.5 mV, which decreased by about 57% and 35%, respectively. Due to the interaction between defects and ultrasonic waves, the energy of the transmitted wave was reduced, which led to the change in the shape and amplitude of the received wave. Due to the different types, sizes, and locations of defects, the received acoustic wave may change into various shapes, but this will not affect the detection of defects. By comparing the waveform amplitude of the received signals with or without defects, the defects are easy to detect. Further, the defect information can be directly distinguished by scanning and detecting the area to be inspected. The air-coupled ultrasonic transducer developed in this study can be used in practical testing applications.

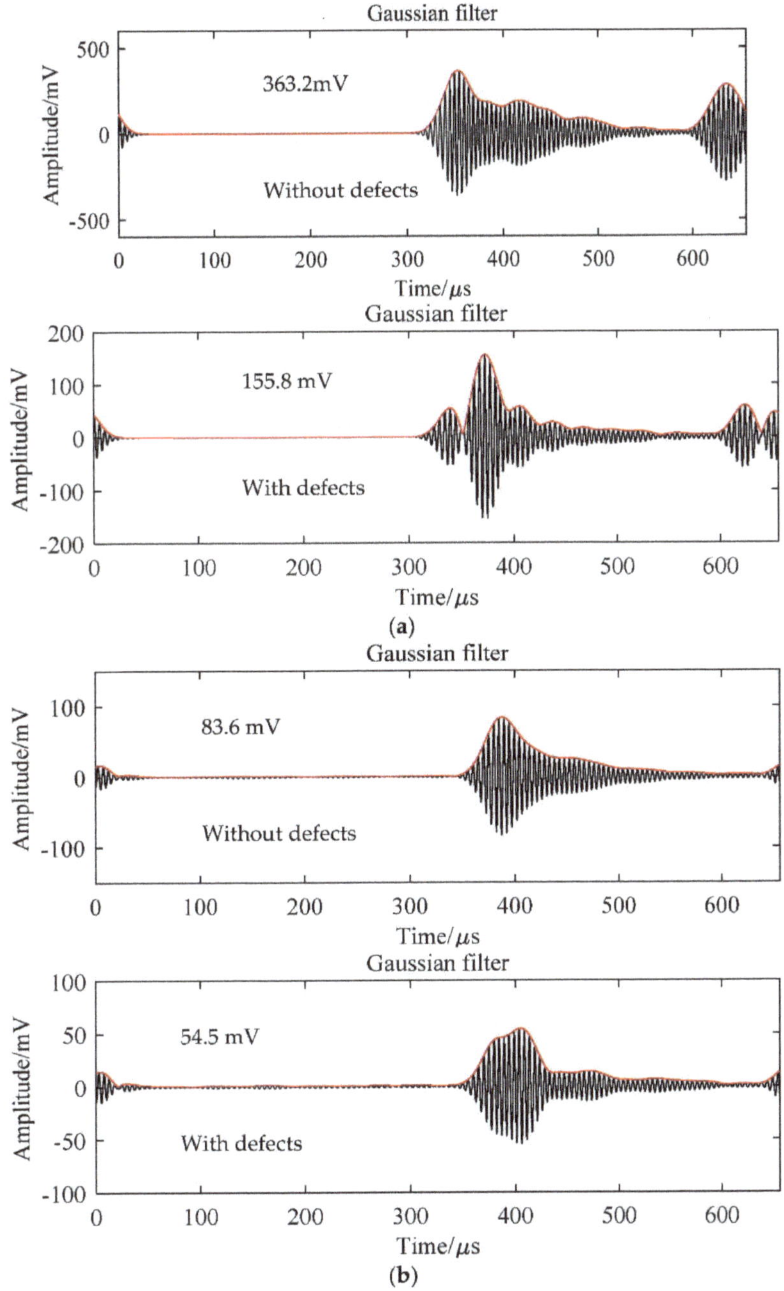

Figure 8. Time–domain filtered and enveloped signal of (**a**) PVC foam, (**b**) CFRP plates with or without defects.

6. Conclusions

This paper introduces a methodology to model, design, fabricate, and optimize air-coupled ultrasonic transducers matching system, which includes the matching layer and bonding layer. The matching layer modeling process and evaluation methods are demonstrated and verified by the self-developed air-coupled ultrasonic transducer with double matching layers. The mechanism of the matching layer components affecting the transducer performance is explained. Additionally, the bonding layer effects on the transducer performance are described. The testing results indicate that the self-developed air-coupled ultrasonic transducer in this study has a 20% higher amplitude than the product on the market and can easily identify defects in non-metallic materials. This study provides the foundation for air-coupled ultrasonic transducer modeling and development.

Author Contributions: Conceptualization, J.Z. and Y.L.; methodology, J.B. and Y.L.; validation, J.B. and Y.L.; investigation, J.B.; data curation, J.B. and Y.L.; writing—original draft preparation, J.B. and Y.L.; writing—review and editing, J.Z.; project administration, J.Z. All authors have read and agreed to the published version of the manuscript.

Funding: Supported by the Opening Project of Shanxi Key Laboratory of Advanced Manufacturing Technology (No. XJZZ201903).

Institutional Review Board Statement: Not applicable.

Informed Consent Statement: Not applicable.

Data Availability Statement: Data sharing not applicable.

Conflicts of Interest: The authors declare no conflict of interest.

References

1. Blomme, E.; Bulcaen, D.; Declercq, F.; Lust, P. Air-coupled ultrasonic evaluation of coated textiles. In Proceedings of the Ultrasonics Symposium, Munich, Germany, 8–11 October 2002.
2. Sukmana, D.D.; Ihara, I. Surface Roughness Characterization through the Use of Diffuse Component of Scattered Air-Coupled Ultrasound. *Jpn. J. Appl. Phys. Part* **2006**, *45*, 4534–4540. [CrossRef]
3. Dang, C.J.; Li, M.X. 1-3 Piezo-Composite materials. *Appl. Acoust.* **1995**, *14*, 2–7.
4. Zhang, Y.Y.; Wang, S.; Liu, D.A.; Zhang, Q.; Wei, W.; Bo, R.; Zhao, X.; Luo, H. Fabrication of angle beam two-element ultrasonic transducers with PMN-PT single crystal and PMN-PT/epoxy 1-3 composite for NDE applications. *Sens. Actuators Phys.* **2011**, *168*, 223–228. [CrossRef]
5. Zhu, J.; Cao, W.W.; Jiang, B.; Zhang, D.S.; Zheng, H.; Zhou, Q.; Shung, K.K. Nano-structured TiO_2 film fabricated atroom temperature and its acoustic properties. *J. Phys. D-Appl. Phys.* **2008**, *41*, 162001. [CrossRef] [PubMed]
6. Ye, Z.Y.; Wu, H.D. Research on the Transducer's Matching Layer Parameters Optimization. *Yadian Yu Shengguang/Piezoelectrics Acoustooptics* **2015**, *37*, 368–372.
7. Toda, M. New type of matching layer for air-coupled ultrasonic transducers. *IEEE Trans. Ultrason. Ferroelectr. Freq. Control.* **2002**, *49*, 972–979. [CrossRef] [PubMed]
8. Kelly, S.P.; Hayward, G.; Alvarez-Arenas, T. Characterization and assessment of an integrated matching layer for air-coupled ultrasonic applications. *IEEE Trans. Ultrason. Ferroelectr. Freq. Control.* **2004**, *51*, 1314–1323. [CrossRef] [PubMed]
9. Alvarez-Arenas, T. Acoustic impedance matching of piezoelectric transducers to the air. *IEEE Trans. Ultrason. Ferroelectr. Freq. Control.* **2004**, *51*, 624–633. [CrossRef]
10. Alvarez-Arenas, T.G. Air-Coupled Piezoelectric Transducers with Active Polypropylene Foam Matching Layers. *Sensors* **2013**, *13*, 5996–6013. [CrossRef] [PubMed]
11. Saito, K.; Nishihira, M.; Imano, K. P2-3 Experimental Study on Intermediumte Layer for Air-Coupled Ultrasonic Transducer with (0–3) Composite Materials(Poster session 2). *Proc. Symp. Ultrason. Electron.* **2006**, *27*, 213–214.
12. Bovtun, V.; Doering, J.; Beck, U.; Erhard, A.; Yakymenko, Y. Ferroelectret non-contact ultrasonic transducers. *Appl. Phys. A* **2007**, *88*, 737–743. [CrossRef]
13. Rymantas, J.K.; Sliteris, R.; Sestoke, J. Air-Coupled Low Frequency Ultrasonic Transducers and Arrays with PMN-32%PT Piezoelectric Crystals. *Sensors* **2017**, *17*, 95.
14. Kazys, R.; Sliteris, R.; Sestoke, J.; Vladisauskas, A. Air-Coupled Ultrasonic Transducers Based on an Application of the PMN-32%PT Single Crystals. *Ferroelectrics* **2015**, *480*, 85–91. [CrossRef]
15. Guo, S.; Li, J.; Wang, H. Design parameters and modal optimization of radial vibration matching layer transducer. *Diam. Abras. Eng.* **2021**, *41*, 68–73.
16. Wang, X.; Wu, H.; Zhang, X.; Zhang, D.; Dong, Z. Investigation of a multi-element focused air-coupled transducer. *AIP Adv.* **2018**, *8*, 095010. [CrossRef]
17. Wu, Q.; Chen, Q.Y.; Lian, G.X.; Wang, X.; Song, X.; Zhang, X. Investigation of an air-coupled transducer with a closed-cell material matching strategy and an optimization design considering the electrical input impedance. *Ultrasonics* **2021**, *115*, 106477. [CrossRef] [PubMed]

18. Song, K.; Kwak, J.; Park, J.; Hur, S. Acoustic Metasurfaces for Efficienr Matching of Non-contact Ultrasonic Transducers. *Smart Mater. Struct.* **2021**, *30*, 085011. [CrossRef]
19. Feng, R. *Ultrasonic Handbook;* Nanjing University Press: Nanjing, China, 2001.
20. Bian, J.C.; Hu, W.X.; Zhou, B.M. Air-coupled 1-3 connectivity composite piezoelectric ultrasonic transducers. *Tech. Acoust.* **2016**, *35*, 717–720.
21. Rose, J.L. *Ultrasonic Guided Waves in Solid Medium;* Cambridge University Press: Cambridge, UK, 2014.

micromachines

MDPI

Article

Effects of pH Values and H_2O_2 Concentrations on the Chemical Enhanced Shear Dilatancy Polishing of Tungsten

Liang Xu [1], Lin Wang [1], Hongyu Chen [1,2,*], Xu Wang [1,2], Fangyuan Chen [1], Binghai Lyu [1,2], Wei Hang [1,2], Wenhong Zhao [1,2] and Julong Yuan [1,2]

1 College of Mechanical Engineering, Zhejiang University of Technology, Hangzhou 310023, China; 2111902045@zjut.edu.cn (L.X.); 2112102013@zjut.edu.cn (L.W.); wx382935877@163.com (X.W.); chenfyupmc@163.com (F.C.); icewater7812@126.com (B.L.); whang@zjut.edu.cn (W.H.); whzhao6666@163.com (W.Z.); jlyuan@zjut.edu.cn (J.Y.)

2 Key Laboratory of Special Purpose Equipment and Advanced Processing Technology, Ministry of Education and Zhejiang Province, Zhejiang University of Technology, Hangzhou 310023, China

* Correspondence: 15222918166@163.com

Abstract: In order to obtain tungsten with great surface qualities and high polishing efficiency, a novel method of chemical enhanced shear dilatancy polishing (C-SDP) was proposed. The effects of pH values and H_2O_2 concentrations on the polishing performance of tungsten C-SDP were studied. In addition, the corrosion behaviors of tungsten in solutions with different pH values and H_2O_2 concentrations were analyzed by electrochemical experiments, and the valence states of elements on the tungsten surface were analyzed by XPS. The results showed that both pH values and H_2O_2 concentrations had significant effects on tungsten C-SDP. With the pH values increasing from 7 to 12, the *MRR* increased from 6.69 μm/h to 13.67 μm/h. The optimal surface quality was obtained at pH = 9, the surface roughness (R_a) reached 2.35 nm, and the corresponding *MRR* was 9.71 μm/h. The *MRR* increased from 9.71 μm/h to 34.95 μm/h with the H_2O_2 concentrations increasing from 0 to 2 vol.%. When the concentration of H_2O_2 was 1 vol.%, the R_a of tungsten reached the lowest value, which was 1.87 nm, and the *MRR* was 26.46 μm/h. This reveals that C-SDP technology is a novel ultra-precision machining method that can achieve great surface qualities and polishing efficiency of tungsten.

Keywords: chemical enhanced shear dilatancy; tungsten; polishing; surface quality

Citation: Xu, L.; Wang, L.; Chen, H.; Wang, X.; Chen, F.; Lyu, B.; Hang, W.; Zhao, W.; Yuan, J. Effects of pH Values and H_2O_2 Concentrations on the Chemical Enhanced Shear Dilatancy Polishing of Tungsten. *Micromachines* **2022**, *13*, 762. https://doi.org/10.3390/mi13050762

Academic Editors: Xiuqing Hao, Duanzhi Duan and Youqiang Xing

Received: 15 April 2022
Accepted: 10 May 2022
Published: 12 May 2022

1. Introduction

Due to the advantages of high melting point, great corrosion resistance, superior electrical conductivity, and high-temperature strength [1–3], tungsten is widely used in integrated circuits, nuclear material, the military industry, medical equipment, and other fields [4–6]. The surface quality of the material has a significant influence on the service performance and service life of the workpiece. For example, as the primary candidate for plasma facing materials (PFMs) in the diverter of the ITER and DEMO fusion reactors, tungsten needs to withstand the impact of high-energy particles. The surface quality will affect its radiation resistance to a certain extent, thus affecting the service life of the nuclear fusion reactor [7]. In semiconductor devices, tungsten is deposited on the device surface by chemical vapor deposition. The surface quality of tungsten film will affect the interconnect performance of devices, and therefore the surface of tungsten film needs to be polished [8]. Ultra-precision polishing is the final processing method to reduce the surface roughness of workpiece, remove the damaged layer, and obtain high surface accuracy and excellent surface quality [9–11]. However, as a typical hard and brittle material, there are great challenges in the ultra-precision polishing of tungsten due to the high hardness, high brittleness, and great wear resistance of material [12,13].

In order to achieve high-efficiency and high-quality polishing, researchers have conducted extensive research on tungsten polishing methods and processes. Poddar et al. [14] used a mixed oxidant composed of H_2O_2 and $Fe(NO_3)_3$ for chemical–mechanical polishing of tungsten and found that the mixed oxidant would generate OH with stronger oxidizing ability, and its removal efficiency was higher than that of a single oxidant. Lim et al. [15] found that the polishing rate could be divided into two regions with increases in $Fe(NO_3)_3$ concentration; when $Fe(NO_3)_3$ concentration was less than 0.1 wt.%, the polishing rate of tungsten increased rapidly, while when the $Fe(NO_3)_3$ concentration was higher than 0.1 wt.%, the polishing rate increased slowly. Han et al. [16] used an alkaline electrolyte containing 0.27 mol/L NaOH for electrochemical polishing of tungsten under the optimal electrode gap width, and surface roughness (R_a) of the polished tungsten reached 7.5 nm. Based on polishing experiments and theoretical analysis, Wang et al. [17] proposed that electrochemical polishing of tungsten should be divided into three stages: corrosion stage, bright stage, and pitting stage. Under the optimal process, surface roughness S_a was only 3.73 nm after polishing for 10 min. Chen et al. [18] proposed a high-efficiency electrochemical polishing method for tungsten surfaces combining forced convection and natural convection. After electrochemical polishing by forced convection for 3 min and natural convection for 3 min, surface roughness R_a was reduced to 17.2 nm. Zhou et al. [19] developed a dynamic electrochemical polishing process using a bi-layer NaOH electrolyte to uniformly polish tungsten microfluidic molds. With the optimized parameters, surface roughness S_a was reduced from 205.98 nm to 4.14 nm after 10 cycles of dynamic electrochemical polishing. Tungsten will be widely used in various fields in the future because of its excellent comprehensive mechanical properties. Therefore, it is particularly important to explore new ultra-precision polishing methods to improve the surface quality and service life of tungsten workpieces.

Shear Dilatancy Polishing (SDP) is a high-efficiency, high-quality, low-cost surface polishing method that has emerged in recent years. The principle is to use viscoelastic material with non-Newtonian fluid properties to prepare a specific shear pad in order to enhance the control of abrasive particles and improve the stress evenness and contact pressure on the workpiece surface based on the shear dilatancy and solidification effects under high-pressure and high-speed conditions to finally achieve high-efficiency and high-quality polishing. The scratching effect of abrasives on the workpiece is the key to achieving material removal [20,21]. Compared with conventional polishing pads, the dilatancy pad can hold more abrasives. Doi et al. [22,23] used viscoelastic materials such as asphalt and potato starch to prepare a specific dilatancy pad, which could reduce the surface defects caused by stress concentrations in local areas during processing. Results showed that under low–medium speed/pressure, the material removal rate of the SiC wafer after the dilatancy pad processing was more than three times of that after metal tin plate processing, the surface scratches after the dilatancy pad processing were lower than 1% of the latter, and the depth of the subsurface damage layer was less than 10% of the latter.

In this study, Chemical enhanced Shear Dilatancy Polishing (C-SDP) as a novel ultra-precision polishing method was proposed to obtain high surface quality tungsten. The effects of pH values and H_2O_2 concentrations of polishing slurry on material removal rate (MRR) and surface roughness (R_a) in the tungsten C-SDP process were studied. In addition, the corrosion behaviors of tungsten in solutions with different pH values and H_2O_2 concentrations were analyzed by electrochemical experiments, and the valence states of elements on the tungsten surface were analyzed by XPS.

2. Principle of Chemical Enhanced Shear Dilatancy Polishing

Figure 1a,b are schematic illustrations of the SDP and C-SDP principles, respectively. In the SDP processing, the abrasives can be trapped in the viscoelastic material to avoid the height difference caused by different abrasive particle sizes and improve the uniformity of force on the workpiece surface. C-SDP is a polishing method with the synergistic effect of mechanical action and chemical action. It selectively removes workpiece surface roughness peaks based on chemical etching of oxidants and efficient mechanical removal of the dilatancy pad, resulting in higher efficiency and greater surface quality of workpieces than can be achieve using SDP.

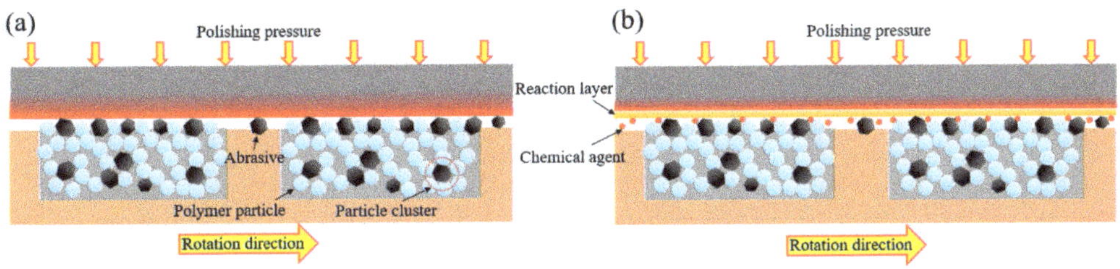

Figure 1. Schematic illustrations of the polishing principle: (**a**) Shear Dilatancy Polishing (SDP); (**b**) Chemical enhanced Shear Dilatancy Polishing (C-SDP).

During C-SDP processing, the chemical polishing slurry contacts the tungsten surface so that a chemical reaction occurs, forming a passivation film with a lower hardness than tungsten. Moreover, relative movement between the dilatancy pad and tungsten workpiece will happen. Under shear force and pressure, viscoelastic material with non-Newtonian fluid properties in the contact area instantly exhibits shear dilatancy and solidification effects, forming a "flexible fixed abrasive tool (FFAT)" in the processing area, which can realize micro-cutting removal of the reaction layer. C-SDP is a synergistic process in which the abrasives remove the passivation film, and the bare surface reacts actively–passively with the polishing slurry to reform the passivation film [24]. In the process of polishing, chemical corrosion of the polishing fluid and mechanical grinding of the abrasives are coupled to achieve material removal at the atomic level and efficient removal of the tungsten surface.

3. Experiments

3.1. Preparation of Dilatancy Pad and Polishing Slurry

Viscoelastic materials, ingredients, and abrasives with a certain mass ratio were evenly mixed by mechanical agitation at 80 °C. After that, the mixture was cooled to room temperature to obtain shear dilatancy material suitable for SDP. In our research, the viscoelastic material was a polymer material with non-Newtonian fluid properties, and fumed silica was used as the ingredient to improve the mechanical properties of viscoelastic material. This viscoelastic material is prone to shear hardening under the action of external forces, which can enhance the holding force of abrasives. The dilatancy pad was obtained by filling the shear dilatancy material in a special polyurethane polishing pad, as shown in Figure 2a. The structure of the polyurethane dilatancy pad is shown in Figure 2b. The polyurethane pad filled with the shear dilatancy material is attached to the rigid layer to achieve a certain rigid support effect. The magnetic layer at the bottom makes the dilatancy pad magnetically adsorbed on the surface of the polishing base plate for easy replacement.

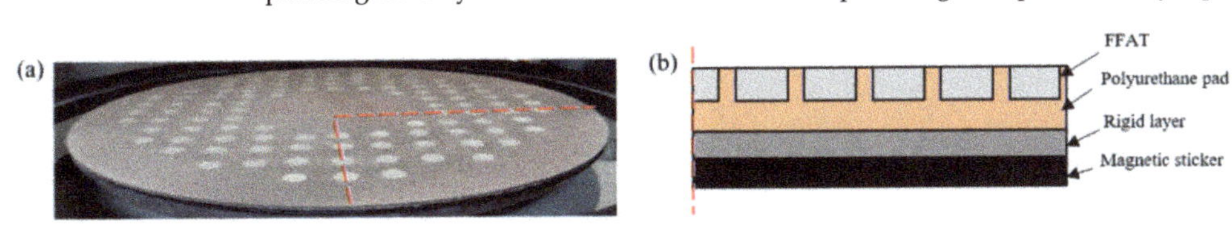

Figure 2. Polyurethane dilatancy pad: (**a**) physical diagram; (**b**) structure diagram.

The polishing base slurry was prepared by deionized water, the dispersant, and the active agent. Diamond micro-powders with a particle size of 0.5 μm were used as the abrasives. H_2O_2 with a concentration of 30 vol.% was used as the oxidant. The pH value of the polishing slurry was adjusted by NaOH. The slurry was continuously stirred for 30 min to ensure that all components were well mixed for the polishing experiments.

3.2. Experimental Process and Conditions

Tungsten workpieces used for the polishing experiments were obtained by a rolling process, which gave them a high density. The characteristics of tungsten are shown in Table 1. In this study, the plane workpiece was taken as the research target. Tungsten samples were 10 mm in diameter and 0.3 mm in thickness. C-SDP polishing experiments were carried out on the experimental device, as shown in Figure 3. During the polishing process, tungsten samples fixed on the fixture rotated along the normal direction with a certain pressure to ensure a uniform polishing of the workpiece surface.

Table 1. Chemical composition and mechanical characteristics of tungsten.

Parameters	Values
Content of W (%)	≥99.99
Density (g/cm^3)	19.35
Mohs hardness	7.5
Tensile strength (MPa)	980–1078
Fracture toughness (MPa·m$^{1/2}$)	5.4

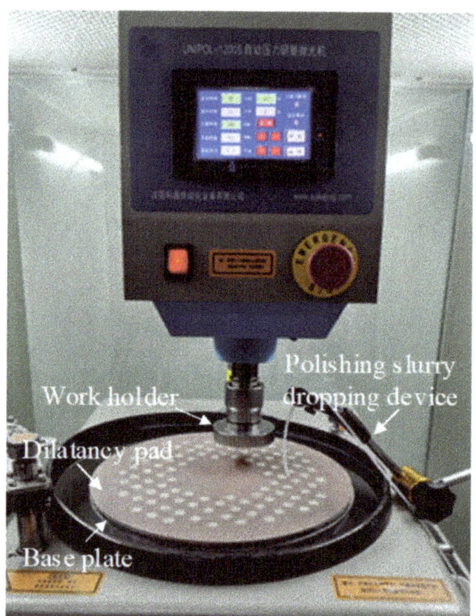

Figure 3. The experimental device of C-SDP.

The experimental conditions are shown in Table 2. In order to achieve better mechanical removal effects of the dilatancy pads, the C-SDP polishing experiments were carried out under the condition of high-pressure and high-speed. Tungsten has a high dissolution rate under alkaline conditions, which can increase the material removal rate. Therefore, the effects of polishing slurries at pH 7, 8, 9, 10, 11, and 12 on the polishing effects of tungsten were studied. Moreover, the effects of H_2O_2 concentrations (0–2.0 vol.%) on the polishing performance of tungsten were also investigated.

Table 2. Experimental conditions.

Parameter	Values
Top plate polishing speed (r/min)	20
Bottom plate polishing speed (r/min)	120
Polishing pressure (kPa)	468
Abrasive particle size (μm)	0.5
Abrasive concentration of slurry (wt.%)	1
Polishing pad	Polyurethane dilatancy pad
Slurry flow rate (mL/min)	3
pH	7; 8; 9; 10; 11; 12
H_2O_2 (vol.%)	0; 0.1; 0.5; 1; 1.5; 2

3.3. Measurement and Testing

The pH values of the polishing slurries were measured by a glass electrode pH meter (PB-10, Sartorius, Germany, resolution: 0.01). The *MRR* of tungsten was measured by a precision balance (ME36S, Sartorius, Germany, resolution: 0.001 mg). The formula for calculating *MRR* is as follows:

$$MRR = \frac{\Delta m}{\rho t s} \quad (1)$$

Δm (g) is the quality difference of the tungsten sample before and after polishing, ρ (g/cm^3) is the density of the tungsten sample, s (cm^2) is the area of the tungsten sample, t (h) is the polishing time, and the unit of *MRR* is μm/h. Each polishing experiment was repeated three times, and the mean value was calculated.

After polishing, the surface roughness (R_a) of tungsten was measured by a 3D profile White Light Interferometer (Super View W1, Chotest, Shenzhen, China), and the sampling range of the White Light Interferometer was 0.5 × 0.5 mm. The surface morphology of the workpiece was observed by a large-field-depth digital microscope (VHX-7000, Keyence, Osaka, Japan). The dynamic potential polarization curves of tungsten in abrasive-free solutions with different pH values and H_2O_2 concentrations were tested by an electrochemical system with a three-electrode cell (CHI760E, CH Instruments, Shanghai, China). The chemical reactions between the components of the polishing slurry and tungsten were analyzed by X-ray photoelectron spectroscopy (ESCALAB 250Xi, Thermo Fisher, Waltham, MA, USA).

4. Results and Discussion

4.1. Rheological Analysis of Shear Dilatancy Material

Rheological testing of shear dilatancy material was performed by a rotational rheometer (MCR302, Anton Paar, Graz, Austria) at a constant testing temperature of 25 °C. During the test, the distance between the plate clamp and rotor (both 25 mm in diameter) was 1 mm. The strain was constant at 0.1%, and frequency sweep tests were performed from 0.1 to 10 Hz. The measurement for each sample was repeated three times in order to quantify the measurement error.

Figure 4a,b respectively show the trends of G' and $tan\delta$ of samples with different abrasive concentrations as the frequency increased. The G' represents the ability of material to store elastic deformation energy, which is used to characterize the elasticity of material. The $tan\delta$ represents the viscoelastic properties of material. When the $tan\delta$ is smaller, the elasticity of material is greater. The frequency ($tan\delta = 1$) is the critical frequency of the solid–liquid phase transition, beyond which the material transforms from a liquid-like state to a solid-like state. It can be clearly found that with increasing frequency, the G' of sample increased, while the $tan\delta$ decreased. The material experienced an obvious shear hardening effect, which met the requirements of shear dilatancy polishing. As shown in Figure 4a,b, the abrasive concentrations significantly affected the rheological properties of the shear dilatancy material. With the increase of abrasive concentrations, the G' of the sample increased, and the $tan\delta$ decreased. The elasticity of the material was improved, and its phase transition frequency decreased. In other words, the material transition to a "flexible fixed abrasive tool" requires a lower polishing speed, which makes the material more prone to transition from liquid-like to solid-like. When the abrasive concentration was 30 wt.%, the G' increased significantly, which was much higher than that of 20 wt.%. Combined with the rheological test results and polishing requirements, VM-30 wt.% Dimond was selected for the preparation of the dilatancy pad.

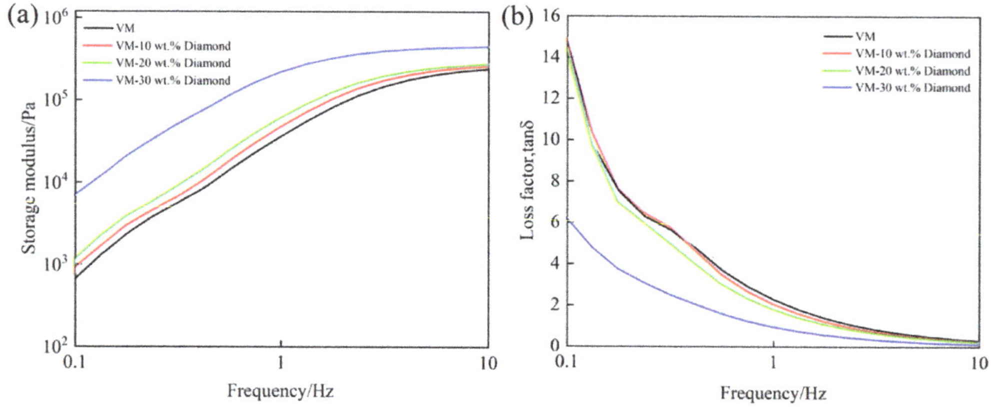

Figure 4. Rheological curves of the shear dilatancy material: (**a**) storage modulus curves; (**b**) loss factor curves.

4.2. Effect of pH Values on the Polishing Performance of Tungsten

pH value is an important component of chemical polishing slurry, which determines the basic polishing environment of C-SDP and directly affects the polishing quality [25]. The effects of pH values on tungsten C-SDP are shown in Figure 5. As shown in Figure 5a, with the increase of pH, the material removal rate of tungsten showed a continuous increasing trend, and it increased from 6.69 μm/h to 13.67 μm/h. In neutral and alkaline environments, the oxide formed on tungsten surfaces is unstable and dissolves in solution at a very low rate to form tungstate ions (WO_4^-) [26,27], which can remove the oxide on the tungsten surface to a certain extent. On the other hand, since the oxide is softer and easier to remove than tungsten, the corresponding mechanical removal effect is also more pronounced. Because the dissolution rate of this oxide is low, the mechanical action of abrasives will remove most of the generated oxide, thereby exposing a new tungsten surface to continue the chemical reaction. In the neutral and alkaline conditions of the polishing slurry, the removal rate increases with the increase of pH, which is also related to the dissolution rate. The material removal rate includes the mechanical removal of abrasives and the dissolution of oxide.

When pH = 7, the material removal was mainly achieved by the mechanical action of diamond abrasives. Because the chemical action was very small at this time, the corrosion effect on the tungsten surface was extremely weak, resulting in a minimum material removal rate. When pH > 7, it was easy for tungsten to react with the alkaline slurry. Micro-convex peaks on the tungsten surface could be oxidized into relatively soft oxides. Under the mechanical grinding of diamond abrasives and the dissolution of alkaline solution, the generated oxides could be easily removed, and thus the material removal rate was improved.

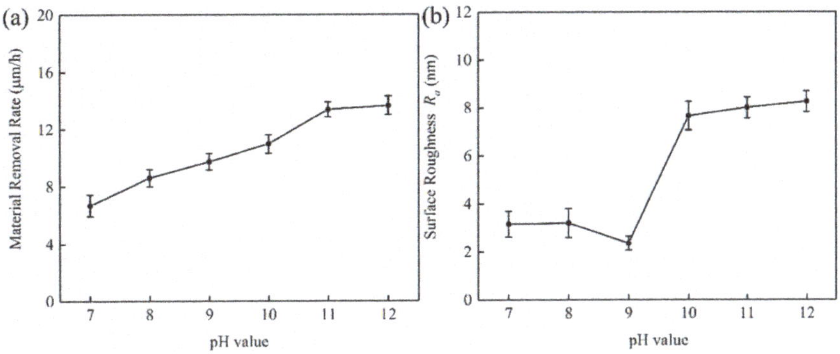

Figure 5. Effect of pH values on the removal rate and surface roughness: (**a**) material remove rate; (**b**) surface roughness.

However, the final tungsten surface quality is determined by the rate of oxide dissolution, the rate of oxide production, and the mechanical action of abrasives. How to control and balance the relationship among the three factors is a key issue that needs to be considered during the polishing process. With the increase of pH values, the surface roughness of the polished tungsten firstly decreased and then increased, as shown in Figure 5b. The surface roughness R_a was the lowest at pH = 9. As the pH increased from 7 to 9, R_a decreased from 3.16 nm to 2.35 nm. However, as the pH sequentially increased to 12, R_a subsequently increased to 8.25 nm.

Figure 6 shows the surface morphologies of tungsten after polishing with different pH values. The tungsten surface after polishing was relatively smooth, as seen in Figure 6a–c. However, the tungsten surface became gradually uneven, as seen in Figure 6d–f, and the corrosion degree of tungsten gradually deepened. It can be inferred that when pH > 9, OH^- in the slurry was very corrosive for tungsten, which likely caused uneven corrosion of micro-convex peaks on the tungsten surface, resulting in poor surface quality after polishing. Figure 7 shows the surface morphologies of tungsten under different pH values. Under strong alkaline conditions (pH = 11 and pH = 12), the tungsten surface was significantly corroded, as shown in Figure 7b,c.

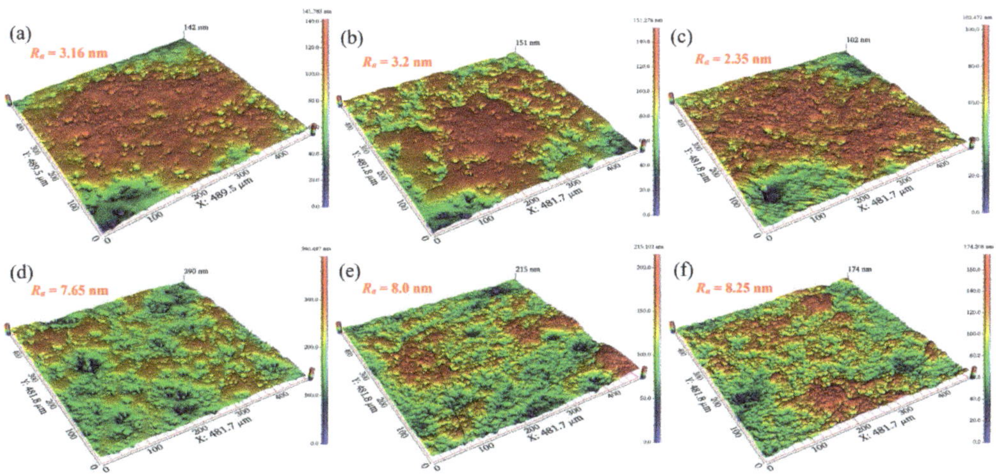

Figure 6. Surface morphologies of polished tungsten under different pH values: (**a**) pH = 7; (**b**) pH = 8; (**c**) pH = 9; (**d**) pH = 10; (**e**) pH = 11; (**f**) pH = 12.

Figure 7. Surface morphologies of polished tungsten under different pH values: (**a**) pH = 9; (**b**) pH = 11; (**c**) pH = 12.

4.3. Effect of H_2O_2 Concentrations on the Polishing Performance of Tungsten

During the polishing process, a passivation film is formed on the tungsten surface due to oxidizing agents. Because of the low hardness of oxide and the weak interface that exists between tungsten and oxide, the passive film is easily removed by the mechanical action of abrasives [28]. The concentrations of H_2O_2 can significantly affect the generation rate of the passivation film on the tungsten surface, which in turn affects the polishing performance of C-SDP [29]. The effects of H_2O_2 concentrations on material removal rate and surface roughness of tungsten are shown in Figure 8. In the experiments, the pH values of polishing slurries with different H_2O_2 concentrations were all 9. As shown in Figure 8a, with the increase of H_2O_2 concentration, the *MRR* of tungsten increased continuously, from 9.71 μm/h to 34.95 μm/h. After the concentration of H_2O_2 increased, the chemical corrosion effect of polishing slurry on the tungsten surface material was enhanced, which could generate faster or thicker soft passivation films. As shown in Figure 8b, the surface roughness of tungsten decreased firstly and then increased, and its surface roughness was the lowest when H_2O_2 concentration was 1 vol.%. As the H_2O_2 concentration increased from 0 to 1 vol.%, R_a decreased from 2.35 nm to 1.87 nm. R_a began to increase, reaching 4.14 nm when the H_2O_2 concentration was 2 vol.%, since the H_2O_2 concentration continued to increase.

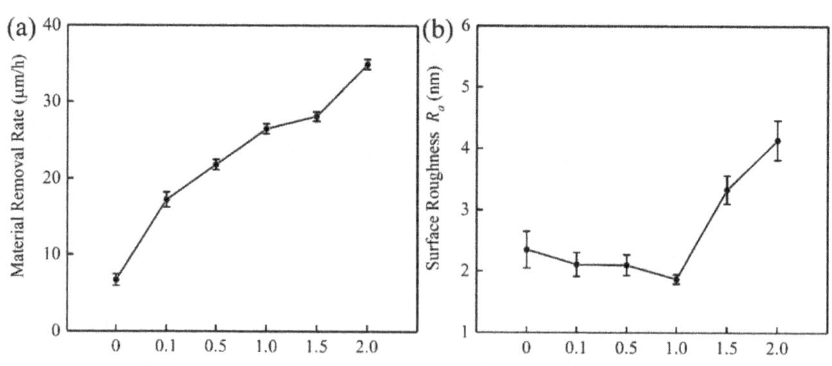

Figure 8. Effect of H_2O_2 concentrations on removal rate and surface roughness: (**a**) material removal rate; (**b**) surface roughness.

Figure 9 shows the surface morphologies of tungsten after polishing with different H_2O_2 concentrations. Compared with the different polishing slurries without H_2O_2, the tungsten surface became smoother after adding 1 vol.% H_2O_2. However, when the concentration of H_2O_2 was higher than 1 vol.%, the surface quality of tungsten became slightly deteriorated due to excessive corrosion.

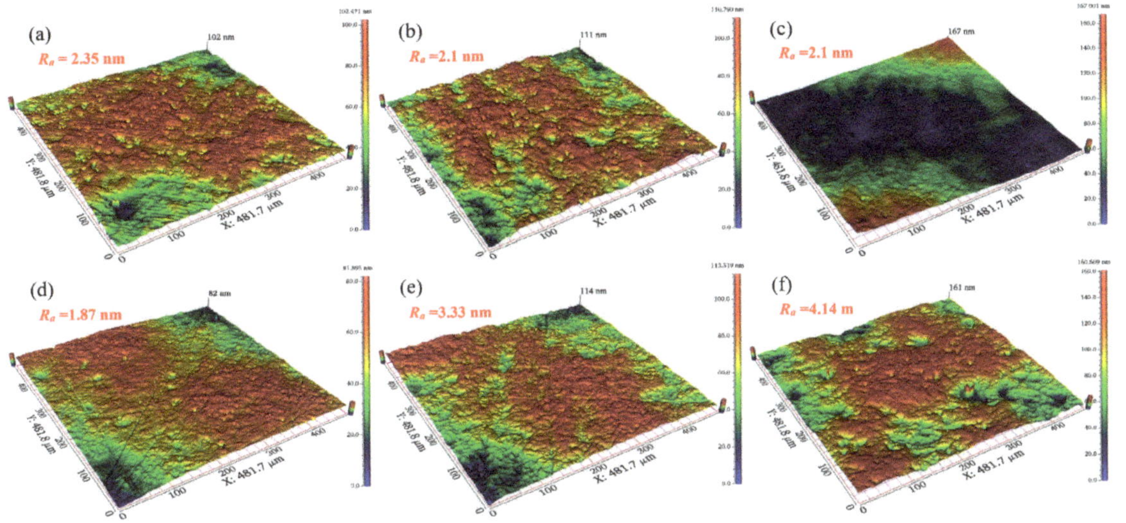

Figure 9. Surface morphologies of polished tungsten under different H_2O_2 concentrations: (**a**) 0 vol.%; (**b**) 0.1 vol.%; (**c**) 0.5 vol.%; (**d**) 1.0 vol.%; (**e**) 1.5 vol.%; (**f**) 2.0 vol.%.

Figure 10 shows the surface morphologies of polished tungsten under different H_2O_2 concentrations. As shown in Figure 10c, when the H_2O_2 concentration was 2 vol.%, a slight orange peel phenomenon and obvious micropores appeared on the tungsten surface. This phenomenon may be related to the fact that the oxidizing property of polishing slurry was too powerful, so that the mechanical removal of abrasives could not keep up with the formation rate of the passivation film. Moreover, the micropores existing in the tungsten were further enlarged. Tungsten samples before and after polishing are shown in Figure 11. Under the optimum parameters of pH = 9 and 1 vol.% H_2O_2 concentration, the surface of the tungsten sample after C-SDP achieved a mirror effect without obvious scratches, pits, or other defects.

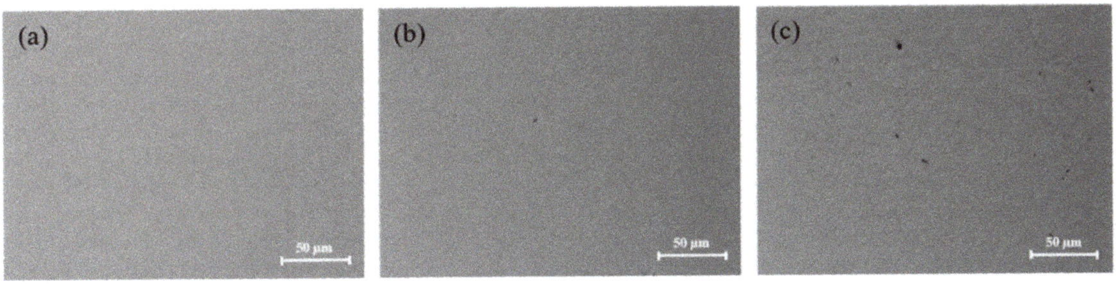

Figure 10. Surface morphologies of polished tungsten under different H_2O_2 concentrations: (**a**) 1.0 vol.% H_2O_2; (**b**) 1.5 vol.% H_2O_2; (**c**) 2.0 vol.% H_2O_2.

4.4. Electrochemical Testing Results

Figure 12a shows the potentiodynamic polarization curves of tungsten in different pH solutions. The composition of electrolyte prepared at different pH values was consistent with that of the polishing slurry except that there was no oxidant. Figure 12b shows the corrosion potential (E_{corr}) and the corrosion current density (I_{corr}) of tungsten at different pH solutions, which could be obtained from the potentiodynamic polarization curves in Figure 12a.

Figure 11. The tungsten samples before and after polishing.

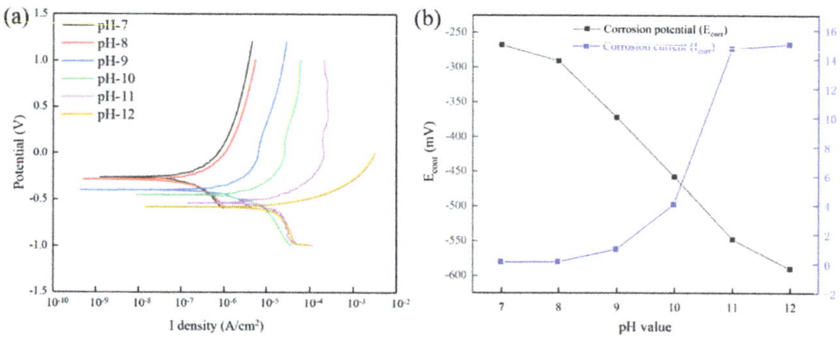

Figure 12. (**a**) Dynamic potential polarization curves at different pH solutions. (**b**) The corrosion potential E_{corr} and the corrosion current density I_{corr} at different pH solutions.

The more negative the corrosion potential is, the more easily the tungsten surface is corroded [30]. The E_{corr} gradually tended to negative values with increasing pH values, as shown in Figure 12b. Among the six pH values, the E_{corr} was the largest (−268 mV) at pH = 7, which indicated that tungsten was the least susceptible to corrosion at this time. The E_{corr} was the smallest (−590 mV) at pH = 12, and tungsten was most easily corroded at this time. These were consistent with the polishing results shown in Figure 6; the surface quality of tungsten was the worst when pH = 12. In electrochemical experiments, the I_{corr} is used to represent the corrosion rate of the workpiece, which can often reflect the change of *MRR* in the polishing process. As shown in Figure 12b, the I_{corr} increased with increases in the pH values, which meant the corrosion rate of tungsten increased gradually. The maximum I_{corr} was 15.01 μA when pH = 12, which was two orders of magnitude higher than that at pH = 7 (0.1185 μA). This phenomenon corresponded to the maximum *MRR* under the condition of pH = 12 shown in Figure 5a. According to the values of I_{corr} and E_{corr}, with increases in pH values, tungsten was more easily corroded, and the corrosion rate increased in the alkaline solution. This phenomenon was consistent with the effect of different pH values on the polishing performance of C-SDP.

Figure 13a shows the potentiodynamic polarization curves of tungsten in different H_2O_2 concentration solutions, and four typical H_2O_2 concentrations of 0, 0.5 vol.%, 1 vol.%, and 2 vol.% were selected. Except for the different concentrations of H_2O_2, the other components of solution were the same as those of the polishing slurry. Figure 13b shows the corrosion potential (E_{corr}) and the corrosion current density (I_{corr}) of tungsten in different H_2O_2 concentration solutions, which could be obtained from the potentiodynamic polarization curves shown in Figure 13a. As shown in Figure 13b, the E_{corr} of tungsten under the H_2O_2 concentration of 1 vol.% (68 mV) was higher than that under the H_2O_2 concentrations of 0 vol.% (−372 mV), 0.5 vol.% (−9 mV), and 2 vol.% (−172 mV). This indicated that when the H_2O_2 concentration was 1 vol.%, the corrosion tendency of the solution for tungsten was the smallest. Between 0 and 1 vol.%, the oxidizing properties of the solution became stronger with the increase of H_2O_2 concentration, and the E_{corr} tended to be more positive. It seems that a dense and thick passivation film gradually formed on the tungsten surface, which reduced the corrosion tendency of tungsten. The E_{corr} decreased at 2 vol.% H_2O_2. This may be due to the high concentration of 2 vol.% H_2O_2, leading to the destruction of the passivation film, which accelerated the dissolution rate of the passivation film in the solution, resulting in a decrease in the corrosion resistance of tungsten. In the four solutions with different H_2O_2 concentrations, the I_{corr} increased with increasing H_2O_2 concentrations, which were 1.003 μA, 4.094 μA, 8.855 μA, and 12.55 μA, respectively, corresponding to the trend of *MRR* in Figure 8a. When the H_2O_2 concentration in solution was 2 vol.%, the I_{corr} was the largest, and it was an order of magnitude higher than that without H_2O_2, which indicated that the passivation film was more likely to be formed on the tungsten surface after the addition of H_2O_2.

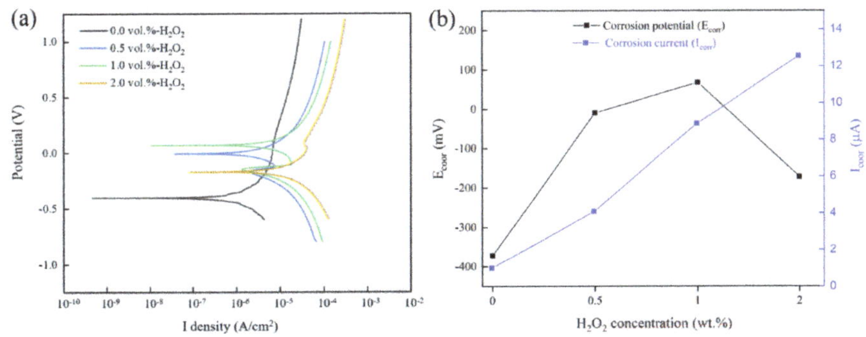

Figure 13. (**a**) Dynamic potential polarization curves at different H_2O_2 concentration solutions. (**b**) The corrosion potential E_{corr} and the corrosion current density I_{corr} at different H_2O_2 concentration solutions.

Figure 14a shows the potentiodynamic polarization curves of tungsten in solutions with different components. Three solutions with different components were an NaOH-based solution (pH = 9), H_2O_2-based solution (1 vol.% H_2O_2), NaOH, and H_2O_2-based solution (pH = 9, 1 vol.% H_2O_2). Figure 14b shows the corrosion potential (E_{corr}) and corrosion current density (I_{corr}) of tungsten in solutions with different components. As shown in Figure 14b, in the solution without H_2O_2 addition, the I_{corr} decreased from 8.855 μA to 1.003 μA, and the E_{corr} changed from 68 V to −372 mV compared with the NaOH and H_2O_2-based solutions. This indicated that H_2O_2 participated in the chemical reaction and could reduce the corrosion of the tungsten surface. After removing NaOH from the solution, the corrosion current I_{corr} on the tungsten surface was the smallest at only 0.454 μA. The result shows that the combined action of NaOH and H_2O_2 can enhance the corrosion rate of tungsten, thereby improving the material removal rate in C-SDP.

It can be seen from the above electrochemical experiments that the chemical corrosion rate and corrosion resistance of tungsten are significantly affected by the chemical agents in the polishing slurry. The etch rate of tungsten can be improved by NaOH and H_2O_2 in the polishing slurry, which is consistent with the material removal rate results shown in Figures 5a and 8a. The chemical composition of the tungsten surface in different solutions is analyzed below, and the chemical reactions that occur during the tungsten C-SDP process are discussed.

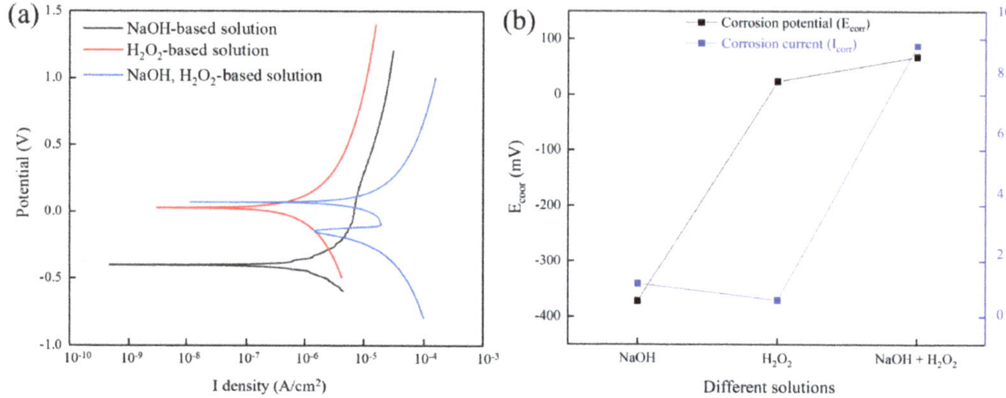

Figure 14. (**a**) Dynamic potential polarization curves at different solutions. (**b**) The corrosion potential E_{corr} and the corrosion current density I_{corr} at different solutions.

4.5. XPS Testing Results

Figure 15 shows the XPS full spectra of the tungsten surface under different conditions. As seen in Figure 15, the main elements on the surface of tungsten samples were C, O, and W. The C element mainly came from the pollutants adsorbed on the tungsten surface during XPS. The presence of the O element under the four different conditions indicates that oxides were produced on the tungsten surface after soaking or polishing. The XPS fine spectra of the tungsten surface in Figure 16 were further analyzed in terms of the element valence.

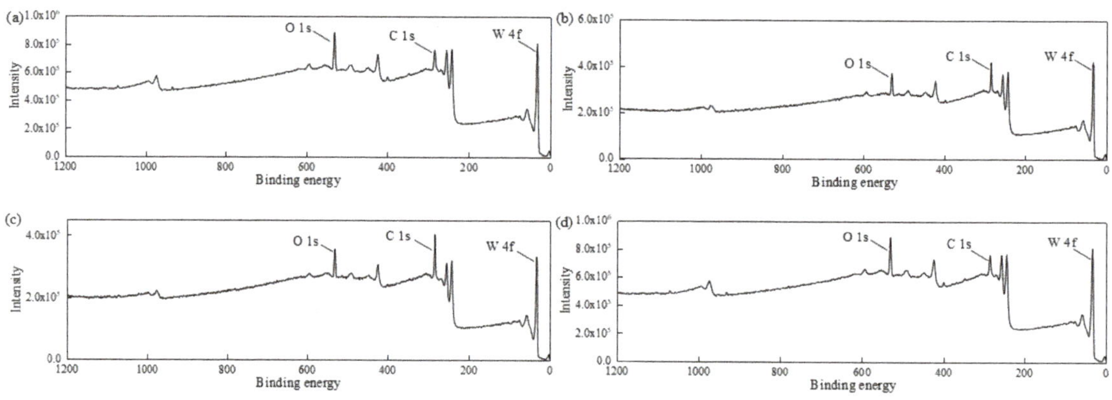

Figure 15. X-ray photoelectron full spectra of the tungsten surface under different conditions: (**a**) pH 9 solution immersion; (**b**) H_2O_2 solution immersion; (**c**) polishing slurry immersion; (**d**) after C-SDP.

Figure 16 shows the XPS fine spectra of the tungsten surface under different conditions. In Figure 16, the deconvolution of the W (4f) spectrum shows four peaks: the two peaks located around 35.5 eV and 37.7 eV can be indexed, respectively, to W $4f_{7/2}$ and W $4f_{5/2}$ of W^{6+}, while the other two peaks at 31.0 eV and 33.2 eV refer, respectively, to W $4f_{7/2}$ and W $4f_{5/2}$ of W [31]. The peaks of the W 4f orbital appeared in pairs, with the peak at the higher binding energy being W $4f_{5/2}$ and the lower one being W $4f_{7/2}$.

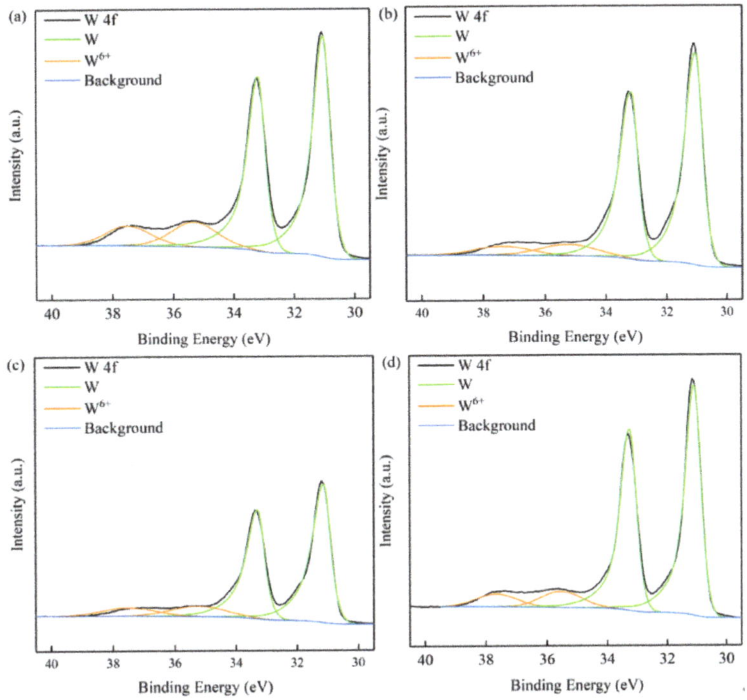

Figure 16. X-ray photoelectron fine spectra of the tungsten surface under different conditions: (**a**) pH 9 solution immersion; (**b**) H_2O_2 solution immersion; (**c**) polishing slurry solution immersion; (**d**) after C-SDP.

Anik et al. investigated the anodic behavior of tungsten at different pH values. When the pH < 1, the main dissolution pathway of oxide is H^+-assisted dissolution. When the pH is between 4 and 6.5, the dissolution of oxide mainly depends on OH^--assisted dissolution. Under strong alkalinity (pH > 12.5), the dissolution is achieved by the slow diffusion of OH^- to the W surface [32]. Lillard et al. believed that under acidic conditions, the oxides formed on tungsten surface could be divided into an inner barrier layer of WO_3 and an outer hydrated layer consisting of WO_3 (H_2O) [26].

It was found that tungsten underwent chemical reactions in pH 9 solution and H_2O_2 solution, and W^{6+} was formed on the tungsten surface, as determined by XPS spectroscopic analysis. Therefore, it is speculated that WO_3 may exist on the tungsten surface. It is worth noting that the equilibrium information in the Pourbaix diagram shows that WO_3 can only be formed stably when pH < 4 [26]. However, this diagram reacts the final product at different pH values and does not involve intermediate products. In addition, it was found by XPS characterization that WO_3 existed in a weakly acidic environment (4.5 < pH < 6.5), and the electrochemical study of tungsten in a weakly alkaline solution (such as pH = 9) showed that the anodic reaction was independent of pH [32]. Weidman et al. believed that the tungsten surface could show limited passivation behavior at neutral and slightly alkaline pH values [33]. Kneer et al. mentioned that a thin protective oxide layer exists on the tungsten surface over almost the entire pH range. When the pH is 4–9, a thin metastable WO_3 film is formed on the tungsten surface [34]. Generally, the oxide film on the tungsten surface will dissolve rapidly under strong alkaline conditions [33]. Combined with the XPS spectra in Figure 16, it can be inferred that NaOH and H_2O_2 react with tungsten, and the main chemical reactions in alkaline C-SDP polishing slurry [35,36] are:

$$W + 6OH^- \rightarrow WO_3 + 3H_2O \quad (2)$$

$$W + 3H_2O_2 \rightarrow WO_3 + 3H_2O \tag{3}$$

$$WO_3 + 2OH^- \rightarrow WO_4^{2-} + H_2O \tag{4}$$

Table 3 analyzes the peak areas of different valence states of tungsten shown in Figure 16. As shown in Table 3, the peak area representing the W element is much larger than that of W^{6+}, which indicates that the content of hexavalent compounds on the tungsten surface is very small, so the oxide film is very thin. Compared with the tungsten surface immersed in pH 9 solution, the tungsten surface immersed in the polishing slurry has a higher W element content. This may be due to a relatively dense passivation film formed by the reaction between H_2O_2 and tungsten in the polishing slurry, preventing further oxidation of the internal tungsten. On the other hand, the passivation film can react with OH^-, leading to a reduction in its thickness. The proportion of the W element on the tungsten surface was also increased after C-SDP, which indicates that the addition of mechanical action can reduce the thickness of the passivation film to a certain extent. In general, the hardness of tungsten oxides tends to decrease as the oxidation state increases. Therefore, in the actual C-SDP process, both the dissolution mode and the chemical state of the oxide play a role in the material removal of the tungsten surface. At present, there are few studies on the chemical corrosion mechanisms of tungsten polishing in alkaline environments, and it is necessary to carry out further systematic research in the follow-up.

Table 3. Analysis results of XPS peak area of tungsten.

Conditions	W^{6+}/W Ratio
pH = 9	0.237
1 vol.% H_2O_2	0.137
Slurry	0.181
C-SDP	0.142

5. Conclusions

In this paper, a novel high-efficiency C-SDP method was proposed to obtain high surface quality tungsten, and the effects of pH values and H_2O_2 concentrations on the polishing performance of tungsten were investigated. The experimental results showed that tungsten C-SDP was significantly affected by pH values and H_2O_2 concentrations. The *MRR* gradually increased with increasing pH values from 6.69 μm/h to 13.67 μm/h. With increasing H_2O_2 concentrations, the *MRR* increased from 9.71 μm/h to 34.95 μm/h. When the pH value was 9 and the H_2O_2 concentration was 1 vol.%, the optimal R_a was 1.87 nm, and the corresponding *MRR* was 26.46 μm/h. This indicates that the C-SDP polishing technique was an effective method to obtain high surface quality tungsten.

The mechanism influences of pH values and H_2O_2 concentrations on tungsten C-SDP were clarified by electrochemical and XPS tests. In alkaline polishing slurries containing H_2O_2, the tungsten surface mainly undergoes an oxidation reaction to form relatively soft tungsten trioxide, which can be quickly removed by abrasives. Then a new surface of the tungsten workpiece is exposed, and the chemical reaction continues, thereby increasing the material removal rate. In the C-SDP polishing slurry, when the alkalinity is too strong or the concentration of oxidant too high, excessive chemical corrosion and poor surface quality will result.

Author Contributions: Conceptualization, L.X. and H.C.; methodology, X.W.; software, L.W.; validation, L.W., W.H. and B.L.; formal analysis, X.W.; investigation, L.W.; resources, L.W.; data curation, H.C.; writing—original draft preparation, L.X. and H.C.; writing—review and editing, W.Z., F.C. and H.C.; visualization, X.W. and L.W.; supervision, J.Y.; project administration, L.X.; funding acquisition, H.C., X.W. and W.Z. All authors have read and agreed to the published version of the manuscript.

Funding: This work is supported by the National Natural Science Foundation of China (Grant Nos. 51905485 and 51805485), and the Joint Funds of the National Natural Science Foundation of China (Grant No. U20A20293) and the Natural Science Foundation of Zhejiang Province (LY21E050011 and LY21E050010).

Conflicts of Interest: The authors declare no conflict of interest.

References

1. Brown, T.; Pitfield, P. Tungsten. In *Critical Metals Handbook*; John Wiley & Sons: Hoboken, NJ, USA, 2014; pp. 385–413. [CrossRef]
2. Dorow-Gerspach, D.; Kirchner, A.; Loewenhoff, T.; Pintsuk, G.; Weißgärberb, T.; Wirtza, M. Additive manufacturing of high density pure tungsten by electron beam melting. *Nucl. Mater. Energy* **2021**, *28*, 101046. [CrossRef]
3. Li, Y.; Du, Z.Y.; Fan, J.L.; Lv, Y.Q.; Lv, Y.Z.; Ye, L.; Li, P.F. Microstructure and texture evolution in warm-rolled fine-grained tungsten. *Int. J. Refract. Hard Met.* **2021**, *101*, 105690. [CrossRef]
4. Wang, G.L.; Xu, Q.; Yang, T.; Xiang, J.J.; Xu, J.; Gao, J.F.; Li, C.L.; Li, J.F.; Yan, J.; Chen, D.P. Application of atomic layer deposition tungsten (ALD W) as gate filling metal for 22 nm and beyond nodes CMOS technology. *ECS J. Solid State Sci. Technol.* **2014**, *3*, P82–P85. [CrossRef]
5. Wang, Z.H.; Zhao, K.X.; Chen, W.M.; Chen, X.D.; Zhang, L.Y. Atomistic modeling of diffusion coefficient in fusion reactor first wall material tungsten. *Appl. Therm. Eng.* **2014**, *73*, 111–115. [CrossRef]
6. Wyngaert, T.V.D.; Elvas, F.; Schepper, S.D.; Kennedy, J.A.; Israel, O. SPECT/CT: Standing on the shoulders of giants, it is time to reach for the sky! *J. Nucl. Med.* **2020**, *61*, 1284–1291. [CrossRef]
7. Chen, H.Y.; Xu, Q.; Wang, J.H.; Li, P.; Yuan, J.L.; Lyu, B.H.; Wang, J.H.; Tokunagad, K.; Yao, G.; Luo, L.M.; et al. Effect of surface quality on hydrogen/helium irradiation behavior in tungsten. *Nucl. Eng. Technol.* 2022, *in press*. [CrossRef]
8. Seo, E.B.; Bae, J.Y.; Kim, S.I.; Choi, H.E.; Son, Y.H.; Yun, S.S.; Park, J.H.; Park, J.G. Interfacial chemical and mechanical reactions between tungsten-film and nano-scale colloidal zirconia abrasives for chemical-mechanical-planarization. *ECS J. Solid State Sci. Technol.* **2020**, *9*, 054001. [CrossRef]
9. Guo, J.; Shi, X.L.; Song, C.P.; Niu, L.; Cui, H.L.; Guo, X.G.; Tong, Z.; Yu, N.; Jin, Z.J.; Kang, R.K. Theoretical and experimental investigation of chemical mechanical polishing of W–Ni–Fe alloy. *Int. J. Extreme Manuf.* **2021**, *3*, 025103. [CrossRef]
10. Wang, X.; Gao, H.; Deng, Q.F.; Wang, J.H.; Chen, H.Y.; Yuan, J.L. Effect of Wetting Characteristics of Polishing Fluid on the Quality of Water-Dissolution Polishing of KDP Crystals. *Micromachines* **2022**, *13*, 535. [CrossRef]
11. Hang, W.; Wei, L.Q.; Debela, T.T.; Chen, H.Y.; Zhou, L.B.; Yuan, J.L.; Ma, Y. Crystallographic orientation effect on the polishing behavior of LiTaO3 single crystal and its correlation with strain rate sensitivity. *Ceram. Int.* **2022**, *48*, 7766–7777. [CrossRef]
12. Omole, S.; Lunt, A.; Kirk, S.; Shokrani, A. Advanced Processing and Machining of Tungsten and Its Alloys. *J. Manuf. Mater. Process.* **2022**, *6*, 15. [CrossRef]
13. Olsson, M.; Bushlya, V.; Lenrick, F.; Ståhl, J. Evaluation of tool wear mechanisms and tool performance in machining single-phase tungsten. *Int. J. Refract. Met. Hard Mater.* **2021**, *94*, 105379. [CrossRef]
14. Poddar, M.K.; Jalalzai, P.; Sahir, S.; Yerriboina, N.P.; Kim, T.; Park, J. Tungsten passivation layer (WO_3) formation mechanisms during chemical mechanical planarization in the presence of oxidizers. *Appl. Surf. Sci.* **2020**, *537*, 147862. [CrossRef]
15. Lim, J.H.; Park, J.H.; Park, J.G. Effect of iron (III) nitrate concentration on tungsten chemical-mechanical-planarization performance. *Appl. Surf. Sci.* **2013**, *282*, 512–517. [CrossRef]
16. Han, W.; Fang, F.Z. Investigation of electropolishing characteristics of tungsten in eco-friendly sodium hydroxide aqueous solution. *Adv. Manuf.* **2020**, *8*, 265–278. [CrossRef] [PubMed]
17. Wang, F.; Zhang, X.Q.; Deng, H. A comprehensive study on electrochemical polishing of tungsten. *Appl. Surf. Sci.* **2019**, *475*, 587–597. [CrossRef]
18. Chen, P.X.; Chen, Y.L.; Lin, H.; Li, H.G.; Li, X. An efficient electrochemical polishing of tungsten with combined forced and natural convections. *Int. J. Adv. Manuf. Technol.* **2021**, *117*, 2819–2834. [CrossRef]
19. Zhou, X.Y.; Wang, F.; Zhang, X.Q.; Deng, H. Electrochemical polishing of microfluidic moulds made of tungsten using a bi-layer electrolyte. *J. Mater. Process. Technol.* **2021**, *292*, 117055. [CrossRef]
20. Qi, H.; Shi, L.W.; Teng, Q.; Hong, T.; Tangwarodomnukund, V.; Liu, G.Y.; Li, H.N. Subsurface damage evaluation in the single abrasive scratching of BK7 glass by considering coupling effect of strain rate and temperature. *Ceram. Int.* **2022**, *48*, 8661–8670. [CrossRef]
21. Hu, W.P.; Teng, Q.; Hong, T.; Saetange, V.; Qi, H. Stress field modeling of single-abrasive scratching of BK7 glass for surface integrity evaluation. *Ceram. Int.* **2022**, *48*, 12819–12828. [CrossRef]
22. Doi, T.K.; Seshimo, K.; Takagi, M.; Ohtsubo, M.; Yamazaki, T.; Nishizawa, H.; Aida, H.; Murakami, S. Development of innovative "dilatancy pad" realizing super high efficiency and high-grade polishing of SiC wide band gap semiconductor substrates. In Proceedings of the International Conference on Planarization/CMP Technology, Kobe, Japan, 19–21 November 2014; pp. 168–173.
23. Doi, T.K.; Seshimo, K.; Yamazaki, T.; Ohtsubo, M.; Ichikawa, D.; Miyashita, T.; Takagi, M.; Saek, T.; Aida, H. Smart polishing of hard-to-machine materials with an innovative dilatancy pad under high-pressure, high-speed, immersed condition. *ECS J. Solid State Sci. Technol.* **2016**, *5*, P598–P607. [CrossRef]

24. Larsen-Basse, J.; Liang, H. Probable role of abrasion in chemo-mechanical polishing of tungsten. *Wear* **1999**, *233*, 647–654. [CrossRef]
25. Shao, Q.; Duan, S.Y.; Fu, L.; Lyu, B.H.; Zhao, P.; Yuan, J.L. Shear Thickening Polishing of Quartz Glass. *Micromachines* **2021**, *12*, 956. [CrossRef] [PubMed]
26. Lillard, R.S.; Kanner, G.S.; Butt, D.P. The nature of oxide films on tungsten in acidic and alkaline solutions. *J. Electrochem. Soc.* **1998**, *145*, 2718–2725. [CrossRef]
27. Shinde, P.A.; Jun, S.C. Review on recent progress in the development of tungsten oxide-based electrodes for electrochemical energy storage. *ChemSusChem* **2020**, *13*, 11–38. [CrossRef] [PubMed]
28. Kaufman, F.B.; Thompson, D.B.; Broadie, R.E.; Jaso, M.A.; Guthrie, W.L.; Pearson, D.J.; Small, M.B. Chemical-Mechanical Polishing for Fabricating Patterned W Metal Features as Chip Interconnects. *J. Electrochem. Soc.* **1991**, *138*, 3460–3465. [CrossRef]
29. Deng, C.B.; Jiang, L.J.; Qin, N.; Qian, L.M. Effects of pH and H_2O_2 on the chemical mechanical polishing of titanium alloys. *J. Mater. Process. Technol.* **2021**, *295*, 117204. [CrossRef]
30. Wang, J.H.; Lyu, B.H.; Jiang, L.; Shao, Q.; Deng, C.B.; Zhou, Y.F.; Wang, J.H.; Yuan, J.L. Chemistry enhanced shear thickening polishing of Ti–6Al–4V. *Precis. Eng.* **2021**, *72*, 59–68. [CrossRef]
31. Lim, G.; Lee, J.H.; Kim, J.; Lee, H.W.; Hyun, S.H. Effects of oxidants on the removal of tungsten in CMP process. *Wear* **2004**, *257*, 863–868. [CrossRef]
32. Anik, M.; Osseo-Asare, K. Effect of pH on the anodic behavior of tungsten. *J. Electrochem. Soc.* **2002**, *149*, B224–B233. [CrossRef]
33. Weidman, M.C.; Esposito, D.V.; Hsu, I.J.; Chen, J.G. Electrochemical stability of tungsten and tungsten monocarbide (WC) over wide pH and potential ranges. *J. Electrochem. Soc.* **2010**, *157*, F179–F188. [CrossRef]
34. Kneer, E.A.; Raghunath, C.; Raghavan, S. Electrochemistry of chemical vapor deposited tungsten films with relevance to chemical mechanical polishing. *J. Electrochem. Soc.* **1996**, *143*, 4095–4100. [CrossRef]
35. Deng, H.; Huang, R.; Liu, K.; Zhang, X.Q. Abrasive-free polishing of tungsten alloy using electrochemical etching. *Electrochem. Commun.* **2017**, *82*, 80–84. [CrossRef]
36. Coetsier, C.M.; Testa, F.; Carretier, E.; Ennahalib, M.; Laborieb, B.; Mouton-arnaudc, C.; Flucherec, O.; Moulina, P. Static dissolution rate of tungsten film versus chemical adjustments of a reused slurry for chemical mechanical polishing. *Appl. Surf. Sci.* **2011**, *257*, 6163–6170. [CrossRef]

micromachines

Article

Integrated Piezoresistive Normal Force Sensors Fabricated Using Transfer Processes with Stiction Effect Temporary Handling

Ni Liu [1], Peng Zhong [2,3], Chaoyue Zheng [2,3], Ke Sun [2], Yifei Zhong [1,*] and Heng Yang [2,3,*]

1 Department of Nephrology, Longhua Hospital, Shanghai University of Traditional Chinese Medicine, Shanghai 200032, China; annie0722@shutcm.edu.cn
2 State Key Laboratory of Transducer Technology, Shanghai Institute of Microsystem and Information Technology, Chinese Academy of Sciences, Shanghai 200050, China; pengzhong@mail.sim.ac.cn (P.Z.); osananajimi@mail.sim.ac.cn (C.Z.); sunke@mail.sim.ac.cn (K.S.)
3 University of Chinese Academy of Sciences, Beijing 100049, China
* Correspondence: sh_zhongyifei@shutcm.edu.cn (Y.Z.); h.yang@mail.sim.ac.cn (H.Y.)

Abstract: Tactile sensation is a highly desired function in robotics. Furthermore, tactile sensor arrays are crucial sensing elements in pulse diagnosis instruments. This paper presents the fabrication of an integrated piezoresistive normal force sensor through surface micromachining. The force sensor is transferred to a readout circuit chip via a temporary stiction effect handling process. The readout circuit chip comprises two complementary metal-oxide semiconductor operational amplifiers, which are redistributed to form an instrumentation amplifier. The sensor is released and temporarily bonded to the substrate before the transfer process due to the stiction effect to avoid the damage and movement of the diaphragm during subsequent flip-chip bonding. The released sensor is pulled off from the substrate and transferred to the readout circuit chip after being bonded to the readout circuit chip. The size of the transferred normal force sensor is 180 μm × 180 μm × 1.2 μm. The maximum misalignment of the flip-chip bonding process is approximately 1.5 μm, and sensitivity is 93.5 μV/μN/V. The routing of the piezoresistive Wheatstone bridge can be modified to develop shear force sensors; consequently, this technique can be used to develop tactile sensors that can sense both normal and shear forces.

Keywords: complementary metal-oxide semiconductor; MEMS; tactile sensor; stiction effect; temporary handling; stiction contact; Au–Si eutectic; flip-chip

Citation: Liu, N.; Zhong, P.; Zheng, C.; Sun, K.; Zhong, Y.; Yang, H. Integrated Piezoresistive Normal Force Sensors Fabricated Using Transfer Processes with Stiction Effect Temporary Handling. *Micromachines* **2022**, *13*, 759. https://doi.org/10.3390/mi13050759

Academic Editor: Jose Luis Sanchez-Rojas

Received: 6 April 2022
Accepted: 9 May 2022
Published: 11 May 2022

1. Introduction

Tactile sensation is a highly desired function in the robotics industry [1–9]. Makihata [5] recently reported that the requirements for tactile sensors include large-area sensing capability, which requires a large number of sensors, rapid response time, and low cost.

Tactile sensor arrays are also crucial sensing elements in pulse diagnosis instruments [10–13], which mimic traditional Chinese doctors and determine pulse signals to assess the health of patients. The tactile sensors used for pulse taking must be organized in a large dense array to cover the entire area around the radial arteries, with a sub-millimeter resolution. The integration of sensors with readout circuits is a crucial technology required for large-area, high-resolution sensing.

Extensive research has been conducted to integrate microelectromechanical systems (MEMS) and readout circuits [14–19]. Monolithic integration features low-electronic parasitics, reduced chip pinout, and small size. However, the strict thermal budget and process compatibility results in complex processes and performance tradeoff, which present various problems. Hybrid integration [20–22], which enables MEMS and complementary metal-oxide semiconductor (CMOS) devices to be optimized independently, is currently the most widely used approach for MEMS and CMOS integration, owing to its short development time, low cost, flexible material selection, and simple fabrication process [23].

Singh et al. [22] proposed a transfer process that can achieve high-density integration by transferring the released MEMS structures onto the readout circuit. Readout circuits can be manufactured using a normal IC foundry and do not undergo etching for release, since the MEMS structures are released before transfer. However, the released MEMS microstructures are movable during the transfer process and can be damaged by shear forces during the bonding and transferring processes. Additionally, the movement of the released MEMS structures decreases the alignment precision of the transferring processes. In a previous study, we proposed a stiction effect temporary handling (SETH) process [24] to temporarily bond the released MEMS structures to the substrates through a stiction effect, which enables temporary handling during the transfer process and reduces alignment errors.

Herein, we present an integrated normal force (force in the z-axis) sensor for pulse diagnosis instruments, wherein the piezoresistive normal force sensor and CMOS readout circuit are integrated using the SETH process. The routing of the piezoresistive Wheatstone bridge can be modified to develop a shear force sensor; consequently, this technique can be used to develop tactile sensors that can sense both normal and shear forces.

2. Design and Fabrication

2.1. Design of the Integrated Normal Force Sensor

In this study, we primarily focused on normal force sensors, since only the force perpendicular to the sensor surface must be measured for pulse taking. The integrated normal force sensor was fabricated by transferring the released force sensor to a CMOS readout circuit chip, as shown in Figure 1. The normal force sensor, which comprises a diaphragm with piezoresistors installed, was fabricated and released through surface micromachining. The diaphragm was suspended by four beams and temporarily attached to the substrate through the stiction effect [25] of surface micromachining to ensure that the normal force sensor did not move during the transfer process, as shown in Figure 1a. The readout circuit chip comprises two CMOS operational amplifiers, which are redistributed to form an instrumentation amplifier. Pads with amorphous silicon/Ti/Au layers on the surface were fabricated on the chip to serve as anchors for the normal force sensor, as shown in Figure 1b. The normal force sensor was then bonded to the readout circuit chip via Au/Si eutectic bonding, as shown in Figure 1c. The diaphragm was transferred to the readout circuit chip after pulling off from the substrate and breaking the suspension beams, as shown in Figure 1d.

The normal force sensor was designed as the sensing element of the pulse diagnosis instrument, which is used in traditional Chinese medicine. A square, flat diaphragm was employed in the normal force sensor, as shown in Figure 2. The size of the low-stress SiN_x diaphragm was 180 μm × 180 μm × 1.2 μm. The polysilicon layer was heavily doped with boron and patterned with piezoresistors and their interconnections. Cr/Pt/Au electrodes were fabricated on top of the polysilicon layer for eutectic bonding and employed as anchors for the diaphragm after transfer. The size of the diaphragm within the electrodes was approximately 120 μm × 120 μm.

Next, the performance of the normal force sensor was simulated. Two piezoresistors were placed perpendicular to the edge of the diaphragm, while two were placed parallel to the edge. Figure 3 depicts the stresses on the piezoresistors, when 500 μN is loaded on the center of the diaphragm. The force 500 μN is equivalent to 260 mmHg pressure on a 120 μm × 120 μm diaphragm, which is slightly higher than the normal blood pressure. The sensitivity of the normal force sensor was calculated to be 34 μV/μN/V, when the longitudinal and transverse gauge factors of boron-doped polysilicon in [26] were used. The normal force sensor is designed to measure pulse signals, whose frequencies are typically lower than 3 Hz. Because the resonant frequency of the sensor is simulated to be as large as 1.11 MHz, as shown in Figure 4, the sensitivities of pulse signals can be considered equal to the DC sensitivity.

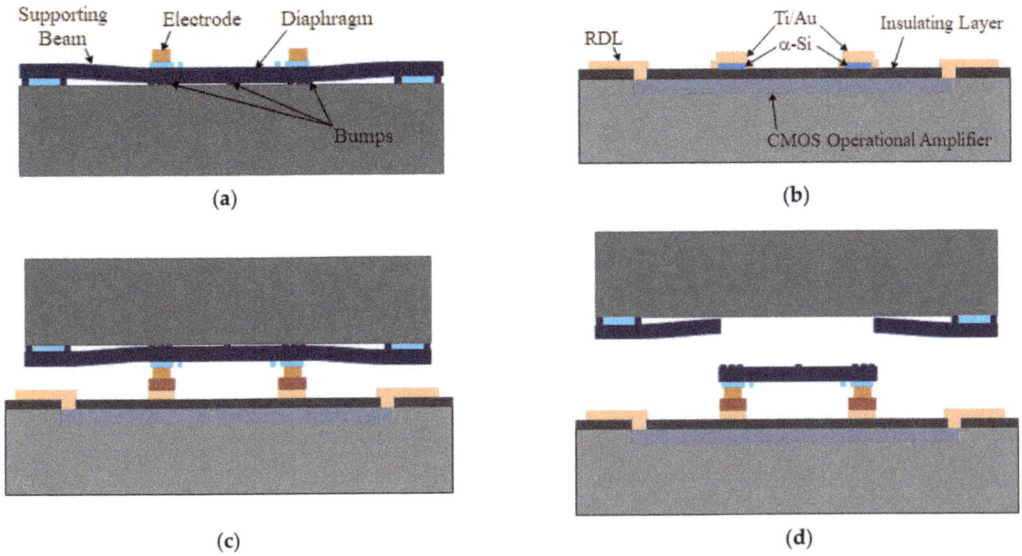

Figure 1. Design process flow of the integrated normal force sensor using stiction effect temporary handling (SETH). (**a**) Normal force sensor is fabricated and temporarily attached to the substrate through the stiction effect; (**b**) readout circuit chip is redistributed, and the pads for eutectic bonding are fabricated; (**c**) normal force sensor is bonded to the readout circuit chip; (**d**) diaphragm is transferred to the readout circuit chip after being pulled off from the substrate and broken from the suspension beams.

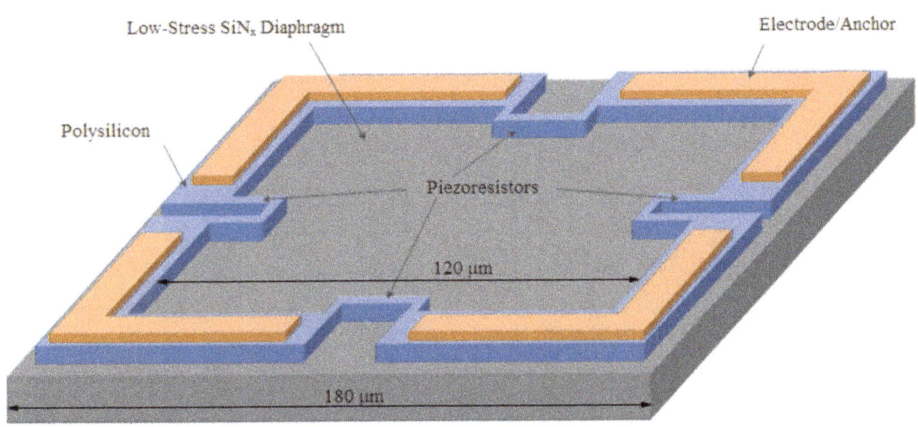

Figure 2. Schematic of the diaphragm of the normal force sensor.

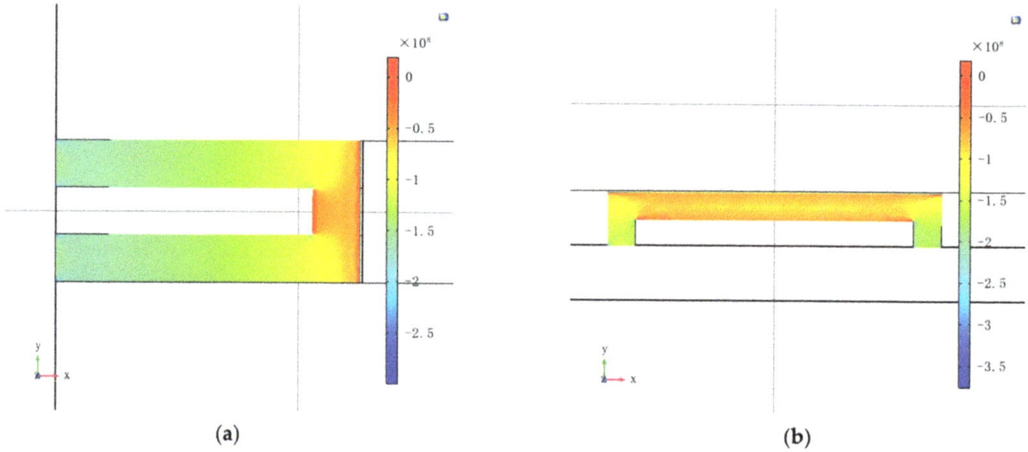

Figure 3. COMSOL simulation results for the normal force sensor. (**a**) Txx along the piezoresistor perpendicular to the edge; (**b**) Tyy along the piezoresistor parallel to the edge.

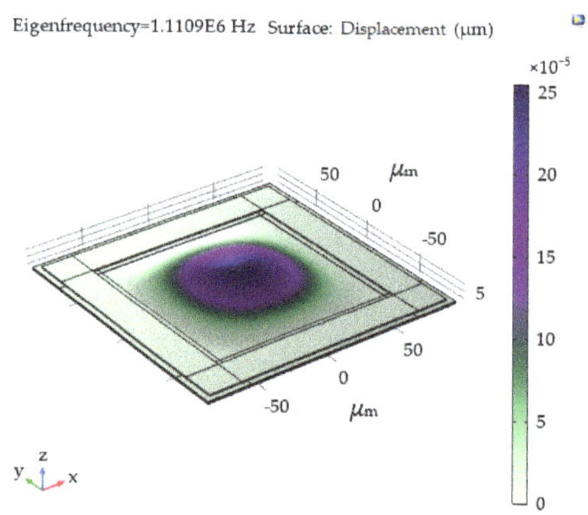

Figure 4. Resonant frequency simulated to be 1.11 MHz by COMSOL.

The diaphragm was attached to the substrate after release to improve the alignment precision of flip-chip bonding. The temporary bonding strength produced due to stiction must be lower than the bonding strength of the flip-chip to ensure that the normal force sensor can be successfully transferred to the readout circuit chip. Bumps were fabricated under the electrode to decrease the stiction area, as shown in Figure 1a. The total area of the bump surface was designed to be 2869 μm^2, which was much lower than the area of the electrodes (9792 μm^2). The normal force sensors can be successfully transferred, even if the temporary bonding strength is equal to the eutectic bonding strength.

To demonstrate integration capability, the normal force sensors were transferred to the CMOS readout circuit chips. The output of the piezoresistive Wheatstone bridge must be amplified using instrumentation amplifiers. Because non-diced wafers of commercial instrumentation amplifiers were unavailable, LMV358 wafers (Yangzhou Genesis Microelectronics Co., Ltd., Yangzhou, China) were employed in our experiments, owing to ease of accessibility. Two LMV358 amplifiers, considered the CMOS version of the LM358 operational amplifier, were redistributed as 2-op amp instrumentation amplifiers [27], as shown in Figure 5. Amplification was determined using external resistors R_1–R_4, which presented resistances of 36, 9.1, 9.1, and 36 kΩ, respectively; the amplification was calculated to be 4.96 using the following equation:

$$V_o = \left(V_{in2} - V_{in1}\left(1 + \frac{R_4}{R_3}\right)\right.\quad(1)$$

where V_{in1} and V_{in2} are the outputs of the Wheatstone bridge. The 3 dB bandwidth exceeded 100 kHz.

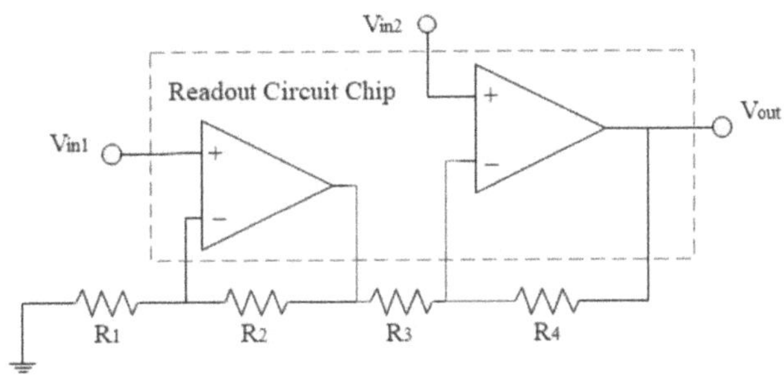

Figure 5. LMV358 is redistributed to serve as a 2-op amp instrumentation amplifier.

2.2. Fabrication

A normal force sensor was fabricated using surface micromachining processes as follows:

(a) A 450 nm thick SiO_2 layer was thermally grown to passivate the substrate. Then, a layer of 800 nm thick low-stress polysilicon was deposited as the sacrificial layer by LPCVD. The polysilicon was subsequently patterned and selectively etched to define the shape of the bumps. Another 200 nm thick layer of low-stress polysilicon was deposited by LPCVD to define the distance between the bumps and substrate. Thereafter, the polysilicon layer was patterned and selectively etched to define the shape of the anchors.

(b) A 1 µm thick silicon-rich SiN_x layer [28] was deposited by LPCVD to serve as the mechanical layer, tuned to reach a low-residual tensile stress of approximately 50 MPa [29,30]. A 300 nm thick LPCVD polysilicon layer was deposited and heavily doped by boron implantation, followed by patterning and selective etching to form the piezoresisitors. Next, a low-stress SiN_x layer of 200 nm thickness was deposited to protect the piezoresisitors.

(c) A composite metal layer of Cr/Pt/Au was sputtered and patterned on the piezoresisitors once the contact windows of the piezoresisitors were etched using the RIE technique. The thicknesses of Cr, Pt, and Au were 50, 100, and 300 nm, respectively. The Pt layer of Cr/Pt/Au prevents the Au–Si alloy formed by the subsequent Au–Si eutectic flip-chip bonding process from penetrating the metal pads.

(d) The SiN_x layer was patterned and selectively etched to form the diaphragm of the tactile sensor and temporarily supported anchors in the silicon nitride diaphragm.

(e) The XeF_2 etching technique was employed to remove the polysilicon sacrificial layer. The released device was subsequently placed in DI water for 24 h and dried at 25 °C for another 24 h to bond the stiction-contact structures temporarily to the substrate using the stiction effect.

Figure 6 illustrates the redistribution flow of the readout circuit chip, as described below:

(a) The composite layers of $SiO_2/SiN_x/SiO_2$ were deposited by PECVD to serve as insulating layers for redistribution. The thickness of each layer was 200 nm. A layer of amorphous silicon of 1 μm thickness was deposited and patterned for subsequent Au–Si eutectic bonding, as shown in Figure 6a.

(b) The contact holes were patterned on the insulating layer and the composite metal layers of Ti/Au were sputtered and patterned to redistribute the operational amplifiers to serve as instrumentation amplifiers, as shown in Figure 6b. The Ti layer was used to decompose native oxide on the surface of amorphous silicon during subsequent Au–Si eutectic bonding [31]. The thicknesses of the Ti and Au layers were 50 and 400 nm, respectively.

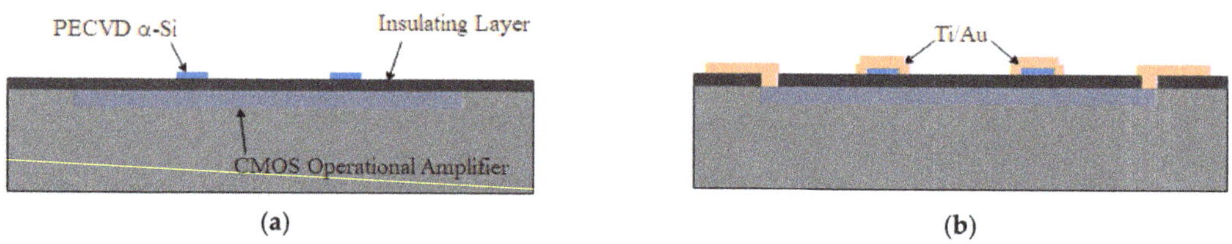

Figure 6. Redistribution flow of the readout circuit chip. (**a**) After composite layers of $SiO_2/SiN_x/SiO_2$ were deposited by PECVD to serve as insulating layers for redistribution, a layer of amorphous silicon was deposited and patterned for subsequent Au–Si eutectic bonding; (**b**) the contact holes were patterned on the insulating layer, and composite metal layers of Ti/Au were sputtered and patterned for redistributing the operational amplifiers to serve as the instrumentation amplifiers.

The released normal force sensor was transferred to the readout circuit chip by flip-chip bonding as follows:

(a) The released normal force sensor was bonded to the readout circuit chip using a flip-chip bonder (FinePlacer Lambda, Fintech, Germany), and the temperature, force, and time required for this process were 380 °C, 20 N, and 300 s, respectively.

(b) The released normal force sensor was subsequently pulled off from the substrate and broken from the suspension beams by applying a pulling force perpendicular to the bonded device. Because the normal force sensors were released before the transfer process, the readout circuit chips did not undergo release etching. This process demonstrates good CMOS compatibility.

3. Results and Discussion

Figure 7a illustrates the released normal force sensor. The interference fingers in the diaphragm and supporting fingers indicated that these structures were temporarily bonded to the substrate due to the stiction effect. The stiction strength during the stiction process was estimated using the longest unattached cantilever to be higher than 7.06 kPa and lower than 22.31 kPa, as shown in Figure 7b.

Figure 8 depicts the integrated normal force sensor, the size of which is approximately equal to those of LMV358, 1070 μm × 640 μm × 525 μm. The maximum alignment error of eutectic bonding was measured to be approximately 1.5 μm, sufficient for the proposed integration process. A Dage Series 4000 bond tester (Nordson DAGE, UK) was used to test the shear strength of Au–Si eutectic bonding. The shear strength of the bonded test structure was approximately 30.74 MPa. The serial resistance of the Au–Si eutectic bonding area is lower than 2 Ω [32], which is much lower than that of polysilicon piezoresistors and can be neglected.

The stress in the transferred normal force sensor caused by the Au–Si eutectic flip-chip bonding process was estimated by comparing the output voltages of the Wheatstone bridge before and after the transfer. The change in the output voltage was in the range of −7.76 to +7.25 mV. Therefore, the stress produced by the Au–Si eutectic flip-chip bonding process was calculated to be in the range of −9.95 MPa to +9.30 MPa, which can be neglected.

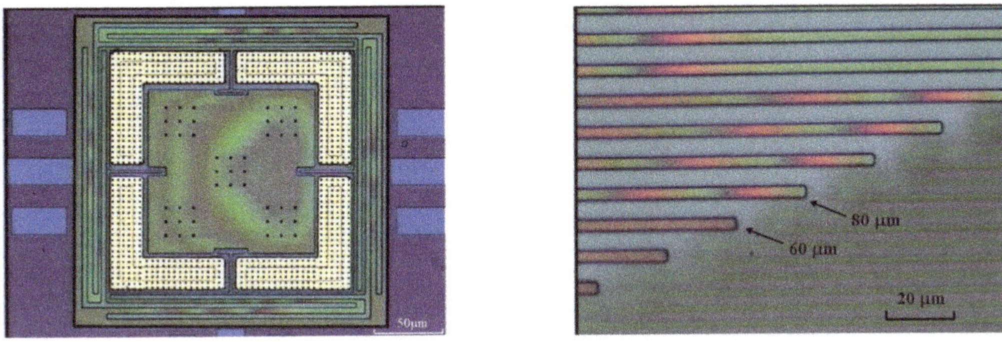

(a) (b)

Figure 7. Optical images of released structures; (**a**) diaphragm supported by four beams; (**b**) cantilever array. Cantilevers longer than 80 μm are colorful due to the interference patterns of uneven gaps, while those shorter than 60 μm exhibit uniform color, which indicates that all cantilevers longer than 80 μm adhered to the substrate. Stiction strength is estimated using the longest unstuck cantilever and shortest stuck cantilever.

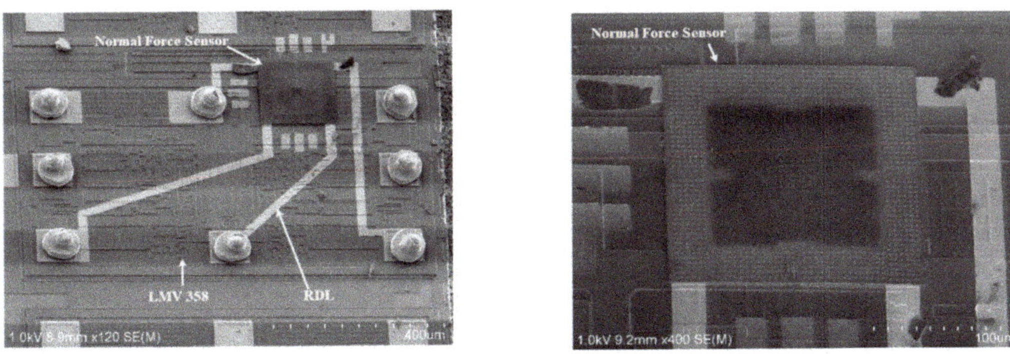

(a) (b)

Figure 8. (**a**) Scanning electron micrographs of the integrated normal force sensor; (**b**) close-up view of the transferred diaphragm.

The integrated normal force sensor was measured using a set of homemade copper wire weights [24]. The source voltage of the Wheatstone bridge was set to 5 V using Agilent E3631A (Agilent, Santa Clara, CA, USA), and the corresponding output voltage of the instrumentation amplifier was recorded using Agilent 34401A (Agilent, USA), when different masses of the beam-shaped copper wire weights were placed on the diaphragm of the transferred normal force sensor under the microscope; Figure 9 presents the measurement results. The sensitivity was calculated to be 93.5 μV/μN/V. The sensitivity of the piezoresistive Wheatstone bridge was calculated to be 18.8 μV/μN/V at an amplification of 4.96. Nonlinearity was approximately 4%, which was quite large and mainly caused by the uncertainty of the point-of-force application. Five sensors were measured. The deviation of sensitivity was less than 20%. System noise was measured at approximately 200 μV.

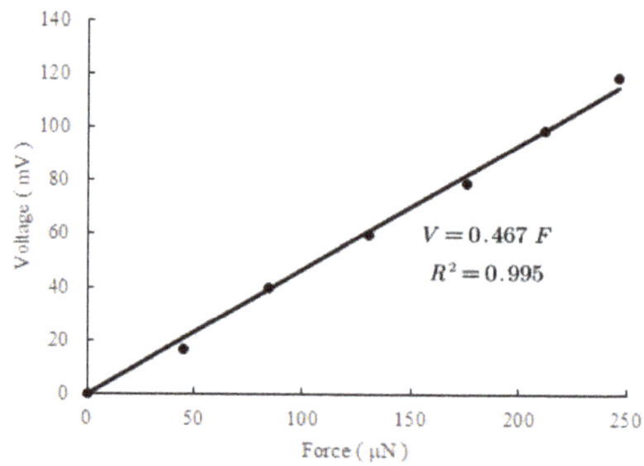

Figure 9. Measurement results of the integrated normal force sensor. Sensitivity was calculated to be 93.5 μV/μN/V.

Shear force sensors can also be developed with flat diaphragms [33] by modifying piezoresistive Wheatstone bridge routing. When shear force in the x direction is applied on the center of the diaphragm, the left side of the diaphragm moves down while the right side moves up, as shown in Figure 10a. The stress of piezoresistor R_1 in Figure 10b is tensile, while that of R_3 is compressive. When the piezoresistors are connected, as in the Wheatstone bridge in Figure 10c, the output is sensitive to shear force and insensitive to normal force. The bumps in the center of the diaphragm can be used as a mesa to improve shear force sensitivity.

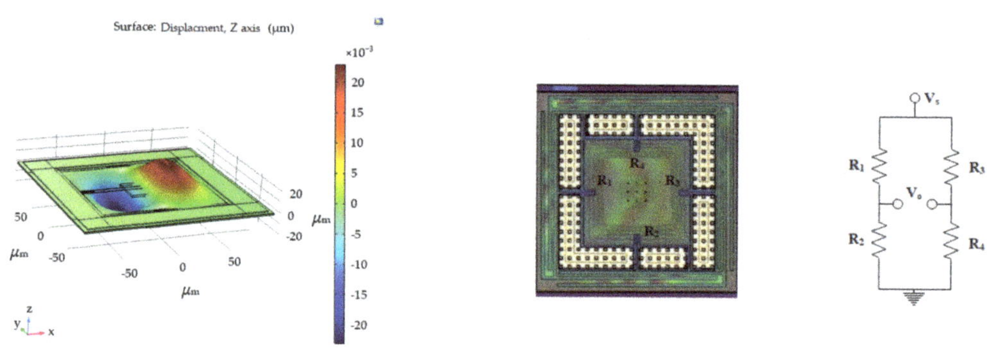

(a) (b) (c)

Figure 10. Shear force sensors developed by modifying routing of piezoresistive Wheatstone bridges. (**a**) When shear force in the x direction is applied on the center of the diaphragm, the left side of the diaphragm moves down while the right side moves up. (**b**) Released shear force sensor. (**c**) Wheatstone bridge of the shear force sensor.

4. Conclusions

Although the integration of sensors with readout circuits is a crucial technology required for large-area, high-resolution tactile sensing, to date, few integrated tactile sensors have been identified, owing to strict thermal budgets and process compatibility. Tactile sensors are typically integrated with the readout circuit in system levels [1–13]. In this study, an integrated piezoresistive normal force sensor was presented. The surface micromachined normal force sensor was transferred to the readout circuit chip, with a temporary stiction effect handling process.

The piezoresistive normal force sensor was manufactured using surface micromachining. The readout circuit chip comprised two CMOS operational amplifiers, which were redistributed to form an instrumentation amplifier. The SETH process was used to transfer the released sensor to the readout circuit chip. Because the MEMS structure and readout circuits were manufactured separately, they were optimized independently. Because the MEMS structures were released before transfer, readout circuits did not undergo etching for release and could be manufactured using a normal IC foundry. These processes feature excellent compatibility with IC chips.

The normal force sensor was designed for pulse diagnosis instruments. The size of the transferred normal force sensor was 180 μm × 180 μm × 1.2 μm. The maximum misalignment in the flip-chip bonding process was approximately 1.5 μm. The sensitivity was measured to be 93.5 μV/μN/V. The routing of the piezoresistive Wheatstone bridge can be modified to develop shear force sensors; hence, this technique can be used to develop tactile sensors, capable of sensing both normal and shear forces.

The size of the integrated normal force sensors is approximately equal to those of the readout circuit chips, because the sensors are significantly smaller and sit on top of the readout circuit chips. In our experiments, the wafers of a very-old-version operational amplifier (LMV358) were employed to verify the technology, owing to ease of accessibility. The LMV358 chip size is approximately 1070 μm × 640 μm × 525 μm, and of approximately 0.5 μm minimum line width. Hence, extremely compact sensors can be developed with modern instrumentational amplifiers.

Author Contributions: N.L. contributed to the design and testing of the device; P.Z. contributed to the fabrication and part of the testing of the device; K.S. contributed to the test idea; C.Z. helped implement part of the fabrication; Y.Z. contributed to the research guidance for the work; H.Y. contributed to the research idea and provided guidance for the work. All authors have read and agreed to the published version of the manuscript.

Funding: This work was funded by the Shanghai Municipal Science and Technology Commission (Project No. 18dz1100600).

Institutional Review Board Statement: The study was conducted in accordance with the Declaration of Helsinki, and approved by the ethical committee of Longhua Hospital (2019LCSY035, 2020.08.27), Shanghai University of Traditional Chinese Medicine.

Informed Consent Statement: Not applicable.

Data Availability Statement: Not applicable.

Acknowledgments: The authors appreciate financial support from the Shanghai Municipal Science and Technology Commission (project 18dz1100600). The authors also appreciate the assistance of engineers at the State Key Laboratory of Transducer Technology and the staff of Longhua Hospital, Shanghai University of Traditional Chinese Medicine.

Conflicts of Interest: The authors declare no conflict of interest.

References

1. Yousef, H.; Boukallel, M.; Althoefer, K. Tactile Sensing for Dexterous in-Hand Manipulation in Robotics—A Review. *Sens. Actuators A* **2011**, *167*, 171–187. [CrossRef]
2. Luo, S.; Bimbo, J.; Dahiya, R.; Liu, H. Robotic Tactile Perception of Object Properties: A Review. *Mechatronics* **2017**, *48*, 54–67. [CrossRef]
3. Sundaram, S.; Kellnhofer, P.; Li, Y.; Zhu, J.; Torralba, A.; Matusik, W. Learning the signatures of the human grasp using a scalable tactile glove. *Nature* **2019**, *569*, 698. [CrossRef] [PubMed]
4. Liu, M.; Zhang, Y.; Wang, J.; Qin, N.; Yang, H.; Sun, K.; Hao, J.; Shu, L.; Liu, L.; Chen, Q.; et al. A star-nose-like tactile-olfactory bionic sensing array for robust object recognition in non-visual environments. *Nat. Commun.* **2022**, *13*, 79. [CrossRef] [PubMed]
5. Makihata, M.; Muroyama, M.; Tanaka, S.; Nakayama, T.; Nonomura, Y.; Esashi, M. Design and Fabrication Technology of Low Profile Tactile Sensor with Digital Interface for Whole Body Robot Skin. *Sensors* **2018**, *18*, 2374. [CrossRef] [PubMed]
6. Li, G.; Zhu, R. A Multisensory Tactile System for Robotic Hands to Recognize Objects. *Adv. Mater. Technol.* **2019**, *4*, 1900602. [CrossRef]

7. Saccomandi, P.; Zollo, L.; Ciancio, A.L.; Schena, E.; Fasano, A.; Oddo, C.M.; Carrozza, M.C.; Camboni, D. Tactile Piezoresistive Sensors for Robotic Application: Design and Metrological Characterization. In Proceedings of the 2017 IEEE International Instrumentation and Measurement Technology Conference (I2MTC), Turin, Italy, 22–25 May 2017; pp. 1–6. [CrossRef]
8. Cirillo, A.; Costanzo, M.; Laudante, G.; Pirozzi, S. Tactile Sensors for Parallel Grippers: Design and Characterization. *Sensors* **2021**, *21*, 1915. [CrossRef]
9. Fujiwara, E.; Wu, Y.T.; dos Santos, M.F.M.; Schenkel, E.A.; Suzuki, C.K. Development of a tactile sensor based on optical fiber specklegram analysis and sensor data fusion technique. *Sens. Actuators A* **2017**, *263*, 677–686. [CrossRef]
10. Chen, J.; Sun, K.; Zheng, R.; Sun, Y.; Yang, H.; Zhong, Y.; Li, X. Three-Dimensional Arterial Pulse Signal Acquisition in Time Domain Using Flexible Pressure-Sensor Dense Arrays. *Micromachines* **2021**, *12*, 569. [CrossRef]
11. Hu, C.; Chung, Y.; Yeh, C.; Luo, C. Temporal and spatial properties of arterial pulsation measurement using pressure sensor array. *Evid. Based Complement. Altern. Med.* **2012**, *2012*, 745127. [CrossRef]
12. Liu, S.; Zhang, S.; Zhang, Y.; Geng, X.; Zhang, J.; Zhang, H. A novel flexible pressure sensor array for depth information of radial artery. *Sens. Actuators A* **2018**, *272*, 92–101. [CrossRef]
13. Chen, C.; Li, Z.; Zhang, Y.; Zhang, S.; Hou, J.; Zhang, H. A 3D wrist pulse signal acquisition system for width information of pulse wave. *Sensors* **2020**, *20*, 11. [CrossRef]
14. Lo, C.C.; Chen, F.; Fedder, G.K. Integrated HF CMOS-MEMS Square-Frame Resonators with On-Chip Electronics and Electrothermal Narrow Gap Mechanism. In Proceedings of the 13th International Conference on Solid-State Sensors, Actuators and Microsystems, Seoul, Korea, 5–9 June 2005; Volume 2, pp. 2074–2077.
15. Tsai, M.H.; Liu, Y.C.; Liang, K.C.; Fang, W. Monolithic CMOS—MEMS Pure Oxide Tri-Axis Accelerometers for Temperature Stabilization and Performance Enhancement. *J. Microelectromech. Syst.* **2015**, *24*, 1916–1927. [CrossRef]
16. Merdassi, A.; Yang, P.; Chodavarapu, V.P. A Wafer Level Vacuum Encapsulated Capacitive Accelerometer Fabricated in an Unmodified Commercial MEMS Process. *Sensors* **2015**, *15*, 7349–7359. [CrossRef]
17. Niklaus, F.; Fischer, A.C. Heterogeneous 3D Integration of MOEMS and ICs. In Proceedings of the 2016 International Conference on Optical MEMS and Nanophotonics (OMN), Singapore, 31 July–4 August 2016; pp. 1–2.
18. Yang, H.S.; Bakir, M.S. Interconnect Technologies for Heterogeneous 3D Integration: CMOS and MEMS. *MRS Proc.* **2010**, *1249*, 1249–1261. [CrossRef]
19. Fedder, G.K.; Howe, R.T.; Liu, T.J.K.; Quevy, E.P. Technologies for Cofabricating MEMS and Electronics. *Proc. IEEE* **2008**, *96*, 306–322. [CrossRef]
20. Basavanhally, N.; Lopez, D.; Aksyuk, V.; Ramsey, D.; Bower, E.; Cirelli, R.; Ferry, E.; Frahm, R.; Gates, J.; Klemens, F.; et al. High-Density Solder Bump Interconnect for MEMS Hybrid Integration. *IEEE Trans. Adv. Packag.* **2007**, *30*, 622–628. [CrossRef]
21. Cohn, M.B.; Böhringer, K.F.; Noworolski, J.M.; Singh, A.; Keller, C.G.; Goldberg, K.Y.; Howe, R.T. Microassembly Technologies for MEMS. *Proc. SPIE Micromach. Microfabr.* **1998**, *3513*, 2–16.
22. Singh, A.; Horsley, D.A.; Cohn, M.B.; Pisano, A.P.; Howe, R.T. Batch Transfer of Microstructures Using Flip-Chip Solder Bonding. *J. Microelectromech. Syst.* **1999**, *8*, 27–33. [CrossRef]
23. Witvrouw, A. CMOS-MEMS Integration: Why, How, and What? In Proceedings of the IEEE ACM International Conference on Computer-Aided Design, San Jose, CA, USA, 5–9 November 2006; pp. 826–827.
24. Zhong, P.; Sun, K.; Zheng, C.; Yang, H.; Li, X. Transfer of Tactile Sensors Using Stiction Effect Temporary Handling. *Micromachines* **2021**, *12*, 1330. [CrossRef]
25. Fearing, R.S. Survey of Sticking Effects for Micro Parts Handling. *IEEE Int. Conf. Intell. Robot. Syst.* **1995**, *2*, 212–217.
26. Kang, T. Evaluation of p-type polysilicon piezoresistance in a full-bridge circuit for surface stress sensors. *Measurement* **2015**, *61*, 243–248. [CrossRef]
27. Kitchin, C.; Counts, L. *Designer's Guide to Instrumentation Amplifiers*, 3rd ed.; Analog Devices: Norwood, MA, USA, 2006.
28. French, P.J.; Sarro, P.M.; Mallee, R.; Fakkeldij, E.J.M.; Wolffenbuttel, R.F. Optimization of a low-stress silicon nitride process for surface-micromachining applications. *Sens. Actuators A* **1997**, *58*, 149–157. [CrossRef]
29. Ni, Z.; Yang, C.; Xu, D.; Zhou, H.; Zhou, W.; Li, T.; Xiong, B.; Li, X. Monolithic Composite "Pressure + Acceleration + Temperature +Infrared" Sensor Using a Versatile Single-Sided "SiN/Poly-Si/Al" Process-Module. *Sensors* **2013**, *13*, 1085–1101. [CrossRef]
30. Li, W.; Li, M.; Wang, X.; Xu, P.; Yu, H.; Li, X. An in-situ TEM microreactor for real-time nanomorphology & physicochemical parameters interrelated characterization. *Nano Today* **2020**, *35*, 100932. [CrossRef]
31. Jing, E.; Xiong, B.; Wang, Y.L. Low-Temperature Au/a-Si Wafer Bonding. *Micromech. Microeng.* **2011**, *21*, 015013. [CrossRef]
32. Liang, H.; Liu, S.; Xiong, B. 3D Wafer Level Packaging Technology Based on the Co-Planar Au-Si Bonding Structure. *J. Micromech. Microeng.* **2019**, *29*, 035010. [CrossRef]
33. Wang, L.; Beebe, D.J. A silicon-based shear force sensor: Development and characterization. *Sens. Actuators A* **2000**, *84*, 33–44. [CrossRef]

micromachines

Article

High Sensitivity Optical Fiber Mach–Zehnder Refractive Index Sensor Based on Waist-Enlarged Bitaper

Na Zhao [1,*], Zelin Wang [1], Zhongkai Zhang [1,*], Qijing Lin [1,2,*], Kun Yao [1], Fuzheng Zhang [1], Yunjing Jiao [1], Libo Zhao [1], Bian Tian [1], Ping Yang [1] and Zhuangde Jiang [1]

1 State Key Laboratory for Manufacturing Systems Engineering, Xi'an Jiaotong University, Xi'an 710049, China; zn2020@xjtu.edu.cn (N.Z.); wzl15086927209@stu.xjtu.edu.cn (Z.W.); yao_kun@outlook.com (K.Y.); xjzfz123@stu.xjtu.edu.cn (F.Z.); 18830970939@163.com (Y.J.); libozhao@xjtu.edu.cn (L.Z.); t.b12@mail.xjtu.edu.cn (B.T.); ipe@xjtu.edu.cn (P.Y.); zdjiang@xjtu.edu.cn (Z.J.)

2 Collaborative Innovation Center of High-End Manufacturing Equipment, Xi'an Jiaotong University, Xi'an 710054, China

* Correspondence: zzkxjtu@outlook.com (Z.Z.); xjjingmi@163.com (Q.L.)

Abstract: A Mach–Zehnder fiber optic sensor with high refractive index response sensitivity was developed. By fabricating a waist-enlarged bitaper structure on the interference arm of a single mode–multimode–single mode (SMS) Mach–Zehnder interferometer (MZI), the spectral contrast and response sensitivity were improved. Subsequently, the response sensitivity was further improved by etching the interference arm. When a beam of light was introduced into the sensor, due to the structural mismatch between the multimode fiber and the normal transmission light, the difference between the low-order mode and the high-order mode was generated in the fiber core and the fiber cladding. In the process of transmission in the sensing arm, due to the different refractive indices of the core and cladding, the optical path difference of the high-order mode and the low-order mode was different, which eventually generated interference fringes. The experimentally measured response sensitivity of SMS MZI in the range of 1.351 RIU to 1.402 RIU is 57.623 nm/RIU; the response sensitivity of a single mode–multimode–bitaper–multimode–single mode (SMBMS) MZI is 61.607 nm/RIU; and the response sensitivity of the etched SMBMS (ESMBMS) MZI is 287.65 nm/RIU. The response sensitivity of the new ESMBMS MZI is three times higher than that of the original SMS MZI. The sensor has the characteristics of compact structure, high sensitivity, easy manufacture, and a wide range of refractive index measurements, and can be used in food processing, pharmaceutical manufacturing and other fields.

Keywords: fiber optic sensor; Mach–Zehnder interferometer; optical fiber waist-enlarged bitaper; the refractive index sensitivity; corrosion

Citation: Zhao, N.; Wang, Z.; Zhang, Z.; Lin, Q.; Yao, K.; Zhang, F.; Jiao, Y.; Zhao, L.; Tian, B.; Yang, P.; et al. High Sensitivity Optical Fiber Mach–Zehnder Refractive Index Sensor Based on Waist-Enlarged Bitaper. *Micromachines* **2022**, *13*, 689. https://doi.org/10.3390/mi13050689

Academic Editors: Xiuqing Hao, Duanzhi Duan and Youqiang Xing

Received: 13 April 2022
Accepted: 27 April 2022
Published: 28 April 2022

Publisher's Note: MDPI stays neutral with regard to jurisdictional claims in published maps and institutional affiliations.

1. Introduction

The refractive index is a commonly used process control index in food production, pharmaceutical development, and other fields. By measuring the refractive index of liquid substances, the composition of the substance can be identified, the concentration determined, and the degree of purity and quality judged. Traditional electronic sensors cannot work in harsh environments, such as high salinity and strong oxidation [1,2]. In recent years, optical fiber sensors had the advantages of excellent anti-electromagnetic interference, good information security, and high precision. The wavelength-dependent type [3] is different from the energy-dependent type [4], which has the advantages of high precision and no interference from light source power.

Optical fiber wavelength-dependent sensors are divided into various types according to their structure, such as the fiber Michelson [5], U-shaped fiber sensor [6,7], coated fiber sensor [8], and Mach–Zehnder interferometer (MZI) [7–16], etc. In 2019, Wang et al. [5] developed a Michelson interferometer by splicing a single-mode fiber and a hollow quartz tube, based on a phase demodulation method, in the range of 1.331 RIU to 1.387 RIU; the refractive index response sensitivity is 8.1498 rad/RIU. The proposed sensor was compact and low-cost, but demodulation analysis was difficult, due to the Fourier analysis of the measurement results. In the same year, Danny et al. [6] proposed a theoretical model of a U-shaped fiber probe, and used the ray tracing method to realize refractive index sensing. In 2020, Wang et al. [7] fabricated a U-shaped double-side polished optical fiber refractive index sensor, with a refractive index response sensitivity of 1541%/RIU from 1.33 RIU to 1.39 RIU. The design and development of the U-shape structure still had certain theoretical and manufacturing difficulties, which needed to be further improved. In 2020, Diegueza et al. [8] reported an optical fiber refractive index sensor coat, with a thin copper film to increase the contrast of the interference fringes, in 0 to 18% glycerol solutions; the response sensitivity is 19 pm/(Glycerol% by weight). Since the development of optical fiber sensing that requires technical support, such as coating, there were certain limitations in production and cost. MZI is small in size, high in sensitivity, and simple in production. Currently, MZI is a hot topic in various research groups.

Since the 1990s, the optical fiber MZI, used to measure the refractive index, began to develop rapidly as a new generation of sensors. In 1994, Liang [9] introduced how to use MZI to measure the refractive index of air. Compared with other methods using double-beam interferometry, it was characterized by convenient operation, stable and reliable data, and easy resolution. In their study of 2014, An et al. [10] measure a humidity response of 0.223 nm/%RH from 35% RH to 85% RH, based on a MZI coated with polyvinyl alcohol material.

In order to further improve the response sensitivity of the sensor, many researchers explored methods to change the structure of the MZI sensing arm, such as adding a fiber taper structure. In 2019, Liao et al. [11] designed an optical fiber MZI based on the fiber taper and bubble structure for ethanol concentration measurement, and record a sensitivity of 28 nm/vol from 0.3 vol to 0.7 vol. In 2019, Vahid et al. [12] fabricated a MZI with ultra-thin sensor arms based on a custom flame-based tapering machine. The sensor's cladding diameter is only 35.5 μm, and the refractive index sensitivity is 415 nm/RIU from 1.332 RIU to 1.384 RIU. In 2019, based on the cascaded up-down taper, Han et al. [13] sandwiched a polarization-maintaining fiber between two single-mode fibers, and they record a refractive index sensitivity of −310.40 dB/RIU from 1.3164 RIU to 1.3444 RIU.

Etching the MZI sensing arm is also a commonly used method in improving the refractive index sensitivity. In 2011, Changping Tang [14] developed a MZI formed by splicing a section of solid-core photonic crystal fiber between two sections of single-mode fiber, and the sensitivity is 70.45 nm/RIU from 1.340 RIU to 1.384 RIU. Then, the coupling degree between the sensor interference light field and the external refractive index is further improved by corrosion, and the sensitivity increases to 198.77 nm/RIU, which is about 2.8 times than before corrosion. In 2019, Huang et al. [15] fabricated a MZI by splicing a photonic crystal fiber between two single-mode fibers, and placed the sensing arm in 40% hydrofluoric acid to reduce the cladding thickness. The experimental results show that the sensitivity increases almost three-fold. In 2019, Haifeng et al. [16] inserted a photonic crystal fiber between two single-mode fibers, and record a sensitivity of 106.19 nm/RIU from 1.333 RIU to 1.381 RIU. After etching the sensor, the cladding diameter is reduced from 250 μm to 112 μm, and the sensitivity improves to 211.53 nm/RIU. In conclusion, an increase in the sensitivity of the refractive index response can be achieved by structural change and cladding etching.

In this paper, based on the principle of double-beam interference, three sensors were designed, fabricated, and compared. The first was a fiber sensor, based on single mode–multimode–single mode (SMS) MZI. The second specific structure was realized by melting a waist-enlarged bitaper in the middle of the sensing arm of the first sensor; the specific structure was single mode–multimode–bitaper–multimode–single mode (SMBMS). The third was realized by etching the sensing arm with hydrofluoric acid on the basis of the second sensing structure, which was referred to as ESMBMS for short. The principle of interference sensing, the fabrication method of the optical fiber cone structure, the analysis of the spectral mode, and the corrosion mechanism of the optical fiber by hydrofluoric acid were analyzed; experiments were designed to measure the sensitivity response of the refractive index of various sensors. Finally, a high-sensitivity refractive index sensor was obtained by making waist-enlarged bitaper and etching. For the convenience of comparison, we summarize the characteristics, advantages, and disadvantages of various sensing structures in Table 1. The comparison shows that the sensor developed in this paper has the advantages of low price, high sensitivity, being simple to make, easy to read, and so on, which has high practical value in the fields of food processing and pharmaceutical production.

Table 1. Comparison of various refractive index sensors.

Structure Type	Monitoring Volume	Range	Sensitivity	Advantages	Insufficient	Ref
Michelson probe	Pattern	1.331RIU to 1.387RIU	8.1498 rad/RIU	Compact and low-cost	The demodulation analysis was difficult	[5]
U-shape fiber probe	Ray tracing method	1.33RIU to 1.39RIU	1541%/RIU	Flexible structural design	Theoretical and manufacturing difficulties, poor repeatability	[6,7]
Optical fiber sensor coat with a thin copper film	Wavelength	0 to 18% glycerol solutions	19 pm/(Glycerol % by weight)	High response sensitivity	The process of optical fiber coating is complex	[8]
MZI coated with polyvinyl alcohol material	Wavelength	35% RH to 85%RH	0.223 nm/%RH	High response sensitivity	The process of optical fiber coating is complex	[10]
MZI based on the fiber taper and bubble structure	Wavelength	0.3 vol to 0.7 vol	28 nm/vol	Wavelength type measurement is not affected by light source, connector, etc.	Low response sensitivity	[11]
MZI with ultra-thin sensor arms	Wavelength	1.332RIU to 1.384RIU	415 nm/RIU	High response sensitivity	Ultra-thin fiber is expensive	[12]
MZI based on polarization-maintaining fiber	Strength	1.3164RIU to 1.3444RIU	310.40 dB/RIU	The price of demodulation equipment is very low	Polarization-maintaining fiber is expensive	[13]
MZI based solid-core photonic crystal fiber	Wavelength	1.340RIU to 1.384RIU	70.45 nm/RIU	Wavelength type measurement is not affected by light source, connector, etc.	Solid-core photonic crystal fiber is expensive	[14]
MZI based photonic crystal fiber	Wavelength	1.333RIU to 1.381RIU	211.53 nm/RIU	Sensitivity can be effectively improved by etching	Photonic crystal fiber is expensive	[16]
ESMBMS MZI	Wavelength	1.351RIU to 1.402RIU	287.65 nm/RIU	Low price, high sensitivity, simple to make, easy to read	The demodulation equipment has not been independently developed.	This paper

2. Principle and Fabrication

2.1. Single mode–Multimode–Single mode (SMS) Mach–Zehnder Interferometer (MZI)

2.1.1. Structural Design

The basic structure of the fiber MZI involved in this paper is SMS MZI, as shown in Figure 1. This MZI included the following parts: the left side was the single mode fiber of the input light, the middle section was a 20 mm long multimode fiber, and the right side was the single mode fiber of the output light. The working principle of MZI is as follows: When light is transmitted from a single mode fiber into a multimode fiber, a part of the cladding mode is excited. Due to the different refractive indices of the core and cladding, there is a certain optical path difference between the core mode and the cladding mode. When light re-entered the single mode fiber from the multimode fiber, the core mode and the cladding mode were coupled and interfered to form interference fringes.

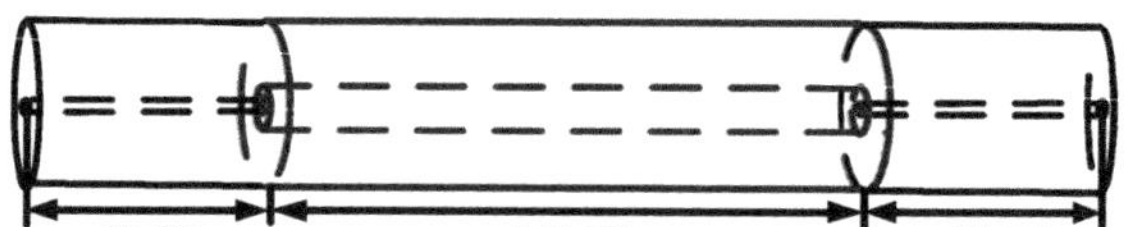

Figure 1. Schematic diagram of the Single mode–Multimode–Single mode Mach–Zehnder Interferometer (SMS MZI).

2.1.2. Principles of Refractive Index Sensing

The occurrence of the interference phenomenon depends on the optical path difference between the two beams. It can be seen from the structure that the core path and cladding path of the MZI sensing arm are of equal length. There is a direct relationship to the frequency spectrum of wavelengths, which is the effective refractive index difference between the different light wave modes. Assuming two different guided modes LP_{0m} and LP_{0n} (m, n are positive integers) are in the multimode fiber, the phase difference between them is:

$$\Delta\varphi = \frac{2\pi(n^m_{eff} - n^n_{eff})L}{\lambda} = \frac{2\pi\Delta n^{m,n}_{eff}L}{\lambda} \tag{1}$$

where L is the length of the fiber and λ is the input wavelength of the light source. $\Delta n^{m,n}_{eff}$ is the effective refractive index difference between $LP_{0\mathrm{m}}$ and $LP_{0\mathrm{n}}$. The equation for the effective refractive index is as follows:

$$n_{eff} = \frac{n \cdot s}{L} \tag{2}$$

where n_{eff} is the effective refractive index, n is the refractive index in the medium, s is the distance traveled by the light, and L is the length of the interference arm. Equation (2) shows that when the refractive index of the medium and L are constant, the mode with the higher order has a larger diffusion angle. That is, the effective refractive index of the high-order mode is greater than that of the low-order mode.

The intensity of the output light can be expressed as:

$$I = I_1 + I_2 + 2\sqrt{I_1 I_2}\cos\Delta\varphi \tag{3}$$

where I is the output light intensity of the MZI, I_1 and I_2 are the light intensity of the guided mode LP_{0m}, and the light intensity of LP_{0n} in the interference core. According to Equation (3), the phase difference equation corresponding to the valley is as follows:

$$\Delta\varphi = (2m+1)\pi \tag{4}$$

Substituting Equation (1) into Equation (4), we obtain:

$$\frac{2\pi\Delta n_{eff}^{m,n}L}{\lambda_m} = (2m+1)\pi \quad (5)$$

where m is an integer, representing the interference order and λ_m is the center wavelength of m-order interference.

According to Equations (4) and (5), Equation (6) is obtained:

$$\delta\lambda_m \approx 2\pi L\delta n_{eff} \quad (6)$$

where $\delta\lambda_m$ is the center wavelength shift of the m-th order interference fringes and δn_{eff} is the change caused by the refractive index of the sucrose solution. Equation (6) demonstrates that when the interference length L is constant, the shift amount of the interference valley wavelength changes with the change of the refractive index of the external liquid. Therefore, the refractive index of the sucrose solution is measured by monitoring the shift in the wavelength of the m-th valley of the MZI.

2.2. *Single mode–Multimode–Bitaper–Multimode–Single mode Mach–Zehnder Interferometer (SMBMS MZI)*

2.2.1. Structural Design

In order to enhance the sensitivity of the fiber MZI, the SMBMS MZI was developed on the basis of the SMS MZI, as shown in Figure 2; a waist-enlarged bitaper was fused to the middle multimode segment. The function of the fusion point in front of the waist-enlarged bitaper was to distribute the light transmitted through the single mode fiber to the multimode fiber core and fiber cladding, while the function of the waist-enlarged bitaper was to redistribute the light transmitted in the core and the cladding, which excited higher-order modes and entered the multimode cladding transmission section behind the waist-enlarged bitaper. The function of the fusion point behind the waist-enlarged bitaper was to couple the light transmitted in the core and the cladding to the output fiber, and through the spectrometer for storage and analysis. In order to verify whether the position of the waist-enlarged bitaper had an effect on the sensor performance, we made a control sensor, and set the position of the waist-enlarged bitaper at one-third of the sensing arm, as shown in Figure 3.

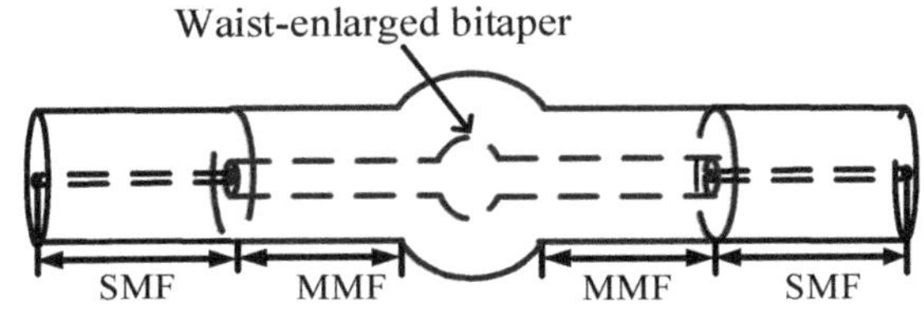

Figure 2. Schematic diagram of the Single mode–Multimode–Bitaper–Multimode–Single mode Mach–Zehnder Interferometer (SMBMS MZI).

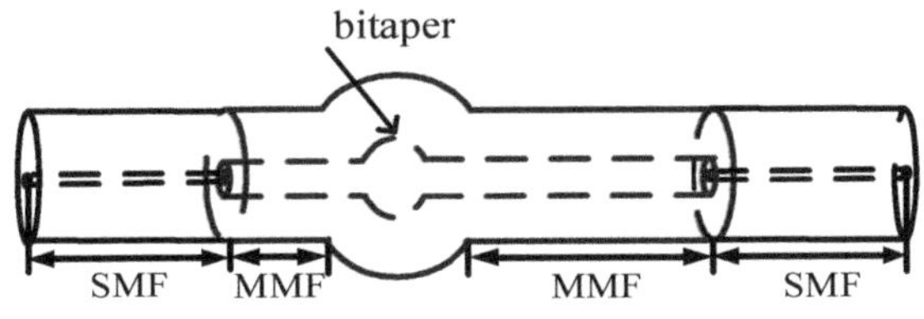

Figure 3. Schematic diagram of the SMBMS MZI with the position of the bitaper set to one third of the sensing arm.

2.2.2. Analysis of Coupling Response Characteristics of Waist-Enlarged Bitaper

The following section focused on analyzing the response characteristics of the optical fiber waist-enlarged bitaper. The inverted taper was composed of a multimode fiber and an inverted taper fiber with a gradually enlarged diameter. The waist-enlarged bitaper part played the role of expanding the beam of the fundamental mode spot. The waist-enlarged bitaper directly affected the propagation response characteristics of the beam in it, as shown in Figure 4. Therefore, we used the waist-enlarged bitaper to make more light waves leak into the cladding and excite higher-order modes. In the following, by explaining the propagation response characteristics of Gaussian beams in tapered fibers, the reasons for the improvement of the coupling efficiency by the waist-enlarged bitaper are theoretically given.

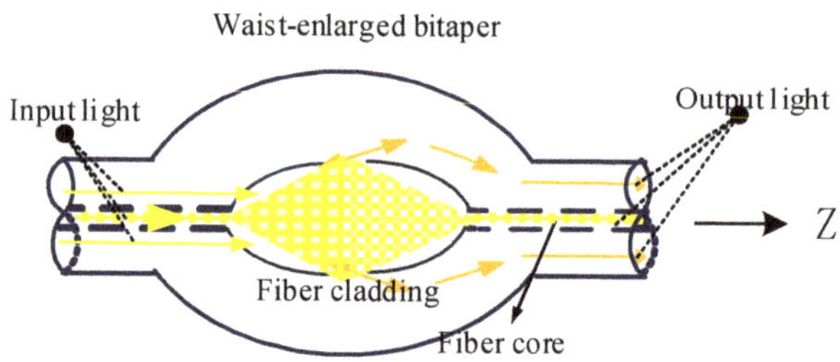

Figure 4. Optical path analysis of waist-enlarged bitaper in SMBMS MZI.

The propagation response characteristics of Gaussian beams in tapered fibers are given by the spot size *ω(z)* and the wavefront curvature radius *R(z)*, used to describe the propagation response characteristics of Gaussian beams [17].

$$\omega^2(z) = \omega^2(0) \cdot [A^2 + (B/k\omega^2(0))^2]^{\frac{1}{2}} \tag{7}$$

$$\frac{1}{R(z)} = \frac{\omega'(z)}{\omega(z)} \tag{8}$$

First, the above two equations can change to the following form:

$$\omega^2(z) = \omega^2(0)[(g(0)/g(z))\cos^2\theta + \frac{1}{g(0)g(z)} \cdot \frac{\sin\theta^2}{k_1^2\omega(0)}] \tag{9}$$

This can be obtained by derivation:

$$\omega^2(z) = a_0 f(z)/(\frac{2\pi n_1}{\lambda})\sqrt{2\Delta} \tag{10}$$

The spot magnification factor M can be expressed as:

$$M = \omega(z)/\omega(0) = f(z) = 1 + c \cdot \exp\gamma(z) \tag{11}$$

Therefore, at the position of the waist-enlarged bitaper, the light spot was enlarged. In a coupled system, the coupling tolerance of the system was only related to the spot size. The larger the spot size, the larger the vertical and horizontal tolerances. In a word, the waist-enlarged bitaper played the role of spot amplification in the optical fiber coupling system, improving the coupling tolerance of the system, and redistributing the spectrum.

2.2.3. Manufacture

Fiber MZI waist-enlarged bitaper is manufactured using a common fusion splicer (FITEL, S178A). In the sensing system, a fiber cleaver (Furukawa, S325) was used to obtain a flat fiber end, and make a waist-enlarged bitaper based on commercial fusion splicers. Multimode optical fiber has a core diameter of 40 μm and a cladding diameter of 125 μm. By increasing the splicing time to 500 ms, and increasing the splicing strength to 155 unit, the waist-enlarged bitaper is obtained. Figure 5 is the photo of the waist-enlarged bitaper observed through a microscope. The diameter of the waist-enlarged bitaper is expanded to 152 μm, and the length of the cone region is 350 μm.

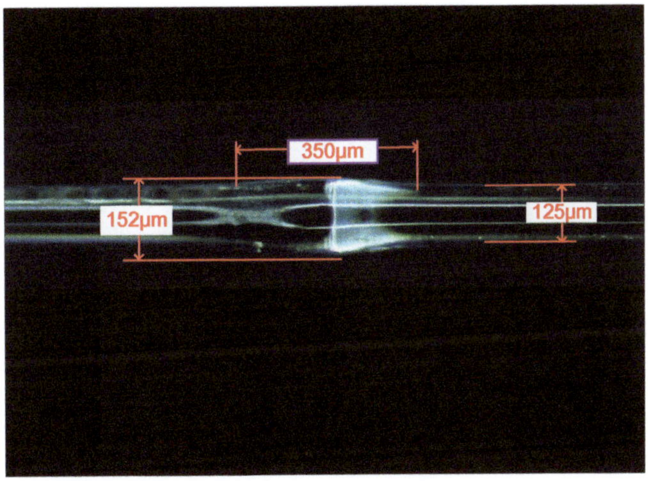

Figure 5. The waist-enlarged bitaper in SMBMS MZI.

2.3. *Etched Single mode–Multimode–Bitaper–Multimode–Single mode Mach–Zehnder Interferometer (ESMBMS MZI)*

2.3.1. Structural Design

Since the refractive index sensitivity response of the SMBMS type structure sensor was still very low, in order to further improve the refractive index sensitivity, the SMBMS type structure sensor was improved. As shown in Figures 6 and 7, the structure of the ESMBMS is based on the structure of the SMBMS, and the middle multimode segment is etched in hydrofluoric acid.

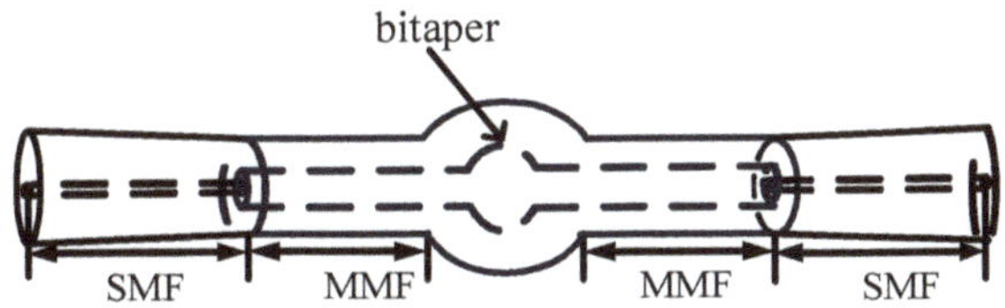

Figure 6. Schematic diagram of the Etched Single mode–Multimode–Bitaper–Multimode–Single mode Mach–Zehnder Interferometer (ESMBMS MZI).

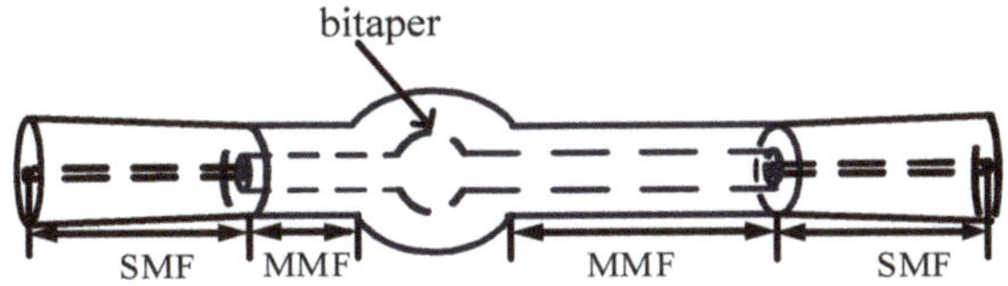

Figure 7. Schematic diagram of the ESMBMS MZI with the position of the bitaper set to one third of the sensing arm.

When the sensing arm was etched, the diameter of the interference arm decreased, which increased the coupling degree between the cladding mode and the ambient refractive index. The change of the refractive index caused the change of the phase difference of the transmitted light in the fiber, and then the change of the refractive index was measured by measuring the change of the interference spectrum. In this experiment, the sensing arm was used to sense the change of refractive index, which was reflected on the spectrum, that is, the wave valley moved. By measuring the drift of the valley, the change in the refractive index was measured.

2.3.2. The Principle of Chemical Corrosion

The etching solution often used in the optical fiber chemical etching method is a hydrofluoric acid solution. The cladding of optical fibers is generally made of silicon dioxide (SiO_2) material. Hydrofluoric acid dissolves silica mainly for the following two reasons: First, hydrofluoric acid solution contains fluoride ions, which combine with silicon ions in SiO_2 to form complex ions. Second, the hydrogen ions in the hydrofluoric acid solution form water with the oxygen ions in the SiO_2. The reaction equation can be expressed as follows:

$$SiO_2 + 6HF = 2H_3O^+ + SiF_6^{2-} \tag{12}$$

In the experiment, we first fixed the sensing unit of the interferometer in a plastic dish, and connected the broadband light source and the spectrometer at both ends. According to past experience, the corrosion rate of 40% hydrofluoric acid at 23 °C is about 2 μm/min [18,19]. After etching the MZI for 20 min, the corrosion surface was relatively smooth.

2.3.3. Manufacture

Figure 5 shows the MZI waist-enlarged bitaper before etching, and Figure 8 shows the bitaper after etching. The 40% hydrofluoric acid solution was directly dropped on the optical fiber sensing arm and etched for 20 min. The diameter of the sensing arm of the 40/125 multimode fiber is reduced to 85 μm. The diameter of the fiber bitaper is reduced from 152 μm to 112 μm, and the length of the waist-enlarged bitaper shows little change.

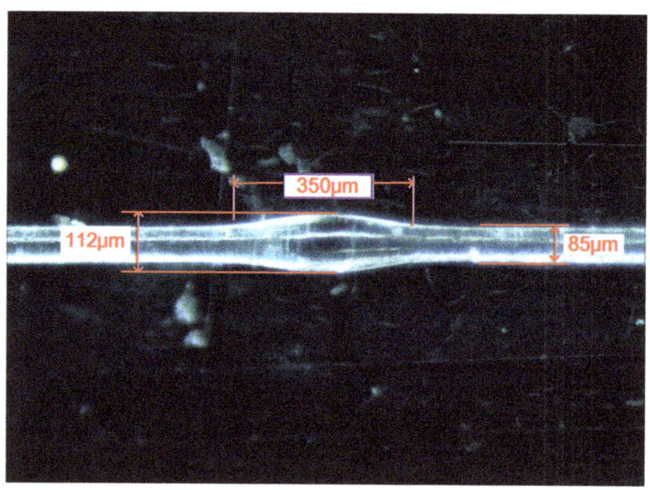

Figure 8. The etched waist-enlarged bitaper in ESMBMS MZI.

3. Spectral Pattern Analysis

There may be multiple modes in the fiber, and due to the different propagation constants between these modes, the optical path difference between the modes will appear under the same transmission length. When these modes are coupled, interference between the modes occurs. Through fabrication experiments, it is found that the position of the waist-enlarged bitaper has little effect on the SMBMS and ESMBMS spectra, so the position of the waist-enlarged bitaper is not distinguished when displaying the spectra. Figure 9 shows the transmission spectra of the MZIs, which show that the density of the interference fringes change due to the addition of waist-enlarged bitaper and corrosion.

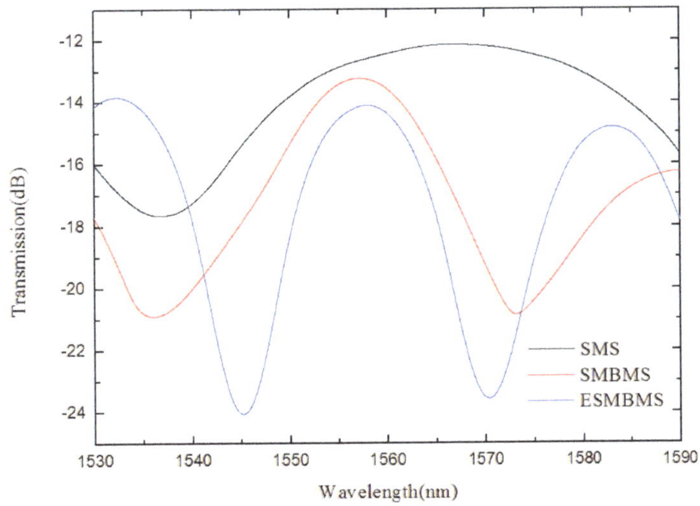

Figure 9. Measured transmission spectra with different L.

According to the transmission spectrum, the higher-order mode is found, and the frequency spectrum obtained after Fourier transform is shown in Figure 10. The main peak amplitudes are located at 0.0199164 nm^{-1}, 0.0300104 nm^{-1}, and 0.0412538 nm^{-1}.

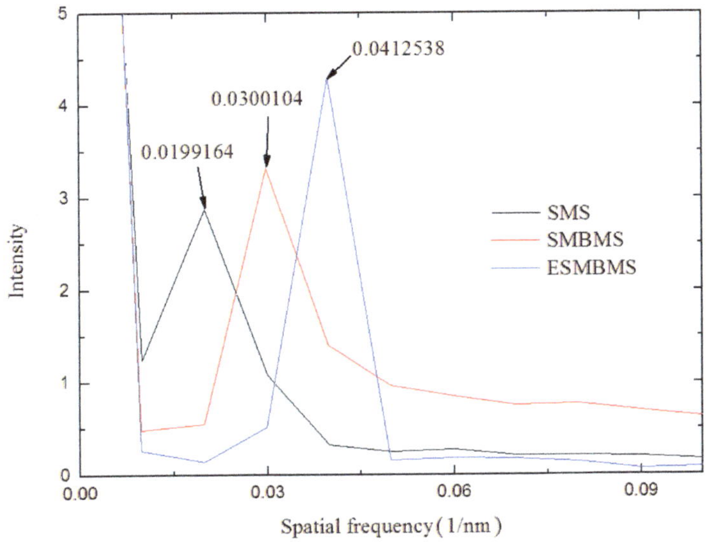

Figure 10. Spatial frequency spectra of the MZIs with different lengths.

Using the Taylor expansion to expand the wavelength the phase ϕ is formulated as:

$$\phi \approx \phi_0 - \frac{2\pi\Delta\lambda}{\lambda^2}\Delta n_{eff} \cdot L \tag{13}$$

where $\Delta\lambda$ is the wavelength difference, ϕ_0 is a initial phase, and λ is the wavelength of spectral valley.

Due to the MZIs spectra corresponding to mathematical cosine patterns, the following equation is obtained:

$$\cos\Delta\phi = \cos(2\pi\xi\Delta\lambda) \tag{14}$$

If ϕ_0 is the initial phase, and we assume it is equal to 0, the spatial frequency ξ [20] is:

$$\xi = \frac{1}{\lambda^2}\Delta n_{eff} \cdot L \tag{15}$$

Considering the modal dispersion, we established the relationship between Δn_{eff} and different modes, based on OptiFiber. The wavelength λ is around 1550 nm, and the lengths of the MZIs are all 20 mm. Through the Fourier transform, the spatial frequency ξ are 0.0199164 nm^{-1}, 0.0300104 nm^{-1}, and 0.0412538 nm^{-1}. Therefore, the parameter Δn_{eff} calculated from Equation (15) are 0.002392458, 0.003604999, and 0.004955613. In other words, as the sensing arm is spliced with waist-enlarged bitaper and etched, higher-order modes are excited, affecting the optical path of transmitted light in the optical fiber core and cladding. Subsequently, the refractive index responses of the three sensors were measured experimentally.

4. Experiment

4.1. The Refractive Index Sensitivity Response Characteristics of SMS MZI

Figure 11 is a schematic diagram of a refractive index sensing system. The measurement system consisted of a broadband light source (BBS, Lightcomm, Shenzhen, China, ASE-CL), a spectrum analyzer (OSA, Anritsu, Kitakyushu, Japan, MS9740A), and a vessel for placing the refractive index solution. The spectrometer resolution was set to 0.02 nm, and the bandwidth of the BBS was 80 nm. In the experiment, sucrose solutions with different concentrations were used as refractive index samples, and their refractive indices are 1.351, 1.369, 1.379, 1.387, 1.394, and 1.402 after being tested by Abbe's refractive index detector. The MZI was uniformly soaked in a sucrose solution of each refractive index for 5 min at a stable room temperature of 23 °C. The wavelength shift was observed and the data were recorded.

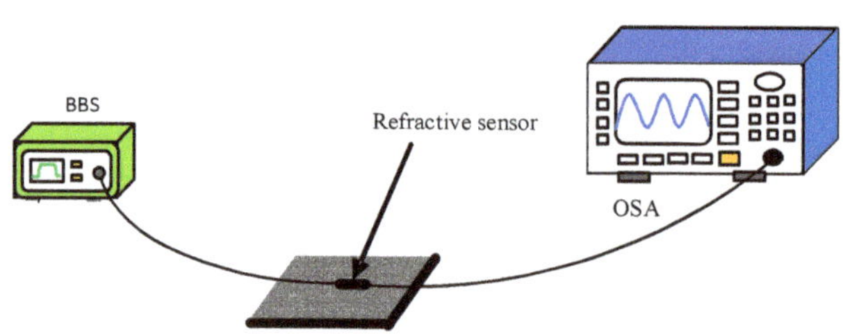

Figure 11. Refractive index sensing experimental device schematic diagram.

The length of the sensing arm of the SMS MZI is about 20 mm. From Figure 12, we see that there is only one wavelength valley in the spectrum of the SMS MZI. We monitored the wavelength of the valley, and Figure 13 shows that the refractive index sensitivity of SMS MZI is 57.623 nm/RIU, and the linearity of the refractive index sensitivity response characteristic is 0.999. Therefore, as the refractive index sensitivity is a positive number, it is seen that the wavelength shifts to the long-wavelength direction; however, the refractive index sensitivity of SMS MZI is too low.

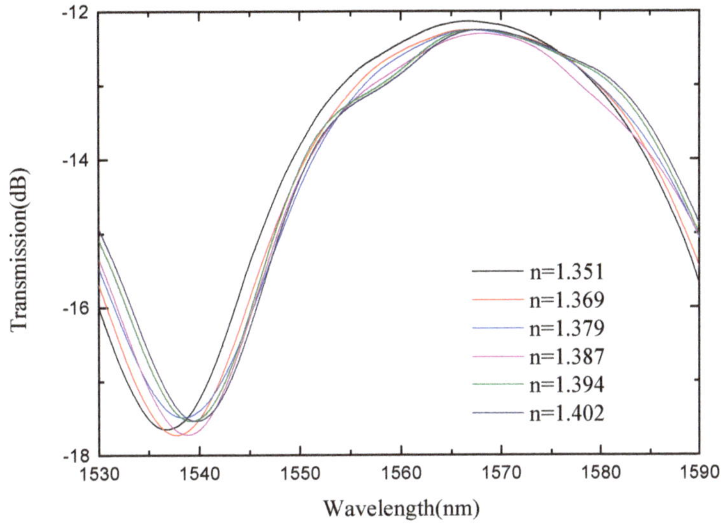

Figure 12. Spectra of SMS MZI at different refractive indices.

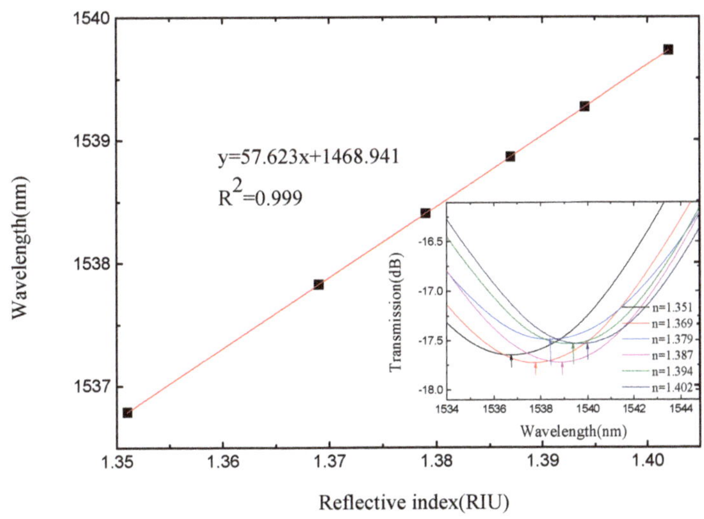

Figure 13. The sensitivity response characteristic diagram of SMS MZI.

4.2. The Refractive Index Sensitivity Response Characteristics of SMBMS MZI

In view of the low refractive index sensitivity of SMS MZI, an improvement was made on the basis of SMS MZI, by melting a waist-enlarged bitaper in the middle of the sensing arm. The photo of the waist-enlarged bitaper is shown in Figure 5. The SMBMS MZI was fabricated, in which the length of the sensing arm of the SMBMS sensor is about 20 mm, the experimental temperature is 23 °C, and the refractive index range is from 1.351 RIU to 1.402 RIU. The sensing arms were uniformly soaked in a sucrose solution for each refractive index for 5 min. The wavelength drift was observed and the data were recorded, shown in Figure 14. At the same time, we changed the position of the waist-enlarged bitaper, and found that the position of the waist-enlarged bitaper had little effect on the spectral shape. Subsequently, we performed refractive index experiments on the sensor with the waist-enlarged bitaper position at one-third of the sensing arm, shown in Figure 15.

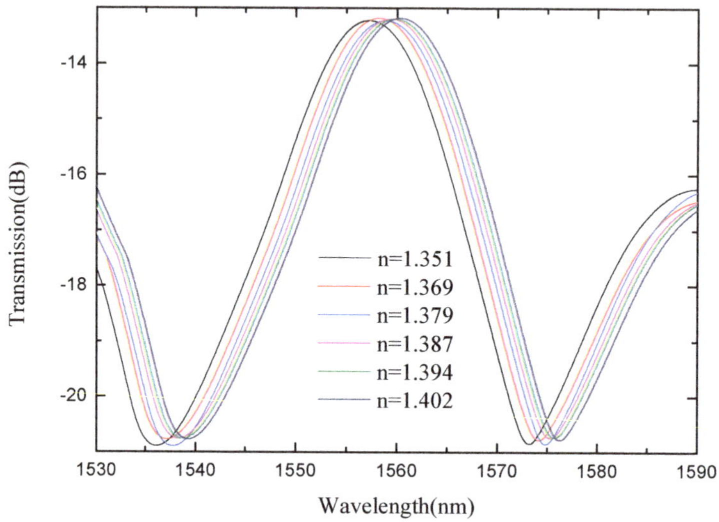

Figure 14. Spectra of SMBMS MZI(1/2L) at different refractive indices.

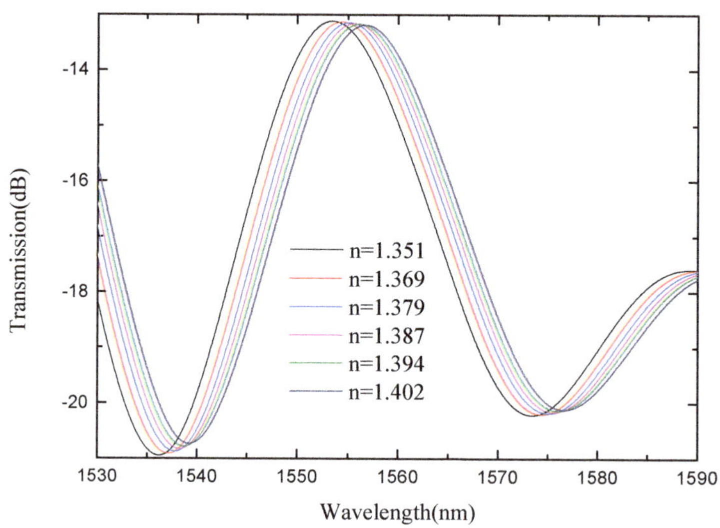

Figure 15. Spectra of SMBMS MZI(1/3L) with the position of the bitaper set to one third of the sensing arm at different refractive indices.

From Figures 14 and 15, we see that there are two valleys within the spectrum of SMBMS MZI. In order to facilitate the comparison with SMS MZI, the valley of the wavelength around 1536 nm was selected for monitoring. Figure 16 shows that the refractive index sensitivity of the SMBMS MZI is 61.607 nm/RIU, and the linearity of the refractive index sensitivity response is 0.999. It can be seen that the wavelength still drifts in the long-wavelength direction, and the refractive index sensitivity of SMBMS MZI improves, compared to SMS MZI. Overall, the sensitivity improvement effect is still not obvious. The reason for this is that the diameter of the sensing arm was relatively thick, and the coupling effect with the refractive index solution was not obvious.

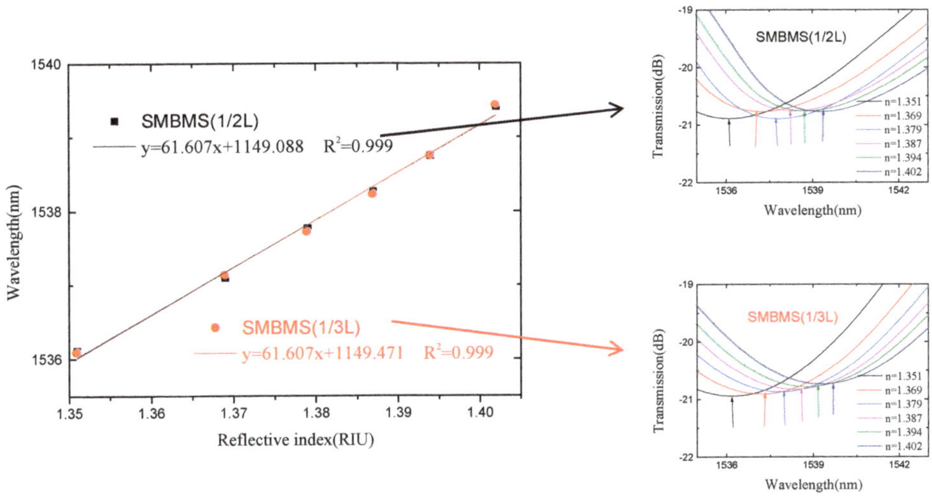

Figure 16. The sensitivity response characteristic diagram of SMBMS MZI.

4.3. The Refractive Index Sensitivity Response Characteristics of ESMBMS MZI

Since the refractive index sensitivity of SMBMS MZI was still low, improvements were made on the basis of SMBMS MZI. The method used was to corrode the sensing arm of SMBMS MZI with hydrofluoric acid for 20 min. The waist-enlarged bitaper is shown in Figure 8. The sensing arm of the fabricated ESMBMS MZI is about 20 mm long, and the fiber diameter is reduced by 40 μm after 20 min of hydrofluoric acid etching at 23 °C. The refractive index of sucrose solution ranges from 1.351 RIU to 1.402 RIU. The sensing arm was immersed in the sucrose solution of each refractive index for 5 min to observe the wavelength and record the data, shown in Figures 17 and 18.

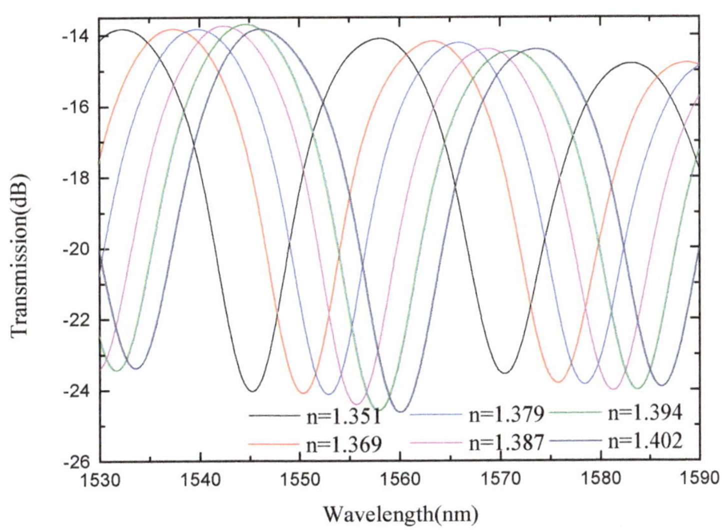

Figure 17. Spectra of ESMBMS MZI(1/2L) at different refractive indices.

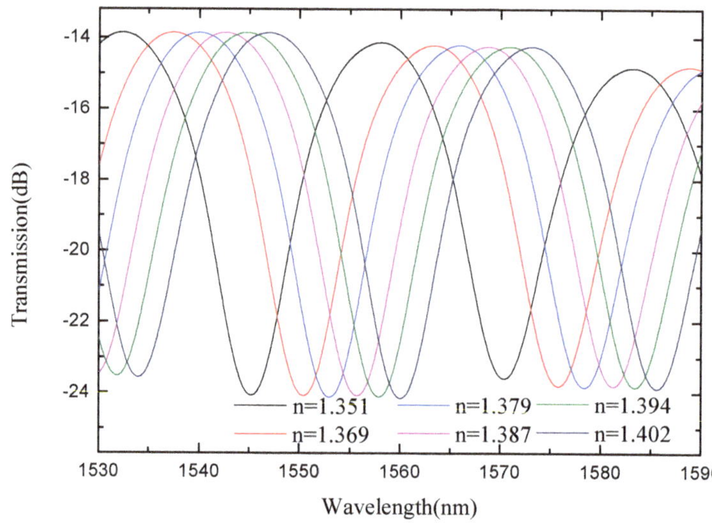

Figure 18. Spectra of ESMBMS MZI(1/3L) with the position of the bitaper set to one third of the sensing arm at different refractive indices.

It can be seen from Figures 17 and 18 that for monitoring the valley with the center wavelength around 1545 nm, the refractive index sensitivity of ESMBMS MZIs is 287.65 nm/RIU, and the linearity of the refractive index sensitivity response is 0.999, as shown in Figure 19. Therefore, it is demonstrated that the wavelength still drifts to the long-wavelength direction, and the refractive index sensitivity of ESMBMS MZI increases almost four-fold compared to SMBMS MZI. Overall, the advantages of ESMBMS MZI refractive index sensitivity response characteristics are obvious. The reason is that the sensing arm cladding of the SMBMS structure was etched with hydrofluoric acid to reduce its diameter, which greatly improved the sensitive response characteristics. Therefore, the refractive index sensitivity of ESMBMS MZI is obviously improved.

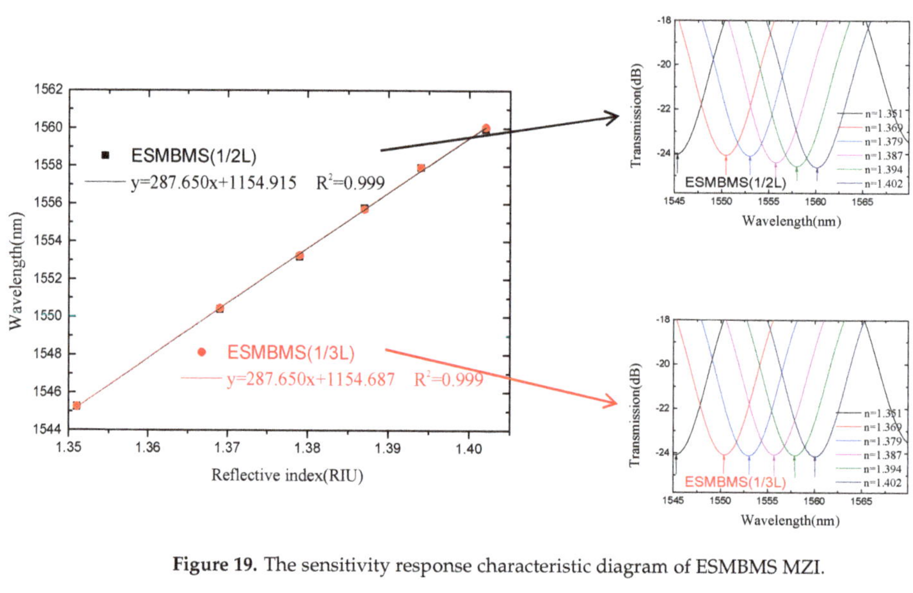

Figure 19. The sensitivity response characteristic diagram of ESMBMS MZI.

5. Conclusions

In this paper, a high-sensitivity MZI was designed and fabricated. The single mode fiber and the multimode fiber were directly spliced by a commercial fusion splicer to form a coupling point. The multimode fiber was in the middle of the sensing arm, and a waist-enlarged bitaper was fused in the middle of the sensing arm, used to realize beam adjustment. Finally, the cladding of the sensing arm was etched with hydrofluoric acid to reduce its diameter, which in turn achieved improved response sensitivity. Fourier transform analysis shows that both the waist-enlarged bitaper and erosion excite higher-order modes, resulting in a larger optical path difference for light transmitted in the optical fiber core and cladding. The refractive index sensitivity responses of these three sensors were experimentally investigated, and the refractive index sensitivity of SMS MZI is 57.623 nm/RIU, with a linearity of 0.9795; the refractive index sensitivity of SMBMS MZI is 61.607 nm/RIU, with a linearity of 0.9545; and the refractive index sensitivity of ESMBMS MZI is 287.65 nm/RIU, with a linearity of 0.9843. It is shown that ESMBMS MZI has a very high refractive index sensitivity and good linearity. As the new ESMBMS MZI, designed in this paper, possesses the advantages of high sensitivity, good linearity, low cost, and simple fabrication, it has a high practical value and good application prospects for future production.

Author Contributions: Conceptualization, Q.L.; data curation, N.Z., Z.W. and F.Z.; formal analysis, Z.Z.; resources, L.Z., B.T. and Z.J.; software, K.Y. and Y.J.; writing—original draft, N.Z.; writing—review & editing, Q.L. and P.Y. All authors have read and agreed to the published version of the manuscript.

Funding: This work was supported by the Postdoctoral Innovative Talent Support Program (Grant No. BX20200274), the National Natural Science Foundation of China (Grant Nos. 51720105016, 52105560), the Post-doctoral Science Foundation (Grant No. 2021M702596), Shaanxi Province Surface Project (Grant No. S2022-JC-YB-1996), and the National Science and Technology Major Project (Grant No. J2019-V-0006-0100).

Conflicts of Interest: The authors declare no conflict of interest.

References

1. Ponzoni, A.; Baratto, C.; Cattabiani, N.; Falasconi, M.; Galstyan, V.; Nunez-Carmona, E.; Rigoni, F.; Sberveglieri, V.; ZambottI, G.; Zappa, D. Metal Oxide Gas Sensors, a Survey of Selectivity Issues Addressed at the SENSOR Lab, Brescia (Italy). *Sensors* **2017**, *17*, 714. [CrossRef] [PubMed]
2. Wang, X.; Zhu, J.; Wen, X.; Wu, X.; Yang, H. Wide range refractive index sensor based on a coupled structure of Au nanocubes and Au film. *Opt. Mater. Express* **2019**, *9*, 3079–3088. [CrossRef]
3. Zhao, N.; Fu, H.; Shao, M.; Yan, X.; Li, H.; Liu, Q.; Gao, H.; Liu, Y.; Qiao, X. High temperature probe sensor with high sensitivity based on Michelson interferometer. *Opt. Commun.* **2015**, *343*, 131–134. [CrossRef]
4. Lin, Q.; Zhao, N.; Yao, K.; Jiang, Z.; Tian, B.; Shi, P.; Chen, F. Ordinary Optical Fiber Sensor for Ultra-High Temperature Measurement Based on Infrared Radiation. *Sensors* **2018**, *18*, 4071. [CrossRef] [PubMed]
5. Wang, J.; Liu, B.; Wu, Y.; Mao, Y.; Zhao, L.; Sun, T.; Nan, T. A novel fiber in-line Michelson interferometer based on end face packaging for temperature and refractive index measurement. *Optik* **2019**, *194*, 163094. [CrossRef]
6. Danny, C.G.; Raj, M.D.; Sai, V.V.R. Investigating the Refractive Index Sensitivity of U-Bent Fiber Optic Sensors Using Ray Optics. *J. Light. Technol.* **2019**, *38*, 1580–1588. [CrossRef]
7. Wang, S.; Zhang, D.; Xu, Y.; Sun, S.; Sun, X. Refractive Index Sensor Based on Double Side-Polished U-Shaped Plastic Optical Fiber. *Sensors* **2020**, *20*, 5253. [CrossRef] [PubMed]
8. Lopez-Dieguez, Y.; Estudillo-Ayala, J.M.; Vázquez, D.J.; Herrera-Piad, L.A. Tip optical fiber refractive index sensor based on a thin copper film. *Supl. Rev. Mex. Fısica* **2021**, *2*, 43–48. [CrossRef]
9. Liang, K. Determination of Refractive Index of Air (or Other Gases) by Mach-Zeder Interferometer. *J. Liaoning Univ. Nat. Sci. Ed.* **1994**, *21*, 64–66.
10. An, J.; Jin, Y.; Sun, M.; Dong, X. Relative Humidity Sensor Based on SMS Fiber Structure with Two Waist-Enlarged Tapers. *IEEE Sens. J.* **2014**, *14*, 2683–2686. [CrossRef]
11. Liao, C.; Zhu, F.; Zhou, P.; Wang, Y. Fiber Taper-Based Mach–Zehnder Interferometer for Ethanol Concentration Measurement. *Micromachines* **2019**, *10*, 741. [CrossRef] [PubMed]
12. Ahsani, V.; Ahmed, F.; Jun, M.B.; Bradley, C. Tapered fiber-optic mach-zehnder interferometer for ultra-high sensitivity measurement of refractive index. *Sensors* **2019**, *19*, 1652. [CrossRef] [PubMed]

13. Han, X.; Liu, C.; Jiang, S.; Leng, S.; Yang, J. Up-down taper based in-fiber Mach-Zehnder interferometer for liquid refractive index sensing. *Sensors* **2019**, *19*, 5440. [CrossRef] [PubMed]
14. Tang, C.; Deng, M.; Zhu, T.; Rao, Y. Photonic crystal fiber based MZ interferometer for refractive index measurement. *J. Optoelectron. Laser* **2011**, *22*, 1304–1308. [CrossRef]
15. Guojia, H.; Shizhang, M.; Lian, W.; Shiping, L.; Wenlin, F. Photonic Crystal Fiber Mach-Zehnder Interferometer Based on Refractive Index Sensitivity of Cladding Corrosion Optimization. *Laser Optoelectron. Prog.* **2019**, *56*, 170629. [CrossRef]
16. Du, H.; Sun, X.; Hu, Y.; Dong, X.; Zhou, J. High Sensitive Refractive Index Sensor Based on Cladding Etched Photonic Crystal Fiber Mach-Zehnder Interferometer. *Photonic Sens.* **2019**, *9*, 126–134. [CrossRef]
17. Fang, Y.; Wu, T.; Tao, S. The characteristics of gaussian beam in up-tapered optical fiber. *J. Beijing Inst. Posts Telecommun.* **1993**, *16*, 71–72.
18. Che, F. Research and Application of Micro-Nano Fiber Grating Refractive IndexSensing. Master's Thesis, Xi'an Shiyou University, Xi'an, China, 2015; p. 29. [CrossRef]
19. Liu, Y.; Che, F.; Fu, H.; Zhao, N.; Shao, M. Research progress of fiber Bragg grating refractive index sensing. *Appl. Optoelectron. Technol.* **2015**, *30*, 1–7. [CrossRef]
20. Pang, F.; Liang, W.; Xiang, W.; Chen, N.; Zeng, X.; Chen, Z.; Wang, T. Temperature-Insensitivity Bending Sensor Based on Cladding-Mode Resonance of Special Optical Fiber. *IEEE Photonics Technol. Lett.* **2008**, *21*, 76–78. [CrossRef]

micromachines

Article

Simultaneous Measurement of Temperature and Refractive Index Using Michelson Interferometer Based on Waist-Enlarged Fiber Bitaper

Na Zhao [1], Zelin Wang [1], Zhongkai Zhang [1,*], Qijing Lin [1,2,*], Kun Yao [1], Liangquan Zhu [1], Bian Tian [1], Libo Zhao [1], Ping Yang [1] and Zhuangde Jiang [1]

1 State Key Laboratory for Manufacturing Systems Engineering, Xi'an Jiaotong University, Xi'an 710049, China; zn2020@xjtu.edu.cn (N.Z.); wzl15086927209@stu.xjtu.edu.cn (Z.W.); yao_kun@outlook.com (K.Y.); zhuliangquan@stu.xjtu.edu.cn (L.Z.); t.b12@mail.xjtu.edu.cn (B.T.); libozhao@xjtu.edu.cn (L.Z.); ipe@xjtu.edu.cn (P.Y.); zdjiang@xjtu.edu.cn (Z.J.)

2 Collaborative Innovation Center of High-End Manufacturing Equipment, Xi'an Jiaotong University, Xi'an 710054, China

* Correspondence: zzkxjtu@outlook.com (Z.Z.); xjjingmi@163.com (Q.L.)

Abstract: An all-fiber temperature and refractive dual-parameter-sensing Michelson interferometer is designed based on the waist-enlarged bitaper. At 5 mm from the fiber end, the waist-enlarged bitaper is manually spliced and the probe is formed. Since the input light encounters the waist-enlarged bitaper, it will excite high-order modes to transmit in the fiber cladding, and there will be an optical path difference between the basic mode and the higher-order mode. The light transmitted in the core and cladding is reflected upon encountering the fiber end face and the interference occurs due to the optical path difference between basic mode and higher-order mode. Changes in temperature and refractive index at the fiber probe can be detected by monitoring the interference fringes. The refractive response sensitivity is −191.06 dBm/RIU from 1.351 RIU to 1.4027 RIU, and the temperature response sensitivity is 0.12 nm/°C from 11 °C to 98 °C. Through the sensitivity matrix equation, the superimposed refractive index and temperature signals can be effectively demodulated. The sensor has the advantages of multi-parameter measurement, compact structure, low cost, easy fabrication and high reliability.

Keywords: fiber-optic sensors; temperature; refractive index; multi-parameter sensing; Michelson interferometer

Citation: Zhao, N.; Wang, Z.; Zhang, Z.; Lin, Q.; Yao, K.; Zhu, L.; Tian, B.; Zhao, L.; Yang, P.; Jiang, Z. Simultaneous Measurement of Temperature and Refractive Index Using Michelson Interferometer Based on Waist-Enlarged Fiber Bitaper. *Micromachines* **2022**, *13*, 658. https://doi.org/10.3390/mi13050658

Academic Editor: Limin Xiao

Received: 15 March 2022
Accepted: 19 April 2022
Published: 20 April 2022

1. Introduction

Temperature and refractive index are important parameters in the biochemical, food processing, pharmaceutical, oil extraction and biochemical measurement fields. Traditional electrical sensors find it difficult to meet the sensing requirements of these fields under severe environmental factors such as corrosion resistance, oxidation resistance and electromagnetic interference resistance. The optical fiber sensor performs sensing measurement based on optical signals, which can overcome the sensing problems such as oxidation resistance and corrosion resistance that electrical sensors cannot deal with, and has the advantages of compact structure, multi-parameter sensing measurement and high sensitivity, which make the optical fiber sensor a multi-parameter sensor.

In recent years, various fiber-optic sensors have been reported, such as long-period fiber gratings [1,2], fiber Bragg gratings [3–6], Mach–Zehnder interferometer [7–11] and Michelson interferometer [12–19], etc. In 2020, JENS HØVIK et al. [1] of Norwegian University developed a wavelength refractive index sensor based on long-period grating, and a response sensitivity of 5078 nm/RIU was measured from 1.33 RIU to 1.34 RIU. Subsequently, Yang H et al. [3] of Nanchang University developed a refractive index sensor using fiber Bragg grating from the perspective of intensity modulation, and a maximum refractive index sensitivity of −134.174 dB/RIU was obtained. In 2021, Wang, S. et al. [12] of Nanyang Technological University in Singapore developed a fiber tip Michelson interferometer, which partially filled with polymer glue in the suspended core fiber, and achieved high-sensitivity temperature sensing of −164 pm/°C from 25 °C to 60 °C. In the same year, Zhao, Y. et al. [13] from Tsinghua University proposed a polarization-maintaining fiber Sagnac temperature sensor based on combination with a fiber-optic Michelson structure, and the temperature sensitivity is 78.984 nm/°C from 31 °C to 35 °C; however, the reduction of the overall structure of the sensor still needs continuous exploration. The above studies have not covered the measurement of temperature and refractive index dual-parameter sensing, so multi-parameter sensing needs further research.

In 2019, Gao, X. et al. [7] of Beijing Jiaotong University measured the refractive index and temperature at the same time based on MZI combined with coreless fiber and few-mode fiber; the refractive index sensitivity from 1.3707 RIU to 1.39809 RIU was 97 pm/RIU, 22.9 pm/RIU and 24.6 pm/RIU, respectively, and the temperature sensitivity from 35 °C to 55 °C was 162 pm/°C, 162 pm/°C and 194 pm/°C, respectively. In order to further improve the temperature and refractive index sensing performance of the sensor, in 2021, Siti Mahfuza, Saimon et al. [6] from Universiti Malaysia proposed a novel fiber-optic sensor, which is composed of a section of single-mode silica rod–single-mode fiber structure cascaded to fiber Bragg grating, by monitoring the transmission spectrum; a sensitivity of 108.07 dBm/RIU in the refractive index range of 1.45 RIU to 1.531 RIU and a sensitivity of 9.31 pm/°C from 35 °C to 85 °C was measured. In the same year, Wu, B. et al. [8] proposed a curved-core-shifted coaxial Mach–Zehnder interferometer with a bending radius of 35.64 mm; the highest refractive index sensitivity in the range of 1.333 RIU to 1.373 RIU was −44.55 nm/RIU, and the highest temperature sensitivity in the temperature range of 25 °C to 60 °C was 0.0799 nm/°C. The above structures can realize temperature and refractive index sensing, but they are transmissive structures that are not easy to transform into probe structures. Therefore, the Michelson interferometers which can be made into probe structures have become the focus of research.

In 2019, Wang, J. et al. [14] fabricated a Michelson interferometer by splicing a single-mode fiber and a hollow silica tube, with a refractive index and temperature sensitivity of 8.1498 rad/RIU from 1.331 RIU to 1.387 RIU, and −0.05 rad/°C from 20 °C to 90 °C. In 2020, Qi K et al. [15] of Harbin Institute of Technology proposed a fiber-optic Michelson interferometer based on a three-microsphere array, which measured a maximum temperature sensitivity of 115.3 pm/°C from 20 °C to 90 °C and a refractive sensitivity of −56.63 nm/RIU in the range of 1.3335 RIU to 1.406 RIU; the fabrication repeatability of the three microsphere structures of the sensor is difficult to control, and the repeatability of the process needs to be further enhanced. In 2021, Zheng J et al. [16] proposed a thin-waist-cone Michelson interferometer, and measured the temperature sensitivity of wavelength demodulation as being 8.4 pm/°C, and the refractive index sensitivity of intensity demodulation as being −145.54 dB/RIU. In the same year, Zhang Y et al. [17] designed a new L-type Michelson interferometer based on the method of flame firing. The refractive index response sensitivity was −131.0 nm/RIU from 1.3430 RIU to 1.3927 RIU, and the temperature response sensitivity was 94.17 pm/°C from 30 °C to 100 °C. The sensor is greatly affected by the position of the flame intensity during the production process, and needs to be further improved in terms of repeatability. The waist-enlarged fiber bitaper can be used as a coupling point [20,21]. Yanhong Liang et al. [22] from Zhejiang University developed a novel optical fiber sensor, based on the waist-enlarged fiber bitaper and the polyvinyl alcohol film, where the refractive index variance ranges from 1.49 to 1.34, the ambient humidity increases from 20%RH to 95%RH, and a sensitivity up to 1.2 dB/%RH can be achieved.

In this paper, a Michelson interferometer based on the waist-enlarged fiber bitaper is developed, which realizes the simultaneous sensing of temperature and the refractive index. The sensor has a compact structure, and a thick waist-cone structure is directly fabricated on the optical fiber by manual welding, which eliminates the process of fiber cutting and alignment welding, and reduces the manufacturing steps of the sensing probe. The spectrum of the optical signal is analyzed, and the mode distribution diagram of the optical fiber section is given. The intensity and wavelength changes of the spectra are analyzed, and the temperature and refractive index are measured using the sensitivity matrix equation.

2. Principle and Design

The schematic diagram of the fiber Michelson interferometer is shown in Figure 1, where the distance L between the waist-enlarged bitaper and the right end face is the length of the sensing arm. When the input light is transmitted from the left to the position of the waist-enlarged bitaper, the higher-order modes are excitation in the cladding. The lower-order fundamental modes continue to propagate in the core. When the light transmitted in the fiber core and fiber cladding encounters the reflective end face, it is reflected and returns along the original path. When the waist-enlarged bitaper is encountered for the second time, the different orders of light are coupled together and form interference fringes.

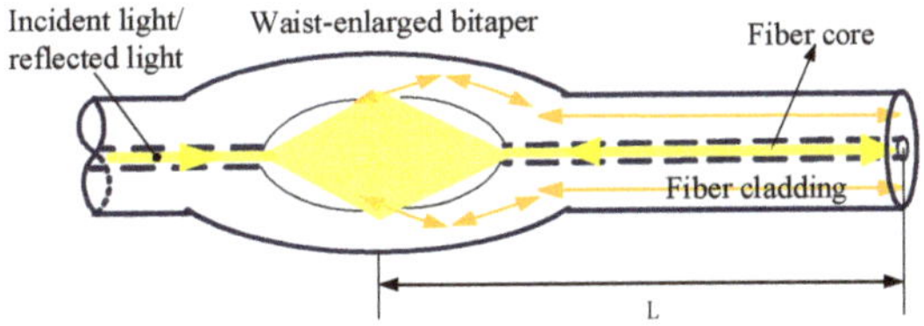

Figure 1. Schematic diagram of Michelson interferometer.

The formation of interference fringes is caused by the optical path difference of the transmitted light in the fiber core and cladding. We can first list the optical path difference equation, as follows [23,24]:

$$\Delta\phi = \frac{4\pi}{\lambda}(n_m - n_{core})L \quad m = 1,2,3\ldots\ldots \tag{1}$$

where n_m is the effective refractive index of the optical fiber cladding, n_{core} is the effective refractive index of the fiber core, L is the length of Michelson interferometer and λ is the wavelength.

If we define the intensities of fiber core mode and fiber cladding mode as I_{core} and $I_m(RI)$, the effective refractive index difference is Δn_{eff} and the initial phase is ϕ_0; the intensity of the transmission spectrum can be defined as [25,26]:

$$I(\lambda) = I_{core} + \sum_m I_m(RI) + \sum_m 2[I_{core} \times I_m(RI)]^{\frac{1}{2}} \cdot \cos\left(\frac{4\pi\Delta n_{eff}L}{\lambda} + \phi_0\right) \tag{2}$$

When the external refractive index changes, the intensity of the reflected light will change. Changes in light intensity are related to changes in the external refractive index. Assuming that the initial phase is equal to 0, the change in the interference intensity can be described as [27]:

$$\begin{gathered}\frac{dI}{dRI} = \sum_m \frac{dI_m(RI)}{dRI} + I_{core} \cdot \sum_m [I_{core} I_m(RI)]^{\frac{1}{2}} \cdot \frac{dI_m(RI)}{dRI} \cdot \cos\left\{\frac{4\pi}{\lambda}[I_{core} - I_m(RI)] \cdot L\right\} + \\ \sum_m [I_{core} I_m(RI)]^{\frac{1}{2}} \cdot \sin\left\{\frac{4\pi}{\lambda}[I_{core} - I_m(RI)] \cdot L\right\} \cdot \frac{8\pi}{\lambda} \cdot L \cdot \frac{dn_m(RI)}{dRI}\end{gathered} \tag{3}$$

The generation of the interference spectrum is mainly caused by the phase difference, the phase condition represented by the cosine phase of Equation (2); therefore, the spectral phase at a certain moment can be expressed as the following equation:

$$\phi = \frac{4\pi n_{core} L}{\lambda} + \phi_0 \tag{4}$$

When the temperature changes, the length and refractive index of the fiber material will change, and the center wavelength will also drift accordingly. Assuming that when the temperature increases by ΔT, the cavity refractive index of the fiber sensor changes from n to $n_{core} + n_{core}\xi\Delta T$, the length changes from L to $L + L\alpha\Delta T$, and the center wavelength λ will drift to $\lambda + \Delta\lambda$, so [23,25]:

$$\phi = \frac{4\pi(n_{core} + n_{core}\xi\Delta T)(L + L\alpha\Delta T)}{\lambda + \Delta\lambda} \tag{5}$$

In the equation: ΔT—temperature change; ξ—thermo-optical coefficient. The change of wavelength with temperature has little to do with the initial phase. In order to simplify the equation, we set the initial phase ϕ_0 as 0; then, from Equations (4) and (5), we can obtain:

$$\Delta\lambda = \lambda\left(\xi\Delta T + \alpha\Delta T + \xi\alpha\Delta T^2\right) \tag{6}$$

Because the wavelength and temperature change is very small, the equation can be approximated to Equation (7):

$$\frac{\Delta\lambda}{\Delta T} \approx \frac{d\lambda}{dT} = \lambda(\xi + \alpha) \tag{7}$$

Equation (7) characterizes the change in wavelength with temperature. The thermal expansion coefficient and the thermo-optic coefficient are constants. Therefore, it can be drawn theoretically that when the monitoring wavelength is constant, the temperature response sensitivity is a constant; in other words, the change of wavelength is approximately linear with the change of temperature, which can be confirmed by the following temperature experiments.

A fusion splicer (Furukawa, S177B, Tokyo, Japan) is used to fabricate the fiber Michelson, and the fiber used for the sensor is a common single-mode fiber with core and cladding diameters of 9 μm and 125 μm, respectively. A fiber cleaver (Furukawa, S325, Tokyo, Japan) is used to cut the fiber end face flat to obtain a reflective end face with high reflectivity. The part between the waist-enlarged bitaper and the reflective end face is the sensing arm. According to the required length of the sensing arm, the position of the waist-enlarged bitaper is selected and the manual welding program is set to make the waist-enlarged fiber bitaper. Through a large number of experimental comparisons, it is found that if the waist size of the waist-enlarged bitaper is too small, the contrast of the generated interference fringes will become smaller; and if the bitaper is too large, the overall intensity of the spectrum will be reduced. Therefore, it is necessary to restrict the size of the waist-enlarged bitaper structure within a certain range, and an intermediate size is selected in the manufacturing process. The parameters in the specific production process are as follows. The discharge cleaning is set to 50 ms, the discharge duration is set to 1100 ms, the discharge intensity is set to 155 xmW and the push distance is set to 140 μm. Due to the high discharge intensity and the large advancing distance, the diameter of the set point will slowly expand. Figure 2 is a 90-times magnified photo of the optical fiber waist-enlarged bitaper under the microscope (Olympus, SZ61, Center Valley, PA, USA). It can be seen from the figure that the diameter of the welding point is enlarged from 125 μm to about 152 μm, and the length of the optical fiber waist-enlarged bitaper is about 350 μm.

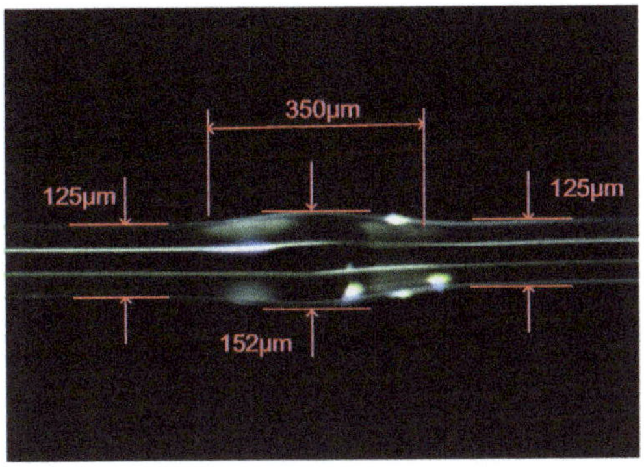

Figure 2. The optical fiber waist-enlarged bitaper under microscope.

Figure 3 shows the interference fringes corresponding to sensors with different lengths. It can be seen from the figure that the contrast of the interference fringes is high, and the interference peaks and interference valleys are clearly visible. By fabricating a large number of sensors, we summarize the relationship between the interference arm length and the interference period per 80 nm spectral range. It can be seen from Figure 4 that as the length of the interference arm increases, the spectrum in the unit spectral range increases. In other words, as the interference arms grow, the interference fringes become denser.

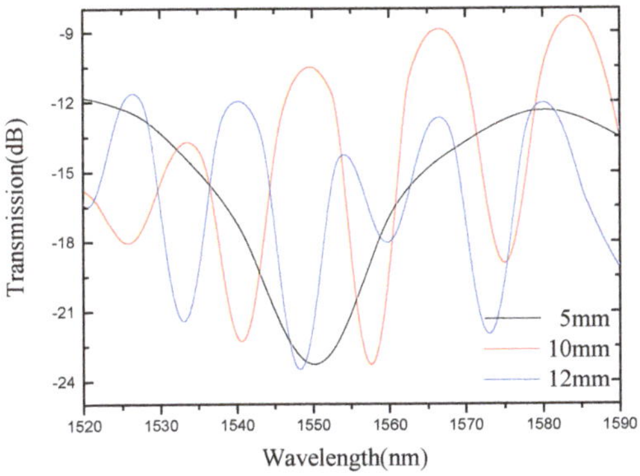

Figure 3. Interferometric spectra of sensors with different lengths.

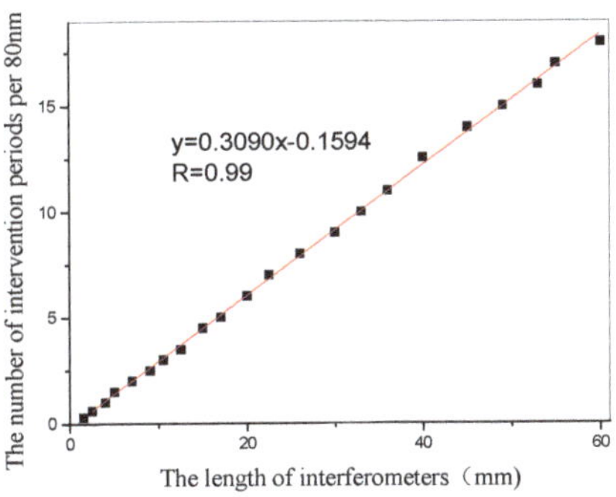

Figure 4. Relationship between interference period and interference arm length in the 80 nm spectral range.

The spectrum is converted to a spatial spectrum by a fast Fourier transform (FFT), as shown in Figure 5. It can be seen that the high-order spectra corresponding to sensors of different lengths have only one main peak. For sensors with interference lengths of 5 mm, 13 mm and 17 mm, the positions of the main peaks are at 0.0204388 nm^{-1}, 0.0573041 nm^{-1} and 0.0712539 nm^{-1}, respectively. Based on Taylor expansion, the phase ϕ can be defined as [24]:

$$\phi \approx \phi_0 - \frac{2\pi\Delta\lambda}{\lambda^2}\Delta n_{eff} \cdot 2L \tag{8}$$

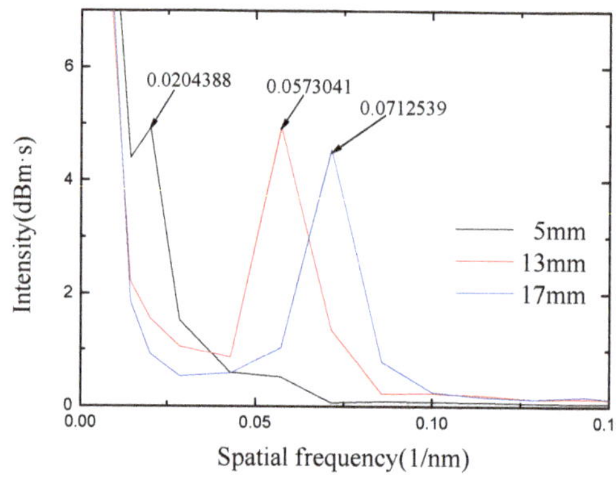

Figure 5. Spatial spectrum analysis of sensors with different lengths.

Meanwhile, the phase part of the interference wavelength can be expressed as a cosine equation, as shown below [28].

$$\cos\phi = \cos(2\pi\xi\Delta\lambda) \tag{9}$$

Assuming the initial phase ϕ_0 is equal to 0, we can obtain the spatial frequency ξ based on Equations (8) and (9), as shown in Equation (10).

$$\xi = \frac{2}{\lambda^2}\Delta n_{eff} \cdot L \tag{10}$$

By transforming Equation (10), it can be obtained that the refractive index difference is as follows.

$$\Delta n_{eff} = \frac{\xi \cdot \lambda^2}{2L} \tag{11}$$

The C + L band is selected in the experiment; the center wavelength is around 1550 nm, and the lengths of the interference arms are 5 mm, 13 mm and 17 mm, respectively. Substituting each parameter into Equation (11), the refractive index difference Δn_{eff} between modes can be calculated to be 0.004910422 dB·s, 0.005295119 dB·s and 0.005034926 dB·s, respectively. The relationship between Δn_{eff} and different modes is analyzed by OptiFiber software, as shown in Table 1. By analyzing the effective refractive index difference between the higher-order mode and the fundamental mode, the experimentally obtained refractive index difference is closest to the difference between LP_{01} and LP_{07}. Among them, LP_{01} is mainly transmitted in the core, and LP_{07} is mainly transmitted in the cladding. The optical mode distribution of the fiber-truncated surface of the fundamental mode LP_{01} and high-order LP_{07} mode given by software simulation is shown in Figure 6.

Table 1. Relationship between mode order and refractive index difference Δn_{eff}.

X	n_{LP0x}	Δn_{eff}
1	1.4658679	0
2	1.4627498	0.0031181
3	1.4625833	0.0032846
4	1.4623038	0.0035641
5	1.4619147	0.0039532
6	1.4614194	0.0044485
7	1.4608213	0.0050466
8	1.4601245	0.0057434
9	1.4593324	0.0065355

(a) Fundamental mode LP_{01} (b) Higher order cladding mode LP_{07}

Figure 6. Optical mode distribution diagram of optical fiber section.

3. Experiment and Discussion

3.1. Refractive Index Response Experiment

As shown in Figure 7, the fiber Michelson interferometer measuring system is composed of an optical spectrum analyzer (OSA, Anritsu, MS9740A, Kanagawa, Japan) and a broadband light source (BBS, Lightcomm, ASE-CL, Shenzhen, China). The resolution of the OSA is set to be 0.02 nm, and the bandwidth of BBS is 80 nm. Sugar solutions with different refractive indices are prepared as samples for measurement, which are set as 1.351 RIU, 1.357 RIU, 1.369 RIU, 1.369 RIU, 1.378 RIU, 1.3872 RIU, 1.3941 RIU and 1.4027 RIU. Further, the refractive index solutions are verified by Abbe refractometer. During the specific experiment, a single concentration of sugar solution is dropped on the sensor head, and the OSA is utilized to record the transmission spectral response. Before the next measurement of the sugar solution, the sensor head is rinsed repeatedly with water and allowed to dry. In addition, the ambient temperature is kept at 11 °C to minimize the cross effects of temperature.

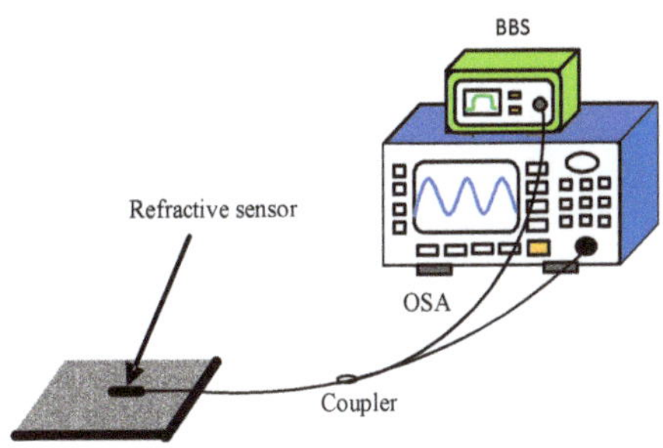

Figure 7. Refractive index sensing experimental device schematic diagram.

As the refractive index of the liquid increases, the intensity and wavelength of the spectral valleys change accordingly, as shown in Figure 8. The blue dots are the measured intensity data, and the black dots are the measured wavelength data. The fitted line is obtained by the least squares fitting method. A linear decrease in the power of the trough can be observed. With the refractive index changes from 1.351 RIU to 1.4027 RIU, an intensity response sensitivity of −191.06 dB/RIU and a wavelength response sensitivity of 5.09 nm/RIU are obtained, as shown in Figure 9.

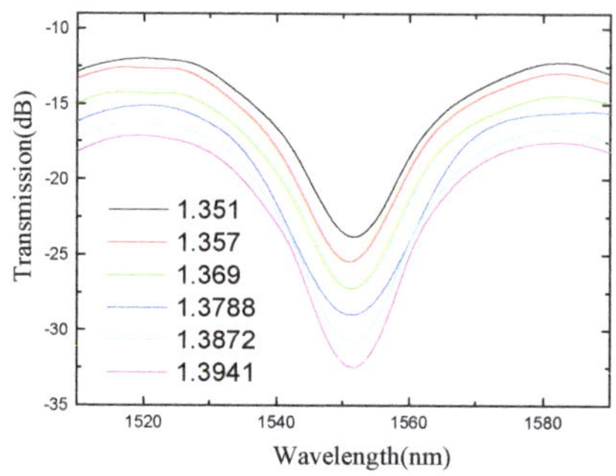

Figure 8. The spectra under different ambient refractive indices.

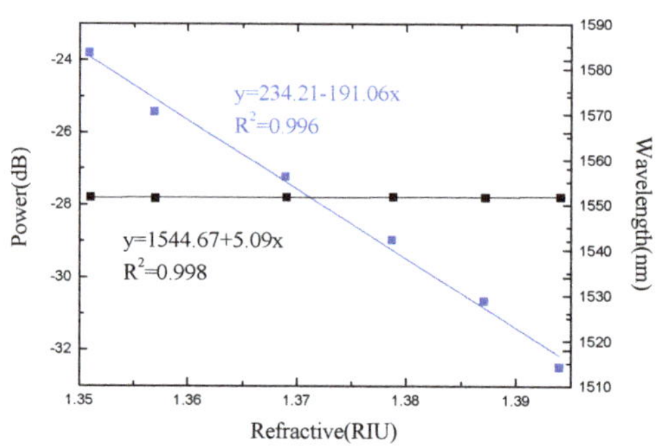

Figure 9. The refractive index response of the Michelson interferometer.

In addition, the change of air humidity can also be attributed to the small change of the air refractive index, so we also experimentally designed a humidity experiment. In the humidity range of 20% to 80%, the interference spectrum of the sensor hardly changes with humidity. This can be attributed to the fact that the refractive index change is too small; the total change of air refractive index with air humidity is less than 0.00005 RIU [29]. The spectral change is too low, and the demodulation equipment can no longer distinguish it.

3.2. Temperature Response Experiment

The schematic diagram of the temperature experimental device is shown in Figure 10. The light emitted by the ASE light source is transmitted into the sensor, and the MS740A spectrometer (Anritsu, MS9740A, Kanagawa, Japan) is used to measure the reflection spectrum of the sensing system. The temperature control system is a high-temperature muffle furnace (Omega, Biel/Bienne, Switzerland), and the temperature control accuracy is 1 °C. As the temperature increases, the spectrum of the fiber Michelson shifts due to the thermo-optic and thermal expansion effects, as shown in Figure 11. It can be seen from the figure that the sensor is sensitive to temperature changes and can be designed as a temperature sensor.

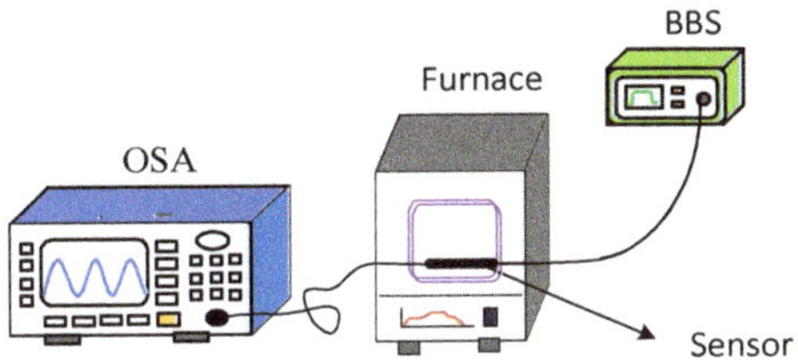

Figure 10. Temperature sensing experimental device schematic diagram.

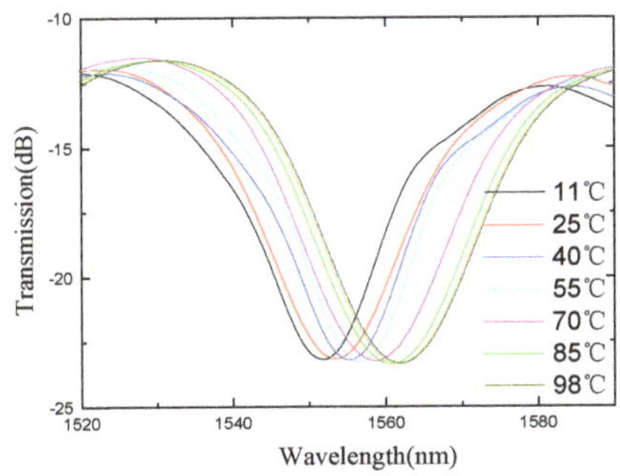

Figure 11. The spectra under different ambient temperatures.

The trough around 1555 nm is monitored and used to analyze the temperature response sensitivity of the fiber Michelson interferometer, as shown in Figure 11. The shift of the trough intensity and wavelength with temperature is analyzed as shown in Figure 12. The blue dots and blue lines are the measured intensity data and the fitted line, and the black dots are the measured wavelength data and the fitted line. Temperature sensitivities are −0.011 nm/°C and –0.009 dB/°C from 30 °C to 90 °C.

$$\begin{bmatrix} \Delta P \\ \Delta\lambda \end{bmatrix} = \begin{bmatrix} K_{nP} & K_{TP} \\ K_{n\lambda} & K_{T\lambda} \end{bmatrix} \begin{bmatrix} \Delta n \\ \Delta T \end{bmatrix} \tag{12}$$

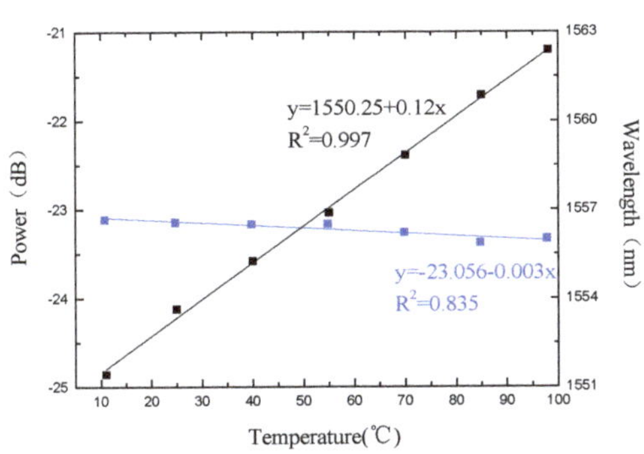

Figure 12. The temperature response of the Michelson interferometer.

In Equation (12), Δn and ΔT are the refractive index and temperature changes, respectively. K_{nP} and $K_{n\lambda}$ are the intensity and wavelength response sensitivities when the refractive index changes. K_{TP} and $K_{T\lambda}$ are the intensity and wavelength response sensitivities when temperature changes. Multiplying both sides of the Equation by the inverse of the sensitivity matrix, Equation (13) can be obtained [30].

$$\begin{bmatrix} \Delta n \\ \Delta T \end{bmatrix} = \begin{bmatrix} K_{nP} & K_{TP} \\ K_{n\lambda} & K_{T\lambda} \end{bmatrix}^{-1} \begin{bmatrix} \Delta P \\ \Delta\lambda \end{bmatrix} \tag{13}$$

The corresponding relationship between the experimentally measured sensitivity coefficient of each parameter and each parameter of the matrix equation is as follows: $K_{nP} = -191.06$ dB/RIU, $K_{n\lambda} = -5.09$ nm/°C, $K_{TP} = -0.003$ dB/RIU, $K_{T\lambda} = 0.12$ nm/°C. Substituting the measurement data into Equation (13), Equation (14) can be obtained. Through the analysis of the spectrum, the changes of the refractive index and temperature value of the detection environment can be obtained based on Equation (14).

$$\begin{bmatrix} \Delta n \\ \Delta T \end{bmatrix} = \begin{bmatrix} -191.06 & 5.09 \\ -0.003 & 0.12 \end{bmatrix}^{-1} \begin{bmatrix} \Delta P \\ \Delta\lambda \end{bmatrix} \tag{14}$$

4. Conclusions

In this paper, a waist-enlarged bitaper Michelson interferometer for multi-parameter sensing is proposed. By comparing the spectra of probes with different lengths, it is found that as the length of the sensor probe increases, the distance between the two valleys of the interference spectrum becomes smaller. The experimental analysis of the temperature-refractive index characteristics of the sensor shows that the sensor has a wavelength response sensitivity of 0.12 nm/°C and an intensity response sensitivity of −0.003 dBm/°C in the range of 11 °C to 98 °C. In the refractive index range of 1.351 RIU to 1.4027 RIU, the wavelength response sensitivity is 5.09 nm/RIU, and the intensity response sensitivity is −191.06 dBm/RIU. The measurement results prove that this type of optical fiber sensor can measure various external parameters according to the change of spectral intensity and wavelength, combined with the sensitivity matrix equation. In the process of food processing and pharmaceutical production, the temperature and refractive index of the product can be obtained through sensor monitoring in a timely manner, and then the product quality can be controlled. At the same time, it can also monitor the temperature and refractive index in real time during oil exploration to obtain the temperature of underground oil and the quality of crude oil. In short, the sensor has the advantages of simple structure, high sensitivity and multi-parameter measurement, and has certain application value.

Author Contributions: Data curation, N.Z. and Z.W.; Formal analysis, L.Z. (Liangquan Zhu); Project administration, L.Z. (Libo Zhao); Resources, Q.L., B.T. and Z.J.; Software, Z.Z. and K.Y.; Writing—original draft, N.Z.; Writing—review & editing, Q.L. and P.Y. All authors have read and agreed to the published version of the manuscript.

Funding: This work is supported by the Postdoctoral Innovative Talent Support Program (Grant No. BX20200274), the National Natural Science Foundation of China (Grant Nos. 51720105016, 52105560), the Post-doctoral Science Foundation (Grant No. 2021M702596), Shaanxi Province Surface Project (Grant No. S2022-JC-YB-1996) and the National Science and Technology Major Project (Grant No. J2019-V-0006-0100).

Conflicts of Interest: The authors declare no conflict of interest.

References

1. Høvik, J.; Yadav, M.; Noh, J.W.; Aksnes, A. Waveguide asymmetric long-period grating couplers as refractive index sensors. *Opt. Express* **2020**, *28*, 23936–23949. [CrossRef] [PubMed]
2. Fukushima, K.; Bui, Q.H.; Nakaya, K.; Soares, M.G.; Wada, A.; Tanaka, S.; Ito, F. EDF ring laser using cascaded-chirped long period fiber grating for temperature measurement. *Opt. Express* **2020**, *28*, 13081–13090. [CrossRef] [PubMed]
3. Yang, H.; Li, Y.; Li, X. Intensity-modulated refractive index sensor based on the side modes of fiber Bragg grating. *Opt. Commun.* **2022**, *505*, 127319. [CrossRef]
4. Laffont, G.; Cotillard, R.; Roussel, N.; Desmarchelier, R.; Rougeault, S. Temperature resistant fiber Bragg gratings for on-line and structural health monitoring of the next-generation of nuclear reactors. *Sensors* **2018**, *18*, 1791. [CrossRef]

5. Kappel, E.; Prussak, R.; Wiedemann, J. On a simultaneous use of fiber-Bragg-gratings and strain-gages to determine the stress-free temperature Tsf during GLARE manufacturing. *Compos. Struct.* **2019**, *227*, 111279. [CrossRef]
6. Saimon, S.M.; Noor, M.Y.M.; Abdullah, A.S.; Salim, M.R.; Ibrahim, M.H.; Azmi, A.I.; Ngajikin, N.H.; Ahmad, M.H.; Othman, A.F. Simultaneous measurement of high refractive index and temperature based on SSRS-FBG. *IEEE Photonics Technol. Lett.* **2021**, *33*, 715–718. [CrossRef]
7. Gao, X.; Ning, T.; Zheng, J.; Li, J.; Pei, L.; Zhang, C.; Liu, L.; He, X. Simultaneous measurement of refractive index, strain, and temperature based on a Mach-Zehnder interferometer with hybrid structure optical fiber. *Appl. Opt.* **2019**, *58*, 8187–8193. [CrossRef]
8. Wu, B.; Bao, H.; Zhou, Y.; Liu, Y.; Zheng, J. Simultaneous measurement of refractive index and temperature based on bent core-offset in-line fiber Mach-Zehnder interferometer. *Adv. Sens. Syst. Appl.* **2021**, *11901*, 72–78. [CrossRef]
9. Marrujo-García, S.; Hernández-Romano, I.; Torres-Cisneros, M.; May-Arrioja, D.A.; Minkovich, V.P.; Monzón-Hernández, D. Temperature-independent curvature sensor based on in-fiber Mach-Zehnder interferometer using hollow-core fiber. *J. Lightwave Technol.* **2020**, *38*, 4166–4173. [CrossRef]
10. Harvey, C.M.; Mühlberger, K.; Fokine, M. Mach-Zehnder interferometer for in-situ non-contact temperature monitoring during thermal processing of an optical fibre. *J. Lightwave Technol.* **2021**, *39*, 7223–7230. [CrossRef]
11. Zhang, H.; Cong, B.; Zhang, F.; Qi, Y.; Hu, T. Simultaneous measurement of refractive index and temperature by Mach–Zehnder cascaded with FBG sensor based on multi-core microfiber. *Opt. Commun.* **2021**, *493*, 126985. [CrossRef]
12. Wang, S.; Yang, Y.; Niu, P.; Wu, S.; Liu, S.; Jin, R.B.; Lu, P.; Hu, X.; Dai, N. Fiber tip Michelson interferometer for temperature sensing based on polymer-filled suspended core fiber. *Opt. Laser Technol.* **2021**, *141*, 107147. [CrossRef]
13. Zhao, Y.; Dai, M.; Chen, Z.; Liu, X.; Gandhi, M.A.; Li, Q.; Fu, H.Y. Ultrasensitive temperature sensor with Vernier-effect improved fiber Michelson interferometer. *Opt. Express* **2021**, *29*, 1090–1101. [CrossRef] [PubMed]
14. Wang, J.; Liu, B.; Wu, Y.; Mao, Y.; Zhao, L.; Sun, T.; Nan, T. A novel fiber in-line Michelson interferometer based on end face packaging for temperature and refractive index measurement. *Optik* **2019**, *194*, 163094. [CrossRef]
15. Qi, K.; Zhang, Y.; Sun, J.; Yi, G. All-Fiber high temperature and refractive index sensor based on three microspheres array Michelson interferometer. *Opt. Laser Technol.* **2020**, *129*, 106300. [CrossRef]
16. Zheng, J.; Liu, B.; Zhao, L.; Mao, Y.; Han, Y.; Nan, T.; Wu, Y.; Wang, J.; Wang, T.; Zhang, Y. Simultaneous transverse load, refractive index, and high-temperature sensor based on michelson interferometer. *Opt. Fiber Technol.* **2021**, *67*, 102686. [CrossRef]
17. Zhang, Y.; Wu, Y.; Han, Y.; Mao, Y.; Wu, J.; Zhao, L.; Tang, R.; Ren, J.; Liu, B. Simultaneous temperature and refractive index sensor based on an L-like Michelson interferometer. *Appl. Opt.* **2021**, *60*, 10101–10108. [CrossRef]
18. Ruan, S. High-Temperature Measurement of a Fiber Probe Sensor Based on the Michelson Interferometer. *Sensors* **2021**, *22*, 289. [CrossRef]
19. Shao, M.; Zhang, R.; Gao, H.; Liu, Y.; Qiao, X.; Lin, Y. A High-Sensitivity Low-Temperature Sensor Based on Michelson Interferometer in Seven-Core Fiber. *IEEE Photonics Technol. Lett.* **2021**, *33*, 1293–1296. [CrossRef]
20. Yu, R.; Wang, C.; Benabid, F.; Chiang, K.S.; Xiao, L. Robust mode matching between structurally dissimilar optical fiber waveguides. *ACS Photonics* **2021**, *8*, 857–863. [CrossRef]
21. Wang, C.; Yu, R.; Debord, B.; Gérôme, F.; Benabid, F.; Chiang, K.S.; Xiao, L. Ultralow-loss fusion splicing between negative curvature hollow-core fibers and conventional SMFs with a reverse-tapering method. *Opt. Express* **2021**, *29*, 22470–22478. [CrossRef] [PubMed]
22. Liang, Y.; Yan, G.; He, S. Enlarged-taper tailored Fiber Bragg grating with polyvinyl alcohol coating for humidity sensing. In Proceedings of the 2015 International Conference on Optical Instruments and Technology: Optical Sensors and Applications, International Society for Optics and Photonics, Beijing, China, 10 August 2015; Volume 9620, p. 962007. [CrossRef]
23. Yao, Q. *Optics Course*, 2nd ed.; Higher Education Press: Beijing, China, 1989; Volume 94, p. 151. ISBN 9787040239157.
24. Bhardwaj, V.; Singh, V.K. Fabrication and characterization of cascaded tapered Mach-Zehnder interferometer for refractive index sensing. *Sens. Actuators A Phys.* **2016**, *244*, 30–34. [CrossRef]
25. Kapany, N.S. Fiber Optics Principles and Applications. In *Chapter E Challenge of Purchasing & Supply Management Leenders*; Academic Press: Cambridge, MA, USA, 1967; Volume 1, p. 564. [CrossRef]
26. Wang, Q.; Wei, W.; Guo, M.; Zhao, Y. Optimization of cascaded fiber tapered Mach-Zehnder interferometer and refractive index sensing technology. *Sens. Actuators B Chem.* **2016**, *222*, 159–165. [CrossRef]
27. Li, H.; Fu, H.; Shao, M.; Zhao, N.; Qiao, X.; Liu, Y.; Li, Y.; Yan, X. In-fiber Mach-Zehnder interferometer based on fiber core etched air-bubble and core diameter mismatch for liquid refractive index sensing. *Acta Phys. Sin.* **2013**, *62*, 214209. [CrossRef]
28. Pang, F.; Liang, W.; Xiang, W.; Chen, N.; Zeng, X.; Chen, Z.; Wang, T. Temperature-insensitivity bending sensor based on cladding-mode resonance of special optical fiber. *IEEE Photonics Technol. Lett.* **2008**, *21*, 76–78. [CrossRef]
29. Wang, Z. Humidity has a bearing on the refractivity of air. *J. Dalian Marit. Univ.* **1986**, *4*, 107–115. [CrossRef]
30. Greub, W.H. *Linear Algebra*; Springer: Berlin/Heidelberg, Germany, 1981. [CrossRef]

micromachines

Review

A Survey of the Influence of Process Parameters on Mechanical Properties of Fused Deposition Modeling Parts

Ge Gao [1,*], Fan Xu [1], Jiangmin Xu [1], Guanghai Tang [1] and Zhenyu Liu [2]

[1] College of Mechanical Engineering, Jiangsu University of Science and Technology, Zhenjiang 212100, China; xufan@stu.just.edu.cn (F.X.); xujiangmin@just.edu.cn (J.X.); tangguanghai@just.edu.cn (G.T.)
[2] Changchun Institute of Optics, Fine Mechanics and Physics, Chinese Academy of Sciences, Changchun 130033, China; liuzy@ciomp.ac.cn
* Correspondence: gaoge@just.edu.cn

Abstract: Due to the availability of materials and low cost for production, fused deposition modeling is becoming the most widely used additive manufacturing (AM) technology. However, the reasonable choice of process parameters for FDM is a significant task that directly affects the performance of the printed part. Therefore, it is necessary to investigate the influences of various process parameters on the quality characteristics of the components. The objectives of this study are to thoroughly review the current state of research that characterizes, estimates the effects of process parameters on mechanical properties, and summarizes the conclusions of existing works. In addition, some general issues of the presented research are summarized, and the need for future development is also emphasized. Finally, the research proposes several areas that deserve further study in this field.

Keywords: additive manufacturing; process parameters; fused deposition modeling; mechanical properties

Citation: Gao, G.; Xu, F.; Xu, J.; Tang, G.; Liu, Z. A Survey of the Influence of Process Parameters on Mechanical Properties of Fused Deposition Modeling Parts. *Micromachines* **2022**, *13*, 553. https://doi.org/10.3390/mi13040553

Academic Editors: Xiuqing Hao, Duanzhi Duan and Youqiang Xing

Received: 12 February 2022
Accepted: 27 March 2022
Published: 30 March 2022

1. Introduction

Fused Deposition Modeling (FDM), also known as Fused Filament Fabrication (FFF), melts thermoplastic filament through a heater and deposits it layer by layer on the platform via a nozzle to form a part. The most significant advantage of FDM is the wide range of molding materials, which includes thermoplastic polymers in general. Sometimes low melting point metals, ceramics, and others materials are also used [1]. Besides, high speed, low cost, pollution-free, and simplicity of the process are also benefits of FDM. Consequently, FDM is emerging as the most widely used and embraced technique of additive manufacturing, which is applied in various fields such as aerospace, automotive, medical, and architecture with rapid growth [2]. However, anisotropic behavior, poor surface quality, and low dimension accuracy are drawbacks of FDM, usually resulting in poor mechanical characteristics of printed components, which dramatically limits the further application of FDM on a large-scale [3].

FDM is a complex process that has a large number of parameters that play different roles in the fabrication. To produce products with good quality and meet requirements for material behavior, it is necessary to evaluate the impact of these parameters on the characteristics. To date, many studies have been conducted to analyze different controllable parameters to achieve desirable properties of parts, including surface roughness [4,5], dimension accuracy [6,7], hardness [8], build time [9,10], and mechanical properties [11–13]. Obviously, mechanical properties are the most fundamental characteristics of FDM printed parts, among which tensile, compressive, and flexural strength are the three most important and concerning properties to the manufacturers and users, which are also the objects of this paper.

FDM involves various parameters that can be classified into three main types: process parameters(raster angle, layer thickness, build orientation, raster width, print speed, infill density, air gap, infill pattern, extrusion temperature); environmental parameters(platform temperature, envelope temperature, humidity, oxygen, etc.); other print parameters(nozzle diameter, material color, filament diameter, etc.). Although these parameters all affect the quality of FDM components, the contributions of which are different. Process parameters are the most commonly analyzed owing to their significant impact on mechanical performance and production efficiency. Actually, several published review papers related to the FDM process parameter are available for interested readers: Gordelier et al. [2], Dey and Yodo [14], Cuan-Urquizo et al. [15], Sheoran et al. [16], Mohamed et al. [17], Popescu et al. [18], Bakır et al. [19], Syrlybayev et al. [20]. These existing literature reviews generally investigate and analyze which process parameters can affect a certain material behavior. Since users of the 3D printer are directly faced with each process parameter, it is necessary and helpful to make them understand how each process parameter affects the quality and characteristics of printed parts at different values. However, to date, no literature review has been reported to explain the influence from the perspective of parameters rather than properties. As a complement, this survey focuses on functions of every process parameter with varying values and discusses the mechanism behind it by amalgamating collusions of existing studies from 2010 to 2021. Some research beyond this range is also included for important topics. This article aims to provide a comprehensive review of the roles of different process parameters in the FDM process, update the recent advances in process parameters optimization for researchers, serve as a resource for newcomers in this field and give directions for anyone wishing to improve the mechanical behaviors of their printed components.

The structure of the paper is organized as follows: Section 2 describes different process parameters and reviews literature related to investigating or improving the mechanical performance of FDM parts. Section 3 contains some key findings of the presented works and highlights concluding results. Section 4 describes difficulties encountered in the improvement of the FDM part characteristics. The last section includes recommendations and perceptions for future work.

2. Process Parameters

The most researched process parameters include air gap, build orientation, extrusion temperature, infill density, infill pattern, layer thickness, raster width, raster angle, and print speed, as shown in Figure 1, which have substantial effects on filament (inter-layer and intra-layer) bonding, and thus influence the mechanical performance of FDM printed components [18]. In addition, interactions of these parameters play a significant role from the perspective of mechanical properties [21,22].

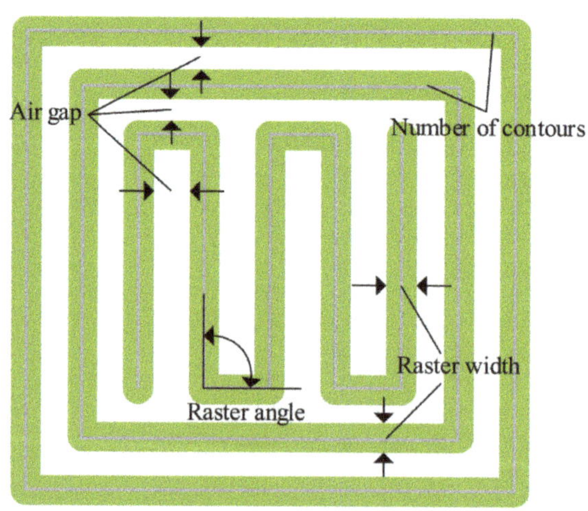

Figure 1. FDM process parameters related to toolpath.

2.1. Build Orientation

Build orientation (or part orientation [23], construction/layer orientation [24]) represents how and in which direction the part is generated on the print platform. In fact, build orientation can represent an arbitrary angle with any value [8,25,26], but in most studies, it is regarded as a certain angle with respect to X, Y, and Z-axis [27,28]. Generally, when test specimens are placed horizontally, vertically, and laterally, the build orientation is named as flat, upright, and on-edge, respectively, which is shown in Figure 2. Flat and on-edge are considered parallel to the print platform, while upright is along the direction of normal of the print platform.

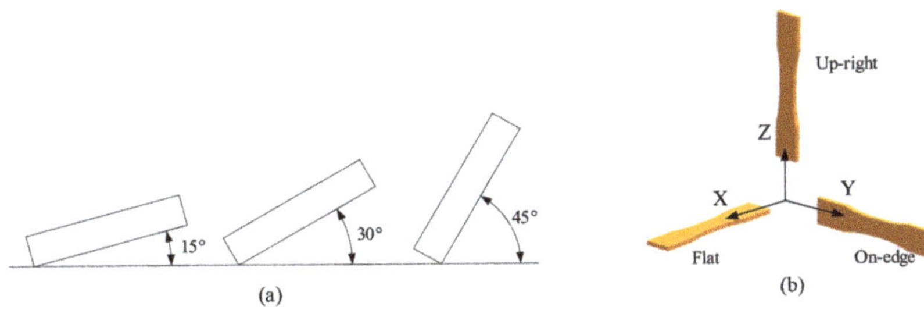

Figure 2. Build orientation: (**a**) arbitrary angle (**b**) certain angle.

The influences of build orientation on the mechanical performance of FDM components have been extensively researched. Different authors investigated the relationship between various materials and building directions. Wang et al. [7] established building factors with various levels based on analysis of variance (ANOVA). The result verified build orientation in the Z-direction to be the most predominant factor for tensile strength. Lee et al. [29] showed that compressive strength of ABS parts was maximum at 0° build orientation. Gorski et.al. [30] noted that tensile strength was maximum at 0° for ABS filaments. Moreover, they found the specimen presented brittle behavior instead of ductile behavior as build orientation increased exceeding certain angles. The conclusions were consistent with Ashtankar et al. [25]. Their study reported that tensile strength of ABS specimen decreased, with the increase of build orientation from 0° to 90°. This trend was also applicable to ultimate compressive strength, which was minimum at 90° orientation. In another study, Hernandez et al. [31] experimentally determined both compressive properties and flexural properties were maximum at 0° build orientation for ABS P430 filaments. Besides, compressive strength was minimum at 45° build orientation. They also deduced that the effect of build orientation on tensile strength of ABS printed parts was insignificant. Bertoldi et al. [32] and Zou et al. [33] experimentally showed that build orientation strongly affected tensile strength and elastic modulus, respectively. Raney et al. [34] evaluated the effects of build orientation and infill density on tensile strength of ABS parts manufactured by a uPrint SE 3D printer, showing that the strength of samples tested against the layers was less than 80% of that tested along the layers.

As for materials other than ABS, Domingo-Espin et al. [35] tested tensile strength of PC parts. This group of researchers proved that tensile strength was maximum at 0° build orientation. Smith and Dean [36] also pointed out that, compared to bulk material, there was a 45 percent decrease in elastic modulus and a 30 to 60 percent decrease in ultimate tensile strength of PC parts depending on orientation. Zaldivar et al. [37] revealed that FDM materials behaved more as laminated composites with macrostructures than isotropic cast resins, consequently tensile strength, failure strain, Poisson's ratio, coefficient of thermal expansion, and modulus all varied significantly depending on the build orientation of PEI dogbones. Taylor et al. [38] analyzed the flexural behavior of PEI parts with varying build orientation and raster angle experimentally and numerically. Results indicated that modulus and yield strength were influenced by an interaction between these two parameters.

In summary, build orientation significantly affected the mechanical properties, which usually played the predominant role when compared to other parameters [39]. For arbitrary angles, in case of other parameters such as air gap and raster angle are kept constant, the 0° orientation is preferable, which shows the highest values for maximum tensile strength, compressive strength, and flexural strength. Consequently, flat or on-edge oriented samples usually exhibit inter-layer failure with higher stiffness and strength performance. On the other side, increasing the angle from the build platform results in microstructures that further reduce the volume fraction of extruded fiber material from the primary load direction resulting in lower strength. That is why upright samples showed inter-layer failure with lower stiffness and strength performance.

2.2. Raster Angle

Raster angle (sometimes called raster orientation [40], layer orientation [41], fiber orientation [42], or even pattern orientation [43]) represents the angle of the filament direction with regard to the X-axis (usually load direction) of the platform. The allowed raster angles can vary from −90° to 90°, and typically used values are 0° (axial), 45°(cross), 90° (transverse), and their combination. For example, −45°/45°(criss-cross) represents the raster printing directions are −45° and 45° alternately for different layers, as shown in Figure 3.

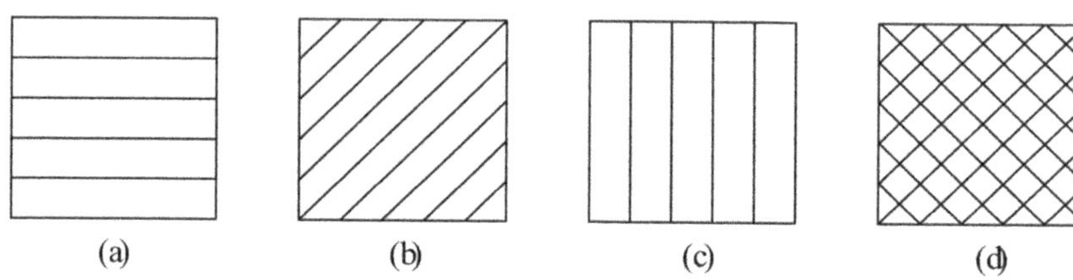

Figure 3. Raster angle: (**a**) 0° (**b**) 45° (**c**) 90° (**d**) −45°/45°.

Ahn et al. [44] applied the Tsai-Wu failure criterion and classical lamination theory to reasonably predict the anisotropic failure model for FDM parts as a function of raster angle. Magalhães et al. [45] suggested that proper choice of raster angles in sandwich specimens could gain in the strength and stiffness of parts, compared to default (45°) FDM configuration. Ziemian et al. [46] and Zhou et al. [47] indicated that the highest tensile strength was obtained at raster angle with 0° for ABS and PP-PC composites, respectively, while the specimens with 90° raster angle exhibited the minimum strength. Es-Said et al. [40] and Garg et al. [48] drove a similar conclusion for flexural strength as well as tensile strength. Moreover, Ziemian et al. [49] further reported that 45° raster specimens in compression were significantly weaker than other raster angles. Based on the analysis of biaxial raster angles, Fatimatuzahraa et al. [50] noted that the structure of 45°/−45°provided better flexural strength than that of 0°/90° of ABS built specimens, despite the almost equivalent tensile strength [51]. A similar conclusion for tensile strength was also driven by Diaconescu et al. [52]. Hart and Wetzel [53] explored the fracture properties of ABS parts with different raster angles. Results confirmed that the elastic-plastic response of the material depended on the raster angle of printed specimens. In contrast, Arbeiter et al. [54] reported that fracture behavior might be not highly dependent on the raster angle by setting ideal processing parameters of PLA samples.

The interaction of build orientation and raster angle can cause strong anisotropy of the FDM parts, therefore these two parameters are generally studied together. Rohde et al. [12] revealed that ABS and PC samples exhibited strong anisotropy as functions of build orientation and raster angle, respectively. Shear moduli were affected by build orientation rather than raster angle for ABS specimens. The lowest values of modulus of rigidity, ultimate shear strength, and yield shear strength were obtained from on-edge configuration specimens. Durgun and Ertan [23] reported that the build orientation had a more significant influence than the raster angle on the mechanical behavior of the resulting fused deposition part. Small raster (e.g., 0° angle) resulted in increased strength resistance in all component positions. Bellini and Güçeri [55] carried out analytical and experimental approaches to study the effect of build orientation and raster angle on flexural strength and tensile strength of ABS material. Balderrama-Armendariz et al. [56] studied elastic properties in torsion of ABS-M30 samples at different build orientations and raster angles. They characterized that build orientation had an insignificant modification of the response of 0.2% yield strength or ultimate shear strength, while the orientation in YXZ with raster at 0° led to improved responses in all measured torsion parameters. Cantrell et al. [57] showed that build orientation and raster angle had a negligible influence on the tensile modulus of ABS specimens. The highest tensile properties and highest shear strength were found in specimens with on-edge orientation and specimens with [+45°/−45°] flat orientation, respectively for PC material. In addition, the shear modulus was almost the same for all specimens with [+45/−45] raster angle regardless of build orientation. Torrado et al. [58] explored the effects of build orientation and raster angle on mechanical anisotropy. The tensile test results exhibited an equivalency between different sample types. Therefore, the authors recommended horizontal specimens printed with a transversal filling due to its higher reliability, higher accuracy, and simplicity of the printing process. Letcher et al. [59] investigated the relationship between layer number, raster angle, and mechanical properties of ABS printed specimens. Results showed that 0° raster orientation yielded the highest strength at each layer number. Furthermore, maximum stress and elastic modulus increased with the increase of the number of layers.

In summary, the relative position of fibers and the axial load causes the specimens to react differently. Raster angles with a higher fraction of specimens oriented along the axis of the load (e.g., 0° orientation) exhibit improved tensile and compressive strength of the part, while those that are offset (e.g., 90° orientation) exhibit reductions in mechanical performance [60–63]. In the former case, fibers themselves withstood most of the applied load, resulting in inter-layer failure. While for the latter case, bonding between adjacent layers and rasters withstood the load, resulting in trans-layer failure, which is much weaker. A similar trend is applicable to the flexural specimen, which can be regarded as one side experiencing compression while the other side experiencing tension when loaded.

2.3. Layer Thickness

Layer thickness (or layer height [64]) represents the thickness of the layer printed by the nozzle tip, as shown in Figure 4. In general, it is smaller than the diameter of the extrusion nozzle (usually one-half), depending on the material and tip size. Layer thickness is directly related to the number of layers printed and hence print time. It has been verified that better accuracy of the component can be achieved by setting lower layer thickness.

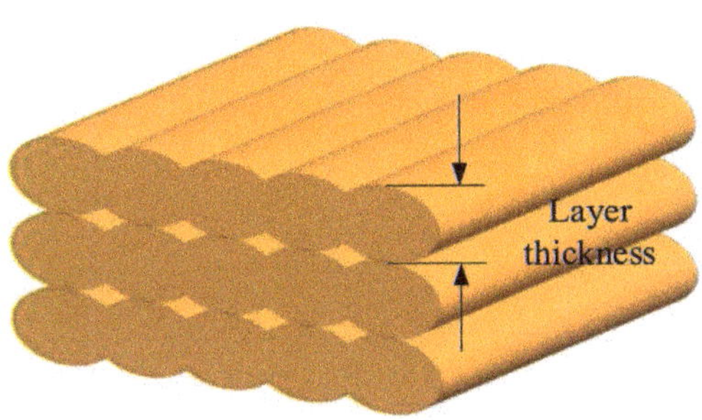

Figure 4. Layer thickness.

Layer thickness is usually studied together with other parameters, most commonly with raster angle. Somireddy et al. [42] researched the influences of raster angle and layer thickness on the flexural behavior using classical laminate theory. Results presented that thinner layer laminates have higher loading capacity and flexural stiffness than thicker ones, except for the maximum deflection. Tymrak et al. [63] quantified the elastic modulus and tensile strength of PLA and ABS parts by comparing different layer thicknesses and bidirectional raster angles. Tests showed that tensile strength dropped with increasing layer thickness. In another study by Rankouhi et al. [62], the mechanical characterization of PLA by varying layer thicknesses and raster angles were analyzed. The maximum elastic modulus and ultimate tensile strength were obtained at lower values of both two factors. Similar results can be obtained for other materials, such as PEEK (Wu et al. [65]) and plaster-based powder (Vaezi and Chua [43]). Garg and Bhattacharya [66] considered layers of different thicknesses and rasters at different angles by simulation and experiment. FE analysis indicated that tensile strength, strain at yield, elongation, and developed stress first decreased with an increase in layer thickness and then increased. Layer thickness, build orientation, and raster angle were evaluated parameters to examine their effects on tensile strength by Nidagundi et al. [67]. Thinner layer thickness, 0° build orientation, and 0° raster angle were optimum for ultimate tensile strength.

In comparison, Rodríguez et al. [24] compared the effect of build orientation, infill density, and layer thickness on the mechanical characteristics of ABS and PLA test components. Regarding ABS, the mechanical strength results barely varied with respect to layer thickness. In contrast, tensile strength of PLA decreased as layer thickness increased. Chacón et al. [27] characterized the effect of layer thickness, build orientation, and print speed to determine the mechanical response of the PLA specimens. They observed that the increased print speed and layer thickness caused ductility to diminish. In addition, the mechanical properties for the upright orientation increased as layer thickness increased and as the print speed decreased, which however were of slight significance for on-edge and flat orientations. Alafaghani et al. [28] demonstrated that mechanical properties were significantly influenced by build orientation, extrusion temperature, and layer thickness; and less significantly on infill pattern, for high infill density specimens, and print speed. To improve the mechanical properties, higher extrusion temperature and larger layer thickness are needed in addition to appropriate build orientation. Carneiro et al. [68] mechanically assessed the influence of raster angle, layer thickness, infill density of PP and GRPP composites. The results showed the infill density had a linear effect on both mechanical properties. Instead, layer thickness had an insignificant effect on the performance of samples. Dong et al. [69] demonstrated that the number of layers was the only dominant factor in improving mechanical strengths of PLA and PLA/wood composites, compared with infill density and layer thickness.

In summary, layer thickness has a different effect on the strength. For a given total height, the thickness of a layer has an inverse proportional relationship with the number of layers. The thinner the layer thickness, the more layers. This response will lead to a high-temperature gradient towards the bottom of the component, which will improve diffusion between adjacent rasters, thus ultimately contributing to the load-bearing and enhancing the strength. In addition, this trend is heightened when at low print speed, which gives a better bonding with the previous layer. On the other hand, an increase in the number of layers also adds to the number of cooling and heating cycles, which in turn gives rise to residual stress accumulation. This behavior can result in distortion and inter-layer cracking, which will reduce the strength. Due to the interaction of these two different influences, in general, a moderate thickness value is obtained as the optimal parameter in some research [70].

2.4. Air Gap

The air gap represents the space between two neighboring printed filaments on the deposited layer. In most cases, the air gap represents the distance between rasters, viz. raster to raster air gap. However, in some research, the air gap is distinguished as raster to contour air gap and contour to contour air gap, respectively. In general, there are three types of air gap, and they are zero, positive and negative. The zero type is generally the default configuration, which places beads just alongside each other. The positive type has a loose place between beads which results in rapid building, while the negative type means that two beads partially overlap the structure, creating a denser component, as shown in Figure 5.

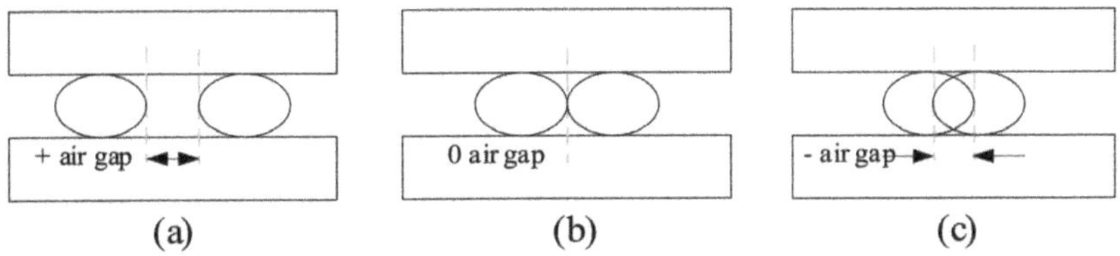

Figure 5. Air gap: (**a**) positive air gap (**b**) zero air gap (**c**) negative air gap.

Rodriguez et al. [71] observed three monofilaments with different air gaps made of ABS. From all arrangements tested, the highest stiffness and tensile strength values were found for the filament with rasters aligned in the loading direction and a small negative air gap. Too et al. [72] characterized that the air gap size had a profound impact on the porosity and compressive strength of FDM built part. With the increasing air gap of the test specimen, compressive strength decreased while porosity increased, respectively. Dawoud et al. [73] researched the impact, flexural and tensile strength of ABS components with different raster angles and air gaps. The air gap with a negative value proved to be the most significant factor in the enhancement of mechanical properties. However, in the case of a positive air gap, varying raster angles seemed to have a more significant effect on tensile strength. Masood et al. [74] presented experimental work on the effect of raster angle, raster width, and air gap on tensile properties of PC. They reported that the air gap was the only dominant parameter influencing tensile properties. This study also found that PC material by FDM had tensile strength in the range of 70 to 80% of the injection molded and extruded PC parts.

In the study of Slonov et al. [75], raster angle, air gap, and raster width on the mechanical properties of samples from PPSF were examined. The authors found that the elastic modulus generally depended on the air gap between rasters, independent of raster angle. On the contrary, the impact strength depended on the raster angle and the adhesion degree between filaments. Hossain et al. [76,77] modified raster width, raster angle, raster to raster air gap, and contour width to improve tensile mechanical properties of PEI material by visual feedback method. Using negative raster to raster air gap led to an average increase in ultimate tensile strength of 16%, compared to the default configuration. Montero et al. [78] examined five process parameters (raster angle, raster width, extrusion temperature, air gap, and color) to understand the ABS properties fabricated by FDM. They observed that the raster angle and air gap influenced tensile strength FDM printed part, while color, extrusion temperature, and raster width had little influence. Moreover, stiffness and shear strength between roads were lower than those measured between layers. Bagsik and Schöppner [79] considered the effect of build orientation, air gap, raster angle, and raster width based on the mechanical data of PEI. Based on their study, the air gap with a negative value contributed to the best results for all directions. With thicker filaments, better mechanical performance could be obtained for the on-edge and upright build direction, while a thinner filament enhanced the strength properties of the flat specimens.

In summary, air gap determines the area of force bearing as well as bonding between filaments. From the perspective of effect, the work of the former one on the mechanical property is more apparent than that of the latter one. In general, the positive air gap results in a loosely packed structure with weak bonding between adjacent filaments, leading to lower strength. In contrast, the negative air gap results in a denser squeezed structure with strong interfacial bonding, significantly improving the strength. Zero air gap may enhance the diffusion between the neighboring rasters, and cause the total bonding area to diminish as well.

2.5. Raster Width

Raster width represents the width of the printed beads or roads for rasters. It depends on the nozzle tip size. Some researchers distinguish contour width from raster width [80,81], as shown in Figure 1. However, in most studies, contour width and raster width are regarded as the same parameter, represented by road width [11,65,82].

Gebisa and Lemu [80] focused on processing parameters, such as contour width, raster angle, contour number, raster width, and air gap, on the effect on the flexural properties of PEI-manufactured parts, which could be arranged as importance: raster width and raster angle > contour width and contour number > air gap. They also found that the effect of a minus air gap could differ between two different materials, which was not recommended for PEI. Ang et al. [83] specified process parameters, namely air gap, raster width, build orientation, number of layers, and infill pattern, on the compressive properties and porosity of ABS scaffold structures. The experiment determined raster width and air gap as the most significant parameters. Moreover, porosity decreased when the air gap decreased or raster width increased. In contrast, compressive strength and modulus increased as raster width increased while the air gap decreased. Rayegani and Onwubolu [84] used the group method of data handling (GMDH) and differential evolution (DE) to quantify the effects of air gap, raster angle, build orientation, and raster width on tensile strength. The investigation showed that negative air gap, as well as smaller raster width, significantly improved tensile strength. Particularly, build orientation played a major role, as could be observed from the results. Onwubolu et al. [85] applied the design of experiment (DOE) to study the main and interaction effects of process variables such as build orientation, raster width, layer thickness, air gap, and raster angle on tensile and strength of ABS components. The maximum tensile strength was obtained with zero build orientation, maximum raster width, raster angle, and negative air gap. In Liu et al. [86], five input process parameters such as build orientation, layer thickness, raster orientation, air gap, and raster width were considered to examine their influence on impact, flexural and tensile strengths. The optimum combination was obtained based on analysis of variance and gray relation analysis. Gkartzou et al. [87] examined the influence of raster width on tensile properties of PLA/Lignin composites. The results showed that specimens with different raster widths had similar tensile strength and Young's modulus.

In summary, larger raster width creates a high temperature near the boding surfaces and a larger bonding area, which may improve the diffusion and lead to stronger bond formation [64]. However, a larger raster can also result in stress accumulation along the width of the part, as well as deterioration in thermal conductivity [88]. On the other hand, smaller raster width will require less production time and material. On the whole, at the intermediate value of the raster width, the higher thermal mass that cools slowly can be achieved, which enhances the bonding between the filaments and thus improves the strength [89].

2.6. Infill Density

The outer region of AM part is usually solid, but the interior area, generally known as the infill, is the inner component covered by the skin, which has different geometries and sizes. Infill density (or infill degree [68], infill ratio [82], infill percentage [90], fill density [88]) refers to the percentage of filament material printed in the given part, where 0% is a shell and 100% is a solid. FDM technology allows users to control the infill density through parameters such as air gap or raster width.

Alvarez et al. [90] observed that the maximum impact resistance, tensile stress, and tensile force were obtained with 100% infill density. Martikka et al. [91] revealed that the increment in infill density enhanced the tensile properties of PLA and PLA/wood composites. Gomez-Gras et al. [92] carried out the Taguchi method to investigate the impact of four process parameters and their intersections—layer thickness, infill density, nozzle diameter, and print speed, on fatigue response. It was concluded that infill density showed the strongest influence in fatigue performance, followed by nozzle diameter and layer thickness, whereas print speed showed no relevant effect in PLA specimens. Aw et al. [93] looked at relating process parameters to tensile properties of CABS/ZnO composites with infill density and infill pattern. Results revealed that tensile strength of CABS composites was little affected by the change of infill density, while the increased infill density caused Young's modulus to increase, resulting in higher stiffness. Line pattern possessed better tensile properties. Kerekes et al. [94] pointed out that with an increase in infill density, Young's modulus, initial yield stress, ultimate strength, and toughness increased, while elongation at break decreased. Layer thickness showed a moderate influence affecting the specimen's properties, where an increasing layer thickness apparently increased Young's modulus, while it decreased elongation at break. Lužanin et al. [95] experimentally analyzed flexural properties depending on the infill density, layer thickness, and raster angle. The researchers reported that layer thickness was the most important parameter affecting flexural force, and the interaction between infill density and raster angle was significant as well. The mechanical effect of printing parameters for carbon fiber-reinforced polyamide was studied by Toro et al. [13]. The most dominant parameter was found to be infill density. Layer thickness and infill pattern played importantly in flexural and tensile behaviors, respectively.

In summary, the mass and strength of FDM produced parts are dependent on the infill density. Lower density requires less print time and material, thus saving cost and reducing the weight. However, more voids are generated within the structure simultaneously, leading to increased porosity. As a result, the dimension of the bonded region between filaments decreases and so as well to the mechanical properties. In contrast, the denser component possesses better mechanical properties but takes much more time to be complete. For example, the specimen built with 100% infill density usually exhibits maximum strength. Generally, infill density ranging from 50% to 98% is recommended, since the improvement in mechanical resistance is countered by longer manufacturing times [90].

2.7. *Infill Pattern*

Infill pattern (or print pattern [93]) represents the way how filaments fill and cross the internal space of the printed part, as shown in Figure 6. Different infill patterns usually have different geometrical layouts and complexity, which will affect print time and the material used.

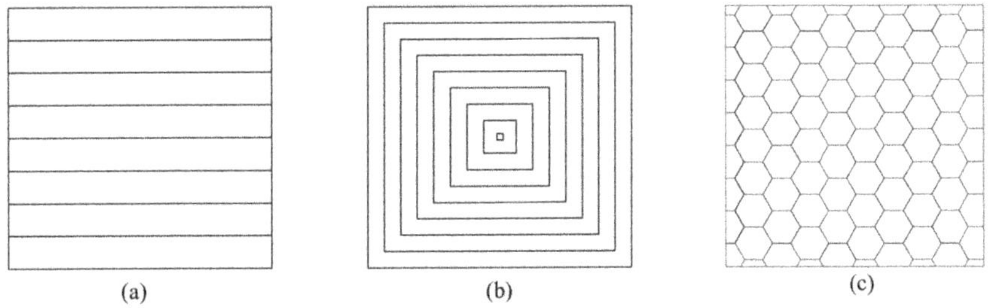

Figure 6. Infill pattern: (**a**) linear (**b**) concentric (**c**) hexagonal.

Many filling patterns are available such as hexagonal (or honeycomb), linear, and diamond, as illustrated by Alafaghani et al. [28], in which the commonly used is the hexagonal pattern. Cho et al. [96] compared the influence of PLA samples with different infill patterns and layer thickness on tensile property. They concluded that layer thickness had a higher effect than infill pattern, and the triangle pattern gave the highest mechanical strength and lowest material consumption. Dave et al. [97] investigated the effect of three process parameters: infill pattern, infill density, and build orientation, on the tensile properties of PLA specimens through a full factorial experiment. ANOVA results indicated that infill density was the most predominant process parameter for tensile strength, compared with infill pattern and build orientation. Fernandez-Vicente et al. [98] found that changes in infill density determined mainly tensile strength of ABS material. At the same time, the influence of the different infill patterns caused a variation of no more than 5% in maximum tensile strength, along with similar behaviors. Akhoundi et al. [99] identified the key factors that influenced tensile and flexural strengths. The input variables, such as infill pattern and infill density, and their relationship with raster angle and void presence, were considered. The result concluded that the highest tensile and flexural strengths were obtained by concentric pattern. They also found that when rasters were deposited at short distances in the Hilbert curve, a high temperature was maintained, which resulted in better fusion and strong bonding between the adjacent rasters. Baich et al. [9] presented the relationship between various infill patterns and different mechanical properties. Statistical analysis revealed that for double-dense infill in all loading conditions, solid infill showed higher strength at the same fabrication cost. Therefore, solid infill was recommended for mechanical applications, in the case of entry-level printers. Moreover, compressive strength increased as the complexity of the infill pattern increased. Nagendra and Prasad [100] revealed significantly linear interactions between infill pattern and other process parameters, such as extrusion temperature, layer thickness, and infill density, on mechanical properties of Nylon/Aramid composite.

In summary, the infill pattern has a complex effect on the mechanical properties of parts produced by FDM owing to a broad spectrum of types. For example, in the hexagonal pattern, each layer lays down on a similar previous layer, the same as the bonding zone. While in the rectilinear pattern, the lay crosses the previous layer at points, which correspond with the bonding zone between each layer. However, the combination of rectilinear patterns in a 100% infill shows higher tensile strength, compared with the honeycomb pattern [98]. Therefore, these results need to be analyzed and explained with caution.

2.8. Print Speed

The print speed (or feed rate [27], print velocity [92], infill speed [101], deposition velocity [102]) represents the speed of the nozzle traveling relative to the print platform. Generally speaking, the lower the print speed, the longer the production time and the better the accuracy of the prints. In comparison, the higher the print speed, the faster parts are produced.

Christiyana et al. [103] produced ABS composite specimens and investigated the role of print speed and layer thickness. It was observed that the maximum flexural and tensile strengths were achieved via setting thinner layer thickness and lower print speed. Similarly, Ning et al. [101] showed that tensile strength of CFRP composites decreased with the increase in print speed. Santana et al. [104] analyzed the factors affecting PLA parts with variations in print speed and extrusion temperature to evaluate the quality of the open-source 3D printer. Based on the value, the print speed and extrusion temperature were irrelevant compared with the mass and modulus of rupture. Kačergis et al. [105] investigated the influence of print speed, platform temperature, and number of layers in the structure printed with PLA and TPU. Experimental results proved that the deformation was strongly influenced by the print speed. By contrast, Li et al. [21] pointed out that air gap played a predominant part in determining tensile strength, followed by layer thickness, and the effect of print speed is the weakest factor. They suggested that smaller values of layer thickness and air gap were preferred if higher tensile strength was needed. Furthermore, print speed could be set relatively higher to improve fabrication efficiency. Lužanin et al. [106] studied the relationship between the maximum flexural force of PLA parts and five process parameters. The input variables were extrusion temperature, infill density, print speed, raster angle, and layer thickness. The optimal parameters setting was maximum levels of infill density and print speed, mid-level of layer thickness, and minimum level of raster angle.

In summary, the effect of print speed on mechanical performance shows a different trend. Generally, lower print speed gives a better bonding and interaction between contiguous filaments, leading to an increase in tensile and flexural strength. However, if the print speed is too slow, the too-long inter-layer cooling time makes just-deposited material cool down at a lower temperature, which disfavors the fusion of the thermoplastics, hence the strength and ductility are affected [107]. On the other hand, rapid print speed could improve the efficiency, but leave not enough time for extrusion materials to plasticize, and the amount of residual stress produced during deposition increases significantly as well [108], which leads to weak mechanical properties. It should be pointed out that the production time is not only affected by print speed but also related to build orientation. Print time decreases as print speed increases for on-edge and flat orientations, while print time remains almost constant for up-right orientation with high-speed values [27].

2.9. Number of Contours

The number of contours (or number of perimeters [109], number of shells [110]) refers to the number of closed roads that are deposited along the edge of the part, as shown in Figure 1. It may range from one to the number of filaments extruded.

Kung et al. [109] studied the influences of three process parameters including number of contours, raster angle, and specimen size. They pointed out that there existed apparent dispersion of the strength for a different number of contours. Interestingly, they also noted that tensile strength of specimens built with 45° is greater than those built with 0°. According to Mahmood et al. [110], there was a positive relationship between tensile strength and number of contours. In addition, a larger cross-section negatively affected tensile strength of a printed part while keeping the other parameters constant. Croccolo et al. [111] experimentally and analytically dealt with the effect of contouring on the static strength and stiffness of ABS parts. They showed that the larger the number of contours, the greater the elastic modulus and stiffness, and thus the higher the maximum strength. Moreover, with the increase of the number of contours, the percentage of elongation to failure decreased. Griffiths et al. [112] performed an experimental investigation on the tensile property of PLA objects. They utilized a full factorial DOE approach considering building orientation, infill density, number of contours, and layer thickness as parameters. The study concluded that the infill density and number of contours were the only significant parameters that should be maximized for optimization. Lanzotti et al. [61] observed the increase in strength with the number of contours and layer thickness. In particular, the strength increased as the raster angle decreased with a rate that was as greater as the layer thickness increased.

In summary, the number of contours impacts the mechanical properties of the part fabricated. When the number of contours increases, the effect is directly seen in the increase in strength. This is owing to the fact that the load is applied directly on the contour rather than the rasters, therefore a growing contour number causes the raster length and number of rasters to decrease, which will lead to improvement in the performance of the part.

2.10. Extrusion Temperature

Extrusion temperature (or print temperature [82], nozzle temperature [113]) refers to the temperature at which the fibers are heated inside the nozzle during the FDM process. It can influence the fluidity and solidification characteristics of the molten material and control the viscosity of filament extruded from the nozzle.

Deng et al. [82] applied an orthogonal test to evaluate the effects of process parameters such as print speed, layer thickness, extrusion temperature, and infill density, on tensile properties of PEEK components. They demonstrated that more micro-pores and slag inclusion were caused by lower print speed and extrusion temperature, leading to lower strength specimens. Aliheidari et al. [113] designed double cantilever beam specimens of ABS and printed at different extrusion temperatures to study the mode-I fracture resistance. Based on critical J-integral value, the authors stated that the higher the temperature was, the greater number of polymer molecules were inter-diffused at the interface, which resulted in higher resistance to fracture. Rinanto et al. [114] optimized extrusion temperature, infill density, and raster angle to produce prototypes with high tensile strength. The optimization combination was 45° of angle, 40% of density, and 210 °C of temperature. Among these three parameters, infill density is the most predominant factor. Sun et al. [115] explored the influence of extrusion temperature and envelope temperature on the quality of bonds between adjacent ABS filaments. Statistical analysis proved that both the envelope temperature and variations in the convective conditions within the printer have substantial influences on the mesostructure and the overall quality of the bond strength between rasters. Leite et al. [116] determined the influence of mechanical properties from layer thickness, extrusion temperature, raster angle, and infill density. The best values reported for the sample were higher infill density and extrusion temperature, and lower layer thickness. Sun et al. [117] demonstrated that increasing platform temperature could enhance the PEEK binding force between layers, making the model more excellent mechanical properties. Moreover, low infill density could also improve the performance of the material. Yang [118] observed a decrease in tensile and flexural properties of WFRPC components with an increase in the extrusion temperature, whose trend is opposite to that of compressive strength.

In summary, the extrusion temperature has an important effect on the crystallinity of the material and polymer filament bonding. Thus, the mechanical performance of printed parts will be affected as well. Higher extrusion temperature of the deposited filament gives better inter-layer fusion, which results in higher mechanical properties. However, too high extrusion temperature may cause material degradation or molding failure during deposition, resulting in dimensional inaccuracy and filament deformation [82]. On the other hand, lower extrusion temperature may prevent the material from melting adequately, leading to nozzle clogging. Both of the two cases above will lead to weak mechanical properties of printed parts.

3. Results and Discussions

In an effort to aggregate thorough information on process parameters of the FDM technique and their influence on mechanical properties, we have summarized the research works in the field concisely. Tables 1–11 give an overview of the parameters and mechanical properties of FDM products intensively investigated in the literature. In most existing research, several parameters are studied together. Therefore, the parameter that plays a major role or authors of the research care about most as the basis for classification. For the case of many parameters included, we attribute it to Table 11 (Others). However, for certain process parameters, there is not much research. Therefore, all studies containing this parameter are grouped into its table. As a consequence, the criteria for the aggregation of these tables are not strictly unique. Since there is much scattered data and information, interested readers are encouraged to review the references provided according to their interests. The key findings of this survey are summarized below:

- The work of different process parameters is coupled and combined to affect the mechanical property of FDM parts, which all have importance and effects. Generally speaking, there exists a parameter playing a dominant role. For example, extrusion temperature, layer thickness, air gap, and print speed can influence the heat transition of the structure, thus affecting the bonding between rasters and the mechanical characteristics. However, extrusion temperature is the most significant factor in determining temperature field variation, followed by layer thickness, print speed, and air gap by order of importance [102,135].
- One process parameter may affect or be affected by several other parameters, directly or indirectly. For instance, layer thickness affects the raster width and print speed. Likewise, the number of layers is related to build orientation and layer thickness in a part. What is more, infill density values significantly have an impact on the print speed, which can be changed by adjusting air gaps and raster width.
- The contribution of a single parameter may be contradictory from different aspects, which should be determined by the final effect. A typical example is raster angle.

Small raster angles (e.g., 0°) will contribute to load-bearing due to filament lying along the loading direction. On the other hand, they will also lead to long rasters, which result in stress accumulation and hence weak bonding [22]. However, the final effect is that a small raster angle ensures the best tensile, compressive and flexural strength, proving that the former one plays a dominant role.

- Optimal parameter values obtained are just in theory, which should be reconsidered and adjusted in practice. According to the conclusion obtained in the former section, thinner layer thickness can help reinforce the tensile strength of the part, which, however, costs more due to more material and time usage for producing [136,137]. Consequently, a compromise needs to be made between improving property and reducing cost.

Table 1. Build direction.

Study	Process Parameters	Mechanical Properties	Materials	Machines
Ashtankar et al. [25]	Build orientation	Tensile strength, compressive strength	ABS	Dimension BST
Lee et al. [29]	Build orientation	Compressive strength	ABS	MIT 3D Printer
Gorski et al. [30]	Build orientation	Tensile strength	ABS	Dimension BST 1200
Hernandez et al. [31]	Build orientation	Compressive strength, tensile strength, flexural strength	ABS	uPrint SE Plus
Zou et al. [33]	Build orientation	Tensile strength, Young's modulus, Poisson's ratio	ABS	Dimension SST 1200 es
Domingo-Espin et al. [35]	Build orientation	Tensile strength, stiffness	PC	Stratasys Fortus 400 mc
Smith and Dean [36]	Build orientation	Tensile strength, modulus	PC	Stratasys Vantage SE
Bagsik et al. [79]	Build orientation	Tensile strength, compressive strength	PEI	Stratasys Fonus 400 mc
Upadhyay et al. [119]	Build orientation	Tensile strength, compressive strength	ABS P400	FDM SST-768
Rohde et al. [12]	Build orientation, raster angle	Shear strength	ABS, PC	Stratasys Fortus 360 mc, Ultimaker 2
Durgun and Ertan [23]	Build orientation, raster angle	Tensile strength, flexural strength.	ABS P430	Dimension BST
Rodriguez et al. [24]	Build orientation, raster angle	Strength, stiffness	ABS	
Bertoldi et al. [32]	Build orientation, raster angle	Tensile strength, modulus, Poisson's ratio,	ABS	Stratasys FDM 1650
Zaldivar et al. [37]	Build orientation, raster angle	Tensile strength, failure strain, modulus, Poisson's ratio, thermal, expansion coefficient	PEI	Stratasys Fortus 400 mc
Taylor et al. [38]	Build orientation, raster angle	Flexural strength	PEI	Stratasys Fortus 400 mc
Bellini and Güçeri [55]	Build orientation, raster angle	Tensile strength, flexural strength	ABS	Stratasys FDM 1650
Balderrama-Armendariz et al. [56]	Build orientation, raster angle	Ultimate shear strength, 0.2%yield strength, shear modulus, fracture strain	ABS	Stratasys Fortus 400 mc
Cantrell et al. [57]	Build orientation, raster angle	Tensile strength, failure strength, Poisson's ratio, modulus	ABS, PC	Stratasys Fortus 360 mc, Ultimaker 2
Raney et al. [34]	Build orientation, infill density	Tensile strength, flexural strength	ABS	uPrint SE Plus
Torrado and Roberson [58]	Build orientation, raster pattern	Tensile strength, anisotropic property	ABS	Lulzbot TAZ 4
Wang et al. [7]	Build direction, layer thickness, deposition style	Tensile strength	ABS P400	Dimension BST
Kamaal et al. [120]	Build direction, infill density, layer thickness	Tensile strength, impact strength	CF/PLA composite	Ypanx Falcon
Tanikella et al. [121]	Building orientation, mass, color	Tensile strength	Ninjaflex, Semi-Flex, HIPS, TGLase, Nylon, ABS, PC	Lulzbot TAZ 3.1 and 4

Table 2. Raster angle.

Study	Process Parameters	Mechanical Properties	Materials	Machines
Es-said et al. [40]	Raster angle	Tensile strength, modulus of rupture, impact resistance	ABS P400	Stratasys FDM 1650
Ahn et al. [44]	Raster angle	Tensile strength	ABS	
Magalhães et al. [45]	Raster angle	Tensile strength, Young's modulus,	ABS P400	Stratasys FDM 2000
Ziemian et al. [46]	Raster angle	Tensile strength, fatigue strength	ABS	Stratasys Vantage-i
Garg et al. [48]	Raster angle	Tensile strength, flexural strength	ABS P400	Stratasys Mojo
Ziemian et al. [49]	Raster angle	Tensile strength, compressive strength, flexural strength, impact strength, fatigue property	ABS	Stratasys Vantage-i
Hart and Wetzel [53]	Raster angle	Fracture property	ABS M30	Lulzbot Taz 6
Arbeiter et al. [54]	Raster angle	Fracture property	PLA	Hage 3DpA2
Carneiro et al. [68]	Raster angle	Tensile strength	PP, Glass/PP composite	Prusa i3
Liu et al. [122]	Raster angle	Tensile property, flexural property	PLA/SCB composite	S1 Architect 3D
Letcher et al. [123]	Raster angle	Tensile strength, flexural strength, fracture property	PLA	MakerBot Replicator 2x
Zhou et al. [47]	Raster angle, layer thickness	Tensile strength	PP/PC composite	LeistritzZSE 18 HPe
Diaconescu et al. [52]	Raster angle, layer thickness	Tensile strength	ABS	MakerBot 2X
Letcher et al. [59]	Raster angle, number of layers	Tensile strength, modulus of elasticity	ABS	MakerBot Replicator 2x
Kung et al. [109]	Raster angle, number of contours, specimen size	Tensile strength	PLA	RepRap 3D printer

Table 3. Layer thickness.

Study	Process Parameters	Mechanical Properties	Materials	Machines
Vaezi and Chua [43]	Layer thickness	Tensile strength, flexural strength	ZP102	Z510/Cx printer
D'Amico et al. [70]	Layer thickness	Tensile strength, flexural strength	ABS	Makerbot 2X
Ayrilmis et al. [124]	Layer thickness	Tensile strength, flexural strength	PLA/wood composite	Zaxe 3D printer
Somireddy et al. [42]	Layer thickness, raster angle	Flexural property	ABS-P430	Stratasys μ printer
Rankouhi et al. [62]	Layer thickness, raster angle	Tensile strength, elastic modulus	ABS	Makerbot Replicator 2x
Wu et al. [65]	Layer thickness, raster angle	Tensile strength, compressive strength, flexural strength	PEEK, ABS P430	Custom-built printer
Garg and Bhattacharyab [66]	Layer thickness, raster angle	Tensile strength	ABS	uPrint SE, Plus and Mojo printers
Knoop et al. [125]	Layer thickness, build orientation	Tensile strength, compressive strength, flexural strength	Nylon	Stratasys Fortus 400 mc
Chacon et al. [27]	Layer thickness, build orientation, print speed	Tensile strength, flexural strength, stiffness	PLA	WitBox desktop 3D printer
Uddin et al. [39]	Layer thickness, build orientation, raster angle	Young's modulus, yield strength, failure strength	ABS	Zortrax M200
Tymrak et al. [63]	Layer thickness, raster angle, color	Tensile strength, elastic modulus	ABS, PLA	A series of open-source3D printers
Dong et al. [69]	Layer thickness, number of layers, infill density	Tensile strength, flexural strength, impact strength	PLA/wood composite	MakerBot Replicator 2x

Table 4. Infill density.

Study	Process Parameters	Mechanical Properties	Materials	Machines
Alvarez et al. [90]	Infill density	Tensile strength, impact resistance	ABS	Makerbot Replicator 2x
Martikka et al. [91]	Infill density	Tensile properties, impact strength	PLA/wood composite	Profi3Dmaker
Aw et al. [93]	Infill density, infill pattern	Tensile property	CABS/ZnO composite	RepRap Mendelmax 1.5
Fernandez-Vicente et al. [98]	Infill density, infill pattern	Tensile strength, Young's modulus	ABS	RepRap Prusa i3
Kerekes et al. [94]	Infill density, layer thickness	Tensile property	ABS-M30	Stratasys uPrint SE Plus
Lužanin et al. [95]	Infill density, layer thickness, raster angle	Flexural strength	PLA	Makerbot Replicator 2
Gomez-Gras et al. [92]	Infill density, layer thickness, nozzle diameter, print speed	Fatigue performance	PLA	Prusa i3
Griffithsa et al. [112]	Infill density, building direction, number of contours, layer thickness	Tensile strength, Young's modulus	PLA	Makerbot Replicator 2

Table 5. Infill pattern.

Study	Process Parameters	Mechanical Properties	Materials	Machines
Ebel et al. [126]	Infill pattern	Tensile strength	PLA, ABS	CB printer, Felix 1.0e
Baich et al. [9]	Infill pattern, infill density	Tensile strength, compressive strength, flexural strength	ABS P430	Stratasys Fortus 200 mc
Cho et al. [96]	Infill pattern, layer thickness	Tensile strength, modulus, yield stress	PLA	
Akhoundi et al. [99]	Infill pattern, infill density	Tensile strength, flexural strength, modulus	PLA	Laboratory FDM 3D printer
Dave et al. [97]	Infill pattern, build orientation, infill density	Tensile strength	PLA	Open-source FDM printer
Vinoth Babu et al. [127]	Infill pattern, layer thickness, infill density	Tensile property, flexural property	CF/PLA composite	Raise 3D V2 N2 Hot end
Zaman et al. [128]	Infill pattern, layer thickness, number of contours, infill density	Compressive strength	PLA, PETG	Makerbot Replicator 2X, Open Edge HDE printer
Nagendra and Prasad [100]	Infill pattern, layer thickness, extrusion temperature, raster angle, infill density	Tensile strength, flexural strength, impact strength, compressive strength	Nylon/Aramid composite	

Table 6. Air gap.

Study	Process Parameters	Mechanical Properties	Materials	Machines
Rodriguez et al. [71]	Air gap	Tensile strength, stiffness	ABS P400	Stratasys FDM1600
Too et al. [72]	Air gap	Compressive strength, porosity	ABS P400	Stratasys FDM1650
Dawoud et al. [73]	Air gap, raster angle	Tensile strength, flexural strength, impact strength	ABS	DIY FDM machine
Masood et al. [74]	Air gap, raster width, raster angle	Tensile strength	PC	Stratasys Vantage
Hossain et al. [76,77]	Air gap, raster angle, contour width, raster width	Tensile strength	PC	Stratasys Fortus 900 mc
Montero et al. [78]	Air gap, raster angle, raster width, extrusion temperature, color	Tensile strength	ABS P400	Stratasys FDM 1650
Bagsik and Schöppner [79]	Air gap, build orientation, raster angle, raster width	Tensile strength	PEI	Stratasys Fortus 400 mc
Ang et al. [83]	Air gap, raster width, build orientation, build layer, build profile	Compressive strength, porosity	ABS	Stratasys FDM 1650

Table 7. Print speed.

Study	Process Parameters	Mechanical Properties	Materials	Machines
Christiyana et al. [103]	Print speed, layer thickness	Tensile strength, flexural strength	ABS/ hydrous magnesium silicate composite	3D protomaker STURDY
Santana et al. [104]	Print speed, extrusion temperature	Flexural strength	PLA	IFSC 3D printer
Li et al. [21]	Print speed, layer thickness, air gap	Tensile strength	PLA	MakerBot Z18
Kačergis et al. [105]	Print speed, number of layers, platform temperature	Deformation	PLA, TPU	Anycubic Prusa i3
Attoye et al. [129]	Print speed, build orientation, extrusion temperature	Young's modulus, yield strength	PLA, ABS	MakerBot
Ning et al. [101]	Print speed, raster angle, extrusion temperature, layer thickness	Tensile strength, Young's modulus, yield strength	CFRP composite	Creatr AM machine

Table 8. Number of contours.

Study	Process Parameters	Mechanical Properties	Materials	Machines
Croccolo et al. [111]	Number of contours, build orientation	Tensile strength, stiffness	ABS-M30	
Lanzotti et al. [61]	Number of contours, layer thickness, raster angle	Tensile strength	PLA	Reprap Prusa I3
Mahmood et al. [110]	Number of contours, infill density, cross-sectional area	Tensile strength	ABS	Makerbot Replicator 2X
Chokshi et al. [130]	Number of contours, layer thickness, infill pattern	Tensile strength, flexural strength	PLA	Prusa MK3S
Gebisa and Lemu [80]	Number of contours, air gap, raster width, raster angle, contour width	Flexural property	PEI	Stratasys Fortus 450
Torres et al. [131]	Number of contours, extrusion temperature, print speed, raster angle, infill density, layer thickness	Tensile strength, fracture property	PLA	MakerBot Replicator2

Table 9. Extrusion temperature.

Study	Process Parameters	Mechanical Properties	Materials	Machines
Aliheidari et al. [113]	Extrusion temperatures	Fracture property	ABS	Felix pro I printer
Sun et al. [117]	Extrusion temperature	Flexural strength	ABS P400	Stratasys FDM 2000
Yang [118]	Extrusion temperature	Tensile property, flexural property, compressive strength	PLA/wood composite	Creator Pro
Rinanto et al. [114]	Extrusion temperature, infill density, raster angle	Tensile strength	PLA	Politeknik ATMI Surakarta FDM Machine
Sun et al. [115]	Extrusion temperature, infill density	Tensile strength	PEEK	High temperature FDM type 3D printer
Abouelmajd et al. [132]	Extrusion temperature, print speed, raster angle	Flexural strength, stiffness	PLA	WANHAO Duplicator 4S
Deng et al. [82]	Extrusion temperature print speed, layer thickness, infill density	Tensile strength, flexural strength impact strength	PEEK	Custom-built FDM equipment
Leite et al. [116]	Extrusion temperature, infill density, raster orientation, layer thickness	Tensile strength, yield strength, modulus of elasticity, elongation at break	PLA	Ultimaker 2 machine

Table 10. Raster width.

Study	Process Parameters	Mechanical Properties	Materials	Machines
Gkartzou et al. [87]	Raster width	Tensile strength, Young's modulus	PLA/ lignin composite	Zmorph 2.0 S
Rajpurohit and Dave [64]	Raster width, layer thickness, raster angle	Tensile property	PLA	Open-source FDM printer
Slonov et al. [75]	Raster width, air gap, raster angle,	Tensile strength, elastic modulus, impact strength	PPSF	Stratasys Fortus 400 mc
Rajpurohit and Dave [89]	Raster width, layer thickness, raster angle	Flexural property	PLA	Open-source FDM printer

Table 11. Others.

Study	Process Parameters	Mechanical Properties	Materials	Machines
Toro et al. [13]	Layer thickness, raster angle, infill pattern, infill density.	Tensile strength, flexural strength	CRF/Nylon composite	Ultimaker 2 Extended +.
Rayegani and Onwubolu [84]	Build orientation, raster angle, raster width, air gap	Tensile strength	ABS	Stratasys Fortus 400 mc
Panda et al. [133]	Layer thickness, raster angle, raster width, air gap	Tensile strength	ABS P400	Fortus 400 mc
Sood et al. [22]	Layer thickness, build orientation, raster angle, raster width, air gap	Tensile strength, flexural strength, impact strength	ABS P400	FDM Vantage SE machine
Panda et al. [26]	Layer thickness, build orientation, raster angle, raster width, air gap	Tensile strength, flexural strength, impact strength	ABS P400	FDM Vantage SE machine
Onwubolu and Rayegani [85]	Layer thickness, build orientation, raster angle, raster width, air gap	Tensile strength	ABS P400	FDM 400 mc machine
Liu et al. [86]	Layer thickness, build orientation, raster angle, raster width, air gap	Tensile strength, flexural strength, impact strength	PLA	MakerBot Replicator2
Giri et al. [134]	Air gap, raster width, layer thickness, build orientation, raster angle, number of contours	Tensile strength	PLA	Customized printer

4. Research Shortcomings and Challenges

This paper reviewed the literature concerned with the effects of various process parameters on mechanical performance by investigating their individual/combined effect. Despite the achievements of the current work, this section describes the major challenges and shortcomings of recent research.

4.1. Diversity of Materials

In most presented research, influences of materials and printers are neglected insignificantly, in fact. From tables, it can be seen that there is a variety of materials for FDM, among which ABS and PLA are the two most widely studied. Other few known materials such as PC [35,57], PEI [37,79], PEEK [138], and Nylon [139] occupy only a small part of the research, not to mention PP [68], PPSF [75], PETG [140,141], or composite materials [93,142]. Therefore, conclusions about process parameters of most studies are obtained from ABS and PLA, which may be not applicable to other materials. For example, negative air gaps are preferred to enhance tensile and flexural behavior for ABS, as demonstrated by multiple works [11,73]. However, for structural materials such as PEI, a minus air gap is not recommended. As this material is processed at high temperature, and zero air gap is sufficient to improve mechanical properties flexural strength by adjusting other parameters, which can reduce the loss of dimensional accuracy and surface quality, caused by the usage of a negative air gap [80]. It should also be noted that materials from different suppliers differ in quality [141]. Moreover, even though the same material from the same source in different colors can lead to variation in properties. For instance, Wittbrodt et al. [143] reported that colors influenced the crystallinity percentage of polymers, and thus impacted the strength, which could not be deemed a low level of significance [44,121]. Therefore, research in a wider variety of materials will contribute to understanding the effect of process parameters better and help overcome shortcomings of FDM.

4.2. Variety of Printers

There exist a wide range of machines from different manufacturers, as presented in the tables. Although samples are from the same material, they may have different properties when printed by other printers [144]. For instance, Tymrak et al. [63] found that ABS parts in a 0° orientation had elastic moduli around 1900 MPa and tensile strengths nearing 30 MPa by RepRap printer, which was higher than that in similar studies from different commercial printers, with moduli varying between 1000 and 1700 MPa, and tensile strengths ranging from 10 to 18 MPa [55,71]. The influence of 3D printers on the mechanical property of FDM parts is definite and obvious. However, there is still a lack of adequate and specific means to measure or evaluate this impact. An effort should be made to identify standard and test methods that could be used to validate FDM machine performance.

4.3. Difference in Results

Since FDM is a complex process, it is difficult to replicate the experiment completely from others, which may lead to different or even opposite conclusions. For example, Dawoud et al. [73] showed that an air gap with a negative value could improve the mechanical property. On the contrary, Mohamed et al. [17] claimed that a positive air gap facilitated the spread of semi-molten materials between the gaps, which led to stronger structures. This phenomenon is more apparent when it involves multi-parameter optimization. As another example, Panda et al. [26,133] investigated process parameters (air gap, build orientation, raster angle, layer thickness, raster width) for mechanical properties of ABS parts. Experiments were conducted using a central composite design and part swarm optimization, respectively. However, the optimum process parameters obtained were different from that by Rayegani et al. [84]. In a word, samples with the optimal combination of parameters may have similar strength to those under the opposite parameters setting. That is why it is difficult to evaluate the role of a specific parameter in a multi-parameter combination.

4.4. Limitation of Research Parameters

It is clear that some of the process parameters are widely studied: infill density, layer thickness, raster angle, build orientation, and air gap. Print speed and raster width also occupy a place in the research field. However, other parameters such as infill pattern, number of contours and extrusion temperature are the least analyzed, which needs more attention. For example, the road width for raster and contour is assumed to have a similar effect on the properties in different studies. However, Gebisa and Lemu [80] concluded that raster width and contour width were two different parameters with completely different influences, which needed to be examined separately. For another instance, raster angle 0° ensures the best mechanical strength, presented by many researchers, while Dave et al. [97] found that samples built with raster angle 90° in Hilbert curve pattern displayed a better result as compared to 0° value. These different results indicate that researchers should spend more time investigating the "ignored" parameters, which may come to a different conclusion or view than before.

4.5. Interaction with Composites Factors

As the characteristics of a pure polymer may not satisfy requirements sometimes, people turn their attention to FDM-based composite materials [145], such as polymer matrix composites [146,147], bio-composites [148,149], nanocomposites [150,151], and fiber-reinforced composites [152,153], which have advantages of high mechanical performance and multi-function. However, the intrinsic properties of different composite materials, such as flow and fiber orientation, solidification behavior, and deformation [142], make it difficult for process parameters optimization related to composite materials. For example, Camineroa et al. [154] examined the influence of fiber volume, layer thickness, and build orientation on the impact properties of continuous fiber-reinforced composites. They noted that the interaction between fiber orientation and build orientation significantly led to different impact strengths for on-edge and flat specimens. In the study of Osman and Atia [155], a significant reduction of tensile modulus was observed for specimens with 45° raster angle, with the increase of rice straw content in the ABS-rice straw composite material. However, this phenomenon was insignificant overall for specimens with a 0° raster angle. In a word, the complicated influences of process parameters on the properties of composites, which are coupled with material factors, remains a big challenge for future research.

5. Summary, Recommendations, and Perspectives

In summary, the research of FDM process parameters is critical for improving the characteristics and quality of parts. Different process parameters may have similar or opposite influences on the mechanical properties and behavior of components, which are also affected by other factors such as materials, printers, experiments, etc. Therefore, a compressive investigation of various process parameters is necessary and helpful. Despite existing research gaps, the future of research on FDM process parameters is the most appealing, and a number of innovative explorations await newcomers in the field. The following contents, though certainly not comprehensive, point out some potential future directions and areas that require attention from the field.

5.1. Condition of Printing

FDM parts are printed in diverse conditions, which inevitably affect the mechanical characteristics of printed samples. The function of environmental parameters such as platform temperature [117], envelope temperature [115,156], humidity [157,158], and oxygen [159]; other print parameters such as nozzle parameter [92] and filament diameter [160], on mechanical behavior has been more or less studied, although not very extensive. In addition, how these factors impact process parameters remains a challenge and only attracts a few researchers' attention. For example, Mohd et al. [161] found that the diameter of the ABS filament increased as it was exposed to prolonged moisture with a certain absorption rate. However, this physical change would not cause nozzle clogging, which would directly affect the print speed. The influence of the FDM process condition could be a potential future research direction in this field.

5.2. Experimental Standard

The current approach to mechanical testing mainly refers to the relevant standards of raw materials and formed parts in their original application fields and utilizes existing standards. There are no specific guidelines for FDM process that prescribe the method of testing mechanical properties. This is one of the reasons that variety can be found when comparing experimental results from different authors. In the existing research, two standards are widely adopted: ASTM and ISO [2,162]. However, some of the standards are intended for materials containing high modulus fibers and are not directly applicable to samples made with FDM process. On the other hand, studies have shown certain composite standards actually improve test consistency on FDM materials [163]. Therefore, a suite of standard test methods should be developed to measure the mechanical property of parts by the FDM process. The authors hope researchers in related fields can work together to solve this urgent and important problem.

5.3. Multi-Parameters Optimization

The properties of FDM built parts exhibit high dependence on process parameters and can be improved by setting parameters at suitable levels. Consequently, experimental approaches are usually adopted to obtain the optimal combination, including Taguchi design [164,165], fractional factorial design [166,167], full factorial design [168,169], face-centered central composites design (FCCCD) [26,79], along with analysis methods such as analysis of variance (ANOVA) [165,166] or signal-to-noise ratio (S/N) [164,170]. Furthermore, some researchers establish the mathematical model between response and parameters (e.g., response surface methodology (RSM) [171,172]) and optimize with various algorithms. For example, particle swarm optimization (PSO) [172,173], artificial neural network (ANN) [134], bacterial foraging optimization(BFO) [26], genetic algorithm (GA) [174], surrogate-based optimization [175], naked mole-rat algorithm (NMRA) [176], and other heuristic optimization methods [177].

Although these optimization methods have achieved satisfactory results, their applicability is limited to some specific problems. In addition, the optimal result may not be achievable in practice, restrained by the parameters setting of the FDM machine. Therefore, exploring new optimization strategies with high efficiency and broad applicability is an attractive prospect. Besides, multi-objective optimization is a more challenging and complex topic [132,165,178], since the optimal result may correspond to multiple parameter combinations. Therefore, there is a need for more research efforts on multi-parameters optimization for the FDM process in the future.

5.4. Post-Processing Technique

Many studies have verified that there are some shortcomings in FDM components that cannot be overcome by only optimizing process parameters. These shortcomings, such as shape distortion, microvoids, uneven fiber distribution, and stairs-stepping effect [179], directly affect the mechanical characteristics of FDM parts. Therefore, post-processing techniques [180], including chemical treatment [181–183], heat treatment [184–186], laser treatment [187,188], and ultrasound treatment [189,190], are often adopted to improve mechanical strength and print quality of parts. However, these treatments may have influences on structural performance as well as process parameters. For instance, heat treatment can enhance the mechanical strength of printed products by improving crystallinity and removing residual stress of polymers [191]. At the same time, this treatment can result in changes in porosity due to annealing temperature as well, which will affect the infill density consequently [185]. Another example is ultrasound treatment. Mohamed et al. [192] used an ultrasonic transducer to improve the surface quality of components with different frequencies, and they observed from the result that the surface roughness was significantly smoother than before, together with a decrease in road width and layer thickness. Therefore, the optimal values obtained from process parameter optimization (classified as pre-processing) may change after post-processing, which needs to be paid more attention to.

5.5. Facing Real Parts

Most studies in the literature focus on "dog bone" samples to analyze the function of process parameters. It should be noted that the conclusion or result obtained from "lab experiment" may not apply to real applications. The review shows that there are only a couple of reports on improving the mechanical performance of a real part. For example, Zaman et al. [128] optimized five process parameters on compressive strength of drilling grid from the aerospace industry using the Taguchi design of experiments. Lee et al. [164] analyzed the relationship between process parameters and elastic performance of a compliant catapult using the Taguchi method. The maximum throwing distance was achieved by setting optimal parameters combination obtained. Since FDM products are ultimately used in practical applications, more research on real objects needs to be carried out, which can be another direction for future research.

5.6. Combination with 4D Printing

4D printed structures can change shape or property by stimulus, showing innovation and smartness, which has attracted unprecedented interest in recent years [193]. With the increasing application of FDM printers for 4D printing, the effect of process parameters on shape memory effect (SME) for smart materials is becoming a research hotspot [194]. For example, Kačergis et al. [105] evaluated the impact of platform temperature, print speed, and number of layers on the behavior of shape-shifting 'hinge' structure. They pointed out the higher print speed and lower platform temperature resulted in a higher deformation angle. In addition, the more active layers, the more time for shape recovery. Rajkumar and Shanmugam [195] analyzed the mechanisms of process parameters, such as infill density, thickness, and print speed, on shape-transformation, based on which they applied the results in manufacturing controllable curved components. In fact, there exist many unknown problems for 4D printing to be investigated, such as material behaviors, shape-shifting effects, and actuation methods [196] for smart and multi-materials obtained through the FDM approach. Therefore, research on the application of the FDM technique in printing 4D structures is exciting and appealing work awaiting further exploration.

Author Contributions: Conceptualization, G.G.; methodology, G.G.; software, G.T.; formal analysis, Z.L.; investigation, G.G., F.X. and G.T.; resources, J.X. and Z.L.; writing—original draft preparation, G.G.; writing—review and editing, G.G. and F.X.; supervision, J.X. All authors have read and agreed to the published version of the manuscript.

Funding: This research received no external funding.

Institutional Review Board Statement: Not applicable.

Informed Consent Statement: Not applicable.

Data Availability Statement: Not applicable.

Acknowledgments: This work is supported by Jiangsu Provincial Double-Innovation Doctor Program. The authors wish to express their appreciation to Lecturer Yumei Wu at Jiangsu University of Science and Technology for her valuable comments.

Conflicts of Interest: The authors declare no conflict of interest.

References

1. Vyavahare, S.; Teraiya, S.; Panghal, D.; Kumar, S. Fused deposition modelling: A review. *Rapid Prototyp. J.* **2020**, *26*, 176–201. [CrossRef]
2. Gordelier, T.J.; Thies, P.R.; Turner, L.; Johanning, L. Optimising the fdm additive manufacturing process to achieve maximum tensile strength: A state-of-the-art review. *Rapid Prototyp. J.* **2019**, *25*, 953–971. [CrossRef]
3. Ngo, T.D.; Kashani, A.; Imbalzano, G.; Nguyen, K.T.; Hui, D. Additive manufacturing (3d printing): A review of materials, methods, applications and challenges. *Compos. B Eng.* **2018**, *143*, 172–196. [CrossRef]
4. Anitha, R.; Arunachalam, S.; Radhakrishnan, P. Critical parameters influencing the quality of prototypes in fused deposition modelling. *J. Mater. Process. Technol.* **2001**, *118*, 385–388. [CrossRef]
5. Horvath, D.; Noorani, R.; Mendelson, M. Improvement of surface roughness on abs 400 polymer using design of experiments (doe). *Mater. Sci. Forum* **2007**, *561–565*, 2389–2392. [CrossRef]
6. Akande, S.O. Dimensional accuracy and surface finish optimization of fused deposition modelling parts using desirability function analysis. *Int. J. Eng. Sci.* **2015**, *4*, 196–202. [CrossRef]
7. Wang, C.C.; Lin, T.; Hu, S. Optimizing the rapid prototyping process by integrating the taguchi method with the gray relational analysis. *Rapid Prototyp. J.* **2007**, *13*, 304–315. [CrossRef]
8. Raju, M.; Gupta, M.K.; Bhanot, N.; Sharma, V.S. A hybrid pso–bfo evolutionary algorithm for optimization of fused deposition modelling process parameters. *J. Intell. Manuf.* **2019**, *30*, 2743–2758. [CrossRef]
9. Baich, L.; Manogharan, G.; Marie, H. Study of infill print design on production cost-time of 3d printed abs parts. *Int. J. Rapid Manuf.* **2015**, *5*, 308–319. [CrossRef]
10. Srivastava, M.; Rathee, S.; Maheshwari, S.; Kundra, T.K. Multi-objective optimisation of fused deposition modelling process parameters using rsm and fuzzy logic for build time and support material. *Int. J. Rapid Manuf.* **2018**, *7*, 25–42. [CrossRef]
11. Mohamed, O.A.; Masood, S.H.; Bhowmik, J.L.; Nikzad, M.; Azadmanjiri, J. Effect of process parameters on dynamic mechanical performance of fdm pc/abs printed parts through design of experiment. *J. Mater. Eng. Perform.* **2016**, *25*, 2922–2935. [CrossRef]
12. Rohde, S.; Cantrell, J.; Jerez, A.; Kroese, C.; Damiani, D.; Gurnani, R.; DiSandro, L.; Anton, J.; Young, A.; Steinbach, D.; et al. Experimental characterization of the shear properties of 3d–printed abs and polycarbonate parts. *Exp. Mech.* **2018**, *58*, 871–884. [CrossRef]
13. Toro, E.V.D.; Sobrino, J.C.; Martínez, A.M.; Eguíaab, V.M. Analysis of the influence of the variables of the fused deposition modeling (fdm) process on the mechanical properties of a carbon fiber-reinforced polyamide. *Procedia Manuf.* **2019**, *41*, 731–738. [CrossRef]
14. Dey, A.; Yodo, N. A systematic survey of FDM process parameter optimization and their influence on part characteristics. *J. Manuf. Mater. Process.* **2019**, *3*, 64. [CrossRef]
15. Cuan-Urquizo, E.; Barocio, E.; Tejada-Ortigoza, V.; Pipes, R.; Rodriguez, C.; Roman-Flores, A. Characterization of the mechanical properties of fff structures and materials: A review on the experimental, computational and theoretical approaches. *Materials* **2019**, *12*, 895. [CrossRef]
16. Sheoran, A.J.; Kumar, H. Fused deposition modeling process parameters optimization and effect on mechanical properties and part quality: Review and reflection on present research. *Mater. Today Proc.* **2020**, *21*, 1659–1672. [CrossRef]
17. Mohamed, O.A.; Masood, S.H.; Bhowmik, J.L. Optimization of fused deposition modeling process parameters: A review of current research and future prospects. *Adv. Manuf.* **2015**, *3*, 42–53. [CrossRef]
18. Popescu, D.; Zapciu, A.; Amza, C.; Baciu, F.; Marinescu, R. FDM process parameters influence over the mechanical properties of polymer specimens: A review. *Polym. Test.* **2018**, *69*, 157–166. [CrossRef]
19. Bakır, A.A.; Atik, R.; Özerinç, S. Mechanical properties of thermoplastic parts produced by fused deposition modeling: A review. *Rapid Prototyp. J.* **2021**, *27*, 537–561. [CrossRef]

20. Syrlybayev, D.; Zharylkassyn, B.; Seisekulova, A.; Akhmetov, M.; Perveen, A.; Talamona, D. Optimisation of strength properties of fdm printed parts—a critical review. *Polymers* **2021**, *13*, 1587. [CrossRef]
21. Li, H.; Wang, T.; Yu, Z. The quantitative research of interaction between key parameters and the effects on mechanical property in fdm. *Adv. Mater. Sci. Eng.* **2017**, *2017*, 1–15. [CrossRef]
22. Sood, A.K.; Ohdar, R.K.; Mahapatra, S.S. Parametric appraisal of mechanical property of fused deposition modelling processed parts. *Mater. Des.* **2010**, *31*, 287–295. [CrossRef]
23. Durgun, I.; Ertan, R. Experimental investigation of fdm process for improvement of mechanical properties and production cost. *Rapid Prototyp. J.* **2014**, *20*, 228–235. [CrossRef]
24. Rodriguez, J.F.; Thomas, J.P.; Renaud, J.E. Design of fused-deposition abs components for stiffness and strength. *J. Mech. Des.* **2003**, *125*, 545–551. [CrossRef]
25. Ashtankar, K.M.; Kuthe, A.M.; Rathour, B.S. Effect of build orientation on mechanical properties of rapid prototyping (fused deposition modelling) made acrylonitrile butadiene styrene (abs) parts. In Proceedings of the ASME 2013 International Mechanical Engineering Congress and Exposition, San Diego, CA, USA, 15–21 November 2013. [CrossRef]
26. Panda, S.K.; Padhee, S.; Anoop Kumar, S.; Mahapatra, S.S. Optimization of fused deposition modelling (fdm) process parameters using bacterial foraging technique. *Intell. Inf. Manag.* **2009**, *1*, 89–97. [CrossRef]
27. Chacón, J.; Caminero, M.; García-Plaza, E.; Núñez, P. Additive manufacturing of pla structures using fused deposition modelling: Effect of process parameters on mechanical properties and their optimal selection. *Mater. Des.* **2017**, *124*, 143–157. [CrossRef]
28. Alafaghani, A.; Qattawi, A.; Alrawi, B.; Guzman, A. Experimental optimization of fused deposition modelling processing parameters: A design-for-manufacturing approach. *Procedia Manuf.* **2017**, *10*, 791–803. [CrossRef]
29. Lee, C.S.; Kim, S.G.; Kim, H.J.; Ahn, S.H. Measurement of anisotropic compressive strength of rapid prototyping parts. *J. Mater. Process. Technol.* **2002**, *187–188*, 248–257. [CrossRef]
30. Górski, F.; Wichniarek, R.; Kuczko, W.; Andrzejewski, J. Experimental determination of critical orientation of abs parts manufactured using fused deposition modelling technology. *J. Mach. Eng.* **2015**, *15*, 121–132.
31. Hernandez, R.; Slaughter, D.; Whaley, D.; Tate, J.; Asiabanpour, B. Analyzing the tensile, compressive, and flexural properties of 3d printed abs p430 plastic based on printing orientation using fused deposition modeling. In Proceedings of the 27th Annual International Solid Freeform Fabrication Symposium, Austin, TX, USA, 12–14 August 2019.
32. Bertoldi, M.; Yardimci, M.A.; Pistor, C.M.; Guceri, S.I.; Sala, G. Mechanical characterization of parts processed via fused deposition. In Proceedings of the 9th Solid Freeform Fabrication Symposium, Austin, TX, USA, 11–13 August 1998.
33. Zou, R.; Xia, Y.; Liu, S.; Hu, P.; Hou, W.; Hu, Q.; Shan, C. Isotropic and anisotropic elasticity and yielding of 3d printed material. *Compos. B. Eng.* **2016**, *99*, 506–513. [CrossRef]
34. Raney, K.; Lani, E.; Kalla, D.K. Experimental characterization of the tensile strength of abs parts manufactured by fused deposition modeling process. *Mater. Today Proc.* **2017**, *4*, 7956–7961. [CrossRef]
35. Domingo-Espin, M.; Puigoriol-Forcada, J.M.; Garcia-Granada, A.-A.; Llumà, J.; Borros, S.; Reyes, G. Mechanical property characterization and simulation of fused deposition modeling polycarbonate parts. *Mater. Des.* **2015**, *83*, 670–677. [CrossRef]
36. Smith, W.C.; Dean, R.W. Structural characteristics of fused deposition modeling polycarbonate material. *Polym. Test.* **2013**, *32*, 1306–1312. [CrossRef]
37. Zaldivar, R.; Witkin, D.; McLouth, T.; Patel, D.; Schmitt, K.; Nokes, J. Influence of processing and orientation print effects on the mechanical and thermal behavior of 3d-printed ultem® 9085 material. *Addit. Manuf.* **2017**, *13*, 71–80. [CrossRef]
38. Taylor, G.; Wang, X.; Mason, L.; Leu, M.C.; Chandrashekhara, K.; Schniepp, T.; Jones, R. Flexural behavior of additively manufactured ultem 1010: Experiment and simulation. *Rapid Prototyp. J.* **2018**, *24*, 1003–1011. [CrossRef]
39. Uddin, M.; Sidek, M.; Faizal, M.; Ghomashchi, R.; Pramanik, A. Evaluating mechanical properties and failure mechanisms of fused deposition modeling acrylonitrile butadiene styrene parts. *J. Manuf. Sci. Eng.* **2017**, *139*, 081018. [CrossRef]
40. Es-Said, O.; Foyos, J.; Noorani, R.; Mendelson, M.; Marloth, R.; Pregger, B. Effect of layer orientation on mechanical properties of rapid prototyped samples. *Mater. Manuf. Process.* **2000**, *15*, 107–122. [CrossRef]
41. Rodríguez-Panes, A.; Claver, J.; Camacho, A. The influence of manufacturing parameters on the mechanical behaviour of pla and abs pieces manufactured by fdm: A comparative analysis. *Materials* **2018**, *11*, 1333. [CrossRef]
42. Somireddy, M.; De Moraes, D.A.; Czekanski, A. Flexural behavior of fdm parts: Experimental, analytical and numerical study. In Proceedings of the 28th Annual International Solid Freeform Fabrication Symposium, Austin, TX, USA, 7–9 August 2017.
43. Vaezi, M.; Chua, C.K. Effects of layer thickness and binder saturation level parameters on 3d printing process. *J. Adv. Manuf. Technol.* **2011**, *53*, 275–284. [CrossRef]
44. Ahn, S.H.; Montero, M.; Odell, D.; Roundy, S.; Wright, P.K. Anisotropic material properties of fused deposition modeling abs. *Rapid Prototyp. J.* **2002**, *8*, 248–257. [CrossRef]
45. Magalhães, L.C.; Volpato, N.; Luersen, M.A. Evaluation of stiffness and strength in fused deposition sandwich specimens. *J. Braz. Soc. Mech. Sci.* **2014**, *36*, 449–459. [CrossRef]
46. Ziemian, S.; Okwara, M.; Ziemian, C.W. Tensile and fatigue behavior of layered acrylonitrile butadiene styrene. *Rapid Prototyp. J.* **2015**, *21*, 270–278. [CrossRef]
47. Zhou, Y.G.; Su, B.; Turng, L.S. Deposition-induced effects of isotactic polypropylene and polycarbonate composites during fused deposition modeling. *Rapid Prototyp. J.* **2017**, *23*, 189. [CrossRef]

48. Garg, A.; Bhattacharya, A.; Batish, A. Failure investigation of fused deposition modelling parts fabricated at different raster angles under tensile and flexural loading. *Proc. Inst. Mech. Eng. Part B J. Eng. Manuf.* **2017**, *231*, 2031–2039. [CrossRef]
49. Ziemian, C.; Sharma, M.; Ziemian, S. Anisotropic mechanical properties of abs parts fabricated by fused deposition modelling. In *Mechanical Engineering*; Gokcek, M., Ed.; IntechOpen: London, UK, 2012; pp. 159–180. [CrossRef]
50. Fatimatuzahraa, A.W.; Farahaina, B.; Yusoff, W. The effect of employing different raster orientations on the mechanical properties and microstructure of fused deposition modeling parts. In Proceedings of the 2011 IEEE Symposium on Business, Engineering and Industrial Applications, Langkawi, Malaysia, 25–28 September 2011. [CrossRef]
51. Jap, N.S.F.; Pearce, G.M.; Hellier, A.K.; Russell, N.; Parr, W.C.; Walsh, W.R. The effect of raster orientation on the static and fatigue properties of filament deposited abs polymer. *Int. J. Fatigue* **2019**, *124*, 328–337. [CrossRef]
52. Diaconescu, C.; Tabacu, S.; Oltean, A. Design and analysis of a fused deposition modelling manufactured part. In Proceedings of the CAR2017 International Congress of Automotive and Transport Engineering—Mobility Engineering and Environment, Pitesti, Romania, 8–10 November 2017. [CrossRef]
53. Hart, K.R.; Wetzel, E.D. Fracture behavior of additively manufactured acrylonitrile butadiene styrene (abs) materials. *Eng. Fract. Mech.* **2017**, *177*, 1–13. [CrossRef]
54. Arbeiter, F.; Spoerk, M.; Wiener, J.; Gosch, A.; Pinter, G. Fracture mechanical characterization and lifetime estimation of near-homogeneous components produced by fused filament fabrication. *Polym. Test.* **2018**, *66*, 105–113. [CrossRef]
55. Bellini, A.; Güçeri, S. Mechanical characterization of parts fabricated using fused deposition modeling. *Rapid Prototyp. J.* **2003**, *9*, 252–264. [CrossRef]
56. Balderrama-Armendariz, C.O.; MacDonald, E.; Espalin, D.; Cortes-Saenz, D.; Wicker, R.; Maldonado-Macias, A. Torsion analysis of the anisotropic behavior of fdm technology. *Int. J. Adv. Manuf. Technol.* **2018**, *96*, 307–317. [CrossRef]
57. Cantrell, J.T.; Rohde, S.; Damiani, D.; Gurnani, R.; DiSandro, L.; Anton, J.; Young, A.; Jerez, A.; Steinbach, D.; Kroese, C. Experimental characterization of the mechanical properties of 3d printed abs and polycarbonate parts. *Rapid Prototyp. J.* **2016**, *23*, 811–824. [CrossRef]
58. Torrado, A.R.; Roberson, D.A. Failure analysis and anisotropy evaluation of 3d-printed tensile test specimens of different geometries and print raster patterns. *J. Fail. Anal. Prev.* **2016**, *16*, 154–164. [CrossRef]
59. Letcher, T.; Rankouhi, B.; Javadpour, S. Experimental study of mechanical properties of additively manufactured abs plastic as a function of layer parameters. In Proceedings of the ASME 2015 International Mechanical Engineering Congress and Exposition (IMECE), Houston, TX, USA, 13–19 November 2015. [CrossRef]
60. Casavola, C.; Cazzato, A.; Moramarco, V.; Pappalettere, C. Orthotropic mechanical properties of fused deposition modelling parts described by classical laminate theory. *Mater. Des.* **2016**, *90*, 453–458. [CrossRef]
61. Lanzotti, A.; Grasso, M.; Staiano, G.; Martorelli, M. The impact of process parameters on mechanical properties of parts fabricated in pla with an open-source 3-d printer. *Rapid Prototyp. J.* **2015**, *21*, 604–617. [CrossRef]
62. Rankouhi, B.; Javadpour, S.; Delfanian, F.; Letcher, T. Failure analysis and mechanical characterization of 3d printed abs with respect to layer thickness and orientation. *J. Fail. Anal. Prev.* **2016**, *16*, 467–481. [CrossRef]
63. Tymrak, B.; Kreiger, M.; Pearce, J.M. Mechanical properties of components fabricated with open-source 3-d printers under realistic environmental conditions. *Mater. Des.* **2014**, *58*, 242–246. [CrossRef]
64. Rajpurohit, S.R.; Dave, H.K. Analysis of tensile strength of a fused filament fabricated pla part using an open-source 3d printer. *Int. J. Adv. Manuf. Technol.* **2019**, *101*, 1525–1536. [CrossRef]
65. Wu, W.; Geng, P.; Li, G.; Zhao, D.; Zhang, H.; Zhao, J. Influence of layer thickness and raster angle on the mechanical properties of 3d-printed peek and a comparative mechanical study between peek and abs. *Materials* **2015**, *8*, 5834–5846. [CrossRef]
66. Garg, A.; Bhattacharya, A. An insight to the failure of fdm parts under tensile loading: Finite element analysis and experimental study. *Int. J. Mech. Sci.* **2017**, *120*, 225–236. [CrossRef]
67. Nidagundi, V.B.; Keshavamurthy, R.; Prakash, C.P.S. Studies on parametric optimization for fused deposition modelling process. *Mater. Today Proc.* **2015**, *2*, 1691–1699. [CrossRef]
68. Carneiro, O.S.; Silva, A.F.; Gomes, R. Fused deposition modeling with polypropylene. *Mater. Des.* **2015**, *83*, 768–776. [CrossRef]
69. Dong, Y.; Milentis, J.; Pramanik, A. Additive manufacturing of mechanical testing samples based on virgin poly (lactic acid)(pla) and pla/wood fibre composites. *Adv. Manuf.* **2018**, *6*, 71–82. [CrossRef]
70. D'Amico, A.A.; Debaie, A.; Peterson, A.M. Effect of layer thickness on irreversible thermal expansion and interlayer strength in fused deposition modeling. *Rapid Prototyp. J.* **2017**, *23*, 943–953. [CrossRef]
71. Rodríguez, J.F.; Thomas, J.P.; Renaud, J.E. Mechanical behavior of acrylonitrile butadiene styrene (abs) fused deposition materials: Experimental investigation. *Rapid Prototyp. J.* **2001**, *7*, 148–158. [CrossRef]
72. Too, M.H.; Leong, K.F.; Chua, C.K.; Du, Z.H.; Yang, S.F.; Cheah, C.M.; Ho, S.L. Investigation of 3d non-random porous structures by fused deposition modelling. *Int. J. Adv. Manuf. Technol.* **2002**, *19*, 217–223. [CrossRef]
73. Dawoud, M.; Taha, I.; Ebeid, S.J. Mechanical behaviour of abs: An experimental study using fdm and injection moulding techniques. *J. Manuf. Process.* **2016**, *21*, 39–45. [CrossRef]
74. Masood, S.H.; Mau, K.; Song, W.Q. Tensile properties of processed fdm polycarbonate material. *Mater. Sci. Forum.* **2010**, *654*, 2556–2559. [CrossRef]
75. Slonov, A.L.; Khashirov, A.A.; Zhansitov, A.A.; Rzhevskaya, E.V.; Khashirova, S.Y. The influence of the 3d-printing technology on the physical and mechanical properties of polyphenylene sulfone. *Rapid Prototyp. J.* **2018**, *24*, 1124–1130. [CrossRef]

76. Hossain, M.S.; Ramos, J.; Espalin, D.; Perez, M.; Wicker, R. Improving tensile mechanical properties of fdm-manufactured specimens via modifying build parameters. In Proceedings of the 2013 International Solid Freeform Fabrication Symposium: An Additive Manufacturing Conference, Austin, TX, USA, 12–14 August 2013. [CrossRef]
77. Hossain, M.S.; Espalin, D.; Ramos, J.; Perez, M.; Wicker, R. Improved mechanical properties of fused deposition modeling-manufactured parts through build parameter modifications. *J. Manuf. Sci. Eng.* **2014**, *136*, 061002. [CrossRef]
78. Montero, M.; Roundy, S.; Odell, D.; Ahn, S.-H.; Wright, P.K. Material characterization of fused deposition modeling (fdm) abs by designed experiments. *Soc. Manuf. Eng.* **2001**, *10*, 1–21.
79. Bagsik, A.; Schöppner, V.; Klemp, E. Fdm part quality manufactured with ultem* 9085. In Proceedings of the 14th International Scientific Conference on Polymeric Materials, Halle, Germany, 15–17 September 2010.
80. Gebisa, A.W.; Lemu, H.G. Investigating effects of fused-deposition modeling (fdm) processing parameters on flexural properties of ultem 9085 using designed experiment. *Materials* **2018**, *11*, 500. [CrossRef]
81. Gebisa, A.W.; Lemu, H.G. Influence of 3d printing fdm process parameters on tensile property of ultem 9085. *Procedia Manuf.* **2019**, *30*, 331–338. [CrossRef]
82. Deng, X.; Zeng, Z.; Peng, B.; Yan, S.; Ke, W. Mechanical properties optimization of poly-ether-ether-ketone via fused deposition modeling. *Materials* **2018**, *11*, 216. [CrossRef] [PubMed]
83. Ang, K.C.; Leong, K.F.; Chua, C.K.; Chandrasekaran, M. Investigation of the mechanical properties and porosity relationships in fused deposition modelling-fabricated porous structures. *Rapid Prototyp. J.* **2006**, *12*, 100–105. [CrossRef]
84. Rayegani, F.; Onwubolu, G.C. Fused deposition modelling (fdm) process parameter prediction and optimization using group method for data handling (gmdh) and differential evolution (de). *Int. J. Adv. Manuf. Technol.* **2014**, *73*, 509–519. [CrossRef]
85. Onwubolu, G.C.; Rayegani, F. Characterization and optimization of mechanical properties of abs parts manufactured by the fused deposition modelling process. *Int. J. Manuf. Eng.* **2014**, *2014*, 598531. [CrossRef]
86. Liu, X.; Zhang, M.; Li, S.; Si, L.; Peng, J.; Hu, Y. Mechanical property parametric appraisal of fused deposition modeling parts based on the gray taguchi method. *Int. J. Adv. Manuf. Technol.* **2017**, *89*, 2387–2397. [CrossRef]
87. Gkartzou, E.; Koumoulos, E.P.; Charitidis, C.A. Production and 3d printing processing of bio-based thermoplastic filament. *Manuf. Rev.* **2017**, *4*, 1. [CrossRef]
88. Elkholy, A.; Kempers, R. Investigation into the influence of fused deposition modeling (fdm) process parameters on the thermal properties of 3d-printed parts. In Proceedings of the 2018 Canadian Society for Mechanical Engineering (CSME) International Congress, Toronto, ON, Canada, 27–30 May 2018. [CrossRef]
89. Rajpurohit, S.R.; Dave, H.K. Flexural strength of fused filament fabricated (fff) pla parts on an open-source 3d printer. *Adv. Manuf.* **2018**, *6*, 430–441. [CrossRef]
90. Alvarez, K.L.; Lagos, R.F.; Aizpun, M. Investigating the influence of infill percentage on the mechanical properties of fused deposition modelled abs parts. *Ing. Investig.* **2016**, *36*, 110–116. [CrossRef]
91. Martikka, O.; Kärki, T.; Wu, Q.L. Mechanical properties of 3d-printed wood-plastic composites. In *Key Engineering Materials*; Trans Tech. Publ. Ltd.: Freienbach, Switzerland, 2018; Volume 777, pp. 499–507. [CrossRef]
92. Gomez-Gras, G.; Jerez-Mesa, R.; Travieso-Rodriguez, J.A.; Lluma-Fuentes, J. Fatigue performance of fused filament fabrication PLA specimens. *Mater. Des.* **2018**, *140*, 278–285. [CrossRef]
93. Aw, Y.Y.; Yeoh, C.K.; Idris, M.A.; Teh, P.L.; Hamzah, K.A.; Sazali, S.A. Effect of printing parameters on tensile, dynamic mechanical, and thermoelectric properties of fdm 3d printed cabs/zno composites. *Materials* **2018**, *11*, 466. [CrossRef] [PubMed]
94. Kerekes, T.W.; Lim, H.; Joe, W.Y.; Yun, G.J. Characterization of process–deformation/damage property relationship of fused deposition modeling (fdm) 3d-printed specimens. *Addit. Manuf.* **2019**, *25*, 532–544. [CrossRef]
95. Lužanin, O.; Movrin, D.; Plančak, M. Effect of layer thickness, deposition angle, and infill on maximum flexural force in fdm-built specimens. *Int. J. Plast.* **2014**, *39*, 49–58.
96. Cho, E.E.; Hein, H.H.; Lynn, Z.; Hla, S.J.; Tran, T. Investigation on influence of infill pattern and layer thickness on mechanical strength of pla material in 3d printing technology. *J. Eng. Sci. Res.* **2019**, *3*, 27–37. [CrossRef]
97. Dave, H.K.; Patadiya, N.H.; Prajapati, A.R.; Rajpurohit, S.R. Effect of infill pattern and infill density at varying part orientation on tensile properties of fused deposition modeling-printed poly-lactic acid part. *Proc. Inst. Mech. Eng. Part C J. Mech. Eng. Sci.* **2021**, *235*, 1811–1827. [CrossRef]
98. Fernandez-Vicente, M.; Calle, W.; Ferrandiz, S.; Conejero, A. Effect of infill parameters on tensile mechanical behavior in desktop 3d printing. 3D Print. *Addit. Manuf.* **2016**, *3*, 183–192. [CrossRef]
99. Akhoundi, B.; Behravesh, A.H. Effect of filling pattern on the tensile and flexural mechanical properties of fdm 3d printed products. *Exp. Mech.* **2019**, *59*, 883–897. [CrossRef]
100. Nagendra, J.; Prasad, M.S. FDM process parameter optimization by taguchi technique for augmenting the mechanical properties of nylon–aramid composite used as filament material. *J. Inst. Eng. (India) Ser. C* **2020**, *101*, 313–322. [CrossRef]
101. Ning, F.; Cong, W.; Qiu, J.; Wei, J.; Wang, S. Additive manufacturing of carbon fiber reinforced thermoplastic composites using fused deposition modeling. *Compos. Part B Eng.* **2015**, *80*, 369–378. [CrossRef]
102. Li, H.; Wang, T.; Sun, J.; Yu, Z. The effect of process parameters in fused deposition modelling on bonding degree and mechanical properties. *Rapid Prototyp. J.* **2018**, *24*, 80–92. [CrossRef]
103. Christiyan, K.J.; Chandrasekhar, U.; Venkateswarlu, K. A study on the influence of process parameters on the mechanical properties of 3d printed abs composite. *IOP Conf. Ser. Mater. Sci. Eng.* **2016**, *114*, 012109. [CrossRef]

104. Santana, L.; Ahrens, C.H.; Netto, A.C.S.; Bonin, C. Evaluating the deposition quality of parts produced by an open-source 3d printer. *Rapid Prototyp. J.* **2017**, *23*, 796–803. [CrossRef]
105. Kačergis, L.; Mitkus, R.; Sinapius, M. Influence of fused deposition modeling process parameters on the transformation of 4d printed morphing structures. *Smart Mater. Struct.* **2019**, *28*, 105042. [CrossRef]
106. Lužanin, O.; Guduric, V.; Ristic, I.; Muhic, S. Investigating impact of five build parameters on the maximum flexural force in fdm specimens–a definitive screening design approach. *Rapid Prototyp. J.* **2017**, *23*, 1088–1098. [CrossRef]
107. Faes, M.; Ferraris, E.; Moens, D. Influence of inter-layer cooling time on the quasi-static properties of abs components produced via fused deposition modelling. *Procedia Cirp.* **2016**, *42*, 748–753. [CrossRef]
108. Zhang, Y.; Chou, K. A parametric study of part distortions in fused deposition modelling using three-dimensional finite element analysis. *Proc. Inst. Mech. Eng. Part B J. Eng. Manuf.* **2008**, *222*, 959–968. [CrossRef]
109. Kung, C.; Kuan, H.C.; Kuan, C.F. Evaluation of tensile strength of 3d printed objects with fdm process on reprap platform. In Proceedings of the 2018 1st IEEE International Conference on Knowledge Innovation and Invention (ICKII), Jeju, Korea, 23–27 July 2018. [CrossRef]
110. Mahmood, S.; Qureshi, A.J.; Goh, K.L.; Talamona, D. Tensile strength of partially filled fff printed parts: Experimental results. *Rapid Prototyp. J.* **2017**, *23*, 122–128. [CrossRef]
111. Croccolo, D.; De Agostinis, M.; Olmi, G. Experimental characterization and analytical modelling of the mechanical behaviour of fused deposition processed parts made of abs-m30. *Comput. Mater. Sci.* **2013**, *79*, 506–518. [CrossRef]
112. Griffiths, C.A.; Howarth, J.; Rowbotham, G.D.A.; Rees, A. Effect of build parameters on processing efficiency and material performance in fused deposition modelling. *Procedia CIRP* **2016**, *49*, 28–32. [CrossRef]
113. Aliheidari, N.; Tripuraneni, R.; Ameli, A.; Nadimpalli, S. Fracture resistance measurement of fused deposition modeling 3d printed polymers. *Polym. Test.* **2017**, *60*, 94–101. [CrossRef]
114. Rinanto, A.; Nugroho, A.; Prasetyo, H.; Pujiyanto, E. Simultaneous optimization of tensile strength, energy consumption and processing time on fdm process using taguchi and pcr-topsis. In Proceedings of the 4th International Conference on Science and Technology (ICST), Yogyakarta, Indonesia, 7–8 August 2018. [CrossRef]
115. Sun, Q.; Rizvi, G.; Bellehumeur, C.; Gu, P. Effect of processing conditions on the bonding quality of fdm polymer filaments. *Rapid Prototyp. J.* **2008**, *14*, 72–80. [CrossRef]
116. Leite, M.; Fernandes, J.; Deus, A.M.; Reis, L.; Vaz, M.F. Study of the influence of 3d printing parameters on the mechanical properties of pla. In Proceedings of the 3rd International Conference on Progress in Additive Manufacturing (Pro-AM 2018), Nanyang Executive Centre, Singapore, 14–17 May 2018. [CrossRef]
117. Sun, X.; Cao, L.; Ma, H.; Peng, G.; Zhanwei, B.; Cheng, L. Experimental analysis of high temperature peek materials on 3d printing test. In Proceedings of the 9th International Conference on Measuring Technology and Mechatronics Automation (ICMTMA), Changsha, China, 14–15 January 2017. [CrossRef]
118. Yang, T.C. Effect of extrusion temperature on the physico-mechanical properties of unidirectional wood fiber-reinforced polylactic acid composite (wfrpc) components using fused deposition modeling. *Polymers* **2018**, *10*, 976. [CrossRef] [PubMed]
119. Upadhyay, K.; Dwivedi, R.; Singh, A.K. Determination and comparison of the anisotropic strengths of fused deposition modeling p400 abs. In *Advances in 3D Printing & Additive Manufacturing Technologies*; Wimpenny, D.I., Pandey, P.M., Kumar, L.J., Eds.; Springer: Singapore, 2017; pp. 9–28. [CrossRef]
120. Kamaal, M.; Anas, M.; Rastogi, H.; Bhardwaj, N.; Rahaman, A. Effect of fdm process parameters on mechanical properties of 3d-printed carbon fibre–pla composite. *Prog Addit Manuf.* **2021**, *6*, 63–69. [CrossRef]
121. Tanikella, N.G.; Wittbrodt, B.; Pearce, J.M. Tensile strength of commercial polymer materials for fused filament fabrication 3d printing. *Addit. Manuf.* **2017**, *15*, 40–47. [CrossRef]
122. Liu, H.; He, H.; Peng, X.; Huang, B.; Li, J. Three-dimensional printing of poly (lactic acid) bio-based composites with sugarcane bagasse fiber: Effect of printing orientation on tensile performance. *Polym. Adv. Technol.* **2019**, *30*, 910–922. [CrossRef]
123. Letcher, T.; Waytashek, M. Material property testing of 3d-printed specimen in pla on an entry-level 3d printer. In Proceedings of the ASME 2014 International Mechanical Engineering Congress and Exposition, Montreal, QC, Canada, 14–20 November 2014. [CrossRef]
124. Ayrilmis, N.; Kariz, M.; Kwon, J.H.; Kitek Kuzman, M. Effect of printing layer thickness on water absorption and mechanical properties of 3d-printed wood/pla composite materials. *Int. J. Adv. Manuf. Technol.* **2019**, *102*, 2195–2200. [CrossRef]
125. Knoop, F.; Schoeppner, V. Mechanical and thermal properties of fdm parts manufactured with polyamide 12. In Proceedings of the 26th Annual International Solid Freeform Fabrication Symposium—An Additive Manufacturing Conference, Austin, TX, USA, 10–12 August 2015.
126. Ebel, E.; Sinnemann, T. Fabrication of fdm 3d objects with abs and pla and determination of their mechanical properties. *RTejournal* **2014**, *2014*. Available online: https://rtejournal.de/wp-content/uploads/2014_Fabrication-of-FDM-3D-objects-with-ABS-and-PLA-and-determination-of-their-mechanical-properties.pdf (accessed on 11 February 2022).
127. Vinoth Babu, N.; Venkateshwaran, N.; Rajini, N.; Ismail, S.O.; Mohammad, F.; Al-Lohedan, H.A.; Suchart, S. Influence of slicing parameters on surface quality and mechanical properties of 3d-printed cf/pla composites fabricated by fdm technique. *Mater. Technol.* **2021**, 1–18. [CrossRef]
128. Zaman, U.K.; Boesch, E.; Siadat, A.; Rivette, M.; Baqai, A.A. Impact of fused deposition modeling (fdm) process parameters on strength of built parts using taguchi's design of experiments. *Int. J. Adv. Manuf. Technol.* **2019**, *101*, 1215–1226. [CrossRef]

129. Attoye, S.; Malekipour, E.; El-Mounayri, H. Correlation between process parameters and mechanical properties in parts printed by the fused deposition modeling process. In *Mechanics of Additive and Advanced Manufacturing*; Kramer, S., Jordan, J., Jin, H., Carroll, J., Beese, A., Eds.; Springer: Cham, Switzerland, 2019; Volume 8, pp. 35–41. [CrossRef]
130. Chokshi, H.; Shah, D.B.; Patel, K.M.; Joshi, S.J. Experimental investigations of process parameters on mechanical properties for PLA during processing in FDM. *Adv. Mater. Processing Technol.* **2021**, 1–14. [CrossRef]
131. Torres, J.; Cole, M.; Owji, A.; DeMastry, Z.; Gordon, A.P. An approach for mechanical property optimization of fused deposition modeling with polylactic acid via design of experiments. *Rapid Prototyp. J.* **2016**, *22*, 387–404. [CrossRef]
132. Abouelmajd, M.; Bahlaoui, A.; Arroub, I.; Zemzami, M.; Hmina, N.; Lagache, M.; Belhouideg, S. Experimental analysis and optimization of mechanical properties of FDM-processed polylactic acid using Taguchi design of experiment. *Int. J. Simul. Multidiscip. Des. Optim.* **2021**, *12*, 30. [CrossRef]
133. Panda, B.N.; Bahubalendruni, M.R.; Biswal, B.B. Comparative evaluation of optimization algorithms at training of genetic programming for tensile strength prediction of FDM processed part. *Procedia Mater. Sci.* **2014**, *5*, 2250–2257. [CrossRef]
134. Giri, J.; Shahane, P.; Jachak, S.; Chadge, R.; Giri, P. Optimization of fdm process parameters for dual extruder 3d printer using artificial neural network. *Mater. Today Proc.* **2021**, *43*, 3242–3249. [CrossRef]
135. Zhang, J.; Wang, X.Z.; Yu, W.W.; Deng, Y.H. Numerical investigation of the influence of process conditions on the temperature variation in fused deposition modeling. *Mater. Des.* **2017**, *130*, 59–68. [CrossRef]
136. Nancharaiah, T. Optimization of process parameters in FDM process using design of experiments. *Int. J. Emerg. Technol.* **2011**, *2*, 100–102.
137. Peng, A.; Xiao, X.; Yue, R. Process parameter optimization for fused deposition modeling using response surface methodology combined with fuzzy inference system. *Int. J. Adv. Manuf. Technol.* **2014**, *73*, 87–100. [CrossRef]
138. Arif, M.F.; Kumar, S.; Varadarajan, K.M.; Cantwell, W.J. Performance of biocompatible peek processed by fused deposition additive manufacturing. *Mater. Des.* **2018**, *146*, 249–259. [CrossRef]
139. Vishwas, M.; Basavaraj, C.K.; Vinyas, M. Experimental investigation using taguchi method to optimize process parameters of fused deposition Modeling for ABS and nylon materials. *Mater. Today. Proc.* **2018**, *5*, 7106–7114. [CrossRef]
140. Szykiedans, K.; Credo, W. Mechanical properties of fdm and sla low-cost 3-d prints. *Procedia Eng.* **2016**, *136*, 257–262. [CrossRef]
141. Vosynek, P.; Navrat, T.; Krejbychova, A.; Palousek, D. Influence of process parameters of printing on mechanical properties of plastic parts produced by fdm 3d printing technology. In Proceedings of the 3rd International Conference on Design, Mechanical and Material Engineering (D2ME 2018), Phuket, Thailand, 27–29 September 2018. [CrossRef]
142. Brenken, B.; Barocio, E.; Favaloro, A.; Kunc, V.; Pipes, R.B. Fused filament fabrication of fiber-reinforced polymers: A review. *Addit. Manuf.* **2018**, *21*, 1–16. [CrossRef]
143. Wittbrodt, B.; Pearce, J.M. The effects of pla color on material properties of 3-d printed components. *Addit. Manuf.* **2015**, *8*, 110–116. [CrossRef]
144. Hodžić, D.; Pandžić, A.; Hajro, I.; Tasić, P. Strength comparison of fdm 3d printed pla made by different manufacturers. *TEM J.* **2020**, *9*, 966–970. [CrossRef]
145. Dewada, S.S.; Telang, A. A review of recently developed polymer composite materials for fused deposition modeling 3D printing. *Mater. Res. Express* **2021**, *8*, 122001. [CrossRef]
146. Kottasamy, A.; Samykano, M.; Kadirgama, K.; Ramasamy, D.; Rahman, M.M.; Pandey, A.K. Optimization of impact energy of copper-polylactic acid (cu-pla) composite using response surface methodology for fdm 3d printing. *J. Adv. Res. Fluid Mech. Therm. Sci.* **2021**, *84*, 78–90. [CrossRef]
147. Nabipour, M.; Akhoundi, B. An experimental study of FDM parameters effects on tensile strength, density, and production time of ABS/Cu composites. *J. Elastomers Plast.* **2021**, *53*, 146–164. [CrossRef]
148. Prasetiyo, A.B.; Sekarjati, K.A.; Tontowi, A.E. Application of rsm method in optimization of 3d printing machine process parameters using biocomposite materials (pmma/hydroxyapatite) to get the highest tension strength. *JEMMME (J. Energy Mech. Mater. Manuf. Eng.)* **2021**, *6*, 119–126. [CrossRef]
149. Rosid, I.A.; Tontowi, A.E. Parameter optimization of customized fdm 3d printer machine for biocomposite material [sago/pmma] using 2k fractional factorial design. *OPSI* **2021**, *14*, 188–196. Available online: http://jurnal.upnyk.ac.id/index.php/opsi (accessed on 11 February 2022). [CrossRef]
150. Karimipour-Fard, P.; Jeffrey, M.P.; Jones Taggart, H. Development, processing and characterization of polycaprolactone/nano-hydroxyapatite/chitin-nano-whisker nanocomposite filaments for additive manufacturing of bone tissue scaffolds. *J. Mech. Behav. Biomed. Mater.* **2021**, *120*, 104583. [CrossRef]
151. Liu, J.; Ye, J.; Momin, F.; Zhang, X.; Li, A. Nonparametric bayesian framework for material and process optimization with nanocomposite fused filament fabrication. *Addit. Manuf.* **2022**, 102765. [CrossRef]
152. Peng, W.A.N.G.; Bin, Z.O.U.; Shouling, D.I.N.G.; Lei, L.I.; Huang, C. Effects of fdm-3d printing parameters on mechanical properties and microstructure of cf/peek and gf/peek. *Chin. J. Aeronaut.* **2021**, *34*, 236–246. [CrossRef]
153. Chen, K.; Yu, L.; Cui, Y.; Jia, M.; Pan, K. Optimization of printing parameters of 3d-printed continuous glass fiber reinforced polylactic acid composites. *Thin-Walled Struct.* **2021**, *164*, 107717. [CrossRef]
154. Caminero, M.A.; Chacón, J.M.; García-Moreno, I.; Rodríguez, G.P. Impact damage resistance of 3d printed continuous fibre reinforced thermoplastic composites using fused deposition modelling. *Compos. Part B Eng.* **2018**, *148*, 93–103. [CrossRef]

155. Osman, M.A.; Atia, M.R. Investigation of abs-rice straw composite feedstock filament for fdm. *Rapid Prototyp. J.* **2018**, *24*, 1067–1075. [CrossRef]
156. Fitzharris, E.R.; Watt, I.; Rosen, D.W.; Shofner, M.L. Interlayer bonding improvement of material extrusion parts with polyphenylene sulfide using the Taguchi method. *Addit. Manuf.* **2018**, *24*, 287–297. [CrossRef]
157. Kim, E.; Shin, Y.J.; Ahn, S.H. The effects of moisture and temperature on the mechanical properties of additive manufacturing components: Fused deposition modeling. *Rapid Prototyp. J.* **2016**, *22*, 887–894. [CrossRef]
158. Kariz, M.; Sernek, M.; Kuzman, M.K. Effect of humidity on 3d-printed specimens from wood-pla filaments. *Wood Res.* **2018**, *63*, 917–922.
159. Lederle, F.; Meyer, F.; Brunotte, G.P.; Kaldun, C.; Hübner, E.G. Improved mechanical properties of 3d-printed parts by fused deposition modeling processed under the exclusion of oxygen. *Prog. Addit. Manuf.* **2016**, *1*, 3–7. [CrossRef]
160. Kristiawan, R.B.; Imaduddin, F.; Ariawan, D.; Arifin, Z. A review on the fused deposition modeling (fdm) 3d printing: Filament processing, materials, and printing parameters. *Open Eng.* **2021**, *11*, 639–649. [CrossRef]
161. Mohd Halidi, S.N.A.; Abdullah, J. Moisture and humidity effects on the abs used in fused deposition modeling machine. *Adv. Mater. Res.* **2012**, *576*, 641–644. [CrossRef]
162. Abbas, T.F.; Othman, F.M.; Ali, H.B. Influence of layer thickness on impact property of 3d-printed pla. *Int. Res. J. Eng. Technol.* **2018**, *5*, 1–4.
163. Forster, A.M. Materials testing standards for additive manufacturing of polymer materials: State of the art and standards applicability. *Natl. Inst. Stand. Technol. US Dept. Commer.* **2015**, 1–54. [CrossRef]
164. Lee, B.H.; Abdullah, J.; Khan, Z.A. Optimization of rapid prototyping parameters for production of flexible abs object. *J. Mater. Process. Technol.* **2005**, *169*, 54–61. [CrossRef]
165. Kafshgar, A.R.; Rostami, S.; Aliha, M.R.M.; Berto, F. Optimization of properties for 3d printed pla material using taguchi, anova and multi-objective methodologies. *Procedia Struct. Integr.* **2021**, *34*, 71–77. [CrossRef]
166. Hikmat, M.; Rostam, S.; Ahmed, Y.M. Investigation of tensile property-based taguchi method of pla parts fabricated by fdm 3d printing technology. *Results Eng.* **2021**, *11*, 100264. [CrossRef]
167. Mazen, A.; McClanahan, B.; Weaver, J.M. Factors affecting ultimate tensile strength and impact toughness of 3d printed parts using fractional factorial design. *Int. J. Adv. Manuf. Technol.* **2022**, *119*, 2639–2651. [CrossRef]
168. Moradi, M.; Aminzadeh, A.; Rahmatabadi, D.; Rasouli, S.A. Statistical and experimental analysis of process parameters of 3d nylon printed parts by fused deposition modeling: Response surface modeling and optimization. *J. Mater. Eng. Perform.* **2021**, *30*, 5441–5454. [CrossRef]
169. Rashed, K.; Kafi, A.; Simons, R.; Bateman, S. Fused filament fabrication of nylon 6/66 copolymer: Parametric study comparing full factorial and taguchi design of experiments. *Rapid Prototyp. J.* **2022**. [CrossRef]
170. Laeng, J.; Khan, Z.A.; Khu, S. Optimizing flexible behaviour of bow prototype using Taguchi approach. *J. Appl. Sci.* **2006**, *6*, 622–630. [CrossRef]
171. Equbal, A.; Sood, A.K.; Equbal, M.I.; Badruddin, I.A.; Khan, Z.A. RSM based investigation of compressive properties of fdm fabricated part. *CIRP J. Manuf. Sci. Technol.* **2021**, *35*, 701–714. [CrossRef]
172. Saad, M.S.; Mohd Nor, A.; Zakaria, M.Z.; Baharudin, M.E.; Yusoff, W.S. Modelling and evolutionary computation optimization on FDM process for flexural strength using integrated approach RSM and PSO. *Prog. Addit. Manuf.* **2021**, *6*, 143–154. [CrossRef]
173. Shirmohammadi, M.; Goushchi, S.J.; Keshtiban, P.M. Optimization of 3d printing process parameters to minimize surface roughness with hybrid artificial neural network model and particle swarm algorithm. *Prog. Addit. Manuf.* **2021**, *6*, 199–215. [CrossRef]
174. Aminzadeh, A.; Aberoumand, M.; Rahmatabadi, D.; Moradi, M. Metaheuristic approaches for modeling and optimization of fdm process. In *Fused Deposition Modeling Based 3D Printing*; Dave, H.K., Davim, J.P., Eds.; Springer: Cham, Switzerland, 2021; pp. 483–504. [CrossRef]
175. Ulu, E.; Korkmaz, E.; Yay, K.; Ozdoganlar, O.B.; Kara, B.L. Enhancing the structural performance of additively manufactured objects through build orientation optimization. *J. Mech. Des.* **2018**, *137*, 111410–111419. [CrossRef]
176. Chohan, J.S.; Mittal, N.; Kumar, R.; Singh, S.; Sharma, S.; Dwivedi, S.P.; Saxena, A.; Chattopadhyaya, S.; Ilyas, R.A.; Le, C.H.; et al. Optimization of fff process parameters by naked mole-rat algorithms with enhanced exploration and exploitation capabilities. *Polymers* **2021**, *13*, 1702. [CrossRef]
177. Liu, B.; Yang, L.; Zhou, R.; Hong, B. Effect of process parameters on mechanical properties of additive manufactured smp structures based on FDM. *Mater. Test.* **2022**, *64*, 378–390. [CrossRef]
178. Patil, P.; Singh, D.; Raykar, S.J.; Bhamu, J. Multi-objective optimization of process parameters of fused deposition modeling (fdm) for printing polylactic acid (pla) polymer components. *Mater. Today Proc.* **2021**, *45*, 4880–4885. [CrossRef]
179. Wickramasinghe, S.; Do, T.; Tran, P. FDM-based 3d printing of polymer and associated composite: A review on mechanical properties, defects and treatments. *Polymers* **2020**, *12*, 1529. [CrossRef] [PubMed]
180. Chohan, J.S.; Singh, R. Pre and post processing techniques to improve surface characteristics of fdm parts: A state of art review and future applications. *Rapid Prototyp. J.* **2017**, *23*, 495–513. [CrossRef]
181. Hambali, R.H.; Cheong, K.M.; Azizan, N. Analysis of the influence of chemical treatment to the strength and surface roughness of FDM. *IOP Conf. Ser. Mater. Sci. Eng.* **2017**, *210*, 012063. [CrossRef]

182. Khosravani, M.R.; Schüürmann, J.; Berto, F.; Reinicke, T. On the post-processing of 3d-printed abs parts. *Polymers* **2021**, *13*, 1559. [CrossRef]
183. Castro-Casado, D. Chemical treatments to enhance surface quality of fff manufactured parts: A systematic review. *Prog. Addit. Manuf.* **2021**, *6*, 307–319. [CrossRef]
184. Jayanth, N.; Jaswanthraj, K.; Sandeep, S.; Mallaya, N.H.; Siddharth, S.R. Effect of heat treatment on mechanical properties of 3d printed pla. *J. Mech. Behav. Biomed. Mater.* **2021**, *123*, 104764. [CrossRef] [PubMed]
185. Singh, S.; Singh, M.; Prakash, C.; Gupta, M.K.; Mia, M.; Singh, R. Optimization and reliability analysis to improve surface quality and mechanical characteristics of heat-treated fused filament fabricated parts. *Int. J. Adv. Manuf. Technol.* **2019**, *102*, 1521–1536. [CrossRef]
186. Chalgham, A.; Ehrmann, A.; Wickenkamp, I. Mechanical properties of fdm printed pla parts before and after thermal treatment. *Polymers* **2021**, *13*, 1239. [CrossRef] [PubMed]
187. Chen, L.; Zhang, X.; Wang, Y.; Osswald, T.A. Laser polishing of cu/pla composite parts fabricated by fused deposition modeling: Analysis of surface finish and mechanical properties. *Polym. Compos.* **2020**, *41*, 1356–1368. [CrossRef]
188. Taufik, M.; Jain, P.K. Laser assisted finishing process for improved surface finish of fused deposition modelled parts. *J. Manuf. Processes* **2017**, *30*, 161–177. [CrossRef]
189. Li, G.; Zhao, J.; Jiang, J.; Jiang, H.; Wu, W.; Tang, M. Ultrasonic strengthening improves tensile mechanical performance of fused deposition modeling 3d printing. *Int. J. Adv. Manuf. Technol.* **2018**, *96*, 2747–2755. [CrossRef]
190. Wu, W.; Jiang, J.; Jiang, H.; Liu, W.; Li, G.; Wang, B.; Zhao, J. Improving bending and dynamic mechanics performance of 3d printing through ultrasonic strengthening. *Mater. Lett.* **2018**, *220*, 317–320. [CrossRef]
191. El Magri, A.; Vanaei, S.; Vaudreuil, S. An overview on the influence of process parameters through the characteristic of 3d-printed peek and pei parts. *High Perform. Polym.* **2021**, *33*, 862–880. [CrossRef]
192. Mohamed, A.S.; Maidin, S.; Mohamed, S.B.; Muhamad, M.K.; Wong, J.H.; Romlee, W.F. Improvement of surface finish by multiple piezoelectric transducers in fused deposition modelling. International Journal on Advanced Science. *Eng. Inf. Technol.* **2016**, *6*, 764–769. [CrossRef]
193. Carrell, J.; Gruss, G.; Gomez, E. Four-dimensional printing using fused-deposition modeling: A review. *Rapid Prototyp. J.* **2020**, *26*, 855–869. [CrossRef]
194. Kafle, A.; Luis, E.; Silwal, R.; Pan, H.M.; Shrestha, P.L.; Bastola, A.K. 3D/4D printing of polymers: Fused deposition modelling (fdm), selective laser sintering (sls), and stereolithography (sla). *Polymers* **2021**, *13*, 3101. [CrossRef] [PubMed]
195. Rajkumar, A.R.; Shanmugam, K. Additive manufacturing-enabled shape transformations via fff 4d printing. *J. Mater. Res.* **2018**, *33*, 4362–4376. [CrossRef]
196. Fu, P.; Li, H.; Gong, J.; Fan, Z.; Smith, A.T.; Shen, K.; Sun, L. 4D printing of polymeric materials: Techniques, materials, and prospects. *Prog. Polym. Sci.* **2022**, *126*, 101506. [CrossRef]

micromachines

Article

A Prognostics Method Based on Back Propagation Neural Network for Corroded Pipelines

Mingjiang Xie, Zishuo Li, Jianli Zhao and Xianjun Pei *

School of Mechanical Engineering, Southeast University, Nanjing 211189, China; mingjiang@seu.edu.cn (M.X.); LiZishuo@sjtu.edu.cn (Z.L.); 220210314@seu.edu.cn (J.Z.)
* Correspondence: xpei@seu.edu.cn

Abstract: A method that employs the back propagation (BP) neural network is used to predict the growth of corrosion defect in pipelines. This method considers more diversified parameters that affect the pipeline's corrosion rate, including pipe parameters, service life, corrosion type, corrosion location, corrosion direction, and corrosion amount in a three-dimensional direction. The initial corrosion time is also considered, and, on this basis, the uncertainties of the initial corrosion time and the corrosion size are added to the BP neural network model. In this paper, three kinds of pipeline corrosion growth models are constructed: the traditional corrosion model, the corrosion model considering the uncertainties of initial corrosion time and corrosion depth, and corrosion model also considering the uncertainties of corrosion size (length, width, depth). The rationality and effectiveness of the proposed prediction models are verified by three case studies: the uniform model, the exponential model, and the gamma process model. The proposed models can be widely used in the prediction and management of pipeline corrosion.

Keywords: pipeline corrosion; BP neural network; uncertainty; corrosion growth model

Citation: Xie, M.; Li, Z.; Zhao, J.; Pei, X. A Prognostics Method Based on Back Propagation Neural Network for Corroded Pipelines. *Micromachines* **2021**, *12*, 1568. https://doi.org/10.3390/mi12121568

Academic Editor: Aiqun Liu

Received: 19 November 2021
Accepted: 15 December 2021
Published: 16 December 2021

1. Introduction

With the global economy and industrialization developing rapidly, the demand for oil and natural gas gradually increases. The primary transportation method for the above two resources is pipeline transportation. However, because of the long-distance transportation of oil and gas pipelines, natural corrosion, third party damage, and other reasons, the pipeline's wall thickness will attenuate, thus affecting the service life of the pipelines [1–4]. Among many failure types, pipeline failure caused by corrosion defects accounts for a large proportion. Many pipelines have been in service for more than ten years, and some of their structures are seriously corroded. Pipeline accidents caused by corrosion defects occur from time to time, becoming a significant threat of the pipeline. These pipeline failures could potentially pollute the environment, waste energy, and threaten public lives and property safety.

To ensure the integrity of corroded pipelines, it is necessary to take corresponding measures to predict the remaining useful life (RUL) of the pipelines [3,5,6]. With the exploration and research of many scientific experts, pipeline corrosion detection and life prediction have been studied a lot. In recent years, breakthroughs have been made in data acquisition, machine learning, and other fields, providing new theoretical support and prognostics methods for the degradation of corrosion defects in pipelines. Continuously improving the accuracy of pipeline corrosion depth prediction and RUL prediction can provide additional benefits to the arrangement of inspection and maintenance actions of pipelines, and further reduce life-cycle costs.

Inline inspection (ILI) tools are widely used to detect and inspect the location and size of pipeline corrosion defects [7–10]. The accuracy of the ILI tool has a great impact on the prediction results. Simple and improved Monte Carlo simulations (MCS) [11–13] are used to calculate the failure probability of a section of corroded pipeline considering the uncertainty of corrosion process. The first-order reliability method (FORM) is also used to evaluate the reliability of pipelines by linearizing the relationship between reliability and parameters of pipelines [14,15]. With the increasing number of variables, MCS and FORM methods can be relatively time-consuming. Aiming at the complex nonlinear relationship between pipeline parameters and corrosion, Ozan [16] utilized artificial neural networks (ANNs) to predict the remaining useful life. Tian [17] took the pipeline's service life and state detection value as the input of the neural network and the life percentage as the output. The neural network model has the key advantage of dealing with nonlinear relationship between pipeline parameters and corrosion growth.

Back propagation (BP) neural network is one kind of artificial neural network that has high prediction accuracy and has been applied to predict the life of pipelines [18,19]. In this paper, a method based on BP neural network is used to simulate and predict corrosion defect growth. The related application of this method has been studied in some literature. Kai et al. [20] used the artificial neural network method to simulate the growth of corrosion defect, and evaluated the structural safety and reliability of pipeline. However, to simplify the structure of the neural networks, they only consider the internal pressure of the corroded pipeline. To assess the risk of the pipeline, Raeihagh et al. [21] established a fuzzy inference system (FIS), and applied the selected factors to the artificial neural networks (ANNs). Ben et al. [22] applied six artificial intelligence models, such as ANN, multivariate adaptive regression splines (MARS), and M5 tree (M5Tree) to study the relationship between the depth of corrosion and probable factors. These studies rarely consider the uncertainty of corrosion size and mostly ignore the initial corrosion time of the pipeline, which may produce inaccurate estimations. What is more, different manufacturing processes at different positions of the pipeline will also affect the corrosion growth of the pipeline. For example, high-speed particles will impact welded nodes [23], which have a relatively high risk. Similarly, only a few pieces of literature have made research on these factors. In this paper, we consider more diversified corrosion parameters, including pipe parameters, corrosion type, service life, corrosion location, corrosion direction, and corrosion size in a three-dimensional direction. In addition, considering the uncertainties of inspection data and initial corrosion time of pipeline, we build an ANN model for the degradation of corroded pipelines and consider the influence of other uncertainty sources to verify the effectiveness of the methodology.

The rest of the paper is organized as follows. Section 2 describes the structure, the modeling process, and the performance assessment of the BP neural network model. Section 3 presents data preprocessing and three prognostics models based on the BP neural network. In Section 4, three case studies with different corrosion growth models are used to demonstrate the effectiveness of the proposed models. Conclusions are presented in Section 5.

2. BP Neural Network Model

Since the relationship between input variables (including pipe properties, corrosion location, corrosion size, corrosion type, etc.) and the corrosion growth is very complex, finding a formula to describe the relationship is difficult. Considering that the BP neural network has strong ability to deal with nonlinear problems, as well as strong self-learning and self-adaptive abilities, a BP neural network is used to predict the corrosion growth of the pipeline. In this section, the structure, the modeling process, and the performance assessment of the BP neural network model will be described.

2.1. Structure of the BP Neural Network

Being composed of many neurons with operation functions, the structure of the BP neural network includes the input layer, hidden layer, and output layer [19], which is shown in Figure 1. The input layer consists of p neurons represented by x_i, $i = 1, 2, 3, \ldots, p$, where p is the number of input variables. The output layer consists of q neurons represented by y_j, $j = 1, 2, 3, \ldots, q$, where q is the number of output variables. Each node of the input layer is connected to all the nodes of the first hidden layer. Each node of the previous hidden layer is connected with all the nodes of the next hidden layer. Similarly, each node of the last hidden layer is connected to the all the nodes of the output layer. Each connection has a weight associated with it. In this paper, 11 input variables (service life, pipe segment length, pipe wall thickness, corrosion type, corrosion location (distance to upstream/downstream girth weld, inner/outer, clock direction), and corrosion size (length, width, depth)) and 3 output variables (corrosion growth coefficients (length, width, depth)) are used to construct the BP neural network. Thus, the BP neural network has 11 input neurons and 3 output neurons, which means $p = 11$, $q = 3$.

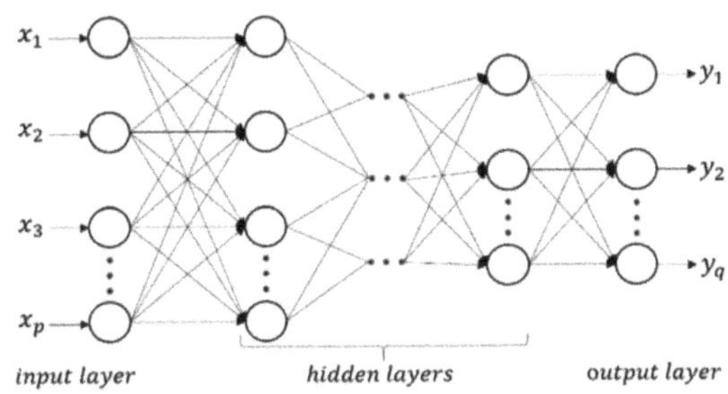

Figure 1. The structure of neural networks.

The output of the neuron in the first hidden layer, u_k^1, $k = 1, 2, \ldots H_1$, where H_1 is the number of neurons in the first hidden layer, is expressed as follows.

$$p_k^1 = \sum_{i=1}^{p} x_i z_{ik}^1 \quad (1)$$

$$u_k^1 = f\left(p_k^1\right) \quad (2)$$

where f is a monotonically increasing function, whose value is within (0, 1). $z^1 = \left[z_{11}^1, z_{21}^1, \ldots, z_{p1}^1, z_{12}^1, z_{22}^1, \ldots, z_{p2}^1, \ldots, z_{1H_1}^1, z_{2H_1}^1, \ldots, z_{pH_1}^1\right]$ is a vector of weights, and the initial values of these weights are all within $[-1, 1]$. Similarly, the output of the neuron in the i-th ($i > 1$) hidden layer, u_k^i, $k = 1, 2, \ldots H_i$, where H_i is the number of neurons in the i-th hidden layer, is expressed as follows,

$$p_k^i = \sum_{t=1}^{H_{i-1}} u_t^{i-1} z_{tk}^i \quad (3)$$

$$u_k^i = f\left(p_k^i\right) \quad (4)$$

where $z^i = \left[z_{11}^i, z_{21}^i, \ldots, z_{H_{i-1}1}^i, z_{12}^i, z_{22}^i, \ldots, z_{H_{i-1}2}^i, \ldots, z_{1H_i}^i, z_{2H_i}^i, \ldots, z_{H_{i-1}H_i}^i\right]$ is also a vector of weights whose initial values are within $[-1,1]$. The output of the model y_j, namely the output of the output neurons, is,

$$l_j = \sum_{s=1}^{H_{last}} u_s^{H_{last}} v_{sj} \quad (5)$$

$$y_j = f(l_j) \quad (6)$$

where $v = \left[v_{11}, v_{21}, \ldots, v_{H_{last}1}, v_{12}, v_{22}, \ldots, v_{H_{last}2}, \ldots, v_{1q}, v_{2q}, \ldots, v_{H_{last}q}\right]$ is a vector of weights within $[-1,1]$, and H_{last} is the number of neurons in the last hidden layer. In this paper, the sigmoid function is selected as the activation functions that can be used in (2), (4), and (6). This function is expressed as follows.

$$f(x) = \frac{1}{1 + e^{(-x)}} \quad (7)$$

The construction of a BP neural network is essentially the process of determining the weights of these connections. When a BP neural network works, it mainly transmits two kinds of data: the forward propagating signal and the back-propagating error. After the input data are obtained, its flow direction is taken from the input layer to the hidden layer, and then to the output layer. Then, the BP algorithm compares the actual outputs with the target outputs and the error is propagated in the opposite direction. The error is shared with each node of each layer, and the weight of each connection is adjusted until the objective function reaches the minimum value by using the back propagation learning rule. Then, the process of establishing the BP neural network is finished.

2.2. Modeling Process

The BP neural network is a data-driven model, and the modeling process is as follows:

- Obtain the database and determine the number of neurons in the input layer and output layer of the BP neural network;
- Randomly sort the collected data (data size = 11,103), and select = 70% of the samples (viz. 7772 data points) as the training samples. Then, the remaining samples are used as the testing samples;
- Train the BP neural network with the training samples and evaluate the performance of the model on the testing samples.

The BP model has two main limitations. The first one is the overfitting problem, which means the trained BP model has pretty high fitting precision on the training set, but has a relatively large prediction error on the testing set. In the proposed method, the target error of the BP model is not set too small, and the redundant samples are deleted. Then, this limitation is avoided. The second limitation is the inherent defect of the BP model and cannot be avoided. In the flat region of the gradient error surface, the variation of the weight is quite small, which makes the convergence of the BP model relatively slow. It spends more time in the training process.

In the process of establishing a BP neural network, the main content is to determine the neural network parameters, including the number of hidden layers, the number of nodes in each hidden layer, the learning rate, the learning objectives, and the frequency of training.

As for the determination of network parameters, we need to determine the number of hidden layers firstly. Then, we can get the number of nodes in each hidden layer according to the empirical formula, where H_i represents the number of neurons in the i-th hidden layer.

$$H_i = 2 \times i + 3 \quad (8)$$

In theory, the BP neural network of three hidden layers has a good fitting result. In this paper, using the collected data and a simple linear growth model, the BP neural network with one, two, three, four, five hidden layers are tested, respectively. The simulation result is the best when the BP neural network has four hidden layers. When the number of hidden layers is too large, such as five, the overfitting problem occurs. So, the number of hidden layers is set as four in this paper. Then, the number of neurons in these four hidden layers are 5, 7, 9, and 11.

Meanwhile, the other parameters of the BP neural network are proposed to use the default value [24]. The parameter selection of the BP neural network is shown in Table 1.

Table 1. Parameter selection of back propagation (BP) neural network.

Type	Parameter	Value
1	Number of hidden layer nodes	5,7,9,11
2	Learning rate	0.1
3	Learning objectives	10-7
4	Frequency of training	1000

Based on the chosen parameters, the initial BP neural network model is built. After training the BP neural network with the training samples, the weights of the connections between neurons are optimized and the final BP neural network structure is determined. Then, the BP neural network is evaluated on the testing samples to verify its validity. After that, the established BP neural network can be used to predict the corrosion growth of the pipeline.

2.3. Performance Assessment

During the BP neural network training and testing process, a measure is needed to be determined to represent the applicability of the model. Expected value and variance are usually adopted in many papers. In this paper, using the corrosion depth of the pipeline in a year as the contrast quantity, we draw the predicted value $\hat{x}_i$ and the actual value x_i on the same picture, and estimate the proximity between the predicted value and the expected actual value by analyzing the shape and trend of the curve. The mean square error (MSE) is selected to represent the performance of the model. The definition of MSE is as follows. The smaller this value is, the better the model is.

$$\text{MSE} = \frac{1}{n}\sum_{i=1}^{n}(\hat{x}_i - x_i)^2 \tag{9}$$

3. The Proposed Models Based on BP Neural Network

In this work, three kinds of pipeline corrosion growth models are constructed and compared. The model 1 is a traditional corrosion model using ANN. The model 2 is a proposed model considering the uncertainties of initial corrosion time and corrosion growth rates. In addition, model 3 is a proposed model which also considers the uncertainties of corrosion length, width, and depth. The traditional corrosion model is set up directly by ANN which is introduced in Section 2. The other two proposed models are introduced in this section.

3.1. Data Preprocessing

The data in this paper mainly come from the inspection and evaluation results of major pipelines by Sinopec pipeline storage and Transportation Co., Ltd. from 2015 to 2017. It records the corrosion type, service life, length of pipe segment, distance to upstream girth weld, size and clock direction of the corrosion, and other information, which is shown in Table 2.

Table 2. Sample of field data.

Sample	1	2	3	4	5
Corrosion type	Circumferential	General	Circumferential	Circumferential	General
Service life (year)	8	8	9	12	18
Length of a pipe segment (m)	12.0	5.7	11.0	11.0	11.9
Distance to upstream girth weld (m)	11.4	2.5	8.3	1.5	11.2
Distance to downstream girth weld (m)	0.6	3.2	2.7	9.5	0.7
Corrosion length (mm)	18.0	80.0	18.0	21.0	30.0
Corrosion width (mm)	87.0	68.0	62.0	75.0	106.0
Clock direction of corrosion	11:00	2:43	6:34	1:54	3:52
Pipe wall thickness (mm)	10.3	10.3	7.1	7.1	8.7
inner or outer	outer	outer	outer	inner	inner
Corrosion percentage	4	7	4	5	4

Before data processing, some field data, such as the corrosion type and the location of the corrosion, cannot be quantified, so they are classified and numbered before processing. Specifically, we classify four types of corrosion, pit corrosion is recorded as 1, general corrosion is recorded as 2, and the circumferential groove is recorded as 3. The corrosion position is marked as 1 for the inner wall and 2 for the outer wall. In industry, defect direction is denoted by clock, namely hour (h) and minute (m). In our model, the clock direction is converted to angle according to Equation (10).

$$\theta = 720 \times \frac{h \times 60 + m}{24 \times 60} \tag{10}$$

where θ is the angle of the corrosion defect. The corrosion depth d of the pipeline can also be calculated by using the corrosion percentage $a\%$ multiplied by wall thickness t, as shown in Equation (11).

$$d = t \times a\% \tag{11}$$

After preprocessing the collected field data, we can use these data as input random variables to construct the neural network model.

3.2. BP Neural Network Model Considering the Uncertainties of Initial Corrosion Time and Corrosion Depth

Because the pipeline does not begin to corrode immediately after being put into use, but takes time to begin to corrode, it is necessary to consider the uncertainty in initial corrosion time. Here, it is assumed that the initial time T_{initial} follows a normal distribution when the pipeline begins to corrode. The detail is as follow [25].

$$T_{\text{initial}} \sim N\left(T_0, std_{T_0}{}^2\right) \tag{12}$$

To construct this proposed ANN model, we first preprocess the input data, which are introduced in last section. Then, we calculate the actual corrosion time T_{actual} using Equation (13). To consider the real situation, the corrosion coefficient O_i related to corrosion rate in a short time is obtained according to the established neural network model. The corrosion amount of the pipeline is calculated and accumulated to obtain the corrosion model of the pipeline. Considering the uncertainties of measured value and actual corrosion rate, we assume that the corrosion coefficient follows the normal distribution represented by Equation (14). Then, the corrosion depth can be calculated using Equation (15).

$$T_{\text{actual}} = T - T_{\text{initial}} \tag{13}$$

$$\widetilde{O}_i \sim N\left(O_i, 2.5 \times 10^{-3} O_i{}^2\right) \tag{14}$$

$$D(t) = D_0 + \sum_{i=1}^{t} \phi\left(\widetilde{O}_i, i\right) \tag{15}$$

where the variable T is the service time of pipeline; O_i represents the output corrosion coefficient of the ith year from the neural network; and D_0 is the initial corrosion depth. The function ϕ represents the relationship between corrosion coefficient and corrosion growth rate, so the corrosion coefficient can be taken as the corrosion rate especially for linear growth corrosion. What is more, to reduce the accidental error, the neural network is trained ten times, and the average of training results is used as the prediction result.

3.3. BP Neural Network Considering the Uncertainties in Corrosion Size

Due to the limitations in inline inspection tools, there exist measurement errors in detected corrosion size. Hence, it is necessary to consider the uncertainties in corrosion size. In this proposed model, in addition to considering the uncertainties mentioned in model 2, the uncertainties in corrosion size (length, width, depth) are also added to the BP neural network model. The corrosion depth, width, and length can then be calculated by Equations (16)–(18).

$$D(t) = D_0 + \sum_{i=1}^{t} \phi\left(\widetilde{O}_D, i\right) \tag{16}$$

$$L(t) = L_0 + \sum_{i=1}^{t} \phi\left(\widetilde{O}_L, i\right) \tag{17}$$

$$W(t) = W_0 + \sum_{i=1}^{t} \phi\left(\widetilde{O}_W, i\right) \tag{18}$$

where D, L, and W represent the corrosion amount of the pipeline in the direction of depth, length, and width, respectively; and D_0, L_0, and W_0 correspond to the initial corrosion depth, length, and width, respectively.

According to a selected sample, we calculate the corrosion coefficient in three corrosion directions firstly. We assume that there are fixed corrosion coefficients $\widetilde{O}_w$ and $\widetilde{O}_l$ in the width and length directions, respectively. Here, we can calculate the corresponding corrosion amounts of $W(t)$ and $L(t)$. Then, the variations of corrosion amount in these two parameters are included in the input data of the BP neural network, and the corrosion coefficient in the depth direction is the output parameter. By substituting the corrosion coefficient into Equation (16), the corrosion depth can be obtained for further risk analysis. In each simulation run, the variations in the corrosion length and width of the test sample over time are added to the sample data. We use the same input data as training data to obtain this proposed model 3 for future comparisons.

4. Case Studies

4.1. General Information

In this section, examples are used to demonstrate the effectiveness of the proposed models. The comparison results for the three corrosion models can be used for the subsequent reliability evaluation and risk analysis of pipelines. Table 3 summarizes the differences among three BP neural network models. The pipeline failure caused by a corrosion defect is mainly because the corrosion depth reaches the critical value. So, in the following case studies, we mainly focus on the growth of corrosion depth rather than length and width. Based on the field data, we investigate three types of corrosion depth growth models. In the first case, the depth of corrosion increases with time linearly. In the second case, corrosion growth follows an exponential distribution. As for the third case, the growth of corrosion depth in each period conforms to the gamma growth process.

Table 3. Three gradually improved models.

Uncertainty	Model 1	Model 2	Model 3
The initial time of corrosion	no	yes	yes
Corrosion depth	no	yes	yes
Corrosion length and width	no	no	yes

4.2. Case Study 1: Uniform Corrosion Hypothesis

The growth of the defect depth is characterized by:

$$d(t) = d_0 + g_d t \tag{19}$$

where d_0 represents the initial corrosion amount and g_d is the growth rate of corrosion depth. g_d is used as the output parameter in the neural network model. When considering the uncertainty, we assume that g_d follows the normal distribution, that the actual corrosion depth growth rate conforms to the theoretical value, and that the variance is 0.05 times the theoretical value.

4.2.1. Traditional Linear Corrosion Growth Model (Model 1)

The BP neural network is used to simulate the corrosion depth of the pipeline, and the results are shown in Figure 2.

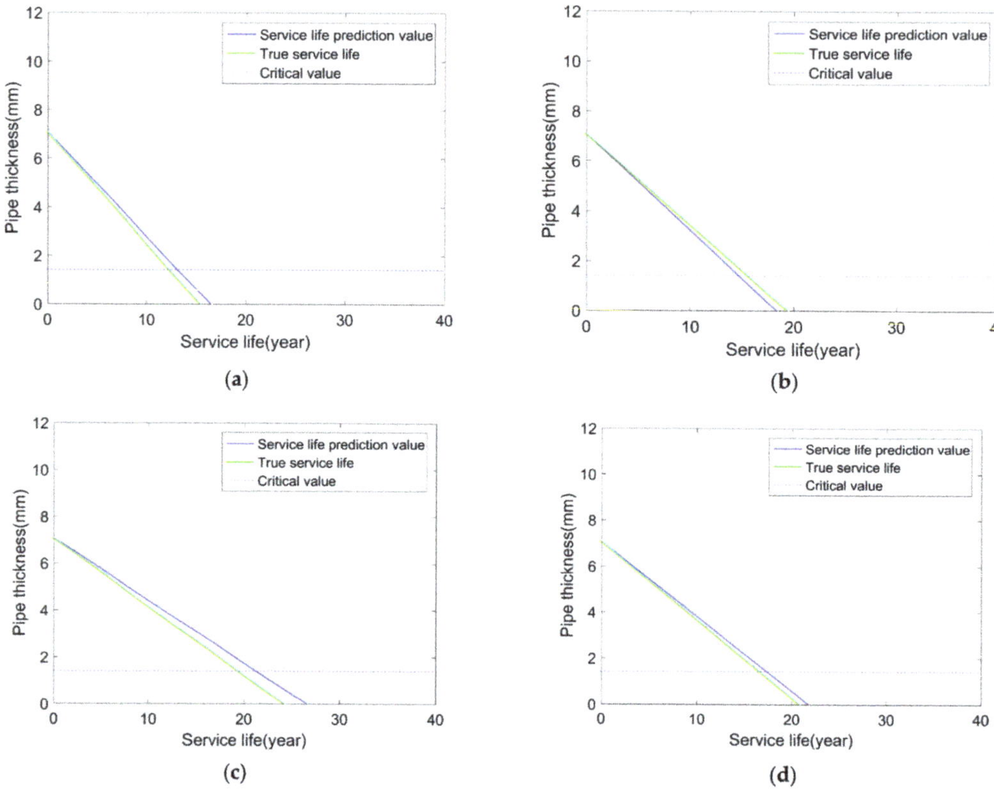

Figure 2. Comparison results of pipeline life prediction: (**a**) sample 1; (**b**) sample 2; (**c**) sample 3; (**d**) sample 4.

As can be seen from the figure, when the BP neural network method is used to predict pipeline corrosion, the prediction results are promising. The predicted corrosion depth growth rates are relatively close to the theoretical growth rate, which can illustrate the great potential of the BP neural network method in predicting pipeline remaining useful life.

4.2.2. Linear Corrosion Growth Model Considering the Uncertainties of Initial Corrosion Time and Corrosion Depth (Model 2)

With the uncertainties of initial corrosion time and corrosion depth in the model 2, and the prediction results of the model 2 is observed and compared with model 1 in Figure 3. The summary of comparison results is shown in Table 4.

After 20 simulation runs of the corresponding network, the service life of the pipeline is shown in Table 4.

To facilitate the comparison, take the absolute value of error for calculation. It can be obtained that the standard deviation of the error of model 2 is 1.1806, and that the standard deviation of the uniform corrosion of model 1 is 1.7084. Thus, it can be concluded that the neural network model 2 is better than the previous model 1. As can be seen from the above figure and table, the prediction results of model 2 are closer to the real values than model 1, which fully illustrates that the proposed model 2 has better performance of predicting pipeline corrosion growth and remaining useful life.

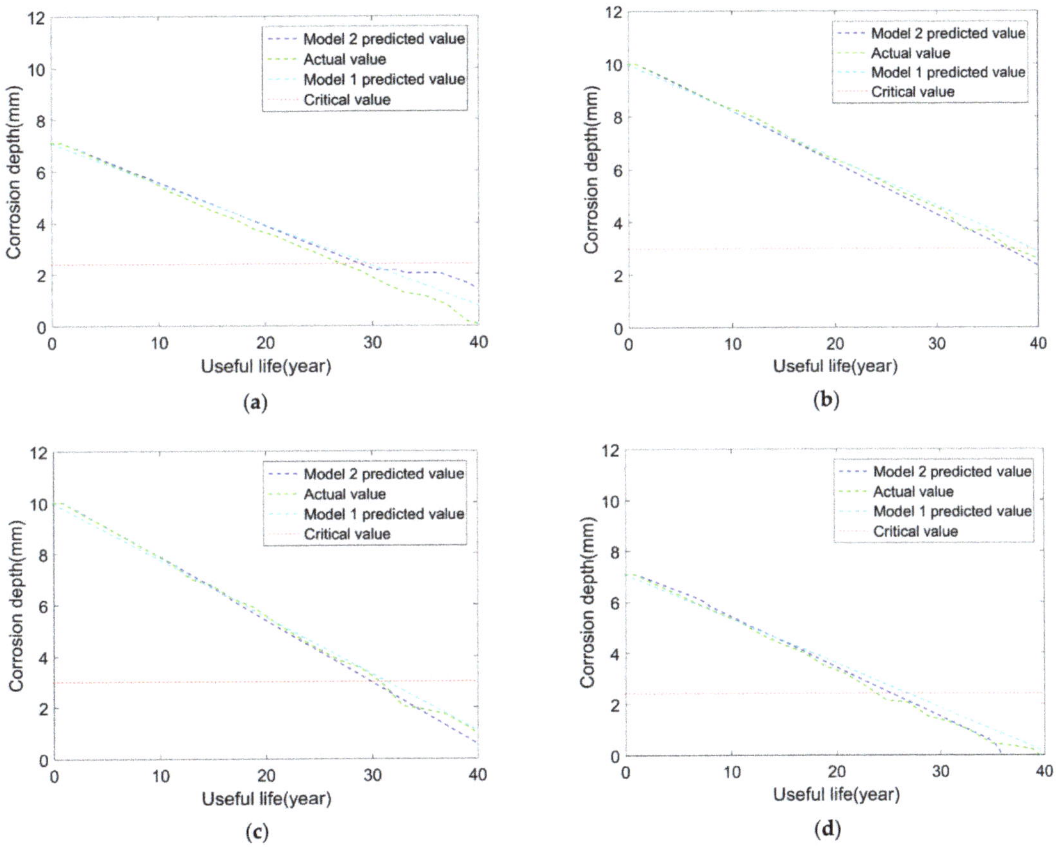

Figure 3. Comparison results of real value, model 1 and model 2: (**a**) sample 1; (**b**) sample 2; (**c**) sample 3; (**d**) sample 4.

Table 4. Analysis of the results of pipeline service life.

No.	Real Service Life	The Outcome of Model 2	The Outcome of Model 1	The Error of Model 2	The Error of Model 1
1	23.9248	25.4248	26.8248	1.5	1.4
2	30.9869	30.5869	32.2869	0.4	1.7
3	20.5586	21.3586	21.9586	0.8	0.6
4	37.4537	36.7537	39.2537	0.7	2.5
5	27.0945	28.8945	29.5945	1.8	0.7
6	25.8102	22.6102	28.6102	3.2	6
7	26.6273	26.0273	28.3273	0.6	2.3
8	35.5698	40.0098	38.0698	4.4302	1.9302
9	37.1356	35.7356	38.4356	1.4	2.7
10	19.6039	18.6039	21.3039	1	2.7
11	32.9597	32.1597	34.0597	0.8	1.9
12	39.095	38.495	40.1	0.6	1.605
13	25.6869	26.6869	26.8869	1	0.2
14	25.0132	26.0132	27.0132	1	1
15	30.8332	30.0332	31.6332	0.8	1.6
16	33.9482	33.0482	35.0482	0.9	2
17	21.4839	20.6389	23.7839	0.845	3.145
18	26.3141	23.1141	29.8141	3.2	6.7
19	21.0338	23.3338	23.6338	2.3	0.3
20	33.499	32.9099	33.1099	0.5891	0.2

4.2.3. Linear Corrosion Growth Model Considering the Uncertainties of Corrosion Size (Model 3)

We build the model 3 based on the previous model 2 and add the uncertainties in corrosion length and width in the BP neural network. The results are shown in Figure 4. Furthermore, to compare the prediction results of the three models more clearly, the mean of squared errors (MSE) between the predicted value and real values is calculated in Table 5.

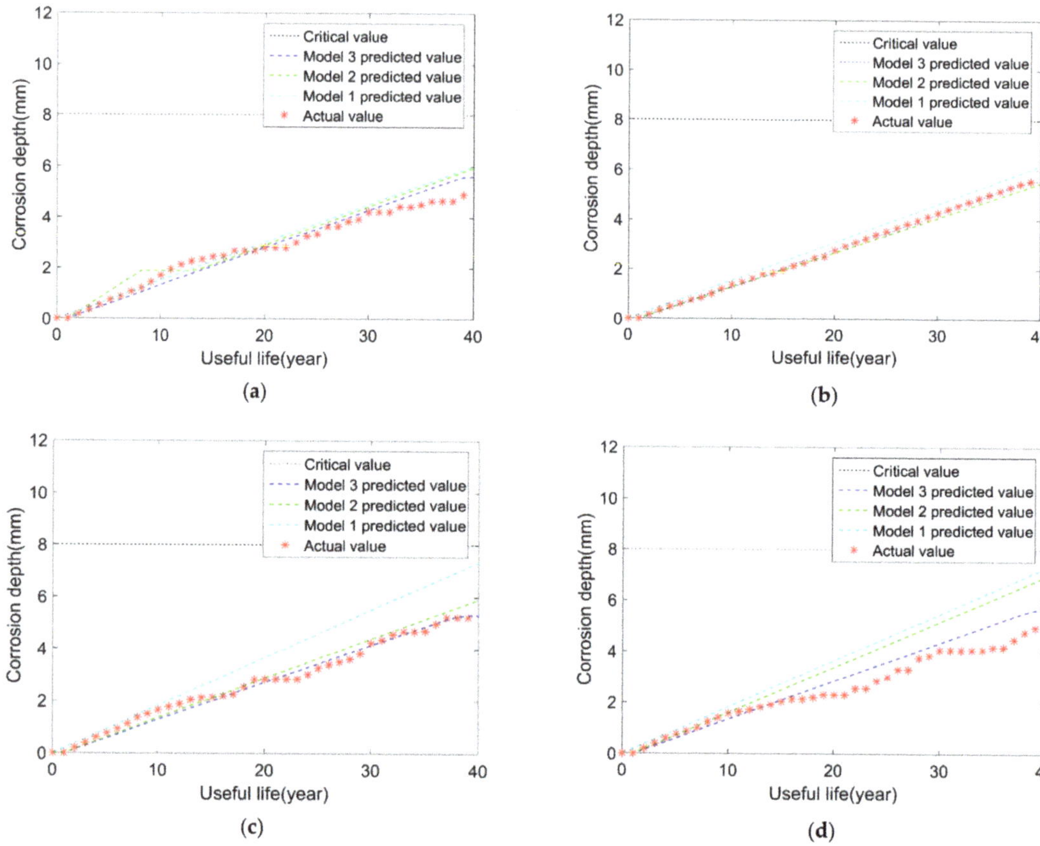

Figure 4. Comparison results of the three models and the real results: (**a**) sample 1; (**b**) sample 2; (**c**) sample 3; (**d**) sample 4.

Table 5. Mean of squared errors (MSE) between the predicted value and real value.

No	The Outcome of Model 1	The Outcome of Model 2	The Outcome of Model 3
1	0.3593	0.2504	0.2026
2	0.2131	0.0846	0.0424
3	0.7162	0.1527	0.1077
4	0.7799	0.6848	0.3004

From the comparison results in the above figures, it can be seen that, after considering the variations in the length and width of corrosion over time, the results of model 3 are closer to the true values compared with model 1 and model 2. At the same time, the above table shows that the MSEs of model 3 are smaller than model 2 and model 1. What is more, the results of model 2 are better than model 1. Considering multiple uncertainty sources, it can be explained that the simulated effects of the above three neural network models are gradually getting closer to reality. Consequently, the proposed BP neural network models produce better results and can make a more accurate pipeline remaining useful life prediction than the traditional BP neural network model (model 1).

4.3. Case Study 2: Exponential Model Hypothesis

The three BP neural network models can also be used to predict the exponential corrosion growth model of the pipeline. The growth of the defect depth is characterized by:

$$d(t) = d_0 + kt^{\gamma} \tag{20}$$

where d_0 is the initial corrosion depth, and k and γ are the parameters of the exponential corrosion growth model. Because the variability of k is usually large, and the variability of γ is usually small [26], the value of γ is selected as 0.5. Then, the model parameter k is used as the output of neural network models.

4.3.1. Traditional Exponential Growth Model (Model 1)

No uncertainty is considered in model 1. When constructing the exponential growth model of the pipeline, it is assumed that the time index of the real corrosion rate obeys a normal distribution with a mean value of 0.5 and a variance of 0.05. At the same time, a proportional coefficient is constructed using real corrosion data. We obtain the simulation results every five years. The corresponding results are shown in Figure 5. The simulated results accord with the trend of actual value, but there are big differences as useful life increases.

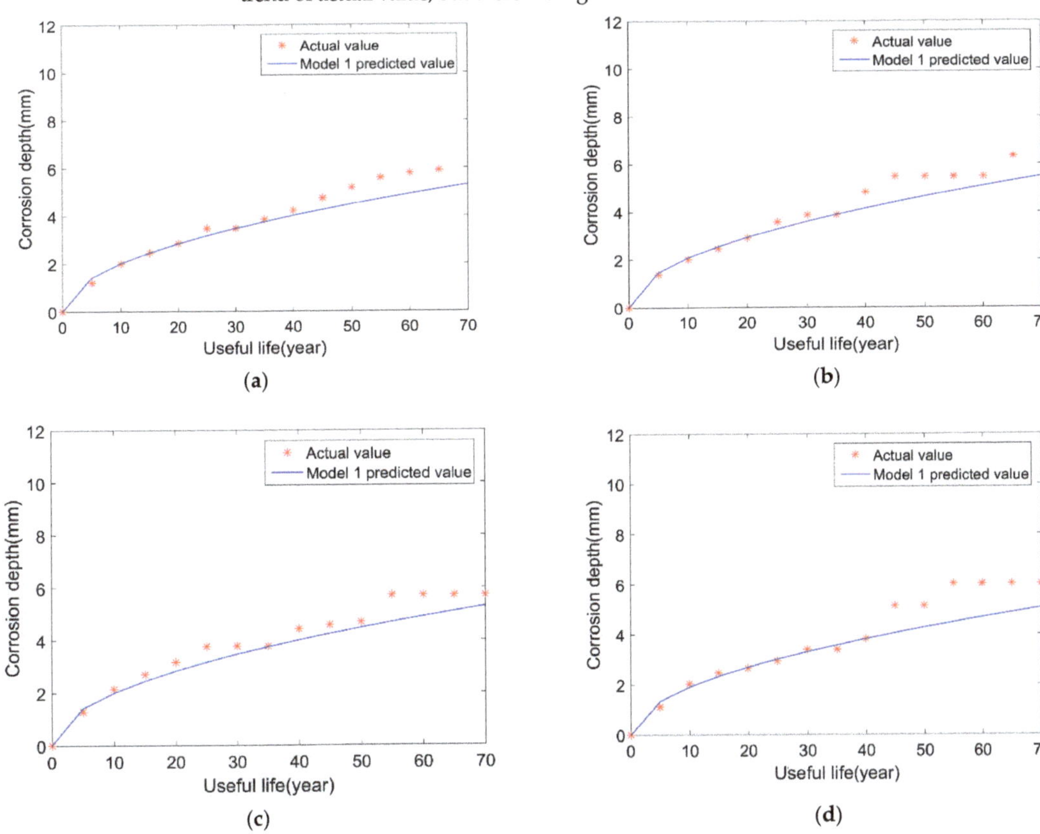

Figure 5. Comparison results of model 1 and the real results: (**a**) sample 1; (**b**) sample 2; (**c**) sample 3; (**d**) sample 4.

4.3.2. Exponential Corrosion Growth Model Considering the Uncertainties in Initial Corrosion Time, Corrosion Index, and Scale Factor (Model 2)

In the proposed model 2, we consider the uncertainties of the initial corrosion time, the corrosion index, and the scale factor in the BP neural network. The simulation results are evaluated every five years, and the results are compared and analyzed, as shown in Figure 6 and Table 6.

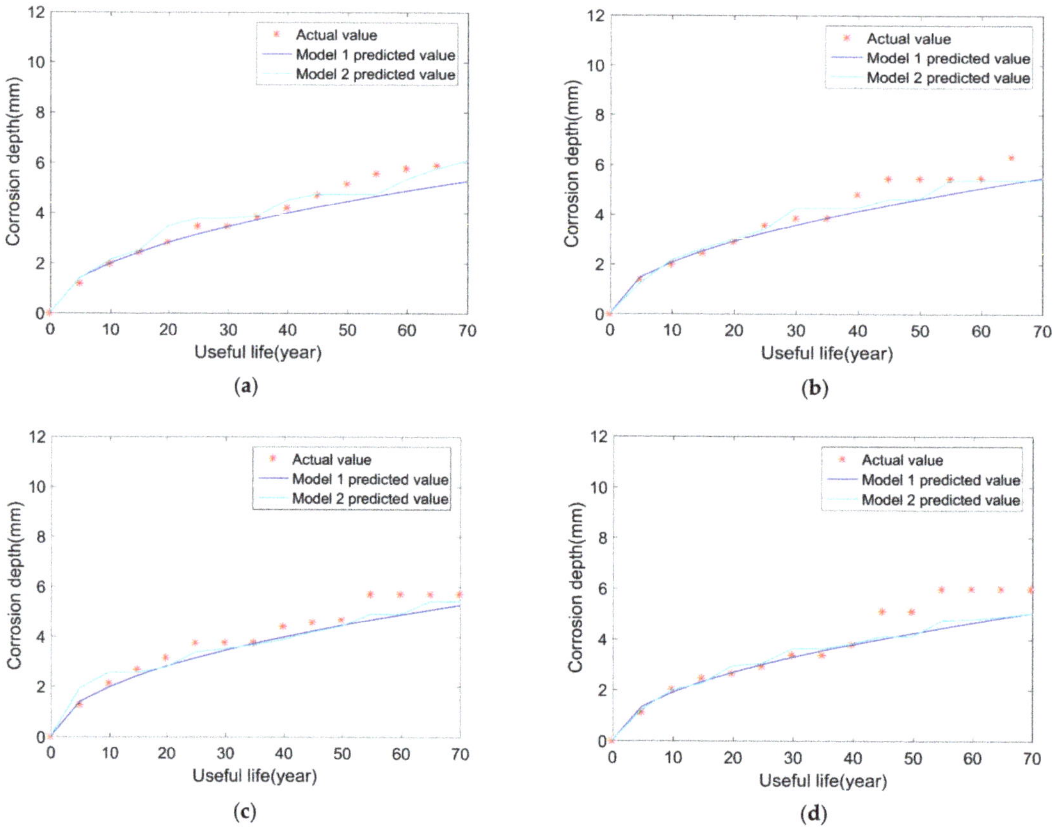

Figure 6. Comparison results of the two models and the real results: (**a**) sample 1; (**b**) sample 2; (**c**) sample 3; (**d**) sample 4.

Table 6. Variance of the difference between the predicted value and real value.

	Sample 1	Sample 2	Sample 3	Sample 4
Model 1	0.2963	0.3689	0.2957	0.4914
Model 2	0.2062	0.3416	0.2770	0.4319

It can be seen from the figures that, among the above four samples, the simulated results of the four figures show that the prediction results generated by model 2 are closer to the true value than model 1. In addition, from Table 6, the standard deviations of model 2 are smaller than model 1, which indicates that model 2 performs better than model 1 in the pipeline corrosion depth growth prediction. However, the prediction results are still not accurate enough for the pipeline remaining useful life prediction.

4.3.3. Exponential Corrosion Growth Model Considering the Uncertainties of Corrosion Size (Model 3)

In addition to considering the uncertainties of initial corrosion time and relevant corrosion depth parameters, the variations in length and width over time are also added to the BP neural network model. Then we compare the results of this new model (model 3) with model 1 and model 2. The results are shown in Figure 7. The differences between the predicted value and real values are also summarized in Table 7.

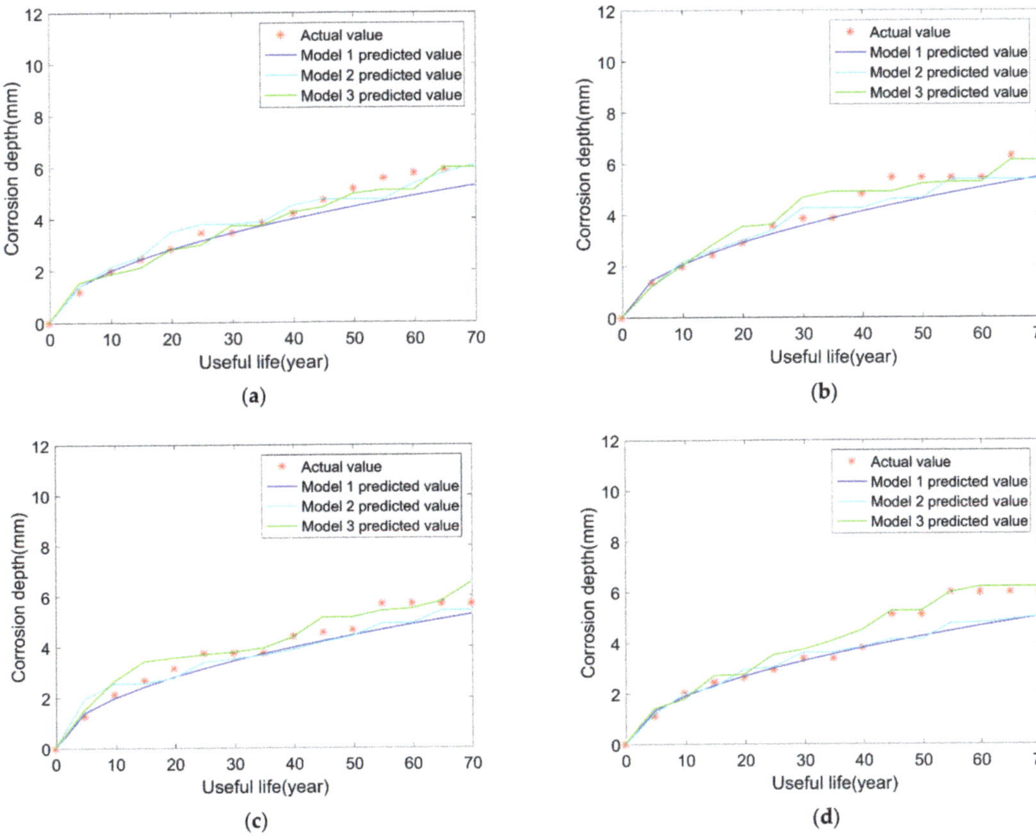

Figure 7. Comparison results of the three models and the real results: (**a**) sample 1; (**b**) sample 2; (**c**) sample 3; (**d**) sample 4.

Table 7. Variance of the difference between the predicted value and real value.

	Sample 1	Sample 2	Sample 3	Sample 4
Model 1	0.2963	0.3689	0.2957	0.4914
Model 2	0.2062	0.3416	0.2770	0.4319
Model 3	0.1664	0.3014	0.2664	0.2429

It also clearly shows that, after considering the changes in the length and width of corrosion over time, the results of model 3 are closer to the true values. In other words, from model 1 to model 3, accuracy and stability are gradually enhanced. Furthermore, compared with model 1 and model 2, the prediction accuracy for model 3 increase a lot, which concludes the uncertainties in corrosion length and width do affect the growth of corrosion depth a lot.

4.4. *Case Study 3: Gamma Distribution Hypothesis*

4.4.1. Traditional Gamma Corrosion Growth Model (Model 1)

In this gamma growth model, it is assumed that the amount of corrosion per year obeys a gamma distribution. The growth of the defect depth is characterized by:

$$d(t) = d_0 + d_g(t) \tag{21}$$

$$F(d_g(t)\Big|\alpha, \beta, t) = \beta^{\alpha t}\left(d_g\right)^{\alpha t-1} exp(-\beta d_g\Big) / \Gamma(\alpha t) \tag{22}$$

where $d_g(t)$ denotes the homogeneous gamma process. The probability density function of $d_g(t)$ is given by Equation (22), where α and β are shape and scale parameters of gammar process, respectively. Moreover, in our paper, the scale parameter β is assumed to be 48 according to the specific value. We use the shape parameter α as the main output parameter of the BP neural network. According to this notion, we establish a BP neural network model, and the results are shown in Figure 8. Furthermore, the prediction results have the same trend with the actual values, but there still are some prediction errors.

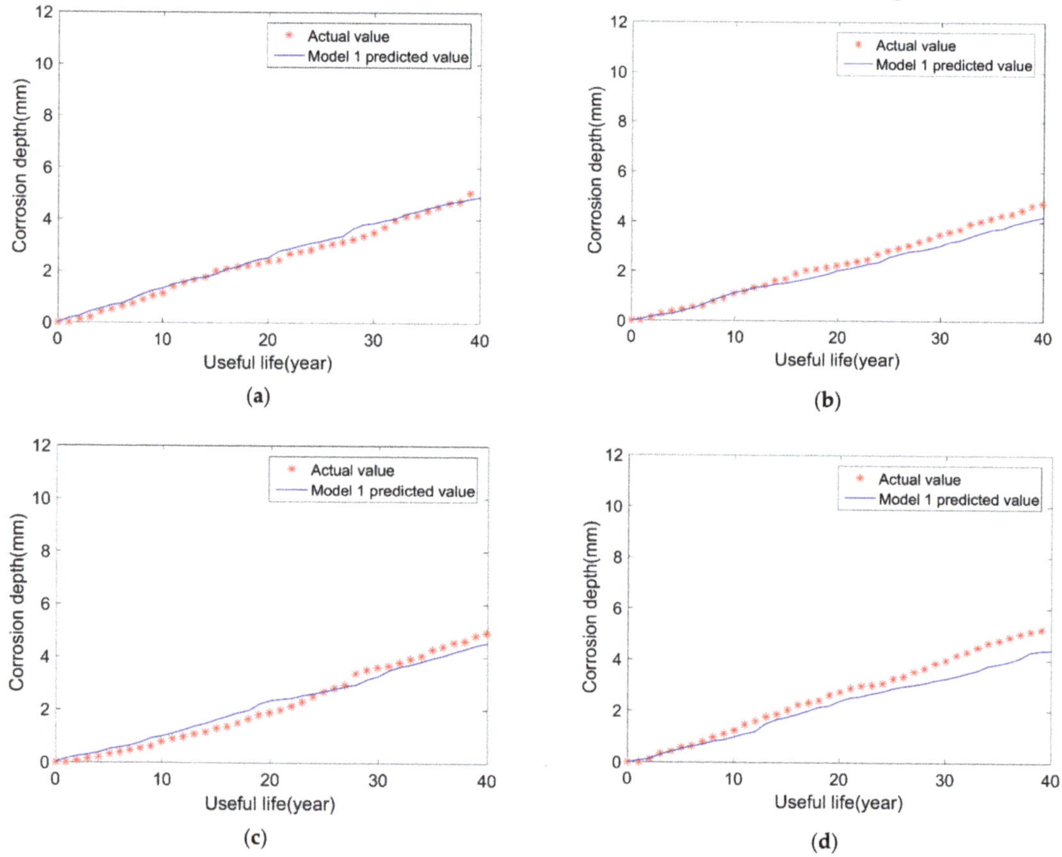

Figure 8. Comparison results of model 1 and the real results: (**a**) sample 1; (**b**) sample 2; (**c**) sample 3; (**d**) sample 4.

4.4.2. Gamma Process Corrosion Growth Model Considering the Uncertainties of Initial Corrosion Time and the Shape Parameters (Model 2)

In this model, the uncertainties of the initial corrosion time and the shape parameters are considered in the BP neural network. We compare this model 2 with model 1 and the actual value. These results are shown in Figure 9.

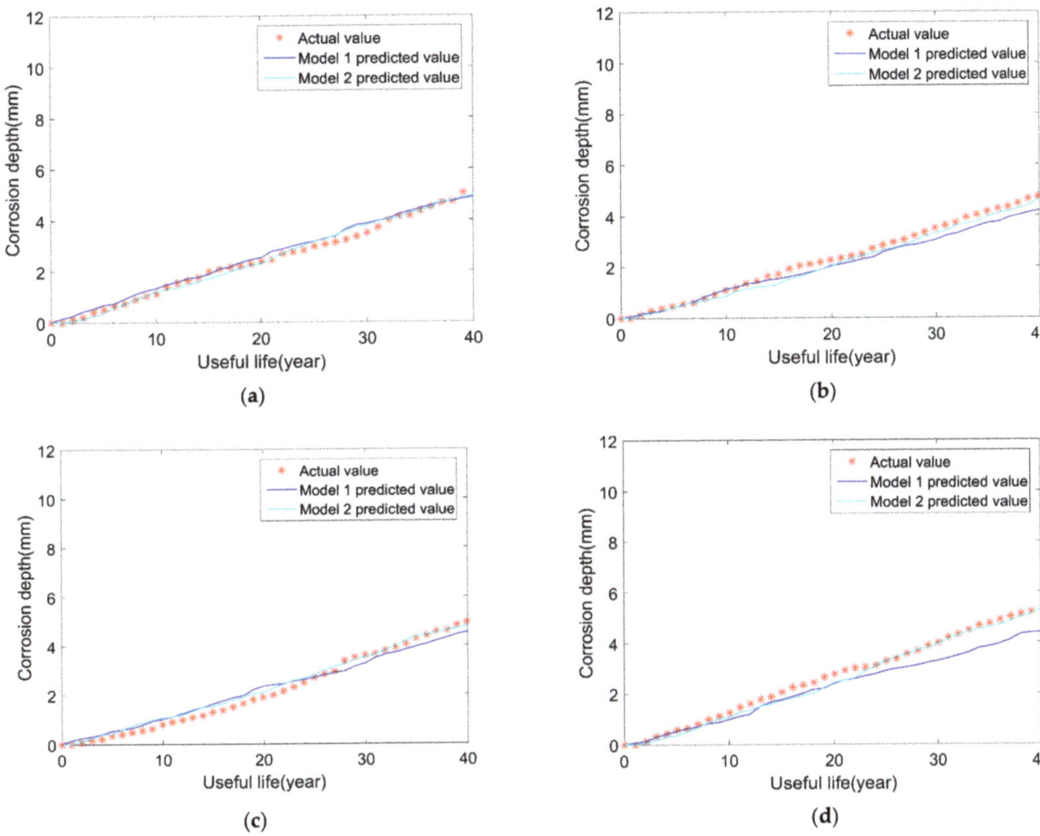

Figure 9. Comparison results of the two models and the real results: (**a**) sample 1; (**b**) sample 2; (**c**) sample 3; (**d**) sample 4.

The tendency of the results is the same as the uniform corrosion model and exponential model. The model which considers uncertainties of initial corrosion time and shape parameters has better prediction results.

4.4.3. Gamma Corrosion Growth Model Considering the Uncertainties of Corrosion Size (Model 3)

The steps are the same as linear and exponential growth models, and we also consider the uncertainties in corrosion size over time in the proposed BP neural network. The comparison results for models 1, 2, and 3 are shown in Figure 10. The differences between the predicted value and real values are also summarized in Table 8.

After considering the uncertainties in the length and width of corrosion over time, the results of model 3 are closer to the actual values. Additionally, from model 1 to model 2 to model 3, accuracy and stability are gradually enhanced. As linear, exponential and Gamma corrosion growth models can be used to describe most corroded the pipelines' degradation working cases. So, we take these three hypothetical growth models as examples, and our proposed BP neural network models' rationality, effectiveness, and universal applicability are verified.

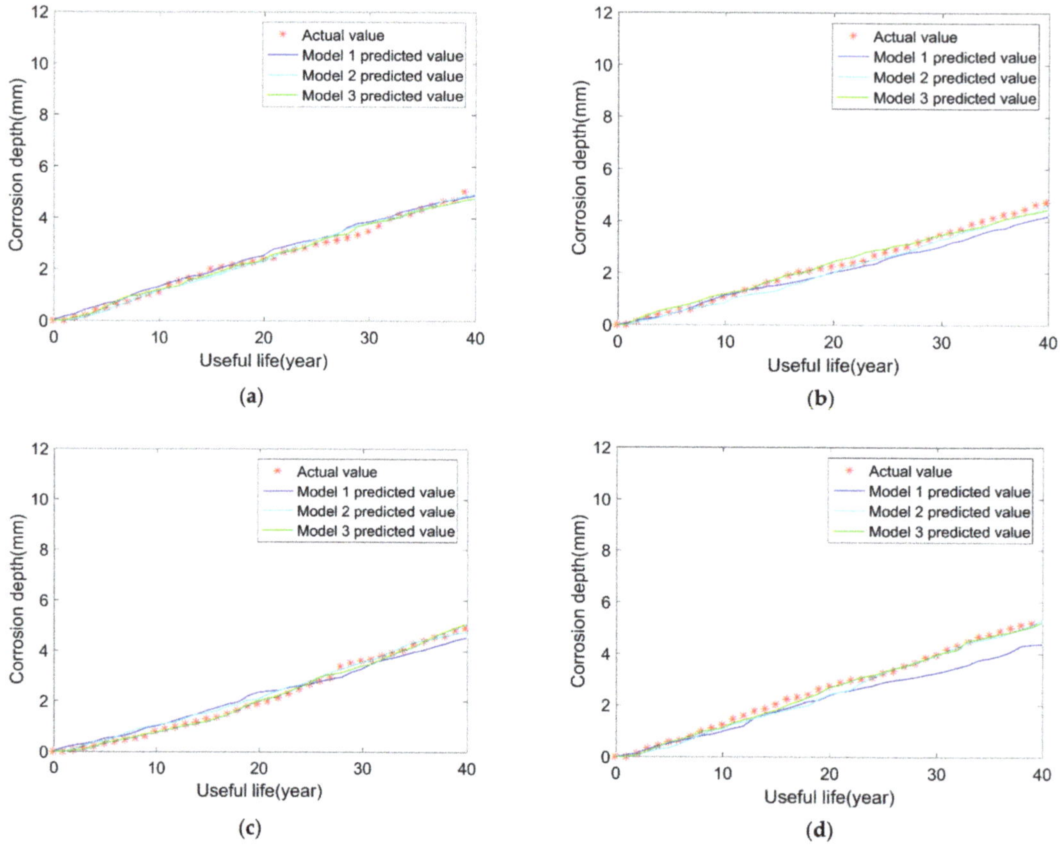

Figure 10. Comparison results of the three models and the real results: (**a**) sample 1; (**b**) sample 2; (**c**) sample 3; (**d**) sample 4.

Table 8. Variance of the difference between the predicted value and real value.

	Sample 1	Sample 2	Sample 3	Sample 4
Model 1	0.1120	0.1681	0.1172	0.3019
Model 2	0.1057	0.1121	0.1028	0.1324
Model 3	0.0880	0.0858	0.0687	0.0754

5. Conclusions

This paper proposed BP neural network models for pipeline useful life prediction considering the uncertainties in initial corrosion time and corrosion size. Furthermore, we use the field data from the Sinopec Pipeline Storage and Transportation Co., Ltd. to demonstrate the effectiveness of the proposed model. We first preprocess the collected field data, and we can use these data (pipe parameters, corrosion location, corrosion size, etc.) as input random variables to construct the neural network model. Three gradually improved models are considered in the pipeline RUL prediction, and the uncertainties are added to each model to make the hypothetical pipeline corrosion situation closer to reality. At the same time, by comparing the results of the neural network with the real values, it can be seen that the results are relatively close, which fully illustrates the effectiveness and the rationality of the BP neural network method in predicting the corrosion degree of pipelines. Three proposed BP neural network models are compared with actual values, a corresponding comparative analysis of the results shows that the model 3 which considers the uncertainties from corrosion initial time and corrosion size produce more accurate prediction results. Lastly, we use three case studies to demonstrate the effectiveness of the proposed models. Three types of corrosion growth models, namely uniform, exponential model, and gamma process models, are applied to the proposed models mentioned above. The comparison results prove that the proposed models have universal applicability to different working conditions.

Author Contributions: Conceptualization, Z.L. and J.Z.; methodology, M.X.; software, Z.L. and M.X.; validation, Z.L., J.Z. and M.X.; formal analysis, M.X. and X.P.; investigation, M.X.; resources, M.X.; data curation, M.X.; writing—original draft preparation, Z.L. and J.Z.; writing—review and editing, M.X.; visualization, M.X. and X.P.; supervision, M.X. and X.P.; project administration, M.X. and X.P.; funding acquisition, M.X. and X.P. All authors have read and agreed to the published version of the manuscript.

Funding: This research was funded by the National Natural Science Foundation of China, grant number 72001039, 71671035, 12102090.

Data Availability Statement: The data presented in this study are available on request from the corresponding author.

Conflicts of Interest: The authors declare no conflict of interest.

References

1. Chio, L.; Wenxing, Z. Statistical Analyses of Incidents on Onshore Gas Transmission Pipelines Based on PHMSA Database. *Int. J. Press. Vessel. Pip.* **2016**, *145*, 29–40.
2. Nessim, M.; Zhou, W.; Zhou, J.; Rothwell, B. Target Reliability Levels for Design and Assessment of Onshore Natural Gas Pipelines. *J. Press. Vessel Technol.* **2009**, *131*, 061701. [CrossRef]
3. Xie, M.; Tian, Z. A Review on Pipeline Integrity Management Utilizing In-Line Inspection Data. *Eng. Fail. Anal.* **2018**, *92*, 222–239. [CrossRef]
4. Askari, M.; Aliofkhazraei, M.; Afroukhteh, S. A Comprehensive Review on Internal Corrosion and Cracking of Oil and Gas Pipelines. *J. Nat. Gas Sci. Eng.* **2019**, *71*, 102971. [CrossRef]
5. Young-Do, J.; Bum, J.A. A Method of Quantitative Risk Assessment for Transmission Pipeline Carrying Natural Gas. *J. Hazard. Mater.* **2005**, *123*, 1–12.
6. Wang, Z.-Q.; Hu, C.-H.; Si, X.-S.; Zio, E. Remaining Useful Life Prediction of Degrading Systems Subjected to Imperfect Maintenance: Application to Draught Fans. *Mech. Syst. Signal Process.* **2018**, *100*, 802–813. [CrossRef]
7. Moon, J.K.; Lee, S.H.; Lee, I.W.; Park, G.S.; Lee, M.H. Defect Depth Estimation Using Magnetic Flux Leakage Measurement for In-Line Inspection of Pipelines. *J. Sens. Sci. Technol.* **2006**, *15*, 328–333.
8. Hector, H.D.; Caspar, G.; Rutger, K.; Richard, L.; Doekle, Y.; Tiedo, T.; Remko, A. Ultrasonic Inline Inspection of a Cement-Based Drinking Water Pipeline. *Eng. Struct.* **2020**, *210*, 110413.
9. Xie, M.; Sutherland, J.; Fang, B.; Gu, B.; Tian, Z. Impact Analysis of Inline Inspection Accuracy on Pipeline Integrity Planning. *J. Press. Vessel Technol.* **2020**, *142*, 061801. [CrossRef]
10. Siraj, T.; Zhou, W. Quantification of Measurement Errors in the Lengths of Metal-Loss Corrosion Defects Reported by Inline Inspection Tools. *J. Press. Vessel Technol.* **2019**, *141*, 061402. [CrossRef]
11. Wenxing, Z. System Reliability of Corroding Pipelines. *Int. J. Press. Vessel. Pip.* **2010**, *87*, 587–595.
12. Bernt, J.L.; Arvid, N.; Ole, E.B.N. Reliability Analysis of Corroding Pipelines by Enhanced Monte Carlo Simulation. *Int. J. Press. Vessel. Pip.* **2016**, *144*, 11–17.
13. Arun, K.C.; Vijay, K. A Bayesian Analysis of Perks Distribution via Markov Chain Monte Carlo Simulation. *Nepal J. Sci. Technol.* **2013**, *14*, 153–166.
14. Changqing, G.; Wenxing, Z. First-Order Reliability Method-Based System Reliability Analyses of Corroding Pipelines Considering Multiple Defects and Failure Modes. *Struct. Infrastruct. Eng.* **2017**, *13*, 1451–1461.
15. Zhou, W.; Gong, C.; Hong, H.P. New Perspective on Application of First-Order Reliability Method for Estimating System Reliability. *J. Eng. Mech.* **2017**, *143*, 04017074. [CrossRef]
16. Ozan, K. Bayesian Learning Based Gaussian Approximation for Artificial Neural Networks. *Turk. J. Forecast.* **2017**, *1*, 54–65.
17. Zhigang, T. An Artificial Neural Network Method for Remaining Useful Life Prediction of Equipment Subject to Condition Monitoring. *J. Intell. Manuf.* **2012**, *23*, 227–237.
18. Weikuan, J.; Dean, Z.; Tian, S.; Shifei, D.; Yuyan, Z.; Chanli, H. An Optimized Classification Algorithm by BP Neural Network Based on PLS and HCA. *Appl. Intell.* **2015**, *43*, 176–191.
19. Zhu, H.; Qi, Y.; Zhang, F.; Song, Y.; Lin, L.; Zhang, J. Calculation Method of Production Pressure Drop Based on BP Neural Network Velocity Pipe String Production in CBM Wells. *IOP Conf. Ser. Earth Environ. Sci.* **2020**, *619*, 012044. [CrossRef]
20. Kai, W.; Lei, H.; Jing, L.; Jing, G. An Optimization of Artificial Neural Network Modeling Methodology for the Reliability Assessment of Corroding Natural Gas Pipelines. *J. Loss Prev. Process Ind.* **2019**, *60*, 1–8.

21. Raeihagh, H.; Behbahaninia, A.; Macki, A.M. Risk Assessment of Sour Gas Inter-Phase Onshore Pipeline Using ANN and Fuzzy Inference System–Case Study: The South Pars Gas Field. *J. Loss Prev. Process Ind.* **2020**, *68*, 104238. [CrossRef]
22. Ben, S.M.E.A.; Keshtegar, B.; Taleb-Berrouane, M.; Abbassi, R.; Trung, N.-T. Advanced Intelligence Frameworks for Predicting Maximum Pitting Corrosion Depth in Oil and Gas Pipelines. *Process Saf. Environ. Prot.* **2021**, *147*, 818–833.
23. Qiao, Q.; Guangxu, C.; Yun, L.; Wei, W.; Haijun, H.; Hui, H. Corrosion Failure Analyses of an Elbow and an Elbow-to-Pipe Weld in a Natural Gas Gathering Pipeline. *Eng. Fail. Anal.* **2017**, *82*, 599–616. [CrossRef]
24. Tawee, C. Investigation Internal Parameters of Neural Network Model for Flood Forecasting at Upper River Ping, Chiang Mai. *KSCE J. Civ. Eng.* **2016**, *20*, 478–484.
25. Mehdi, R.; Hamid, R.M.; Sareh, G. Divergence Instability of Pipes Conveying Fluid with Uncertain Flow Velocity. *Phys. A Stat. Mech. Its Appl.* **2018**, *491*, 650–665.
26. Bruce, R.E.; Yasuhiro, M. Reliability-Based Service Life Assessment of Concrete Structures in Nuclear Power Plants: Optimum Inspection and Repair. *Nucl. Eng. Des.* **1997**, *175*, 247–258.

micromachines

MDPI

Article

Effect of Wetting Characteristics of Polishing Fluid on the Quality of Water-Dissolution Polishing of KDP Crystals

Xu Wang [1,*], Hang Gao [2,*], Qianfa Deng [1], Jinhu Wang [1], Hongyu Chen [1] and Julong Yuan [1]

1 Ultra-Precision Machining Center, Key Laboratory of Special Purpose Equipment and Advanced Processing Technology of Ministry of Education, College of Mechanical Engineering, Zhejiang University of Technology, Hangzhou 310024, China; qfdeng@zjut.edu.cn (Q.D.); wangjinhu@zjut.edu.cn (J.W.); hychen@zjut.edu.cn (H.C.); jlyuan@zjut.edu.cn (J.Y.)

2 Key Laboratory for Precision and Non-Traditional Machining Technology of Ministry of Education, Dalian University of Technology, Dalian 116024, China

* Correspondence: wx382935877@163.com (X.W.); hanggao4187@126.com (H.G.)

Abstract: KDP crystals constitute the only laser-frequency conversion and electro-optical switches that can be used in laser systems for inertial confinement fusion. However, KDP crystals are difficult to produce because of their inherent softness, brittleness, water-solubility, and temperature sensitivity. The authors' group developed a water-dissolution polishing method in previous studies to obtain near-damage-free KDP surfaces. In this article, the effect of the wetting characteristics of the water dissolution polishing fluid on the crystal surface—a factor rarely considered in the usual process optimization—on the polished surface quality was comprehensively studied. The mean radius of micro water droplets at 5 wt.% and 7.5 wt.% water content was approximately 0.6 nm and 1.2 nm, respectively. Theoretically, the smaller micro water droplet size is beneficial to the polished surface quality. When the water content was 5 wt.%, due to the poor wetting characteristics of the polishing fluid, surface scratches appeared on the polished surface; when the water content was 7.5 wt.%, the effects of the wetting characteristics and the radius of the micro water droplets reached a balance, and the polished surface quality was the best (Ra 1.260 nm). These results confirm that the wetting characteristics of the polishing fluid constitute one of the key factors that must be considered. This study proves that the wetting characteristics of the polishing fluid should be improved during the optimization process of polishing fluid composition when using oil-based polishing fluids for ultra-precision polishing.

Keywords: KDP crystals; water-dissolution polishing; wetting characteristics; surface quality

Citation: Wang, X.; Gao, H.; Deng, Q.; Wang, J.; Chen, H.; Yuan, J. Effect of Wetting Characteristics of Polishing Fluid on the Quality of Water-Dissolution Polishing of KDP Crystals. *Micromachines* **2022**, *13*, 535. https://doi.org/10.3390/mi13040535

Academic Editors: Youqiang Xing, Xiuqing Hao and Duanzhi Duan

Received: 10 March 2022
Accepted: 27 March 2022
Published: 29 March 2022

1. Introduction

Potassium dihydrogen phosphate (KH_2PO_4 and KDP) crystals have excellent nonlinear electro-optical properties and currently constitute the only material that can be used as electro-optical switching and frequency doubling conversion elements in laser-induced inertial confinement fusion (ICF) technology [1]. High-power ICF laser devices require a high number of large-size, ultra-high-quality KDP crystal elements. For instance, this is the case of the National Ignition Facility [2,3], which requires a large diameter-thickness ratio (410 mm $\times$ 410 mm $\times$ 10 mm), high surface quality (surface roughness root mean square (rms) $\leq$ 5 nm, transmission wave-front aberration less or equal than $\lambda/6$ PV, and high laser-induced damage threshold ($\geq$15 J/cm ^{2}) for more than 600 pieces of KDP components. KDP crystals are materials that are extremely difficult to process owing to their soft and brittle nature, strong anisotropy, and extreme sensitivity to processing temperature [4]. Moreover, KDP crystals are soluble in water (solubility of 33 g/100 g H_2O at 25 °C) [5], and prolonged exposure to air will absorb moisture and damage the processed surface, further increasing the difficulty of ultra-precision processing and storage of KDP crystals.

Researchers have explored multiple techniques to process KDP crystals. Single-point diamond turning is the most important technique to process KDP components in engineering, producing KDP crystals with a high surface quality. However, micro-waviness appears on the surface of these crystals [6,7]. Such micro-waviness can have an impact on the laser-induced damage threshold of the KDP samples [8]. Moreover, the mechanical processing can produce subsurface damage that affects the optical performance of the KDP elements [9,10].

Ultra-precision grinding is also a commonly used technique in the processing of large-size optical components [11,12]. Qu et al. [13,14] successfully applied an ultra-precision grinding method to obtain ultra-smooth KDP crystal surfaces. Yin et al. [15] developed a magnetorheological polishing technique with controlled fluid-crystal temperature difference to improve the polishing quality and efficiency. Shi et al. [16] developed a combined magnetorheological and ion beam polishing treatment to be applied on the surface after fly-cutting. The ion beam was used to remove residual iron ions from the surface after magnetorheological polishing, which increased the crystal laser-induced damage threshold. Gao et al. [17] proposed an abrasive-free jet polishing method for KDP crystals to address surface particle residues and thermal effects. Despite the significant progress achieved, problems such as abrasive grain embedding, surface atomization, iron powder embedding, and high cost remain. Ultra-precision processing of large-size KDP crystal elements is still one of the limitations of ICF engineering.

The research group authoring this paper developed an oil-based, abrasive-free polishing fluid with an "water-in-oil" structure exploiting the water-soluble nature of KDP crystals, and verified the feasibility of this water-soluble processing method by obtaining a non-destructive and ultra-smooth surface via a continuous polishing method only through the physical dissolution of micro water droplets in the polishing fluid [18]. A flattened polishing with surface roughness rms of 2.182 nm and flatness of 22.013 μm (KDP crystal size of 100 mm × 100 mm × 10 mm) was achieved using an indefinite eccentric motion [19]. Meanwhile, in combination with computer-controlled optical surfacing technology, which is widely employed for large-size optical component processing, a tool influencing the function of the water-dissolution numerical control polishing method, was implemented by using small tools to process large-size KDP crystals [20]. Thus, the micro waviness on the surface after single-point diamond turning processing was successfully removed, reducing the surface roughness rms to values below 3 nm [21]. The laser-induced damage thresholds of KDP crystals after processing by different methods were also compared, and the morphology of laser damage points was analyzed. It was concluded that the proposed water-dissolution polishing could enhance the laser-induced damage thresholds of KDP crystals [22].

In a previous study, the effect of water content of oil-based water-dissolution polishing fluid on the surface roughness of KDP crystals was also studied. We found that the surface roughness after polishing does not always decrease with the decrease of the water content of polishing fluid, and that the surface roughness becomes worse when the water content of polishing fluid is too low (5 wt.%). However, theoretically, the lower the water content is, the smaller the size of micro water droplets in the polishing fluid becomes, and the better the surface quality should be. In addition, we also experimentally found that there is a significant difference in the wetting characteristics of the crystal surface among polishing fluids with different water contents. Furthermore, the effect of wetting characteristics on the quality of ultra-precision polishing has been rarely reported.

To analyze the factors affecting the quality of KDP water-dissolution polishing, the effect of the wetting characteristics of polishing fluid on the crystal surface—a factor rarely considered in the usual process optimization—on the polishing quality was investigated in the current study.

2. Correlation between Surface Roughness and Water Content during the Water-Dissolution Polishing Process of KDP Crystals

2.1. Selective Removal Mechanism of Water-Dissolution Polishing

The selective removal mechanism of water-dissolution, ultra-precision polishing of KDP crystals that can produce ultra-smooth surfaces is shown in Figure 1. In the oil-based polishing fluid used, the micro water droplets are wrapped in surfactants and dispersed in the oil-phase mother liquor to form a "water-in-oil" structure. In the polishing area, the polishing pad is in contact with the crystal surface material under a certain pressure. Under such a mechanical action, the "water-in-oil" structure at the rough peak of the polishing pad and the crystal surface are squeezed and deformed by friction, and the crystal material at the high point is dissolved and removed by the released water. The dissolved products are then carried away from the crystal surface by the mechanical action of the polishing pad and the flowing action of the polishing fluid, while at the low concave part of the crystal surface, the polishing pad is not in direct contact with the crystal material, and the "water-in-oil" structure is in a stable state, so no dissolution phenomenon occurs. This achieves selective removal of the crystal surface roughness peak and reduces the crystal surface roughness [18].

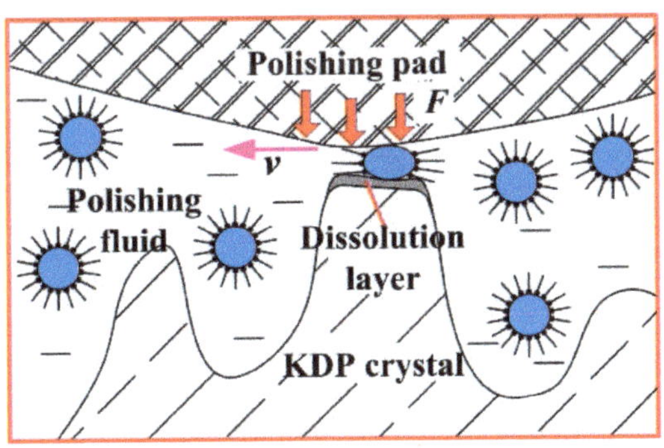

Figure 1. Illustration of the selective removal of material from a KDP crystal.

2.2. Analysis of the Effect of Water Content of Water-Dissolution Polishing Fluid on the Polished Surface Roughness

The size of the micro water droplets in the "water-in-oil" polishing fluid is directly related to the polished surface roughness. If the size of the micro water droplets is too large, the ultra-smooth surface with nanoscale roughness will not be formed. The micro water droplet size in the polishing fluid with different water contents was measured by a dynamic light scattering (DLS) instrument (model no.: ALV-NIBS); the results are shown in Figure 2. The micro water droplet radius was basically distributed within 10 nm, which proves the feasibility of the above described removal mechanism. Note also that the micro water droplet size in polishing fluid increases with the increase of water content in the polishing fluid. The average values of micro water droplet radius when the polishing fluid contained 5 wt.%, 7.5 wt.%, and 10 wt.% water were approximately 0.6 nm, 1.2 nm, and 3.8 nm, respectively. Theoretically, the lower the water content of the polishing fluid, the smaller the size of the micro water droplets—which should be more favorable for the selective removal of crystal surface roughness peaks—and the lower the surface roughness after polishing.

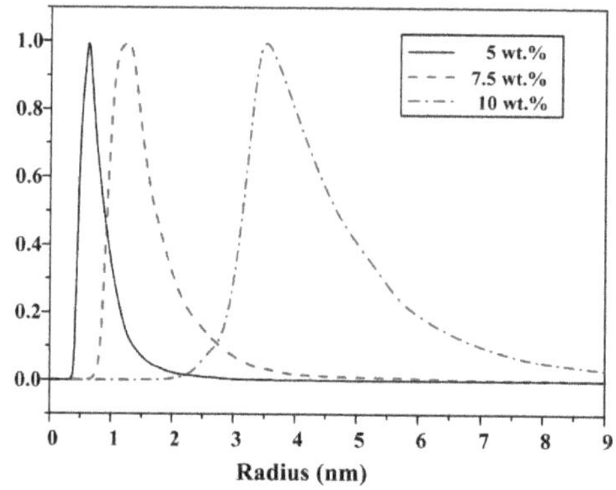

Figure 2. Radius distribution curve of micro water droplets in different polishing fluids.

In a previous study, we analyzed the variation pattern of the surface roughness of KDP crystals after polishing with the water content of the water-dissolution polishing fluid (as shown in Figure 3). When the water content was greater than 7.5 wt.%, the surface roughness decreased significantly with the decrease of water content, as expected. However, when the water content was reduced to 5 wt.%, the surface roughness value after polishing was higher than that at 7.5 wt.% [21], and it did not keep decreasing with the decrease in water content, which was inconsistent with the theoretical analysis based on the micro water droplet size.

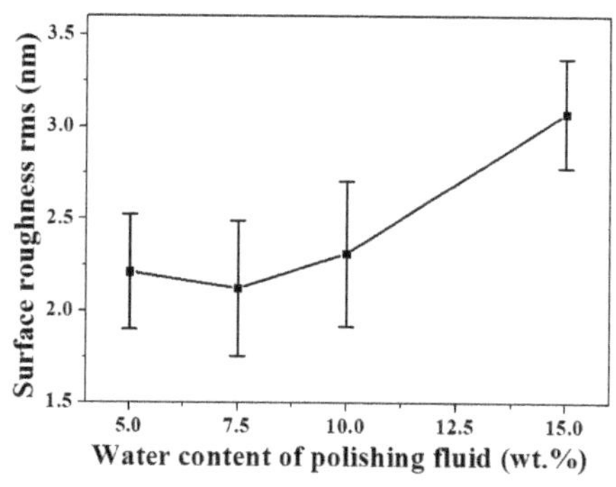

Figure 3. Variations in surface roughness with water content of polishing fluid (the data used to plot this curve are from [21]).

3. Analysis of the Effect of Polishing Fluid Viscosity on the Surface Quality of Water-Dissolution Polishing

In the traditional polishing process, it is generally assumed that a lower viscosity of the polishing fluid will be beneficial to obtain a higher quality surface. Gutmann et al. [23] considered that a high viscosity polishing fluid cannot be easily transported to the processing area between the polishing pad and the polished surface compared to a low viscosity polishing fluid, which will lead to a lubrication state in the polished area that is not conducive to processing and affects the quality of the polished surface, with defects such as scratches possibly occurring. Mullany et al. [24] assumed that a lower viscosity of polishing fluid leads to a higher friction, which can improve the material removal rate of polishing, and is conducive to obtaining an ultra-smooth surface.

Changes in the water content of the polishing fluid during the water-dissolution polishing of KDP crystals can also cause changes in viscosity. In this study, an NDJ-1 rotary viscometer was used to measure the viscosity of a polishing fluid with different levels of water content. The measurement results are shown in Figure 4. The viscosity of the polishing fluid increased with the increase in water content. According to the results of Gutmann, Mullany, and others, the viscosity of polishing fluid is lower when the water content is lower, which should theoretically facilitate the processing, but is not consistent with the actual results when KDP crystals are processed by water-dissolution ultra-precision polishing methods. Therefore, the difference in viscosity is not the reason for the degradation in polished surface quality when the water content is too low.

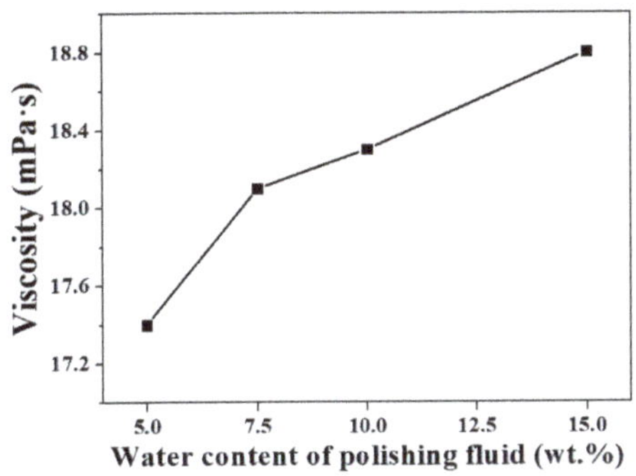

Figure 4. Variations in the viscosity of polishing fluid with water content.

4. Analysis of the Effect of the Wetting Characteristics of Polishing Fluid on the Surface Quality of Water-Dissolution Polishing

In actual processing, the surface of KDP crystals and a polishing pad with a certain degree of elasticity do not perfectly fit each other. A large number of small-size gaps exists between the rough peak of the crystal surface and the polishing pad (as shown in Figure 5). Selective localized water-dissolution at the rough peaks of the crystal surface can only occur when the polishing fluid is able to enter these gaps sufficiently. During the polishing process, if the polishing fluid cannot be guaranteed to fully enter the small-size gaps between the polishing pad and the KDP crystal, the rough peak will lack sufficient polishing fluid to ensure that material removal occurs. Thus, even direct contact friction between the polishing pad and the crystal surface will occur locally, which will have a negative effect on the quality of the polished surface and directly affect the acquisition of an ultra-smooth surface.

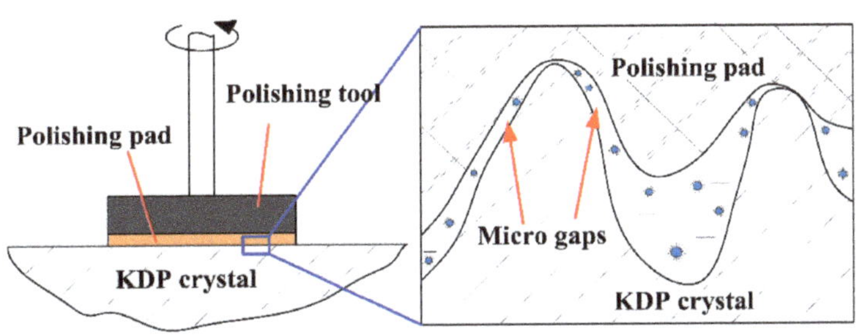

Figure 5. Illustration of micro gaps between polishing pad and KDP crystal.

Given that the size of the rough peak on the KDP crystal surface is extremely small (micro-nanometer scale), the gaps between the polishing pad and the surface can be approximated as micro-nanometer scale pipes, demanding the consideration of the effect of fluid capillarity. To evaluate the ability of the polishing fluid to enter the pore space between the polishing pad and the crystal surface, denoted as E_a, with reference to the formula for calculation of the column rise height h of the fluid in the capillary, E_a can be considered to be positively correlated with h:

$$E_a \propto h = \frac{2\gamma \cos\theta}{\rho g r} \tag{1}$$

where γ is the surface tension, θ is the contact angle, ρ is the density of the fluid, and r is the radius of the capillary. Therefore, to analyze E_a with a certain pore size, the surface tension, contact angle, and density of the polishing fluid need to be examined and analyzed.

The surface of the polyurethane polishing pad was covered with pores and channels. It provided a certain storage capacity for polishing fluid, which could quickly penetrate into the surface layer of the polishing pad. Thus, the contact angle of the polishing fluid on the surface of the polishing pad can be approximated to be close to 0°. Therefore, in the present study, we assumed that the contact angle of the water-dissolution polishing fluid on the surface of KDP crystal was the main parameter impacting E_a. The contact angle can be a direct measurement of the wettability of a fluid on a solid surface, i.e., the ability and tendency of a fluid to spread out on a solid surface. It is the angle between the tangent line of the liquid–gas interface and the solid–liquid interface at the junction of the solid–liquid–gas phase (as shown in Figure 6). The contact angle satisfies the Young equation:

$$\cos\theta = (\gamma_{SG} - \gamma_{SL})/\gamma_{LG} \tag{2}$$

where γ_{LG} is the gas–liquid interfacial tension, γ_{SG} is the solid–gas interfacial tension, and γ_{SL} is the solid–liquid interfacial tension. When $\theta = 0°$, the liquid can completely wet the solid surface (the contact angle of ethanol on the surface of KDP crystal is 0°); when $\theta < 90°$, the liquid can wet the solid surface; and when $\theta > 90°$, the solid surface is hydrophobic, and the liquid cannot wet the solid surface.

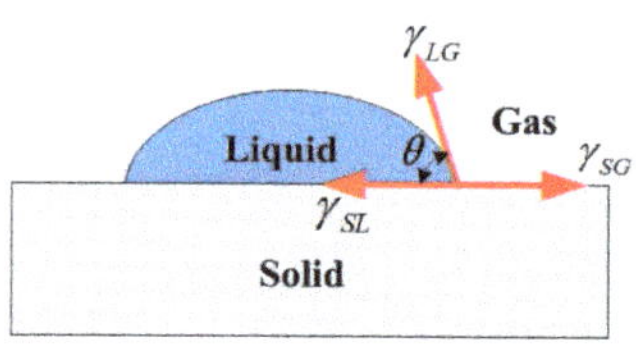

Figure 6. Illustration of the contact angle.

If the contact angle of the polishing fluid on the surface of a KDP crystal is large, the polishing fluid will not easily enter the processing area between the polishing pad and the crystal. Therefore, to evaluate the wettability of the polishing fluid with different water contents on the surface of KDP crystals, the contact angle was measured. The surface roughness of the KDP crystal in the test was 3–5 nm, which was close to the surface state in the ultra-precision polishing process; the droplet volume of the polishing fluid was 5 μL to avoid the influence of the droplet's self-weight on the contact angle.

The test results of the contact angles of polishing fluid on the KDP surface are shown in Figure 7. The wettability of the polishing fluid on the KDP surface was the worst when the water content was 0 wt.%, and the contact angle reached 52.6°, but it could still wet the KDP crystal surface. The contact angle was 46.2° when the water content of polishing fluid was 5 wt.%. The contact angle of polishing fluid on the surface of the KDP crystal decreased with the increase of water content, and the decreasing trend was linear. When the water content increased to 20 wt.%, the contact angle decreased to 19.1°.

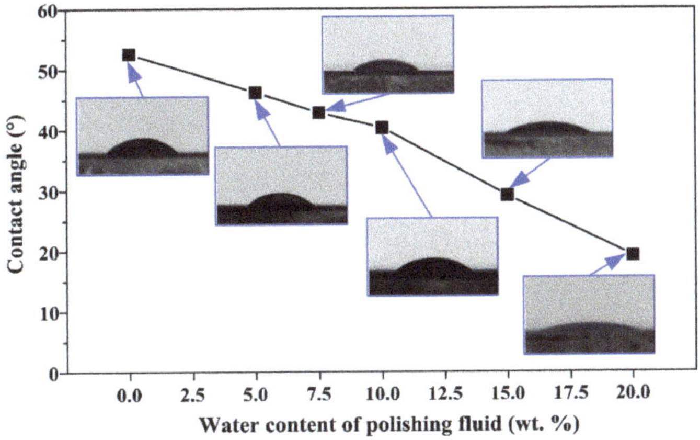

Figure 7. Contact angles of polishing fluid with different water contents on the KDP surface.

According to Equation (1), in addition to the contact angle, the polishing fluid density and surface tension are also important factors affecting its access to the polishing area. In this study, we used a DSA100 surface tension meter to measure the surface tension of polishing fluids with different water contents through the "hanging drop method". Therefore, the surface tension was determined by the radius of curvature of the bottom of the hanging drop and the shape factor of the droplet when the droplet reached static-force (tension to gravity) equilibrium. The surface tension of polishing fluids with different water contents is listed in Table 1.

Table 1. Surface tension of polishing fluids with different water contents.

Water Content	Measuring Results	Density (g/cm^3)	Surface Tension (mN/m)
0 wt.%		0.820	24.78
5 wt.%	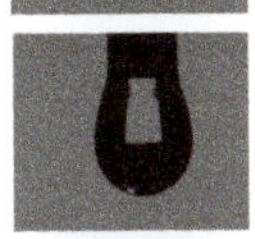	0.842	24.90
7.5 wt.%	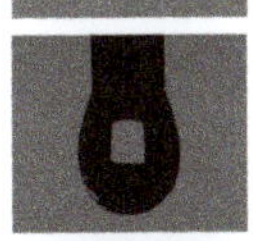	0.849	25.02
10 wt.%	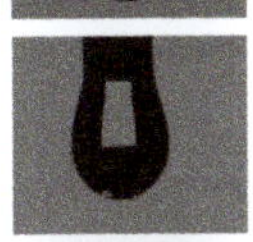	0.856	25.29
15 wt.%	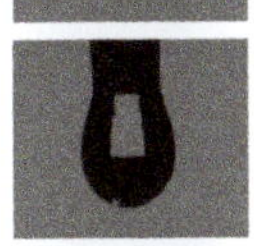	0.867	25.58
20 wt.%		0.887	25.73

The surface tension of water at room temperature was 72.7 mN/m, and the surface tension of other organic components of the polishing fluid was approximately 10–40 mN/m. Note from Table 1 that only a very small increase of the surface tension of polishing fluid occurred with the increase of water content; this tension was basically maintained at approximately 25 mN/m; when the water content increased from 0% to 20%, the density of the polishing solution slightly increased from 0.82 g/cm^3 to 0.887 g/cm^3. The changes in contact angle were then relatively large. The variation of the value of cosθ was much greater than that of γ and ρ. Thus, it can be concluded that the surface tension and density have much less influence on the ability of the polishing fluid to enter the pore space between the polishing pad and the crystal surface than the contact angle.

Substituting the experimentally measured contact angle, surface tension, and density into Equation (1), we can easily conclude that the lower the water content of the polishing fluid, the worse its ability to enter the pore space between the polishing pad and the crystal surface. This means that in actual water-dissolution polishing, the use of polishing micro-emulsions with a too low water content will lead to the polishing fluid to enter the processing area with some difficulty, and even with local "unlubrication-friction" of the polishing pad and crystal, the surface quality of the polished KDP crystal will be affected.

In addition to E_a, the wetting work reflects the firmness of the solid–liquid interfacial bonding, and can also characterize the ability of polishing fluid to wet the KDP crystal surface. The wetting work W_a is calculated according to Equation (3). It can be concluded that the larger its value, the stronger the solid–liquid interfacial bonding, and the easier to wet the solid surface. Table 2 lists the wetting work W_a calculated according to the results of surface tension and contact angle measurements. It is evident that W_a increases basically linearly with the increase of water content of polishing fluid, which is consistent with the variation trend of E_a. This further proves that the water content of the water-dissolution polishing fluid directly affects the wettability of the processed surface, thus affecting the polishing quality.

$$W_a = (\gamma_{SG} + \gamma_{LG}) - \gamma_{SL} = \gamma_{LG}(1 + \cos\theta) \tag{3}$$

Table 2. Wetting work of polishing fluids with different water contents.

Water Content	Contact Angle (°)	Surface Tension (mN/m)	Wetting Work (mN/m)
0 wt.%	52.6	24.78	39.83
5 wt.%	46.2	24.90	42.13
7.5 wt.%	42.8	25.02	43.38
10 wt.%	40.3	25.29	44.58
15 wt.%	29.1	25.58	47.93
20 wt.%	19.1	25.73	50.04

To verify the above analysis, the surface of KDP crystals polished with a polishing fluid containing 7.5 wt.% water was polished for 20 min with a polishing fluid containing 5 wt.% water. To ensure that the observations of the polished surface were representative, five points on the KDP surface were observed with an optical microscope (Olympus MX40, Tokyo, Japan), as shown in Figure 8.

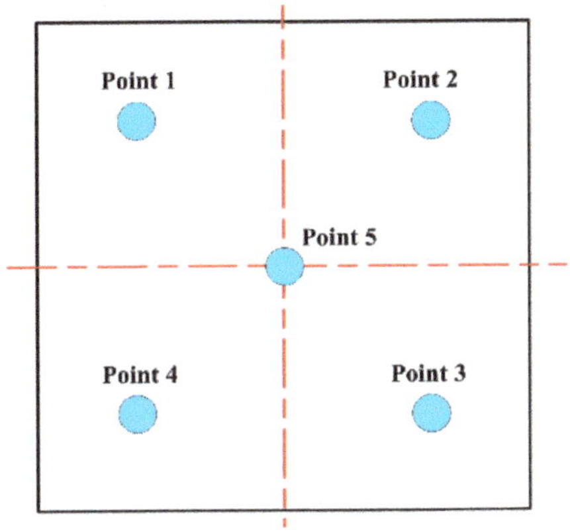

Figure 8. Schematic showing the five observing points on the KDP crystal.

For the water-dissolution polishing method, the material is removed by water dissolution with micro water droplets in the polishing fluid, and a super smooth surface without scratches was obtained by polishing with a polishing fluid containing 7.5 wt.% water (as shown in Table 3). Under the condition of sufficient supply of polishing fluid, the surface quality of the polished KDP with polishing fluid containing 5 wt.% water crystal was not further improved. Moreover, microscopic scratches appeared locally on the crystal surface that caused damage to the polished surface.

Table 3. KDP surface after polishing with different water contents of polishing fluid.

	5 wt.%	7.5 wt.%
Point 1		
Point 2		

Table 3. *Cont.*

	5 wt.%	7.5 wt.%
Point 3		
Point 4	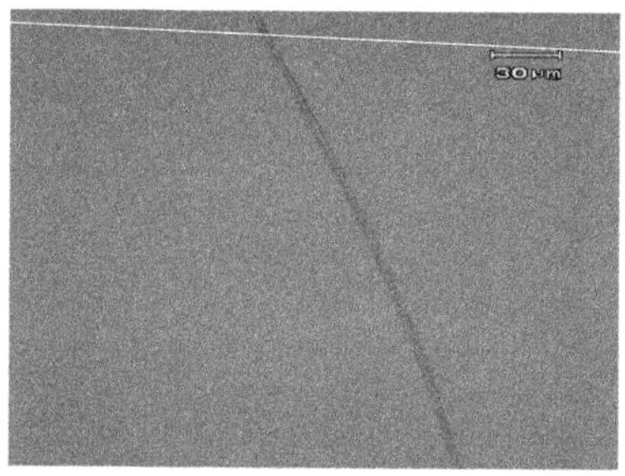	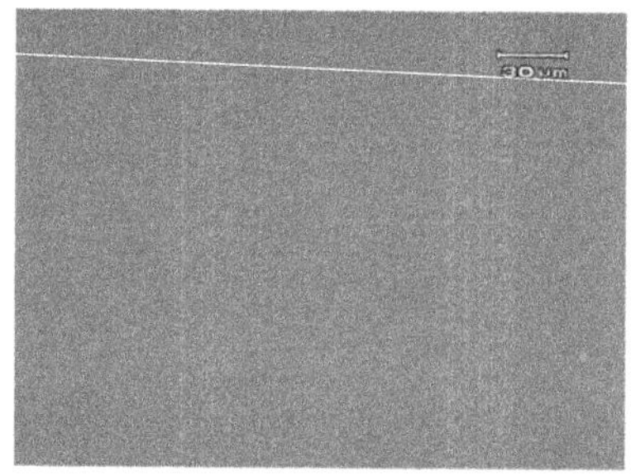
Point 5		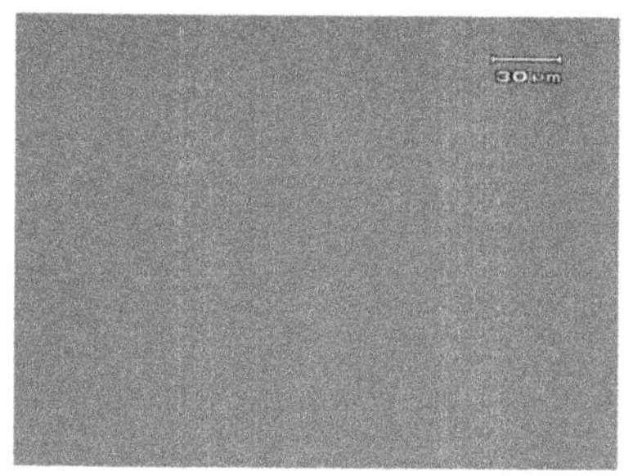

The surface roughness of KDP crystal was measured with a three-dimensional profilometer (ZYGO Newview 5022, Middlefield, CT, USA) over an area of 0.35 × 0.26 mm (the measuring results are shown in Figure 9). When the water content was greater than 7.5 wt.%, the surface roughness decreased significantly with the decrease in water content. The surface roughness value was Ra 1.918 nm and Ra 2.652 nm after being processed with a polishing fluid with water content of 10 wt.% and 15 wt.%, respectively. The surface roughness value decreased to Ra 1.260 nm after being processed with a polishing fluid with water content of 7.5 wt.%, while the surface roughness value increased to Ra 1.660 nm after being processed with a polishing fluid with water content of 5 wt.%. When the water content was lower than 7.5 wt.%, continuing to reduce the water content of the polishing fluid would lead to a decrease in surface quality.

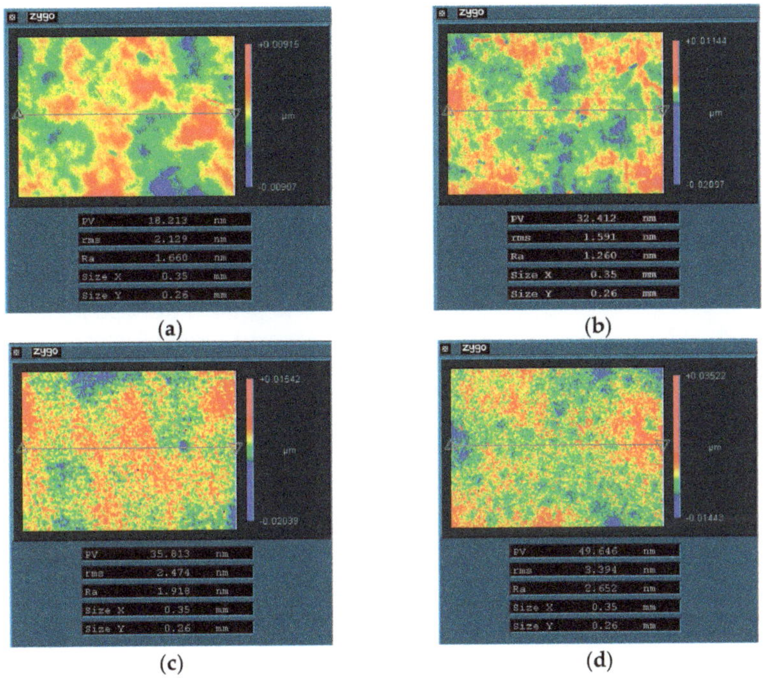

Figure 9. Surface roughness of KDP crystal after polishing with different water contents of polishing fluid: (**a**) 5 wt.%, (**b**) 7.5 wt.%, (**c**) 10 wt.%, and (**d**) 15 wt.%.

These experimental results are consistent with the previous analysis. When using oil-based polishing fluid, if the wettability of the polishing fluid on the surface of the workpiece is too low, a significantly negative impact on the polishing quality is caused. The polished surface quality of KDP crystals was jointly determined by the micro water droplet size and the wetting characteristics of the polishing fluid on the KDP surface. The wettability of the polishing liquid on the KDP surface was enhanced with the increase in water content, and it was easier for the water-dissolution polishing fluid with high water content to enter the micro-porosity between the polishing pad and the crystal surface. However, when the water content was greater than 7.5 wt.%, the size of micro water droplets in the polishing fluid also increased substantially, thereby departing from achieving a high-quality surface. At this time, the surface quality after polishing still decreased with the decrease of water content. However, when the water content was reduced to 5 wt.%, the surface quality deteriorated due to the reduced wettability of the polishing solution on the KDP surface, even though the mean micro water droplet radius was reduced from 1.2 nm at 7.5 wt.% to 0.6 nm. When the water content was 7.5 wt.%, the wetting characteristics of the polishing fluid and the effect of the micro water droplet radius reached a balance, and the surface quality was optimized after polishing.

5. Conclusions

When using a "water-in-oil" polishing fluid for ultra-precision polishing of KDP crystals, theoretically, the lower the water content, the easier to obtain lower surface roughness. In this study, we analyzed the anomalous phenomenon that the surface quality decreases when the water content of the polishing fluid is too low in the process of water-dissolution polishing of KDP crystals. The conclusions are as follows:

1. The radius distribution of micro water droplets in water-dissolution polishing fluid was within 10 nm, and it increased significantly with the increase in water content.

The mean value of the micro water droplet radius was approximately 0.6 nm when the polishing fluid contained 5 wt.% of water, it was approximately 1.2 nm with a water content of 7.5 wt.%, and it was approximately 3.8 nm when the water content was 10 wt.%.

2. The viscosity of the polishing fluid increased with the increase of water content; theoretically, a lower viscosity should be more conducive to polishing the surface quality. Thus, the polishing fluid viscosity was not considered as a key factor affecting the quality of water-dissolution polishing of KDP crystals.
3. By measuring the contact angle and surface tension, it was concluded that the wetting characteristics of the polishing solution on the crystal surface were significantly affected by the water content. The higher the water content, the better the polishing fluid wet the crystal surface, and the easier to enter the micro-porosity between the polishing pad and the crystal surface, which is theoretically beneficial to obtain a high-quality surface.
4. The surface roughness value of KDP was Ra 1.260 nm after being processed with a polishing fluid containing 7.5 wt.% water, while the surface roughness value increased to Ra 1.660 nm after being processed with a polishing fluid with a water content of 5 wt.%, and the surface became clearly scratched. Concerning the KDP crystal water-dissolution polishing method, a low water content of the polishing solution led to a decrease in the wettability, making it difficult to enter the gaps between the polishing pad and the crystal surface, and thus leading to a degradation in the quality of the polished surface. The polished surface quality of KDP crystals was jointly determined by the micro water droplet size and the wetting characteristics of the polishing fluid on the KDP surface. When the water content was 7.5 wt.%, the wetting characteristics of the polishing fluid and the effect of micro water droplet radius were balanced and the best polished surface quality was achieved.
5. When using oil-based polishing fluids for ultra-precision processing, in addition to conventional factors such as viscosity, size of polishing particles (abrasive particles, microdroplets, etc.), and polishing speed, the wetting characteristics of the polishing fluid on the processed surface constituted one of the key factors that must be considered. In the future, we will further optimize the polishing fluid composition and incorporate appropriate additives to obtain a water-dissolution polishing fluid with good microemulsion radius and wetting characteristics to further improve the surface quality of KDP crystals after polishing. The present study also proved that the wetting characteristics of the polishing fluid should be improved during the optimization process of polishing fluid composition when using oil-based polishing fluids for ultra-precision polishing.

Author Contributions: Conceptualization, X.W. and H.G.; methodology, J.Y.; software, Q.D.; validation, J.W.; formal analysis, X.W. and H.G.; investigation, X.W. and H.G.; resources, X.W., H.G., and J.Y.; data curation, X.W., J.W. and H.C.; writing—original draft preparation, X.W.; writing—review and editing, H.G. All authors have read and agreed to the published version of the manuscript.

Funding: This research was funded by the Natural Science Foundation of Zhejiang Province (Grant No. LY21E050010) and the National Natural Science Foundation of China (Grant No. 51805485, 51775511, 52175442, 51905485).

Conflicts of Interest: The authors declare no conflict of interest.

References

1. Zhang, Z.; Wang, H.; Quan, X.; Pei, G.; Tian, M.; Liu, T.; Long, K.; Li, P.; Rong, Y. Optomechanical analysis and performance optimization of large-aperture KDP frequency converter. *Opt. Laser Technol.* **2019**, *109*, 633–642. [CrossRef]
2. Zylstra, A.B.; Nora, R.; Patel, P.; Hurricane, O. Model validation for inferred Hot-Spot conditions in National Ignition Facility experiments. *Phys. Plasmas* **2021**, *28*, 122703. [CrossRef]
3. Lindl, J.D.; Atherton, L.; Amednt, P.; Batha, S. Progress towards ignition on the National Ignition Facility. *Nucl. Fusion* **2011**, *51*, 0940249SI. [CrossRef]

4. Hang, W.; Wei, L.; Debela, T.; Chen, H.; Zhou, L.; Yuan, J.; Ma, Y. Crystallographic orientation effect on the polishing behavior of $LiTaO_3$ single crystal and its correlation with strain rate sensitivity. *Ceram. Int.* **2022**, *48*, 7766–7777. [CrossRef]
5. Zhou, X.; Zheng, W.; Xu, D.; Luo, T.; Zhang, Z.; Wang, X. Solubility measurement and thermodynamics modelling for potassium dihydrogen phosphate in a water-ethanol system from 293.2 to 323.2 K. *Fluid Phase Equil.* **2020**, *512*, 112533. [CrossRef]
6. Chen, D.; Zhang, S.; Liu, J.; Zha, C.; Pan, R. Morphological analysis of KDP-crystal workpiece surfaces machined by ultra-precision fly cutting. *Materials* **2020**, *13*, 432. [CrossRef]
7. An, C.; Feng, K.; Wang, W.; Xu, Q.; Lei, X.; Zhang, J.; Yao, X.; Li, H. Interaction mechanism of thermal and mechanical field in KDP fly-cutting process. *Micromachines* **2021**, *12*, 855. [CrossRef]
8. Pang, Q.; Shu, Z.; Kuang, L.; Xu, Y. Effect of actual frequency features generated in machining process on the temperature and thermal stress of potassium dihydrogen phosphate crystal. *Mater. Today Commun.* **2021**, *29*, 102984. [CrossRef]
9. Liu, Q.; Liao, Z.; Cheng, J.; Xu, D.; Chen, M. Mechanism of chip formation and surface-defects in orthogonal cutting of soft-brittle potassium dihydrogen phosphate crystals. *Mater. Des.* **2021**, *198*, 109327. [CrossRef]
10. Zhang, Y.; Hou, N.; Zhang, L. Understanding the formation mechanism of subsurface damage in potassium dihydrogen phosphate crystals during ultra-precision fly cutting. *Adv. Manuf.* **2019**, *7*, 270–277. [CrossRef]
11. Yang, Y.; Ji, Y. Experimental study on grinding damage control of optical materials. *Diam. Abras. Eng.* **2020**, *40*, 84–88.
12. Chen, H.; Xu, Q.; Wang, J.; Li, P.; Yuan, J.; Lyu, B.; Wang, J.; Tokunaga, K.; Yao, G.; Luo, L.; et al. Effect of surface quality on hydrogen/helium irradiation behavior in tungsten. *Nucl. Eng. Technol.* 2022, *in press*. [CrossRef]
13. Qu, M.; Jin, T.; Xie, G.; Cai, R. Developing a novel binderless diamond grinding wheel with femtosecond laser ablation and evaluating its performance in grinding soft and brittle materials. *J. Mater. Process. Technol.* **2020**, *275*, 116359. [CrossRef]
14. Qu, M.; Xie, G.; Jin, T.; Cai, R.; Lu, A. Realization of high efficiency and low damage machining of anisotropic KDP crystal by grinding. *Precis. Eng.* **2019**, *55*, 464–473. [CrossRef]
15. Yin, Y.; Zhang, Y.; Dai, Y.; Xiao, Q.; Tie, G. Novel magneto-rheological finishing process of KDP crystal by controlling fluid-crystal temperature difference to restrain deliquescence. *CIRP Ann. Manuf. Technol.* **2018**, *67*, 587–590. [CrossRef]
16. Shi, F.; Qi, X.; Dong, W.; Zhu, Z. Improvement of surface laser damage resistance of KDP crystal under combined machining process. *Opt. Eng.* **2019**, *57*, 121911. [CrossRef]
17. Gao, W.; Wei, Q.; Ji, J.; Sun, P.; Ji, F.; Wang, C.; Xu, M. Theoretical modeling and analysis of material removal characteristics for KDP crystal in abrasive-free jet processing. *Opt. Express* **2019**, *27*, 6268–6282. [CrossRef]
18. Gao, H.; Wang, B.; Guo, D.; Li, Y. Experimental study on abrasive-free polishing for KDP crystal. *J. Electrochem. Soc.* **2010**, *157*, 853–856. [CrossRef]
19. Cheng, Z.; Gao, H.; Liu, Z.; Guo, D. Investigation of the trajectory uniformity in water dissolution ultraprecision continuous polishing of large-sized KDP crystal. *Int. J. Extrem. Manuf.* **2020**, *4*, 45101. [CrossRef]
20. Wang, X.; Gao, H.; Yuan, J. Experimental investigation and analytical modelling of the tool influence function of the ultra-precision numerical control polishing method based on the water dissolution principle for KDP crystals. *Precis. Eng.* **2020**, *65*, 185–196. [CrossRef]
21. Wang, X.; Gao, H.; Chen, Y.; Guo, D. A water dissolution method for removing micro-waviness caused by SPDT process on KDP crystals. *Int. J. Adv. Manuf. Technol.* **2016**, *85*, 1347–1360. [CrossRef]
22. Chen, Y.; Gao, H.; Wang, X.; Guo, D.; Liu, Z. Laser induced damage of potassium dihydrogen phosphate (KDP) optical crystal machined by water dissolution ultra-precision polishing method. *Materials* **2018**, *11*, 419. [CrossRef] [PubMed]
23. Gutmann, R.; Price, D.; Neyrick, J.; Saino, C.; Permana, D.; Muraka, S. CMP of copper-polymer interconnect structures. *CMP-MIC Conf.* **1998**, *19*, 257–266.
24. Mullany, B.; Byrne, G. The effect of slurry viscosity on chemical-mechanical polishing of silicon wafers. *J. Mater. Process. Technol.* **2003**, *132*, 28–34. [CrossRef]

micromachines

Article

Influence of Annealing Temperature on Optical Properties of Sandwiched ZnO/Metal/ZnO Transparent Conductive Thin Films

Qijing Lin [1,2,3,†], Fuzheng Zhang [2,†], Na Zhao [2,*] and Ping Yang [2]

1 Collaborative Innovation Center of High-End Manufacturing Equipment, Xi'an Jiaotong University, Xi'an 710054, China; xjjingmi@163.com
2 State Key Laboratory of Mechanical Manufacturing Systems Engineering, Xi'an Jiaotong University, Xi'an 710049, China; xjzfz123@stu.xjtu.edu.cn (F.Z.); ipe@xjtu.edu.cn (P.Y.)
3 School of Mechanical and Manufacturing Engineering, Xiamen Institute of Technology, Xiamen 361021, China
* Correspondence: zn2020@xjtu.edu.cn
† These authors should be considered as co-first authors.

Abstract: Two sandwiched ZnO/Metal/ZnO transparent conductive thin films, 50nm ZnO/Cu/50nm ZnO (abbreviated as ZnO(Cu)) and 50nm ZnO/Ti/Cu/Ti/50nm ZnO (abbreviated as ZnO(Ti/Cu)) were deposited by magnetron sputtering technology. The comparative analysis of experiment results shows that the introduction of the Ti layer is beneficial to the overall properties of ZnO(Ti/Cu) thin film compared to ZnO(Cu) thin film with the same metal layer thickness. The effect of the annealing temperature on the performance of the two film systems was studied. Although the carrier concentration did not always increase with annealing temperature, the sheet resistances did decrease due to the obvious increase of mobility. The transmittance of ZnO(Cu) thin films increases with annealing temperature, while that of ZnO(Ti/Cu) films increases at first and then decreases. The optical band gap of ZnO(Cu) thin films increases with temperature, but is lower than that of ZnO(Ti/Cu) thin films, whose bandgap first increases with temperature and then decreases. The figure of merit of the ZnO(Ti/Cu) film is better than that of ZnO(Cu), which shows that the overall performance of ZnO(Ti/Cu) films is better, and annealing can improve the performance of the film systems.

Keywords: optical properties; sheet resistance; transparent conductive thin film; annealing temperature; ZnO/Metal/ZnO

Citation: Lin, Q.; Zhang, F.; Zhao, N.; Yang, P. Influence of Annealing Temperature on Optical Properties of Sandwiched ZnO/Metal/ZnO Transparent Conductive Thin Films. *Micromachines* **2022**, *13*, 296. https://doi.org/10.3390/mi13020296

Academic Editors: Xiuqing Hao, Duanzhi Duan and Youqiang Xing

Received: 27 January 2022
Accepted: 11 February 2022
Published: 13 February 2022

1. Introduction

As a transparent conductive oxide, zinc oxide (ZnO) is a wide-bandgap semiconductor with high transmittance in the visible range [1,2], which has led to its broad application in the fields of flat panel displays, camera tubes, solar cells, organic light-emitting diodes, liquid crystal displays and so on [3,4]. However, pure ZnO has a high resistivity and cannot meet the requirements of transparent conductive films due to two mutually restrictive factors: high transmittance and low resistivity in the visible light range. Sandwiched ZnO/Metal/ZnO multilayer thin film combining metal with a semiconductor can effectively improve the electrical conductivity without reducing the optical transmittance [5] because the metal in the middle layer can conduct electrons in order to reduce the resistivity. Various metal-sandwich ZnO multilayers, such as ZnO/Ag/ZnO [6], ZnO/Au/ZnO [7], ZnO/Al/ZnO [8] and ZnO/Cu/ZnO [9–12], have been developed. Due to the positive effect of copper in improving electrical conductivity and optical properties, ZnO(Cu) sandwiched multilayer has attracted much interest.

Various technologies, such as magnetron sputtering [11,12], atomic layer deposition [13] and sol-gel [14], have been used to fabricated ZnO(Cu) sandwiched films on different substrates, such as glass [11–14] and flexible polyethylene naphthalate [15]. Simulation and experimental methods were used to study the influence of Cu and ZnO layer thickness on the performance of ZnO(Cu) sandwiched film. Both theoretical analysis and experimental results prove that the electrical conductivity and optical properties of oxide transparent conductive films with a Cu interlayer depend considerably on its thickness [10–18]. Generally speaking, electrical conductivity will be improved with the increase of Cu layer thickness because of its excellent conductivity, but the transmittance will increase at first, reach the maximum value at a certain thickness, and then decrease sharply because of the absorption in the Cu layer. Although ZnO is one of the best choices for the semiconductor layer, oxygen chemisorption on its surface and grain boundaries of ZnO result in higher resistivity [17]. Therefore, the electrical properties of pure ZnO are unsuitable. To solve this problem, ZnO films are usually treated by annealing to improve the stability of the film by releasing strain energy and improving the crystal shape [19]. Previous studies [16,18] have shown that the annealing atmosphere and temperature affect the properties of the ZnO/Cu/ZnO multilayers.

Reducing the absorption of light in the metal layer is an effective measure to improve the properties of metal-sandwich ZnO multilayers. However, ultrathin Cu films (i.e., less than 10 nm) are susceptible to oxidation and corrosion, which significantly affect their electrical and optical properties [15]. One solution is to cover the layer with a protective, ultrathin film with stronger reducibility. Previous research has demonstrated that a continuous, ultrathin Cu/Ti bilayer film in which the Ti film acts as a protective film can improve the performance and stability of the transparent conductive electrode [19–21].

In this paper, two ZnO-based transparent conductive thin-film systems, consisting of either sandwiched Cu or Ti/Cu/Ti metal layers, were fabricated using magnetron sputtering technology and were annealed at temperatures from 100 °C to 400 °C in an Ar atmosphere.

Comparative experiments and characterization analysis, such as microstructure analysis, sheet resistance and optical properties, were investigated, and the influence of annealing temperature on performance was studied.

2. Experiment

Two ZnO transparent conductive thin-film systems (50nm ZnO/Cu/50nm ZnO and 50nm ZnO/Ti/Cu/Ti/50nm ZnO) were fabricated on glass substrates using magnetron sputtering technology. The sputtering equipment (Explorer14, Seattle, WA, USA) had three target guns, allowing it to perform RF sputtering of ZnO and the DC sputtering of Cu and Ti (both purity grades were 99.99%) in an Ar atmosphere with a base pressure of 2×10^{-5} Pa. Then, the fabricated thin-film systems were annealed in a high temperature furnace for 30 min at temperatures from 100 °C to 400 °C in an Ar atmosphere.

The crystallization structures of the two ZnO transparent conductive thin-film systems created at the various annealing temperatures were determined using X-ray diffraction (XRD) (Xpert Pro, Rotterdam, The Netherlands). In order to evaluate the performance of the two film systems, the four-point probe method and ultraviolet and visible spectrophotometer (Shimadzu UV-3600, Kyoto, Japan) were used to measure the sheet resistance and transmission spectra, respectively.

3. Results and Discussion

The transmission spectra of unannealed ZnO/10nm Cu/ZnO, ZnO/20nm Cu/ZnO and ZnO/5nm Ti/10nm Cu/5nm Ti/ZnO, as measured by ultraviolet and visible spectrophotometer, are presented in Figure 1. It can be noted that the transmittance of ZnO/10nm Cu/ZnO is higher than that of ZnO/20nm Cu/ZnO. This is similar to previous experiment results showing that the transmittance of ZnO(Cu) thin-film systems decreases as Cu film thickness increases [10,16,17]. As listed in Table 1, the transmittance of ZnO/5nm Ti/10nm Cu/5nm Ti/ZnO is also higher than that of ZnO/10nm Cu/ZnO in the wavelength range of 470–780 nm. The reasons for this phenomenon need to be analyzed in combination with the characteristics of Cu and Ti. It has been proven that the reflective effect and light absorption of Cu are the two main influencing factors of transmittance reduction, and they have a different degree of influence at short and long wavelengths [10,22–24]. That is, the reflective effect of the Cu film layer is the main cause for the transmittance reduction in the short-wavelength region, while light absorption affects the long-wavelength region. To be specific, in short wavelengths (less than 470 nm), the extinction coefficients of Cu and Ti are close, which means that the introduction of Ti layers has little effect on the attenuation of light absorption. However, because Ti has a higher refractive index than Cu, the reflection effect is more obvious, which leads to the transmittance of ZnO/5nm Ti/10nm Cu/5nm Ti/ZnO thin film being lower than that of ZnO/10nm Cu/ZnO in this region. In the wavelength range of 470–780 nm, since the extinction coefficient of Ti is significantly lower than that of copper, the addition of the Ti layer is beneficial to transmittance by reducing the effect of light absorption, as shown in Table 1. It is worth mentioning that the transmittance of the sandwiched ZnO/Metal/ZnO film is lower than that of the ZnO single film with the same thickness due to the reflective effect and light absorption of the metal interlayer. This has been confirmed by previous study [16].

Table 1. The refractive index and extinction coefficient of ZnO, Cu and Ti at different wavelengths [24].

Wavelength (nm)	Refractive Index			Extinction Coefficient	
	ZnO	Cu	Ti	Cu	Ti
387.44	2.264	1.231	1.500	2.068	2.12
413.27	2.181	1.185	1.590	2.208	2.17
442.79	2.122	1.168	1.680	2.363	2.25
476.85	2.078	1.152	1.750	2.504	2.34
516.58	2.045	1.12	1.810	2.603	2.47
539.04	2.032	1.038	1.860	2.592	2.56
563.55	2.019	0.826	1.920	2.602	2.67
590.38	-	0.468	2.010	2.809	2.77
619.9	-	0.272	2.110	3.326	2.88
652.53	-	0.214	2.220	3.667	2.99
688.78	-	0.213	2.360	4.043	3.11
729.294	-	0.223	2.540	4.433	3.23
826.53	-	0.26	2.980	5.26	3.32

In order to further compare the electric conductivity of three thin films, their sheet resistances are exhibited in Table 2. ZnO/20nm Cu/ZnO thin film has the minimum sheet resistance, while ZnO/10nm Cu/ZnO has the maximum. The reason for the change of sheet resistance can be explained intuitively. As a semiconductor material, the ZnO thin film has a high resistivity in the ZnO/Metal/ZnO film structure, hence, the overall conductivity of the film structure depends largely on the sandwiched metal layer introduced to improve the conductivity. Therefore, the increase of the metal layer thickness greatly reduces the sheet resistances of the ZnO/Metal/ZnO multilayer film structure. On the other hand, because the conductivity of Cu is superior to that of Ti, the resistance of thin film with 20nm Cu is less than that of 5nm Ti/10nm Cu/5nm Ti, although the overall thicknesses of both of the metal layers is 20 nm.

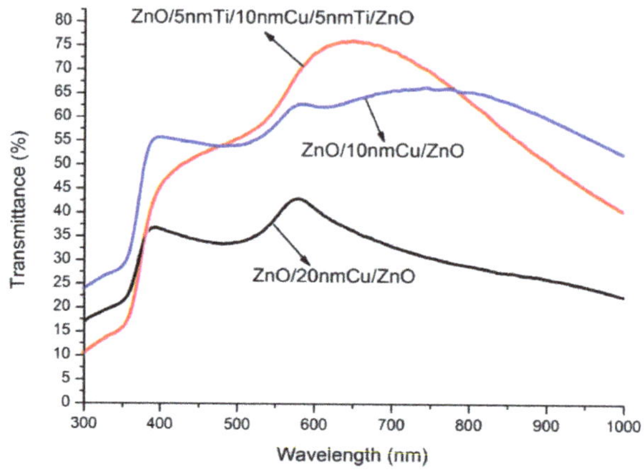

Figure 1. The optical transmission spectra of three unannealed ZnO transparent conductive thin-film systems.

Table 2. Sheet resistance and figure of merit of the three ZnO transparent conductive thin films.

Figure	Sheet Resistance (Ω/sq)	Figure of Merit
ZnO/10nm Cu/ZnO	10.1	0.00181
ZnO/20nm Cu/ZnO	5.5	0.00004
ZnO/5nm Ti/10nm Cu/5nm Ti/ZnO	6.6	0.01060

It is expected that both optical transmittance and electric conductivity of transparent conductive films are maximized as they are two important parameters for transparent conductive thin film. In fact, optical transmittance and electric conductivity are two mutual constraints: improvement of optical transmittance (or electrical conduction) will lead to a decrease in electrical conduction (or optical transmittance). Therefore, an evaluation parameter named figure of merit was introduced by Haacke to evaluate the performance of transparent conductive film. The expression is defined as [25]:

$$\Phi_{TC} = \frac{T^{10}}{R_S} \tag{1}$$

where T is the peak transmittance and R_S is the sheet resistance. As listed in Table 2, the Φ_{TC} of ZnO/20nm Cu/ZnO is the minimum, which indicates that the overall performance of ZnO/20nm Cu/ZnO is the worst although it has the minimum sheet resistance. ZnO/5nm Ti/10nm Cu/5nm Ti/ZnO has the best performance.

It is known that the metal layer thickness has a significant effect on the performance of ZnO/Metal/ZnO transparent conductive thin films, which is also reflected in these films. Obviously, the figure of merit for ZnO/20nm Cu/ZnO is lower than that of ZnO/10nm Cu/ZnO. This indicates that the overall performance of ZnO/20nm Cu/ZnO is inferior to that of ZnO/10nm Cu/ZnO. The main reason is that, although an increase in thickness of the Cu layer is beneficial to the improvement of conductivity, it greatly reduces the transmittance. On the other hand, although the metal layer thickness of ZnO/5nm Ti/10nm Cu/5nm Ti/ZnO is also increased, its overall performance is improved, which is reflected in the increase of the figure of merit. It shows that the introduction of the Ti layer is helpful to the improvement of overall performance. This indicates that the Ti layer plays an important role in improving the overall performance of the conductive films.

To further study the performance evolution of ZnO transparent conductive thin film under various annealing conditions, two film systems, ZnO/20nm Cu/ZnO (abbreviated as ZnO(Cu)) and ZnO/5nm Ti/10nm Cu/5nm Ti/ZnO (abbreviated as ZnO(Ti/Cu)) were annealed at different temperature. Figures 2 and 3 present the XRD spectra of the two film systems after annealing. As shown in Figure 2, each spectrum has two distinct diffraction peaks. The first peaks, belonging to ZnO, are hexagonal wurtzite, which indicates that ZnO films grow preferentially toward (002) orientation. The second peaks, corresponding to Cu(111), obviously indicate that Cu films grow toward (111) preferred orientation. The intensity of the Cu(111) peak increases with the temperature. This shows that annealing can improve the crystallization quality of the Cu layer. However, in Figure 3 it is observed that the ZnO hexagonal wurtzite polycrystalline structure with the (002) preferred orientation have been formed, and their peak intensities increase with the temperature. However, the diffraction peaks that can represent Cu are not obvious, which indicates that Cu does not crystallize very well. This is different from the XRD patterns of ZnO(Cu) thin films.

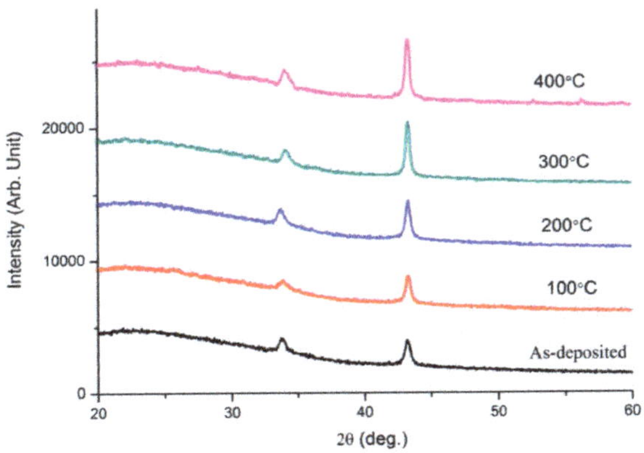

Figure 2. XRD spectra of ZnO(Cu) film system with varying annealing temperature.

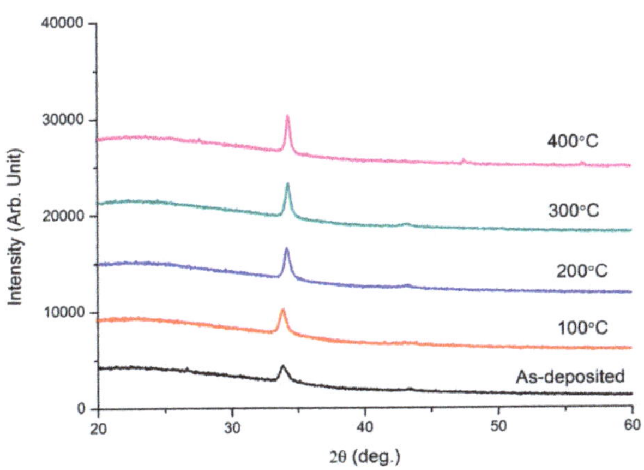

Figure 3. XRD spectra of ZnO(Ti/Cu) film system with varying annealing temperature.

The electrical properties of the two annealed thin-film systems are presented in Figure 4. The electrical conductivity of both film systems was improved with the increase of temperature. In detail, the sheet resistances of ZnO(Ti/Cu) thin films decreased from 6.59 Ω/sq to 5.19 Ω/sq when annealing temperature was raised from room temperature to 400 °C; the decrease of sheet resistance was smaller than that of ZnO(Cu) thin films, whose sheet resistance decreased from 5.54 Ω/sq to 2.68 Ω/sq. Their mobility shows an increasing trend with the temperature. The carrier concentration of ZnO(Cu) thin films tends to increase continuously while that of ZnO(Ti/Cu) thin films increases from room temperature to 300 °C and then decreases. The reasons for this change in electrical properties can be explained simply as follows. The annealing process is conducive to the improvement of the quality and the degree of crystallization of the thin film (as obviously shown in the XRD patterns in Figure 3). It leads to the decrease of free electron scattering, which induces the increase of the carrier concentration and mobility [26]. Further, for ZnO(Cu) thin films, the increase of the carrier concentration and mobility leads to the decrease of sheet resistance. On the other hand, for ZnO(Ti/Cu) thin films, the particle size of the nanomaterials will increase obviously as the annealing temperature goes up to 300 °C, after which the carrier concentration will finally decrease under the influence of the quantum size effect. If the rate of carrier concentration decrease is lower than that of mobility increase, the conductivity of the thin films is improved.

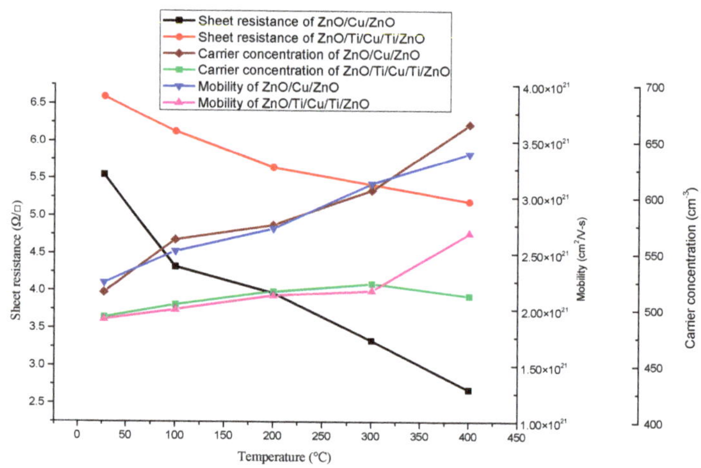

Figure 4. Electrical properties of two ZnO transparent conductive thin films with varying annealing temperatures.

The transmission spectra of two annealed film systems are presented in Figures 5 and 6. For the ZnO(Cu) film system, there are two crests in the spectra at wavelengths of about 380 nm and 580 nm. The wavelength shift of peak transmittance around the center wavelength of 380 nm is not obvious. However, the center wavelength of peak transmittance in long wavelengths shifts towards the longer wavelength regions. It can be noted that all the optical transmittances are lower than 60% in the range of wavelengths shown in the figure. It is also observed that there is only one crest in the spectrum, and all the peak optical transmittances are higher than 75% for ZnO(Ti/Cu) multilayers annealed at different temperatures. In order to clearly exhibit the relation between transmittance and temperature, the maximum transmittances of the two annealed multilayer film systems are presented in Figure 7. It is noted obviously that the transmittance of ZnO(Ti/Cu) thin films is higher than that of ZnO(Cu) thin films. In detail, the transmittance of ZnO(Cu) thin films increases with the annealing temperature. In particular, when the temperature increases from 300 °C to 400 °C, there is a large improvement in transmittance. But for ZnO(Ti/Cu) thin films, the transmittance shows a trend of increasing at first and then decreasing. That is, the multilayer has the maximum transmittance (about 87%) when the annealing temperature is 300 °C. Although annealing can improve the optical properties of the sandwiched ZnO/Metal/ZnO films to a certain extent, the reflective effect and light absorption of the metal layers still exist. Therefore, the average visible transmittance of sandwiched ZnO/Metal/ZnO films after annealing is lower than that of ZnO single film with the same thickness.

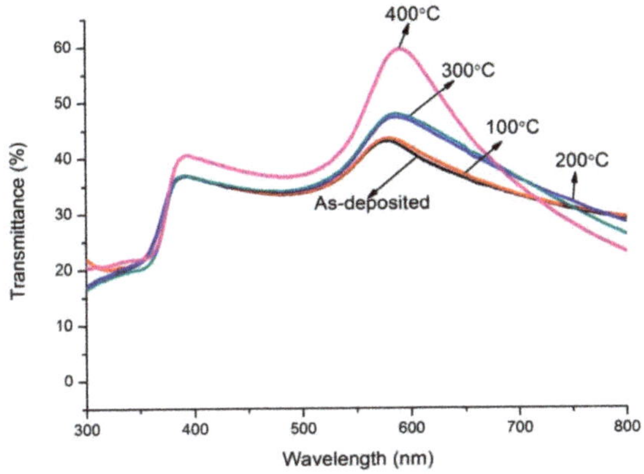

Figure 5. Optical transmission spectra of the ZnO(Cu) film system with varying annealing temperatures.

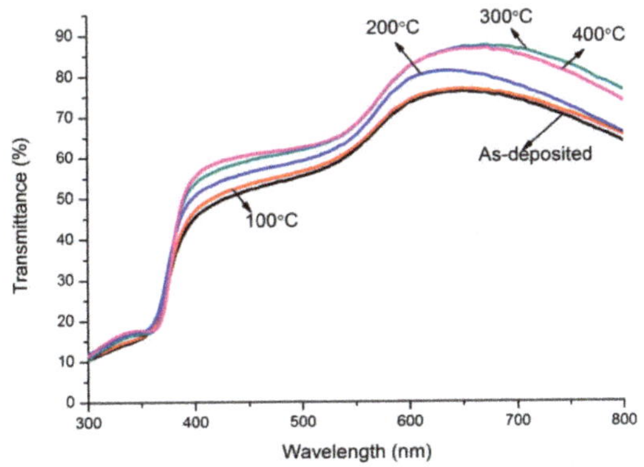

Figure 6. Optical transmission spectra of the ZnO(Ti/Cu) film system with varying annealing temperatures.

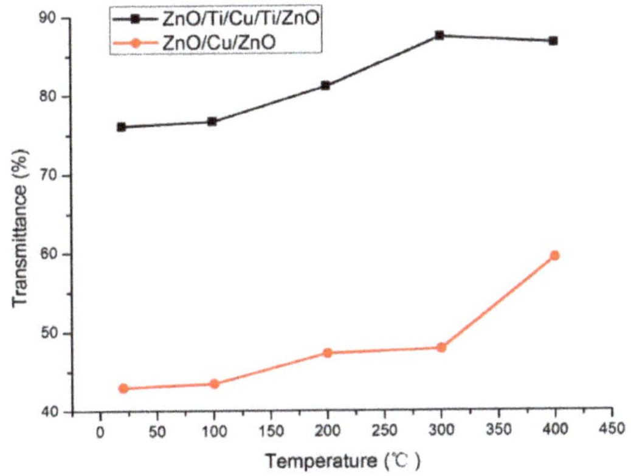

Figure 7. The maximum transmittance of the two multilayers with different annealing temperatures.

According to the transmittance spectra in Figures 5 and 6, the relationship curve between the $(\alpha hv)^2$ and hv of the two multilayer film systems with different annealing temperatures can be obtained, as shown in Figure 8, where α is the absorption coefficient and h is the photon energy. For ZnO/Metal/ZnO thin-film materials, the relationship between the α and hv can be expressed as [12,18,27–30]:

$$(\alpha hv)^2 = \Lambda(hv - E_g)$$

where Λ is a constant and E_g is the optical bandgap. Normally, E_g is extrapolated from the linear part of the relationship curve to the hv axis, as shown in Figure 8. The optical bandgaps of two annealed film systems are presented in Figure 9. The optical bandgap of ZnO obtained from ZnO(Cu) thin films (3.15~3.19 eV) is lower than that of ZnO(Ti/Cu): 3.20~3.23 eV, which varies near the theoretical value (~3.2 eV) [31,32]. In particular, the bandgap of ZnO(Cu) thin film increases with temperature, while that of ZnO(Ti/Cu) first increases and then decreases. The increase of bandgap after annealing is caused by the Burstein–Moss migration effect [33], which is related to the increase of carrier concentration in the film (Figure 4). The increased carriers fill in the lower energy level of the conduction band, making the valance electrons transfer to the higher energy level, thus increasing the bandgap width. As to the ZnO(Ti/Cu) thin films, the decrease of the bandgap between 300 °C and 400 °C is due to the increase of the particle size of the film, whose quantum size effect reduces the carrier concentration, leading to the bandgap narrowing.

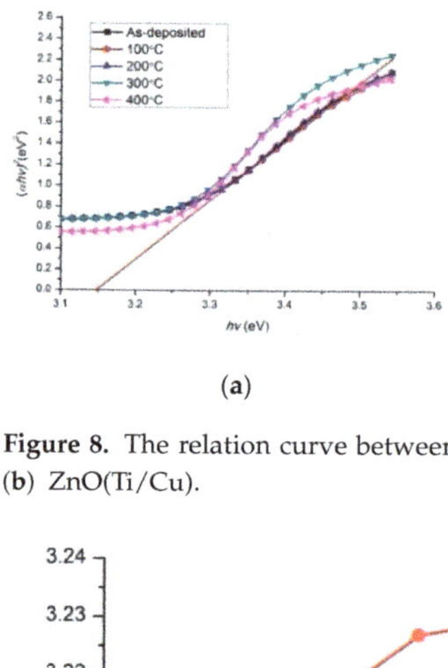

(a)

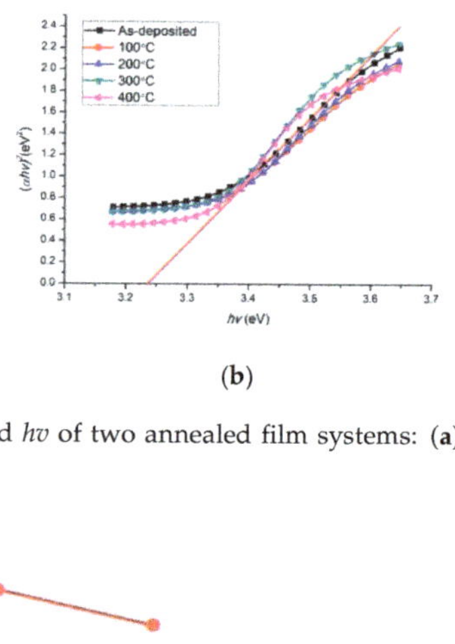

(b)

Figure 8. The relation curve between $(\alpha hv)^2$ and hv of two annealed film systems: (**a**) ZnO(Cu); (**b**) ZnO(Ti/Cu).

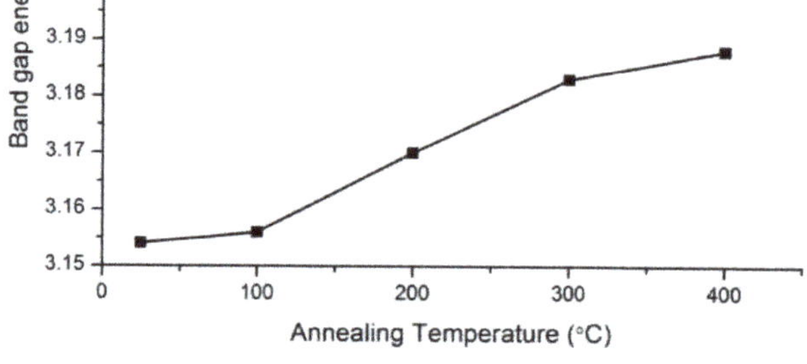

Figure 9. Bandgap energy of two annealed film systems.

The figures of merit of the ZnO(Cu) and ZnO(Ti/Cu) film systems after annealing are presented in Figure 10. The figure of merit of ZnO(Ti/Cu) is higher than that of ZnO(Cu). This shows that the performance of ZnO(Ti/Cu) is superior to that of ZnO(Cu). To be more specific, the higher annealing temperature increases the figure of merit for ZnO(Ti/Cu), but not obviously for ZnO(Cu). In other words, annealing has greatly improved the performance of ZnO(Ti/Cu). The figure of merit is highest when the annealing temperature is 300 °C although its conductivity is worse than thin film annealed at 400 °C. This means that the ZnO(Ti/Cu) thin film annealed at 300 °C has the best overall performance [17].

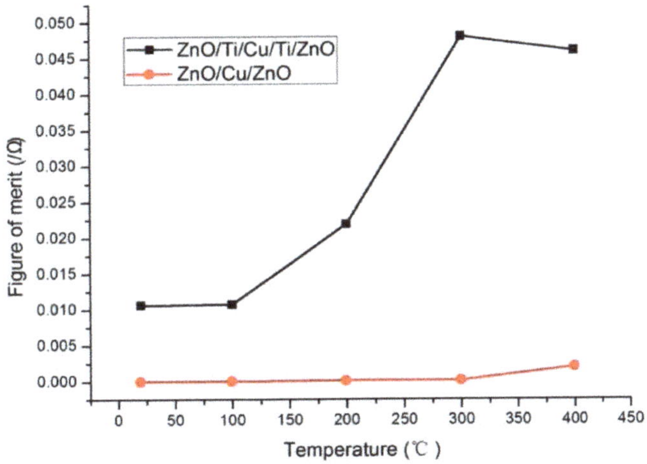

Figure 10. The figures of merit of the ZnO(Cu) and ZnO(Ti/Cu) thin films with varying annealing temperatures.

4. Conclusions

Two metal-sandwiched ZnO transparent conductive thin-film systems (50nm ZnO/Cu/ 50nm ZnO and 50nm ZnO/Ti/Cu/Ti/50nm ZnO) were fabricated, and the comparative analysis of their properties shows that the introduction of Ti layers can improve the overall performance of the film while maintaining the same overall metal layer thickness. The influence of the annealing temperature on the performance of two film systems was studied. The XRD patterns show that ZnO films have the (002) preferred orientation in both film systems, and the Cu films have [111] crystal orientation in 50nm ZnO/Cu/50nm ZnO films but do not crystallize very well in 50nm ZnO/Ti/Cu/Ti/50nm ZnO films. Annealing improved the conductivity of the two films due to a combination of changes in carrier concentration and mobility (although carrier concentration did not always increase with annealing temperature). The optical transmittance of ZnO(Ti/Cu) thin films with different annealing temperatures were higher than that of ZnO(Cu) thin films. The transmittance of ZnO(Cu) thin films increased with the annealing temperature, while that of ZnO(Ti/Cu) thin films increased from 100 °C to 300 °C and then decreased with higher temperatures. The bandgap of ZnO obtained from ZnO(Cu) thin films increased with temperature but remained lower than that of ZnO(Ti/Cu), which first increased and then decreased with temperature. The figure of merit of the ZnO(Ti/Cu) thin film is higher than that of ZnO(Cu), which indicates that the performance of ZnO(Ti/Cu) thin films is superior to ZnO(Cu). By comparing figures of merit, it can be determined that annealing can improve the performance of the film systems.

Author Contributions: Conceptualization, Q.L.; Data curation, N.Z.; Methodology, Q.L.; Software, F.Z.; Writing—original draft, Q.L.; Writing—review & editing, P.Y. All authors have read and agreed to the published version of the manuscript.

Funding: The authors would like to thank the financial supports by the National Key Research and Development Program (Grant No. 2021YFB3201803), the National Natural Science Foundation of China (Grant Nos. 51720105016, 52105560), and the Fundamental Research Funds for the Central Universities(No. xpt012021018).

Data Availability Statement: Not applicable.

Acknowledgments: The authors appreciate the support from the International Joint Laboratory for Micro/Nano Manufacturing and Measurement Technologies.

Conflicts of Interest: The authors declare no conflict of interest.

References

1. Lin, Y.-S.; Tseng, W.-C. Effect of Al Nanoparticles on the Microstructure, Electrical, and Optical Properties of AZO/Al/AZO Trilayer Thin Film. *J. Electron. Mater.* **2011**, *41*, 437–441. [CrossRef]
2. Yu, M.; Wang, G.; Zhao, R.; Liu, E.; Chen, T. Improved interfacial wetability in Cu/ZnO and its role in ZnO/Cu/ZnO sandwiched transparent electrodes. *J. Mater. Sci. Technol.* **2020**, *37*, 123–127. [CrossRef]
3. Zaabat, S.; Challali, F.; Chakaroun, M.; Boudine, B.; Boudrioua, A. Investigation and Comparison of Electrical and Optical Properties of RF Sputtered CU Doped ZnO and ZnO/Cu/ZnO Multilayer Films. *J. New Technol. Mater.* **2019**, *9*, 45–50. [CrossRef]
4. Namoune, A.; Touam, T.; Chelouche, A. Thickness, annealing and substrate effects on structural, morphological, optical and waveguiding properties of RF sputtered ZnO thin films. *J. Mater. Sci. Mater.* **2017**, *28*, 12207–12219. [CrossRef]
5. Liu, X.Y.; Li, Y.A.; Liu, S.; Wu, H.-L.; Cui, H.-N. ZnO/Cu/ZnO multilayer films: Structure optimization and investigation on photoelectric properties. *Thin Solid Films* **2012**, *520*, 5372–5377. [CrossRef]
6. Liang, Y.C.; Deng, X.S. Microstructure evolution and optical properties of c-axis-oriented ZnO thin films incorporated with silver nanoisland layers. *Ceram. Int.* **2014**, *40*, 1687–1692. [CrossRef]
7. He, X.X.; Wang, W.J.; Li, S.H.; Liu, Y.L.; Zheng, W.Q.; Shi, Q.; Luo, X. Experimental and theoretical analysis of ZnO/Au/ZnO transparent conducting thin films. *Vacuum* **2015**, *120*, 17–21. [CrossRef]
8. Lee, B.-S.; Kim, C.-I.; Joo, Y.-H. The Optical and Electrical Characteristics for ZnO/Al/ZnO Multilayer Films Deposited by RF Sputtering. *J. Nanoelectron. Optoelectron.* **2015**, *10*, 402–407. [CrossRef]
9. Li, W.Y.; Jiang, L.X.; Yin, G.L.; Wang, Y.-Y.; Yu, Z.; He, D.-N. Preparation and characterization of ZnO/Cu/ZnO transparent conductive films. *Rare Met.* **2013**, *32*, 273–277. [CrossRef]
10. Lin, Q.; Yang, S.; Jing, W.; Wang, C.; Jiang, Z. Numerical Simulation on Optical Characteristics of ZnO/Cu/ZnO Thin Film. *J. Nanosci. Nanotechnol.* **2016**, *16*, 873–877. [CrossRef]
11. Sahu, D.R.; Huang, J.L. Properties of ZnO/Cu/ZnO multilayer films deposited by simultaneous RF and DC magnetron sputtering at different substrate temperatures. *Microelectron. J.* **2007**, *38*, 299–303. [CrossRef]
12. Wang, L.; Chen, W.D.; Ling, L. Investigation of the optical and electrical properties of ZnO/Cu/ZnO multilayer structure for transparent conductive electrodes by magnetron sputtering. *J. Mater. Sci. Mater. Electron.* **2016**, *28*, 3458–3466. [CrossRef]
13. Wang, T.; Ma, H.P.; Yang, J.G.; Zhu, J.-T.; Zhang, H.; Feng, J.; Ding, S.-J.; Lu, H.-L.; Zhang, D.W. Investigation of the optical and electrical properties of ZnO/Cu/ZnO multilayers grown by atomic layer deposition. *J. Alloys Compd.* **2018**, *744*, 381–385. [CrossRef]
14. Wang, M.; Ji, J.; Luo, S.; Jiang, L.; Ma, J.; Xie, X.; Ping, Y.; Ge, J. Sol-gel-derived ZnO/Cu/ZnO multilayer thin films and their optical and electrical properties. *Mater. Sci. Semicond. Process.* **2016**, *51*, 55–59. [CrossRef]
15. Ionescu, M.; Bensebaa, F.; Luan, B.L. Study of optical and electrical properties of ZnO(Cu) multilayers deposited on flexible substrate. *Thin Solid Films* **2012**, *525*, 162–166. [CrossRef]
16. Sahu, D.R.; Huang, J.L. Dependence of film thickness on the electrical and optical properties of ZnO-Cu-ZnO multilayers. *Appl. Surf. Sci.* **2006**, *253*, 915–918. [CrossRef]
17. Song, S.; Yang, T.; Lv, M.; Li, Y.; Xin, Y.; Jiang, L.; Wu, Z.; Han, S. Effect of Cu layer thickness on the structural, optical and electrical properties of AZO/Cu/AZO tri-layer films. *Vacuum* **2010**, *85*, 39–44. [CrossRef]
18. Li, W.; Zhong, J.; Zhang, K.; Wang, Y.-Y.; Yin, G.; He, D.-n. Effects of Annealing on Properties of ZnO/Cu/ZnO Transparent Conductive Film. *J. Mater. Eng.* **2015**, *43*, 44–48.
19. Lee, J.H.; Yeo, B.W.; Park, B.O. Transparent conducting ZnO:Al, In and Sn thin films deposited by the sol-gel nethod. *Thin Solid Films* **2003**, *126*, 94–99. [CrossRef]
20. Sahu, D.; Huang, J.-L. The properties of ZnO/Cu/ZnO multilayer films before and after annealing in the different atmosphere. *Thin Solid Films* **2007**, *516*, 208–211. [CrossRef]
21. Ghosh, D.S.; Chen, T.L.; Pruneri, V. Ultrathin Cu-Ti bilayer transparent conductors with enhanced figure-of-merit and stability. *Appl. Phys. Lett.* **2010**, *96*, 091106. [CrossRef]
22. Park, J.S.; Kim, T.S.; Son, K.S.; Lee, E.; Jung, J.S.; Lee, K.-H.; Maeng, W.-J.; Kim, H.-S.; Kim, E.S.; Park, K.-B.; et al. Ti/Cu bilayer electrodes for SiNx-passivated Hf-In-Zn-O thin film transistors: Device performance and contact resistance. *Appl. Phys. Lett.* **2010**, *97*, 590. [CrossRef]

23. Han, Z.; Zhu, P.; Xu, L.; Kannan, C.S.; Guo, S.; Liu, J.; Koppala, S.; Ju, S. Electrochemical properties of the IrO_2-Ta_2O_5 coated anodes with Al/Ti and Cu/Ti layered composites substrates. *J. Alloys Compd.* **2018**, *769*, 210–217. [CrossRef]
24. Palik, E.D. *Handbook of Optical Constants of Solids*; Academic Press: Cambridge, MA, USA, 1985.
25. Haacke, G. New figure of merit for transparent conductors. *J. Appl. Phys.* **1976**, *47*, 4086–4089. [CrossRef]
26. Lee, G.; Yamamoto, Y.; Kourogi, M.; Ohtsu, M. Blue shift in room temperature photoluminescence from photo-chemical vapor deposited ZnO films. *Thin Solid Films* **2001**, *386*, 117–120. [CrossRef]
27. Montero, J.; Herrero, J.; Guillén, C. Preparation of reactively sputtered Sb-doped SnO_2 thin films: Structural, electrical and optical properties. *Sol. Energy Mater. Sol. Cells* **2010**, *94*, 612–616. [CrossRef]
28. Sivaramakrishnan, K.; Alford, T.L. Metallic conductivity and the role of copper in ZnO/Cu/ZnO thin films for flexible electronics. *Appl. Phys. Lett.* **2009**, *94*, 052104. [CrossRef]
29. Sivaramakrishnan, K.; Theodore, N.D.; Moulder, J.F.; Alford, T.L. The role of copper in ZnO/Cu/ZnO thin films for flexible electronics. *J. Appl. Phys.* **2009**, *106*, 063510. [CrossRef]
30. Minemoto, T.; Negami, T.; Nishiwaki, S.; Takakura, H.; Hamakawa, Y. Preparation of $Zn_{1-x}Mg_xO$ films by radio frequency magnetron sputtering. *Thin Solid Films* **2000**, *372*, 173–176. [CrossRef]
31. Choopun, S.; Vispute, R.D.; Yang, W.; Sharma, R.P.; Venkatesan, T. Realization of Band Gap above 5.0 eV in Metastable Cubic-Phase $Mg_xZn_{1-x}O$ Alloy Films. *Appl. Phys. Lett.* **2002**, *80*, 1529–1531. [CrossRef]
32. Tiwari, A.; Bhosle, V.M.; Ramachandran, S.; Sudhakar, N.; Narayan, J.; Budak, S.; Gupta, A. Ferromagnetism in Co doped CeO_2: Observation of a giant magnetic moment with a high Curie temperature. *Appl. Phys. Lett.* **2006**, *88*, 160. [CrossRef]
33. Lin, Q.; Zhao, N.; Jiang, Z.; Meng, Q.; Chen, F. Electrical and optical properties of metal-sandwiched ZnO/Ti/Cu/Ti/ZnO transparent conductive thin film. *Micro Nano Lett.* **2018**, *13*, 1511–1515. [CrossRef]

micromachines

Article

Research on the High Temperature and High Pressure Gold-Plated Fiber Grating Dual-Parameter Sensing Measurement System

Na Zhao [1], Zhongkai Zhang [1,*], Qijing Lin [1,2,*], Kun Yao [1], Liangquan Zhu [1], Yi Chen [3], Libo Zhao [1], Bian Tian [1], Ping Yang [1] and Zhuangde Jiang [1]

1 State Key Laboratory for Manufacturing Systems Engineering, Xi'an Jiaotong University, Xi'an 710049, China; zn2020@xjtu.edu.cn (N.Z.); yao_kun@outlook.com (K.Y.); zhuliangquan@stu.xjtu.edu.cn (L.Z.); libozhao@xjtu.edu.cn (L.Z.); t.b12@mail.xjtu.edu.cn (B.T.); ipe@xjtu.edu.cn (P.Y.); zdjiang@xjtu.edu.cn (Z.J.)

2 Collaborative Innovation Center of High-End Manufacturing Equipment, Xi'an Jiaotong University, Xi'an 710054, China

3 Aeronautics and Astronautics Engineering College, Air Force Engineering University, Xi'an 710038, China; chenyikgd@163.com

* Correspondence: zzkxjtu@outlook.com (Z.Z.); xjjingmi@163.com (Q.L.)

Abstract: In electrohydrostatic drive actuators, there is a demand for temperature and pressure monitoring in complex environments. Fiber Bragg grating (FBG) has become a promising sensor for measuring temperature and pressure. However, there is a cross-sensitivity between temperature and pressure. A gold-plated FBG is proposed and manufactured, and an FBG is used as a reference grating to form a parallel all-fiber sensing system, which can realize the simultaneous measurement of pressure and temperature. Based on the simulation software, the mechanical distribution of the pressure diaphragm is analyzed, and the fixation scheme of the sensor is determined. Using the demodulator to monitor the changes in the reflectance spectrum in real-time, the pressure and ambient temperature applied to the sensor are measured. The experimental results show that the temperature sensitivity of gold-plated FBG is 3 times that of quartz FBG, which can effectively distinguish the temperature changes. The pressure response sensitivity of gold-plated FBG is 0.3 nm/MPa, which is same as the quartz FBG. Through the sensitivity matrix equation, the temperature and pressure dual-parameter sensing measurement is realized. The accuracy of the temperature and pressure measurement is 97.7% and 99.0%, and the corresponding response rates are 2.7 ms/°C and 2 ms/MPa, respectively. The sensor has a simple structure and high sensitivity, and it is promising to be applied in health monitoring in complex environments with a high temperature and high pressure.

Keywords: optical fiber sensor; gold-plated fiber Bragg grating; pressure; temperature sensing

Citation: Zhao, N.; Zhang, Z.; Lin, Q.; Yao, K.; Zhu, L.; Chen, Y.; Zhao, L.; Tian, B.; Yang, P.; Jiang, Z. Research on the High Temperature and High Pressure Gold-Plated Fiber Grating Dual-Parameter Sensing Measurement System. *Micromachines* **2022**, *13*, 195. https://doi.org/10.3390/mi13020195

Academic Editor: Limin Xiao

Received: 2 January 2022
Accepted: 25 January 2022
Published: 27 January 2022

1. Introduction

The electro-hydrostatic actuator (EHA) is a highly integrated actuator [1,2], which can be widely used in the field of hydraulic transmission and control. However, the EHA structural system is complex and the working environment is harsh. There are defects, such as poor control accuracy, the difficulty in predicting failures, and low maintainability, which limit its operational reliability on major equipment. Monitoring the important working parameters of the temperature and pressure of the whole machine in the EHA system through sensing technology to form a feedback and regulation mechanism, can improve the control accuracy and reliability of the EHA.

At present, the reported electrical pressure and temperature integrated sensors are mainly used for normal temperature and low pressure monitoring [3–5], and there are few reports on integrated sensors that can be used in high temperature and high pressure harsh environments. Compared with the conventional electrical sensors, optical fiber sensors [6,7] have attracted attention for their stable chemical properties, anti-electromagnetic interference, compact structure, low cost, diverse functions, good insulation performance, and light weight. However, the research of optical fiber sensors mostly focuses on single-parameter measurement. To date, the optical fiber multi-parameter integrated sensors that meet the special needs of extreme environments are still problems that need to be tackled.

The related research of optical fiber temperature sensing mainly focuses on, for instance, the Mach–Zehnder interferometer [8–10], Michelson interferometer [11,12], Fabry–Perot interferometer [13,14], long period grating [15,16], and fiber Bragg grating (FBG) [17–37]. Although the above-mentioned various sensors can be directly used for temperature measurement, there are some shortcomings. For example, the structure of the Mach–Zehnder interferometer cannot be made into a probe structure; the structure of the Fabry–Perot interferometer is poor in stability; and the large measuring point size of the long-period grating leads to inaccurate measurements. In contrast, FBG has attracted much attention due to its small size, low cost, easy manufacturing, and high maturity. In 2003, NASA [17] adopted a distributed quartz fiber grating sensing system and installed the fiber sensing network on the X-38 space shuttle to realize the real-time monitoring of the temperature of the composite fuel tank. To date, the FBG sensor based on quartz fiber formed a relatively stable industrialized temperature detection system, with a temperature response sensitivity of about 0.01 nm/°C [18–20]. In 2021, Keith M.Alcock et al. [21] used the optical fiber sensor to measure the temperature of a lithium-ion battery, realizing the miniaturized installation of the optical fiber sensor. In the same year, Angela Brindisi et al. [22] equipped a fiber optic sensor on a small landing gear, based on the mechanical sensing performance of FBG, to evaluate whether there was a hard landing and the degree of the hard landing. In order to improve the sensitivity, people have made various attempts, including corrosion, coating, and other methods. Finally, it was found that corrosion does not help much to improve the temperature sensitivity of the grating, and it is easy to introduce physical interference, such as the refractive index and humidity. Therefore, optical fiber coating has become the focus of attention. Additionally, gold-plated optical fiber has the advantages of high expansion coefficients, good adhesion of gold atoms to optical fibers, and a mature manufacturing industry. Compared with quartz FBG, gold-plated FBG has a higher temperature response sensitivity, and gold-plated FBG also responds to strain, which can realize high temperature and high pressure measurements. In 2015, Monaghan et al. [23] used a metalized packaging method to improve the temperature sensitivity of FBG. In 2017, Liu Yanchao et al. [24] proposed a method for the in situ detection of lithium-ion batteries by pasting gold-plated fiber Bragg grating (FBG) sensors during the production of lithium-ion batteries. In 2019, Wu Hao et al. [25] used gold-plated grating and quartz grating cascades to realize the temperature and strain sensing measurements; the temperature response sensitivity was 26.5 pm/°C in the range of 30–70 °C, and the strain response sensitivity was 1.19 pm/με up to 400 με. In the same year, I. Laarossi et al. [26] used gold-plated gratings to measure the temperature and strain; the measurement response sensitivity was 1.10 pm/με and 3.7 pm/°C, respectively. In 2021, Yanjun Zhang et al. [27] conducted research on gold-plated FBG; the sensitization characteristics of the sensor were theoretically analyzed, and the response characteristics of the sensor were studied. To date, gold-plated FBG is mostly used in plasmon resonance [28–30] and battery biomarkers [31,32], and there have been few studies on the temperature and pressure.

Pressure value monitoring in EHA health monitoring is also very important. The pressure range of the pure optical fiber sensing structure is generally in the order of kPa [33–36], and with the help of a cantilever beam and diaphragm structure, it can reach the order of MPa [37–41]. In 2013, Lijun Li et al. [37] developed the FBG pressure sensor in order to meet the needs of coal mine production safety, with a sensitivity of 0.5983 nm/MPa in the range of 0 to 7.15 MPa. In 2016, Wang Hui et al. [38] designed a fiber grating pressure sensor combined with a cantilever beam, and the pressure sensitivity reached 3×103 nW/MPa in the measurement range of 0–6 MPa. In the same year, Hongtao Zhang [39] proposed a high-sensitivity pressure sensor based on FBG wavelength detection to measure the downhole pressure in oil and gas wells, and the sensitivity of the sensor can reach 230.9 pm/MPa in the range of 0 to 20 MPa. In 2017, Yiping Wang et al. [40] developed a high-sensitivity pressure sensor using phase shifted FBGs, and achieved a high-sensitivity measurement of 418.8 MHz/MPa in the sensing range of 0–4 MPa. In 2019, Zhen'an Jia et al. [41] proposed an FBG pressure sensor using a composite structure; the composite structure included a square diaphragm, a steel truss, and a vertical beam, and the pressure sensitivity of the sensor was 622.71 pm/MPa in the range of 0 to 2 MPa.

The above studies mostly focused on the measurement of the temperature or pressure, but the temperature and pressure responses of the sensor were all wavelength types. The combined effect of the temperature and pressure causes cross-interference between the data. How to overcome the multi-parameter interference and meet the needs of temperature and pressure sensor monitoring in a complex environment is the key point of later research. In order to achieve the simultaneous response to pressure and temperature, Nan Wang [42] of the PLA Naval Armament Department integrated and multiplexed the optical fiber pressure sensors and temperature compensation gratings to achieve rapid temperature compensation at low and normal temperatures, and complete 8 MPa large-scale high-precision pressure sensing; the response sensitivity was 0.15 nm/MPa. In 2019, Wenhua Wang et al. [43] proposed a Fabry–Perot interferometer and FBG cascaded fiber optic pressure sensor, and measured a pressure response of 0–1 MPa and a temperature response from 5.6 to 26.4 °C. In 2021, Qinggeng Fan et al. [44] designed a high-sensitivity square diaphragm pressure sensor based on FBG, and conducted theoretical and experimental verifications. The experimental results show that the pressure sensitivity of the sensor is 3.402 pm/kPa, in the range of 0–200 kPa, and the temperature response sensitivity is 19.29 pm/°C at 20–55 °C; this structure is suitable for low pressure measurement. In the existing papers, the research mostly focuses on the low-pressure and low-temperature section; however, the single-parameter measurement of temperature can already reach a high temperature [45]. However, for the temperature measurement under high pressure, the current research generally stays at room temperature, and there is less measurement and monitoring in the high-temperature and high-pressure complex environments.

In this paper, a high-temperature-resistant pressure diaphragm-type FBG temperature and pressure dual-parameter sensor is developed. The mechanical characteristics of the pressure sensitive diaphragm are simulated, the position range of the sensor on the pressure diaphragm is guided, and the working principle of the sensor is discussed. Based on the temperature and pressure characteristics of the gold-plated grating, which is different from the quartz grating, combined with the sensitivity matrix equation, the purpose of constructing a temperature and pressure dual-parameter sensor monitoring on the sensing probe is realized. The experimental results show that the optical fiber sensor has great potential in the simultaneous sensing and measurement of temperature and pressure.

2. Principle and Design

The preparation of gold-plated FBG is based on the photosensitive characteristics of optical fibers to produce FBG. By using ultraviolet light, some specific optical waveguide structures can be written into the optical fiber. After forming the optical fiber optical waveguide device using electron beam evaporation or the magnetron sputtering system gold plating method, the outer surface of the FBG can be gold-plated. FBG is packaged with a material with a large thermal expansion coefficient to improve the temperature sensitivity. The experimental design of the temperature and pressure sensing probe structure design is shown in Figure 1a. The overall diameter of the probe is 11 mm and the height is 5 mm. The sensing probe is divided into three parts, including a hexagonal nut-type cap for fixing the elastic diaphragm and the threaded connection structure below; the diaphragm is used to sense the pressure and the grating is pasted on it. The connection with the pressure supply device is based on a threaded connection end, which cooperates with a hexagonal nut to form a fixation to the diaphragm, and the physical diagram of the developed sensing probe is shown in Figure 1b,c. When the fluid pressure acts on the circular diaphragm, the diaphragm is deformed. The slight stretching and deformation of the fiber grating will cause the center wavelength of the fiber grating to shift, and the center wavelength shift will reflect the fluid pressure value. The gold-coated FBG and reference quartz FBG used in the experiments have the different grating pitches of 523.52 nm and 528.40 nm, respectively. Different grating pitches ensure that they have different center wavelengths, which is convenient for data analysis. In addition, they are all single-mode fibers with a fiber diameter of 250 μm and a FBG length of 300 mm. Considering the two-level difference of the grating length and diameter, the parallel connection of the gratings is selected for sensor installation, as shown in Figure 1d. The center wavelength of the reflection spectrum of the dual FBG is different, and the two sensing probes can be analyzed separately and the reflection spectrum can be coupled through the fiber coupler.

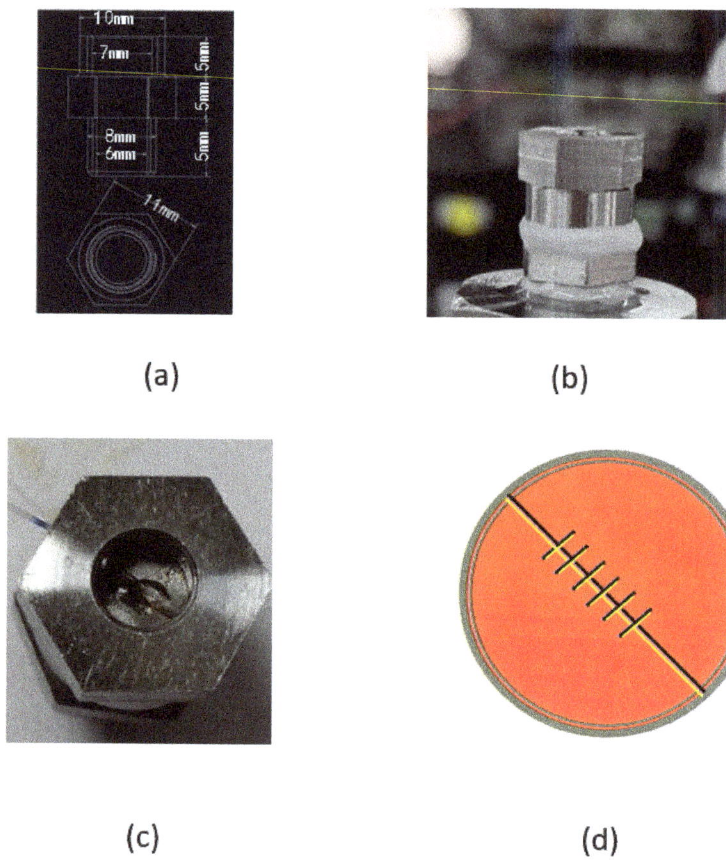

Figure 1. Design and manufacture of the sensor probe. (**a**) Structural design drawing of the sensor probe; (**b**) Physical map of the sensor probe; (**c**) Hexagonal nut; (**d**) Schematic diagram of the installation of parallel grating sensors.

Combined with the working environment of the pressure sensor, and based on ANSYS software, a mechanical simulation of a circular pressure diaphragm with a diameter of 1.5 cm, a thickness of 2 mm, and a material of 0Cr17Ni12Mo2 (AISI316) are carried out. As shown in Figure 2, the stress distribution in the linear part of the diaphragm at the middle diameter of 1 cm is relatively uniform, so that the force part of the FBG is concentrated on the center of the diaphragm as much as possible to obtain a more uniform stress distribution. When fluid pressure acts on the circular diaphragm, the diaphragm is deformed, and the slight stretching and deformation of the FBG will cause the center wavelength of the fiber grating to shift, and the center wavelength shift will reflect the fluid pressure value.

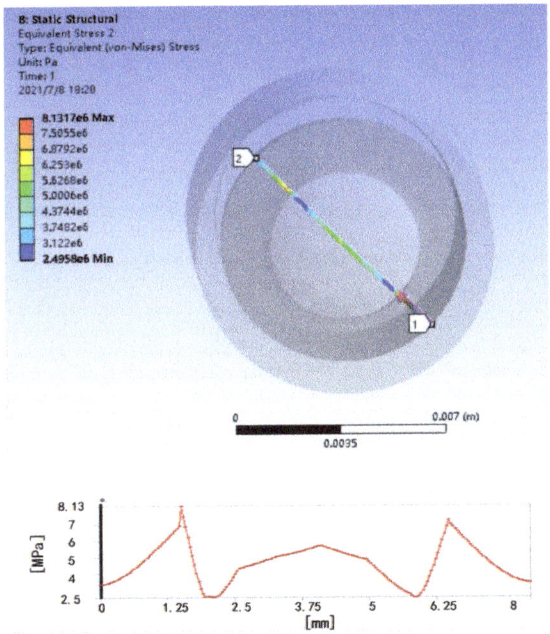

Figure 2. Force analysis of the pressure diaphragm.

The sensor measurement of the temperature and pressure environments can be obtained by designing a dual grating parallel structure with different response sensitivities. In terms of the temperature measurement, for the FBG structure, the thermo-optical effect and thermal expansion effect affect the change of the optical path difference. Therefore, when the ambient temperature changes, the length and effective index of the FBG will change. The optical path difference can be expressed as the following formula [46]:

$$\lambda_b = 2n_{eff}\Lambda \tag{1}$$

Among them, Λ is the grating pitch and n_{eff} is the effective refractive index of the core. When the ambient temperature acts on the FBG, the reflection spectrum will drift. In the formula, the coefficient of the thermal expansion and the optical path are constants. It is easy to obtain a reflection peak that is proportional to the temperature difference. In other words, the reflection peak changes with the outside temperature. We can obtain the ambient temperature by monitoring the frequency spectrum.

$$\frac{d\lambda_b}{dT} = 2\left(n_{eff}\frac{d\Lambda}{dT} + \Lambda\frac{dn_{eff}}{dT}\right) \tag{2}$$

The change of the grating pitch caused by the thermal expansion effect [47] is:

$$\frac{d\Lambda}{dT} = \alpha\cdot\Lambda \tag{3}$$

The change in the effective refractive index of the fiber caused by the thermo-optic effect is:

$$\frac{dn_{eff}}{dT} = n_{eff}\cdot\varepsilon \tag{4}$$

So the temperature sensitivity [48] is:

$$K_T = \frac{d\lambda_b}{dT} \cdot \frac{1}{\lambda_b} = \alpha + \varepsilon \quad (5)$$

$$\alpha = \frac{d\Lambda}{dT} \cdot \frac{1}{\Lambda} \quad (6)$$

$$\varepsilon = \frac{dn_{eff}}{dT} \cdot \frac{1}{n_{eff}} \quad (7)$$

In the formula, K_T is the temperature sensitivity, α is the thermal expansion coefficient corresponding to the optical fiber material, and ε is the thermo-optic coefficient. Due to the difference in the doping composition and doping concentration, the expansion coefficient α and the thermo-optic coefficient ε of the various optical fibers are quite different, so the temperature response sensitivity will also be different.

In terms of the pressure, among all the external factors that cause the FBG wavelength shift, the most direct is the mechanical parameter. This is because no matter whether the grating is stretched or squeezed, it will cause the change of the grating period Λ, and the elasto-optical effect of the fiber itself makes the effective refractive index n_{eff} also change with the change of the external stress state. Therefore, the use of FBG can be made into an optical fiber stress–strain sensor, where the wavelength shift caused by the stress can be uniformly described by Equation (8):

$$\Delta\lambda_b = 2n_{eff}\Delta\Lambda + 2\Delta n_{eff}\Lambda \quad (8)$$

where $\Delta\Lambda$ is the deformation of the fiber grid under stress, and Δn_{eff} is the elastic-optical effect of the fiber.

Differentiate both sides of the expression (1) of the center wavelength to obtain the following formula:

$$d\lambda_b = 2n_{eff}d\Lambda + 2\Lambda dn_{eff} \quad (9)$$

Divide both ends of Equation (9) by the terms on both sides of Equation (1) to obtain the following equation:

$$\frac{d\lambda_b}{\lambda_b} = \frac{dn_{eff}}{n_{eff}} + \frac{d\Lambda}{\Lambda} \quad (10)$$

Since the change of the refractive index of the optical fiber material is less affected by the stress, the influence of the refractive index, n, can be ignored. The above formula can be simplified to:

$$\frac{d\lambda_b}{\lambda_b} = \frac{d\Lambda}{\Lambda} = \frac{\Delta L}{L} \quad (11)$$

L represents the total length of the optical fiber, and ΔL represents the longitudinal expansion and contraction of the optical fiber.

In addition, since the grating is fixed on the elastic diaphragm, the change in the length of the grating ΔL is mainly affected by the elastic diaphragm. Therefore, the pressure response sensitivity of the two types of sensors is relatively similar from a theoretical point of view.

3. Experiment and Discussions

When conducting temperature experiments, keep the pressure constant and only change the temperature of the thermostat. The experimental temperature device is shown in Figure 3. The optical fiber grating demodulator used in this experiment adopts the U.S. MICRON OP ICS company SI 155; its minimum resolution is 0.02 nm, and the demodulator integrates a coupler, ASE light source (1510 nm–1590 nm), and signal demodulation system. The OMEGA thermometer has an accuracy of 0.5 °C. The length of the quartz fiber grating and the gold-plated grating in the experiment are both 3 mm, the wavelengths of the center after the structure is fixed are 1537.16 nm and 1551.48 nm, and the reflectivity is 90% and 30%, respectively.

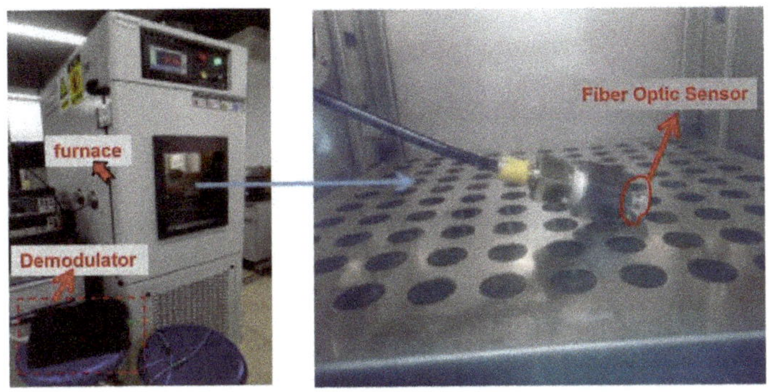

Figure 3. The actual photograph for the temperature measurement system.

In order to study the temperature response, the FBG is placed in a thermostat. The experiment carried out two temperature rising and falling experiments. The two temperature response curves can overlap well, and the four temperature response curves of the rising and falling temperatures can also overlap well, indicating that the sensor probe has good repeatability, as shown in Figures 4 and 5. During the heating process, the center wavelength of the FBG decreases linearly with the increase in temperature. Similarly, during the cooling process, the center wavelength of the fiber grating decreases linearly with the decrease in temperature.

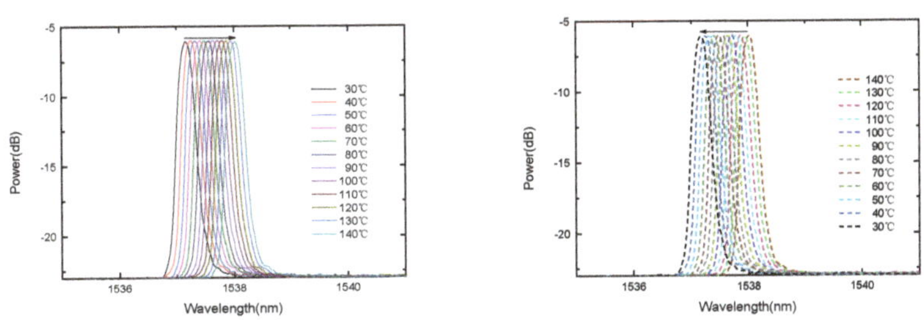

(**a**) Heating (**b**) Cooling

Figure 4. The change of the FBG center wavelength with the temperature.

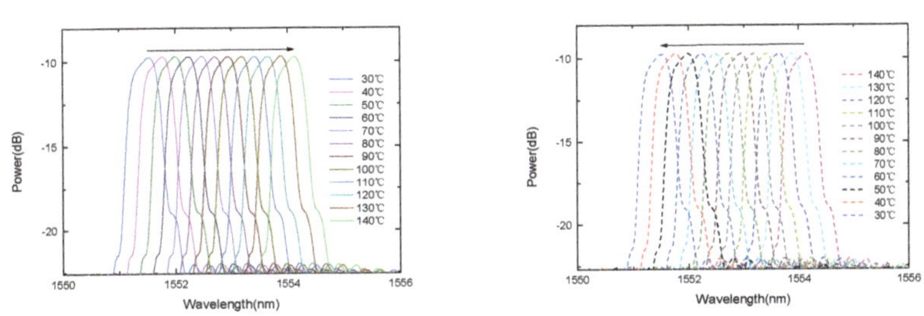

(**a**) Heating (**b**) Cooling

Figure 5. The change of the gold-plated FBG center wavelength with the temperature.

Figure 6 is a diagram of the pressure response of the FBG pressure sensor based on the data collected in Table 1, which is the temperature experimental data measured at

30 °C to 120 °C. The points in the figure are the measured data, and the fitting line is obtained by the least squares linear fitting. From the results of the experimental data analysis, the temperature response sensitivity of the sensor during the two heating processes are 0.009 nm/°C and 0.027 nm/°C, respectively. Combining the resolution of the spectrometer and the sensitivity of the gold-plated FBG, the sensor has a temperature resolution of 0.8 °C. Based on the measured wavelength, the sensitivity coefficient, and standard temperature value, the maximum temperature error is 2.5 °C, so the measurement accuracy δ can be obtained through the accuracy measurement equation as follows: δ = 100% − 2.5/(140 − 30) × 100%= 97.7%.

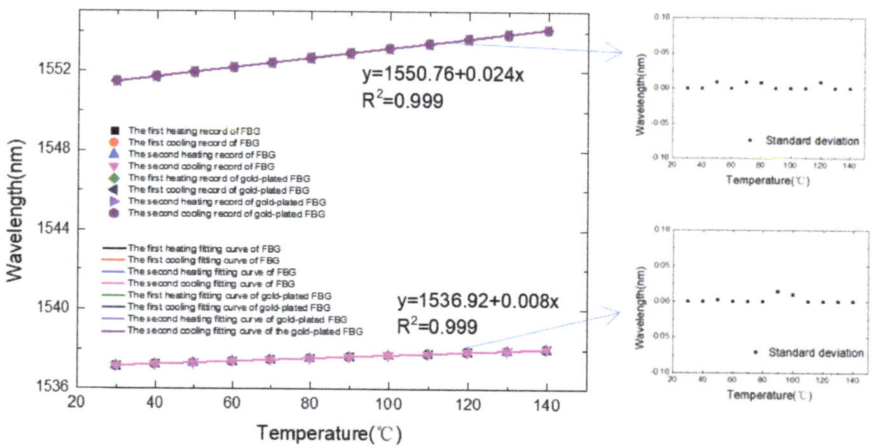

Figure 6. Fitting curve of the temperature response sensitivity of the double grating sensor probe and the error bars.

A pressure measurement system was designed, as shown in Figure 7. The probe was placed at the impulse tube of the pressure gauge. As the external pressure changes, the pressure diaphragm is deformed, and the grating is also deformed, which in turn causes the pitch of the grating to change. Therefore, the FBG pressure sensing system can demodulate the spectrum corresponding to the FBG at different pressure. The pressure source used in the experiment was the CW-600T pressure calibrator, which was connected with the sensor by the threaded interface (M20 × 1.5). The sensor, coupler, ASE light source, and demodulation system were connected to each other by optical fibers. The pressure gauge for monitoring the pressure source of the fiber grating pressure sensor was produced by the Xi'an Instrument Factory, with a range of 0–60 MPa and an accuracy of 0.01-level standard pressure gauge. The ASE light source emitted a beam of broadband light that entered the sensor probe through the coupler, and the modulation system demodulated the center wavelength of the FBG. The circular diaphragm was deformed due to the pressure, and the center wavelength at this time was less than the unpressurized center wavelength.

Table 1. Wavelength data collected during the temperature rise and fall.

	T/°C	Wavelength Corresponding to the First Temperature Change of the FBG/nm	Wavelength Corresponding to the Second Temperature Change of the FBG/nm	Wavelength Corresponding to the First Temperature Change of the Gold-Plated FBG/nm	Wavelength Corresponding to the Second Temperature Change of the Gold-Plated FBG/nm
	30	1537.16	1537.16	1551.48	1551.48
	40	1537.24	1537.24	1551.72	1551.72
	50	1537.32	1537.32	1551.96	1551.96
	60	1537.4	1537.4	1552.2	1552.2
	70	1537.48	1537.46	1552.44	1552.44
	80	1537.56	1537.56	1552.68	1552.68
↑	90	1537.64	1537.64	1552.94	1552.90
	100	1537.72	1537.72	1553.16	1553.16
	110	1537.8	1537.8	1553.4	1553.4
	120	1537.86	1537.88	1553.64	1553.64
	130	1537.96	1537.96	1553.88	1553.88
	140	1538.04	1538.04	1554.12	1554.12
	140	1538.04	1538.04	1554.12	1554.12
	130	1537.96	1537.96	1553.88	1553.88
	120	1537.88	1537.88	1553.64	1553.64
	110	1537.8	1537.8	1553.4	1553.4
	100	1537.72	1537.72	1553.16	1553.18
	90	1537.64	1537.64	1552.92	1552.92
↓	80	1537.54	1537.56	1552.68	1552.68
	70	1537.48	1537.48	1552.44	1552.44
	60	1537.4	1537.4	1552.2	1552.2
	50	1537.32	1537.34	1551.96	1551.96
	40	1537.24	1537.24	1551.72	1551.72
	30	1537.16	1537.16	1551.48	1551.48

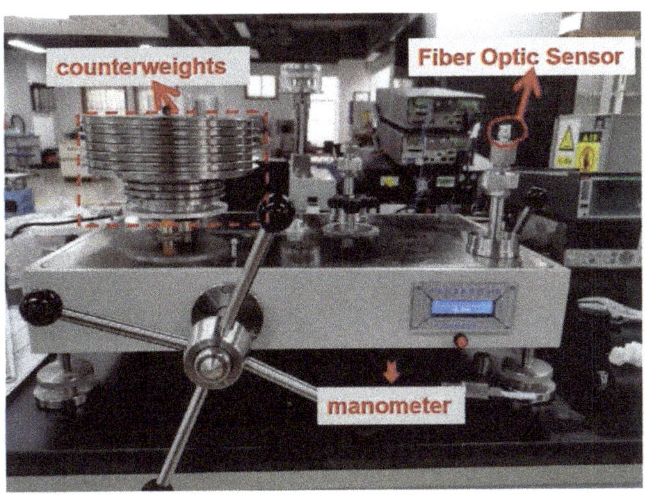

Figure 7. The actual photograph for the pressure measurement system.

The experiment carried out two compression processes and monitored the peaks at 1537.04 nm and 1551.60 nm, respectively. During the boost process, the center wavelengths of the FBGs decreases linearly with the increase in the pressure, as shown in Figures 8 and 9. Similarly, during the depressurization process, the center wavelengths of the FBGs increase linearly as the pressure decreases.

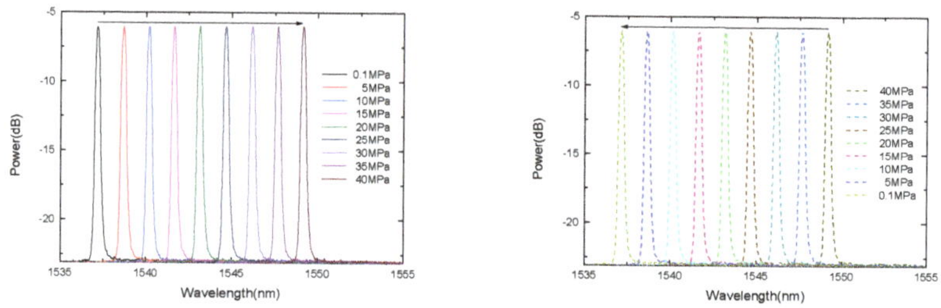

(a) Rise (b) Reduction

Figure 8. The change of the FBG center wavelength with pressure.

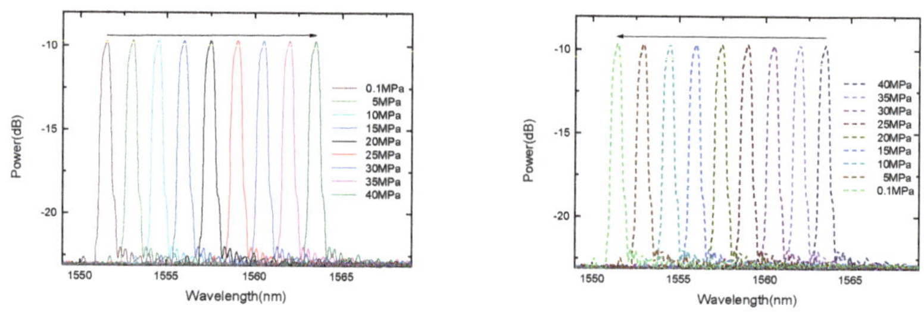

(a) Rise (b) Reduction

Figure 9. The change of the gold-plated FBG center wavelength with pressure.

Table 2 is the experimental data collected under the different temperature environments, and Figure 10 is the FBG pressure response graph. The experimental results show that the pressure response sensitivity of the FBG and gold-plated FBG is similar, which is 0.3 nm/MPa at 0.1 MPa to 40 MPa. The sensor has good linearity, and the response curves of the two pressure rise-and-fall processes can be well overlapped. Combining the resolution of the spectrometer and the sensors' sensitivity, the gold-plated FBG has a temperature resolution of 0.8 °C. Based on the measured wavelength, sensitivity coefficient, and standard pressure value, the maximum temperature error is 0.4 MPa, so the measurement accuracy can be obtained through the accuracy measurement equation as follows: $\delta = 100\% - 0.4/(40 - 0.1) \times 100\% = 99.0\%$.

The response time of the temperature and pressure is a key parameter of the sensor, especially when the sensor is used in some extreme environments. Its influence mainly includes the following four parameters: the elastic deformation speed of the diaphragm structure, the change speed of the pressure to be measured, the change of the refractive index of the optical fiber material and the grid with the temperature and pressure, and the response time of the detector. A fast response experiment was carried out, and the response time of the FBG was measured by the time constant, which is defined as the time taken when the temperature or pressure rises to 63.2% of the steady-state value, that is, the collected signal rises from the initial value to 63.2%. The results show that the response times of the temperature and pressure are 2.7 ms/°C and 2 ms/MPa, respectively.

Table 2. Wavelength data collected during the pressure rise and fall.

	P/MPa	Wavelength Corresponding to the First Pressure Change of the FBG/nm	Wavelength Corresponding to the Second Pressure Change of the FBG/nm	Wavelength Corresponding to the First Pressure Change of the Gold-Plated FBG/nm	Wavelength Corresponding to the Second Pressure Change of the Gold-Plated FBG/nm
	0.1	1537.24	1537.04	1551.6	1551.5
	5	1535.66	1535.66	1549.96	1550.08
	10	1534.08	1534.18	1548.54	1548.48
	15	1532.68	1532.78	1546.98	1546.92
↑	20	1531.18	1531.18	1545.48	1545.49
	25	1529.73	1529.66	1544	1543.98
	30	1528.14	1528.14	1542.58	1542.54
	35	1526.66	1526.68	1541	1540.99
	40	1525.16	1525.18	1539.54	1539.38
	40	1525.16	1525.16	1539.5	1539.48
	35	1526.66	1526.72	1540.98	1541.04
	30	1528.16	1528.16	1542.6	1542.58
	25	1529.66	1529.74	1544.06	1544.08
↓	20	1531.24	1531.16	1545.48	1545.38
	15	1532.66	1532.66	1547	1547.08
	10	1534.16	1534.16	1548.54	1548.44
	5	1535.72	1535.66	1550.04	1550.04
	0.1	1537.16	1537.28	1551.58	1551.38

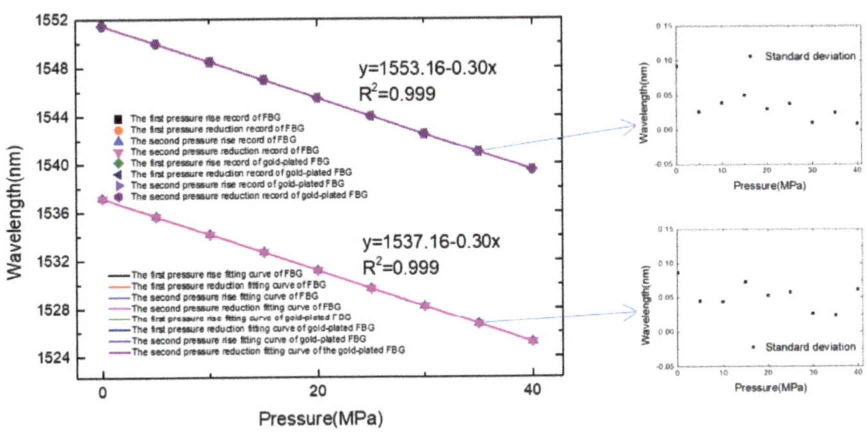

Figure 10. Fitting curve of the pressure response sensitivity of the double grating sensor probe and the error bars.

As shown in Figure 11, connect the sensor probe to the pressure gauge so that the sensor probe can sense the pressure signal, and at the same time place the sensor in the thermostat, so that the sensor can sense temperature and pressure information at the same time. Based on the dual-parameter sensor probe, the temperature and pressure are measured at the same time, and the spectral data at different moments are obtained, as shown in Figure 12.

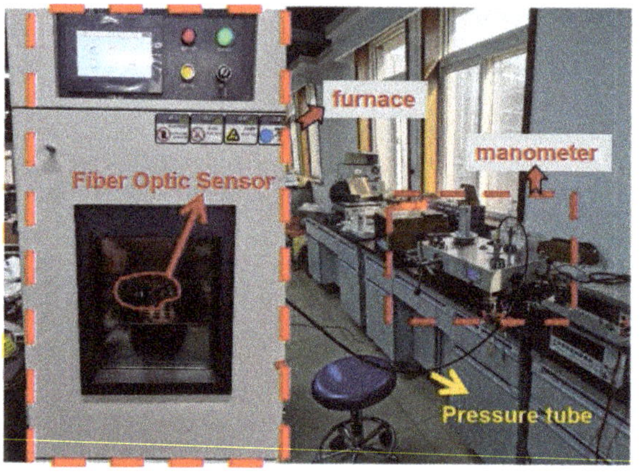

Figure 11. Dual-parameter sensing experiment platform.

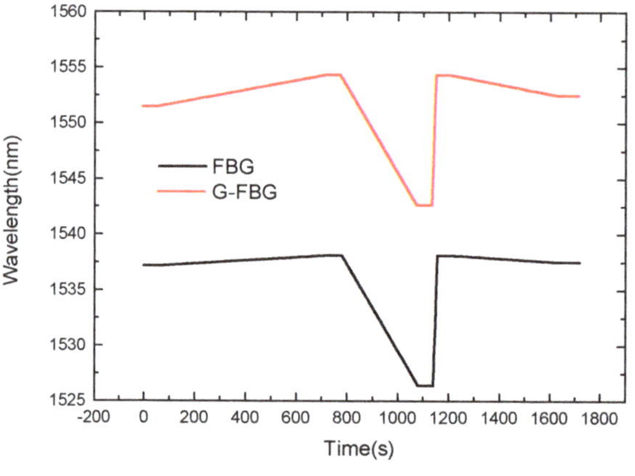

Figure 12. Measure the spectra with temperature and pressure.

When the temperature and pressure act on the sensor, the sensitivity matrix equation can be used to achieve the dual-parameter differential measurement, such as Equation (12), where k_x is the temperature and pressure response sensitivity of the two sensors, λ^0 is the initial wavelength, T is the temperature to be measured in the experiment, and the subscripts 1 and 2 are used to distinguish two different sensor structures. The wavelength λ can be expressed as follows:

$$\begin{bmatrix} \lambda_1 \\ \lambda_2 \end{bmatrix} = \begin{bmatrix} \lambda_1^0 \\ \lambda_2^0 \end{bmatrix} + \begin{bmatrix} k_1 & k_2 \\ k_3 & k_4 \end{bmatrix} \begin{bmatrix} T \\ P \end{bmatrix} \tag{12}$$

$$\begin{bmatrix} \lambda_1 - \lambda_1^0 \\ \lambda_2 - \lambda_2^0 \end{bmatrix} = \begin{bmatrix} k_1 & k_2 \\ k_3 & k_4 \end{bmatrix} \begin{bmatrix} T \\ P \end{bmatrix} \tag{13}$$

By multiplying the reciprocal matrix and combining the formula, the temperature and pressure parameters can be obtained as follows:

$$\begin{bmatrix} k_{T1} & k_P \\ k_{T2} & k_P \end{bmatrix}^{-1} \begin{bmatrix} \lambda_1 - \lambda_1^0 \\ \lambda_2 - \lambda_2^0 \end{bmatrix} = \begin{bmatrix} k_{T1} & k_P \\ k_{T2} & k_P \end{bmatrix}^{-1} \begin{bmatrix} k_{T1} & k_P \\ k_{T2} & k_P \end{bmatrix} \begin{bmatrix} T \\ P \end{bmatrix} \tag{14}$$

$$\begin{bmatrix} T \\ P \end{bmatrix} = \begin{bmatrix} k_{T1} & k_P \\ k_{T2} & k_P \end{bmatrix}^{-1} \begin{bmatrix} \lambda_1 - \lambda_1^0 \\ \lambda_2 - \lambda_2^0 \end{bmatrix} \tag{15}$$

The initial wavelengths λ_1^0 and λ_2^0 of the detection wavelengths in the experiment are 1537.04 nm and 1551.60 nm, respectively. The temperature response sensitivities, k_{Tx}, of the two sensing structures are 0.008 nm/°C and 0.024 nm/°C, respectively, and the pressure response sensitivities k_{Px} are both 0.3 nm/MPa. Substituting each parameter into formula (15) can obtain formula (16), which is used to measure the environmental parameters. Specifically, with the help of the matrix method, the temperature and pressure values at different time points can be detected at the same time, as shown in Figure 13.

$$\begin{bmatrix} T \\ P \end{bmatrix} = \begin{bmatrix} 0.008 & 0.3 \\ 0.024 & 0.3 \end{bmatrix}^{-1} \begin{bmatrix} \lambda_1 - 1537.04 \\ \lambda_2 - 1551.60 \end{bmatrix} \tag{16}$$

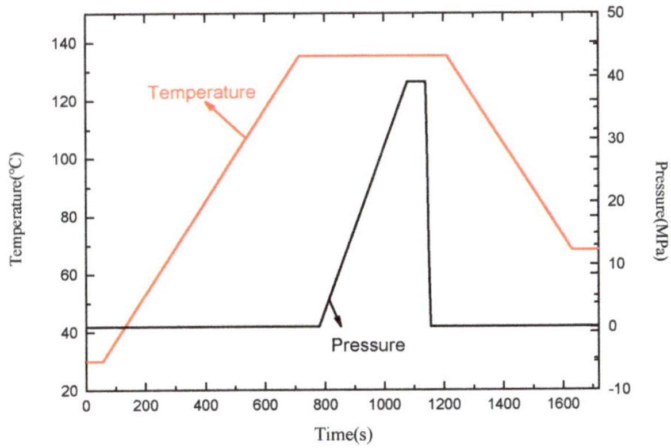

Figure 13. The temperature and pressure conditions to be measured based on the spectrum analysis of the sensor probe.

4. Conclusions

This paper proposes and manufactures an all-optical fiber sensor system based on the parallel structure of gold-plated FBG and quartz FBG, which can simultaneously measure temperature and pressure. As the temperature and pressure sensitivity of the two sensor structures are different, we can measure the temperature and pressure by monitoring the response of the wavelength in real-time based on the sensitivity matrix equation. The experimental results show that the pressure response sensitivity of the quartz FBG and gold-plated FBG are both 0.3 nm/MPa. The temperature sensitivity of the gold-plated FBG is 0.024 nm/°C with a resolution of 0.8 °C, and the sensitivity of the quartz FBG is 0.008 nm/°C with a resolution of 0.067 MPa, which can distinguish the temperature and pressure changes well. The sensor probe has the advantages of a simple structure, easy production, small size, high sensitivity, and dual-parameter measurement, which can be applied to monitor the running status of the EHA.

Author Contributions: Conceptualization, N.Z., Q.L. and P.Y.; Data curation, N.Z. and Z.Z.; Formal analysis, N.Z. and L.Z. (Libo Zhao); Project administration, B.T. and Z.J.; Resources, Y.C., L.Z. (Libo Zhao), B.T. and Z.J.; Software, K.Y. and L.Z. (Liangquan Zhu); Writing-original draft, N.Z.; Writing-review & editing, Q.L. and P.Y. All authors have read and agreed to the published version of the manuscript.

Funding: This work is supported by the Postdoctoral Innovative Talent Support Program (Grant No. BX20200274), the National Natural Science Foundation of China (Grant Nos. 52105560, 51890884, 51720105016), the Postdoctoral Science Foundation (Grant No. 2021M702596), and the National Science and Technology Major Project (Grant No. J2019-V-0006-0100).

Conflicts of Interest: The authors declare no conflict of interest.

References

1. Zhang, H.; Ding, L.; Zhang, W.; Li, C. Performance analysis of an electro-hydrostatic actuator with high-pressure load sensing based on fuzzy PID. *Mech. Sci.* **2021**, *12*, 529–537. [CrossRef]
2. Lei, Z.; Qin, L.; Wu, X.; Jin, W.; Wang, C. Research on Fault Diagnosis Method of Electro-Hydrostatic Actuator. *Shock Vib.* **2021**, *2021*, 6688420. [CrossRef]
3. Oh, B.; Park, Y.-G.; Jung, H.; Ji, S.; Cheong, W.H.; Cheon, J.; Lee, W.; Park, J.-U. Untethered soft robotics with fully integrated wireless sensing and actuating systems for somatosensory and respiratory functions. *Soft Robot.* **2020**, *7*, 564–573. [CrossRef] [PubMed]
4. Wang, L.; Zhu, R.; Li, G. Temperature and strain compensation for flexible sensors based on thermosensation. *ACS Appl. Mater. Interfaces* **2019**, *12*, 1953–1961. [CrossRef] [PubMed]
5. Zhang, L.; Yang, M.-Z.; Liang, X.-F.; Zhang, J. Oblique tunnel portal effects on train and tunnel aerodynamics based on moving model tests. *J. Wind Eng. Ind. Aerodyn.* **2017**, *167*, 128–139. [CrossRef]
6. Jiang, C.L.; Yan, T.J.; Song, M.C.; Shi, J.L. Application of Fiber Bragg Grating Sensing Technology in Key Equipment Monitoring of Offshore Platform. In Proceedings of the 2020 2nd International Conference on Industrial Artificial Intelligence (IAI), Shenyang, China, 24–26 July 2020; pp. 1–6. [CrossRef]
7. Chen, R.; He, W.; Xie, H.; Liu, S. Monitoring the strain and stress in FDM printed lamellae by using Fiber Bragg Grating sensors. *Polym. Test.* **2021**, *93*, 106944. [CrossRef]
8. Antonio-Lopez, J.E.; Eznaveh, Z.S.; LiKamWa, P.; Schülzgen, A.; Amezcua-Correa, R. Multicore fiber sensor for high-temperature applications up to 1000 °C. *Opt. Lett.* **2014**, *39*, 4309–4312. [CrossRef]
9. Zhao, N.; Lin, Q.; Jing, W.; Jiang, Z.; Wu, Z.; Yao, K.; Tian, B.; Zhang, Z.; Shi, P. High temperature high sensitivity Mach-Zehnder interferometer based on waist-enlarged fiber bitapers. *Sens. Actuators A Phys.* **2017**, *267*, 491–495. [CrossRef]
10. Liu, T.; Wang, J.; Liao, Y.; Yang, L.; Wang, S. Splicing point tapered fiber Mach-Zehnder interferometer for simultaneous measurement of temperature and salinity in seawater. *Opt. Express* **2019**, *27*, 23905–23918. [CrossRef]
11. Zhou, J.; Wang, Y.; Liao, C.; Sun, B.; He, J.; Yin, G.; Liu, S.; Li, Z.; Wang, G.; Zhong, X.; et al. Intensity modulated refractive index sensor based on optical fiber Michelson interferometer. *Sens. Actuators B Chem.* **2015**, *208*, 315–319. [CrossRef]
12. Zhao, N.; Fu, H.; Shao, M.; Yan, X.; Li, H.; Liu, Q.; Gao, H.; Liu, Y.; Qiao, X. High temperature probe sensor with high sensitivity based on Michelson interferometer. *Opt. Commun.* **2015**, *343*, 131–134. [CrossRef]
13. Wu, Y.; Li, H.; Yan, G.; Meng, F.; Zhu, L. Simultaneous measurement of refractive index and temperature based on asymmetrical Fabry-Perot interferometer. *Microw. Opt. Technol. Lett.* **2019**, *61*, 2190–2195. [CrossRef]
14. Zhang, C.; Fu, S.; Tang, M.; Liu, D. Femtosecond laser fabricated all-multicore-fiber parallel Fabry-Perot interferometers for dual-parameter sensing. In Proceedings of the Optical Fiber Communication Conference and Exhibition, San Diego, CA, USA, 8–12 March 2020. [CrossRef]
15. Eid, M.M.; Rashed, A.N.Z. Numerical simulation of long-period grating sensors (LPGS) transmission spectrum behavior under strain and temperature effects. *Sens. Rev.* **2021**, *41*, 192–199. [CrossRef]
16. Zhao, Y.; Liu, S.; Luo, J.; Chen, Y.; Fu, C.; Xiong, C.; Wang, Y.; Jing, S.; Bai, Z.; Liao, C. Torsion, refractive index, and temperature sensors based on an improved helical long period fiber grating. *J. Lightwave Technol.* **2020**, *38*, 2504–2510. [CrossRef]
17. Bennett, T.; Fox, R. Design, Development & Flight Testing of the NASA X-38 7500 ft2 Parafoil Recovery System. In Proceedings of the 17th AIAA Aerodynamic Decelerator Systems Technology Conference and Seminar, Monterey, CA, USA, 19–22 May 2003. [CrossRef]
18. Bremer, K.; Alwis, L.S.M.; Zheng, Y.; Roth, B.W. Towards mode-multiplexed fiber sensors: An investigation on the spectral response of etched graded index OM4 multi-mode fiber with Bragg grating for refractive index and temperature measurement. *Appl. Sci.* **2020**, *10*, 337. [CrossRef]
19. Hu, W.; Li, C.; Cheng, S.; Mumtaz, F.; Du, C.; Yang, M.; Farhan, M. Etched multicore fiber Bragg gratings for refractive index sensing with temperature in-line compensation. *OSA Contin.* **2020**, *3*, 1058–1067. [CrossRef]
20. Wu, F.; Jiang, D.; He, W.; Yang, C.; Yao, J. Surface displacement measurement based on fiber Bragg grating under narrow gap conditions. *J. Mech. Eng.* **2007**, *43*, 231–234. [CrossRef]
21. Alcock, K.M.; Grammel, M.; González-Vila, Á.; Binetti, L.; Goh, K.; Alwis, L.S. An accessible method of embedding fibre optic sensors on lithium-ion battery surface for in-situ thermal monitoring. *Sens. Actuators A Phys.* **2021**, *332*, 113061. [CrossRef]
22. Brindisi, A.; Vendittozzi, C.; Travascio, L.; Di Palma, L.; Ignarra, M.; Fiorillo, V.; Concilio, A. A Preliminary Assessment of an FBG-Based Hard Landing Monitoring System. *Photonics* **2021**, *8*, 450. [CrossRef]
23. Monaghan, T.; Capel, A.J.; Christie, S.; Harris, R.; Friel, R. Solid-state additive manufacturing for metallized optical fiber integration. *Compos. Part A Appl. Sci. Manuf.* **2015**, *76*, 181–193. [CrossRef]
24. Liu, Y.; Fang, J.; Xu, C.; Wei, B.; Guan, Y.; Fan, M.; Yan, X.; Gao, C. The feasibility of gold-plated fiber Bragg grating sensors for in-situ detection of lithium-ion batteries. *Prog. Laser Optoelectron.* **2017**, *54*, 105–111. [CrossRef]
25. Wu, H.; Lin, Q.; Jiang, Z.; Zhang, F.; Li, L.; Zhao, L. A temperature and strain sensor based on a cascade of double fiber Bragg grating. *Meas. Sci. Technol.* **2019**, *30*, 065104. [CrossRef]

26. Laarossi, I.; Roldán-Varona, P.; Quintela-Incera, M.A.; Rodríguez-Cobo, L.; López-Higuera, J.M. Ultrahigh temperature and strain hybrid integrated sensor system based on Raman and femtosecond FBG inscription in a multimode gold-coated fiber. *Opt. Express* **2019**, *27*, 37122–37130. [CrossRef] [PubMed]
27. Zhang, Y.; Gao, H.; Zhang, L.; Liu, Q.; Fu, X. Embedded gold-plated fiber Bragg grating temperature and stress sensors encapsulated in capillary copper tube. *Opto-Electron. Eng.* **2021**, *48*, 200195. [CrossRef]
28. Luo, J.; Liu, G.S.; Zhou, W.; Hu, S.; Chen, Y.; Luo, Y.; Chen, Z. A graphene-PDMS hybrid overcoating enhanced fiber plasmonic temperature sensor with high sensitivity and fast response. *J. Mater. Chem. C* **2020**, *8*, 12893–12901. [CrossRef]
29. Liu, F.; Zhang, X.; Li, K.; Guo, T.; Ianoul, A.; Albert, J. Discrimination of bulk and surface refractive index change in plasmonic sensors with narrow bandwidth resonance combs. *ACS Sens.* **2021**, *6*, 3013–3023. [CrossRef]
30. Arumona, A.E.; Garhwal, A.; Youplao, P.; Amiri, I.S.; Ray, K.; Punthawanunt, S.; Yupapin, P. Electron cloud spectroscopy using micro-ring Fabry–Pérot sensor embedded gold grating. *IEEE Sens. J.* **2020**, *20*, 10564–10571. [CrossRef]
31. Dell'Olio, F.; Conteduca, D.; Ciminelli, C.; De Palo, M. Design of a new ultracompact resonant plasmonic multi-analyte label-free biosensing platform. *Sensors* **2017**, *17*, 1810. [CrossRef]
32. Westbrook, P.S.; Feder, K.S.; Kremp, T.; Taunay, T.F.; Monberg, E.; Puc, G.; Ortiz, R. Multicore optical fiber grating array fabrication for medical sensing applications. *Int. Soc. Optics Photonics* **2015**, *9317*, 93170C. [CrossRef]
33. Schenato, L.; Aneesh, R.; Palmieri, L.; Galtarossa, A.; Pasuto, A. Fiber optic sensor for hydrostatic pressure and temperature measurement in riverbanks monitoring. *Opt. Laser Technol.* **2016**, *82*, 57–62. [CrossRef]
34. Liu, Y.; Jing, Z.; Li, R.; Zhang, Y.; Liu, Q.; Li, A.; Zhang, C.; Peng, W. Miniature fiber-optic tip pressure sensor assembled by hydroxide catalysis bonding technology. *Opt. Express* **2020**, *28*, 948–958. [CrossRef] [PubMed]
35. Cheng, X.; Dash, J.; Gunawardena, D.; Htein, L.; Tam, H.-Y. Silicone Rubber Based Highly Sensitive Fiber-Optic Fabry–Perot Interferometric Gas Pressure Sensor. *Sensors* **2020**, *20*, 4927. [CrossRef] [PubMed]
36. Liu, Y.; Jing, Z.; Liu, Q.; Li, A.; Teng, C.-A.; Cheung, Y.; Lee, A.; Tian, F.; Peng, W. Differential-pressure fiber-optic airflow sensor for wind tunnel testing. *Opt. Express* **2020**, *28*, 25101–25113. [CrossRef] [PubMed]
37. Lijun, L.; Xu, Z.; Bin, T.; Jing, S.; Xuemei, Y.; Yongzheng, L. A kind of mirco-fiber Bragg grating mine pressure sensor. *J. China Coal Soc.* **2013**, *38*, 2084–2088. [CrossRef]
38. Hui, W.; Yang, Y.; Bing, L. Demodulation method of fiber grating pressure sensor based on dense wavelength division multiplexer. *Prog. Laser Optoelectron.* **2016**, *53*, 220–226. [CrossRef]
39. Zhang, H.; Song, W.; Wang, Z.; Guo, B.; Zhang, G.; Liu, L. Numerical and experimental studies of high-sensitivity plug-in pressure sensor based on fiber Bragg gratings. *Opt. Eng.* **2016**, *55*, 096104. [CrossRef]
40. Wang, Y.; Wang, M.; Xia, W.; Ni, X.; Wu, D. Optical fiber Bragg grating pressure sensor based on dual-frequency optoelectronic oscillator. *IEEE Photonics Technol. Lett.* **2017**, *29*, 1864–1867. [CrossRef]
41. Jia, Z.; Fan, Q.; Feng, D.; Yu, D.; Zhao, X.; Yang, K. Design and investigation of the fiber Bragg grating pressure sensor based on square diaphragm and truss-beam structure. *Opt. Eng.* **2019**, *58*, 097109. [CrossRef]
42. Nan, W. Design and implementation of a structure-matched conformal temperature-compensated marine fiber grating pressure sensor. *Electron. World* **2020**, *10*, 130–134.
43. Wang, W.; Zhou, X.; Wu, W.; Chen, J.; He, S.; Guo, W.; Gao, J.; Huang, S.; Chen, X. Monolithic structure-optical fiber sensor with temperature compensation for pressure measurement. *Materials* **2019**, *12*, 552. [CrossRef]
44. Fan, Q.; Feng, D.; Yong, Z. Highly sensitive FBG pressure sensor based on square diaphragm. *Optik* **2021**, *225*, 165559. [CrossRef]
45. Zhao, X.; Dong, M.; Zhang, Y.; Luo, F.; Zhu, L. Simultaneous Measurement of Temperature and Refractive Index Using High Temperature Resistant Pure Quartz Grating Based on Femtosecond Laser and HF Etching. *Materials* **2021**, *14*, 1028. [CrossRef] [PubMed]
46. Rollinson, C.M.; Wade, S.A.; Dragomir, N.M.; Baxter, G.W.; Collins, S.F. Strain and temperature response of transmission dips at 2/3 of the Bragg wavelength in type Iand IIA fiber Bragg gratings. In Proceedings of the 17th International Conference on Optical Fibre Sensors, Bruges, Belgium, 23–27 May 2005; Volume 5855, pp. 310–313. [CrossRef]
47. Li, H.; Ma, L.; Zhao, Q.; Zhang, F.; Wang, C.; Liu, X. Rapid response of high precision fiber bragg grating based temperature sensor. In Proceedings of the 2018 IEEE 3rd Optoelectronics Global Conference (OGC), Shenzhen, China, 4–7 September 2018; pp. 64–67. [CrossRef]
48. Yucel, M.; Ozturk, N.F.; Goktas, H.H.; Gunduz, A.E. Design of a Fiber Bragg Grating based temperature sensor. In Proceedings of the 2016 24th Signal Processing and Communication Application Conference (SIU), Zonguldak, Turkey, 16–19 May 2016; pp. 669–672. [CrossRef]

MDPI
St. Alban-Anlage 66
4052 Basel
Switzerland
Tel. +41 61 683 77 34
Fax +41 61 302 89 18
www.mdpi.com

Micromachines Editorial Office
E-mail: micromachines@mdpi.com
www.mdpi.com/journal/micromachines

www.ingramcontent.com/pod-product-compliance
Lightning Source LLC
Chambersburg PA
CBHW061757210926
51850CB00006B/583
9783036577913